PHYSICS

D1613892

Growth Rhythms
and
The History
of
The Earth's Rotation

Growth Rhythms
and
The History
of
The Earth's Rotation

Edited by

G. D. Rosenberg

and

S. K. Runcorn

*Department of Geophysics, School of Physics,
University of Newcastle upon Tyne*

A Wiley–Interscience Publication

JOHN WILEY & SONS

London · New York · Sydney · Toronto

Copyright © 1975 by John Wiley & Sons, Ltd.

All rights reserved.

No part of this book may be reproduced by any means, nor translated, nor transmitted into a machine language without the written permission of the publisher.

Library of Congress Cataloging in Publication Data:

Interdisciplinary Winter Conference on Biological Clocks and Changes in the Earth's Rotation: Geophysical and Astronomical Consequences, Newcastle upon Tyne, 1974.
Growth rhythms and the history of the Earth's rotation.

   'A Wiley–Interscience publication.'
   Includes bibliographies.
     1. Earth—Rotation—Addresses, essays, lectures.
2. Biology—Periodicity—Addresses, essays, lectures.
I. Rosenberg, Gary, D., ed. II. Runcorn, S. K., ed.
III. Title.

QB633.I57   1974   525'.3   74-18096
ISBN 0 471 73616 3

Printed in Great Britain by
William Clowes and Sons Ltd., London, Colchester and Beccles

PREFACE

> Science needs all kinds of people ... since each person can ask different questions and see different worlds.
>
> ABRAHAM H. MASLOW
> *Motivation and Personality*

More than 30 people have contributed to the writing of this book. They include astronomers, molluscan physiologists, geophysicists, palaeontologists, archaeologists and geologists. In addition, each manuscript offers a different scientific philosophy. Some are problem-centred theses, others are papers of technique. Some are experiments in scientific abstraction, others show an interest in the uniqueness of scientific data. Some are well-polished, and have the aspect of being scientifically 'correct', while others of equal importance will seem to be bold 'hunches'.

The molluscan physiologist may not find an immediate interest in the astronomical or geophysical papers. The astronomer will not immediately see the relevance of a paper on molluscan physiology to his discipline. The orthodox scientist may dispute the roughly hewn papers, the scientific pioneer may overlook the conservative orthodox work.

However, the diverse disciplines and philosophies represented in this book are unified by an important theme: the rate of the Earth's rotation is not what it used to be. Many different disciplines provide evidence for this hypothesis, and the idea is important to cosmology, Earth and life history. The synthesis of many different philosophies is one type of scientific progress.

Of course, each of the papers has merit on its own, in addition to its relevance to determining the history of the Earth's rotation. For example, ancient astronomical observations preserve a concern for astrology, with eclipses associated with the death of a Chinese Empress and even the collapse of a dynasty. Astronomical and weather records preserved in similar texts also provide much information on living conditions (growing seasons, climate, etc.) within the past 3000 years. The biological papers add to our knowledge of growth processes and biological clocks in invertebrates and in plants. This information can be used in palaeoecological studies.

This book is intended to be an introduction to the methods of determining the history of the Earth's rotation, and it is also aimed at surveying the potential of the work for other sciences. It is hoped that the book will supplement undergraduate reading as well as serving as a reference for research and postgraduate seminars.

Preface

The manuscripts published here were originally presented at the Interdisciplinary Winter Conference on Biological Clocks and Changes in the Earth's Rotation: Geophysical and Astronomical Consequences. The Conference was held at the School of Physics, University of Newcastle upon Tyne, 8–10 January 1974 and we wish to thank many people for their help with the Conference and for their assistance in preparing this volume.

The Conference sessions were chaired by Dr. George Farrow, University of Glasgow, Mr. Leslie Morrison, Royal Greenwich Observatory, Dr. Colin Scrutton, University of Newcastle and Dr. Don Tarling, University of Newcastle. Colin Scrutton and Don Tarling also generously advised in scientific matters regarding organization of the Conference and of the book.

Several colleagues assisted with review of the manuscripts. Mr. W. F. Mavor and Marion Turner organized administrative and secretarial matters for the Conference. They were helped by the secretarial staff of the School of Physics.

Marion Turner also helped prepare portions of this volume. Mrs. E. Thompson and Dorothy Hewett prepared slides for the Conference, and many of the drawings for the text. The efficient and patient co-operation of several staff-members of the Publisher also rendered us invaluable aid (but unfortunately their code of ethics does not allow us to mention them by name).

The Scientific Affairs Division of NATO provided Rosenberg with a NATO Senior Fellowship.

We are grateful to all of the above for their assistance.

G. D. ROSENBERG
S. K. RUNCORN

May, 1974

CONTRIBUTING AUTHORS

DR. K. D. ALDRIDGE, Institute of Earth and Planetary Physics, University of Alberta, Edmonton, Alberta T6G 2EI, Canada

DR. RICHARD M. BARKER, Eel Brook, Yarmouth Co., Nova Scotia, Canada

PROFESSOR WILLIAM B. N. BERRY, Professor of Palaeontology and Curator of Palaeozoic Invertebrates, Museum of Palaeontology, University of California, Berkeley, California 94720, USA

DR. ROBERT W. BUDDEMEIER, Department of Oceanography, University of Hawaii, 2525 Correa Road, Honolulu, Hawaii 96822, USA

DR. GEORGE R. CLARK II, Department of Geology, University of New Mexico, Albuquerque, New Mexico 87131, USA

DR. P. J. F. COUTTS, State Archaeologist, Archaeological and Aboriginal Relics Office, 213 Lonsdale Street, Melbourne 3000, Australia

MR. G. T. CREBER, Botany Department, Birkbeck College, Malet Street, London WC1E 7HX

PROFESSOR K. M. CREER, Head, Department of Geophysics, University of Edinburgh, 6 South Oswald Road, Edinburgh EH9 2HX, Scotland

MR. JOHN W. DOLMAN, School of Physics, University of Newcastle upon Tyne, Newcastle upon Tyne NE1 7RU, England

MR. K. A. ERIKSSON, Department of Geology, University of the Witwatersrand, Jan Smuts Avenue, Johannesburg, South Africa

DR. JOHN W. EVANS, Department of Biology, Memorial University of Newfoundland, St. John's, Newfoundland, Canada

DR. JOHN GRIBBIN, Assistant Editor, Nature—Times News Service, Nature, 4 Little Essex Street, London WC2R 3LF, England

PROFESSOR C. A. HALL, JR., Chairman, Geology Department, University of California, Los Angeles, Los Angeles, California 90024, USA

DR. ROGER G. HIPKIN, Department of Geophysics, University of Edinburgh, 6 South Oswald Road, Edinburgh EH9 2HX, Scotland

PROFESSOR J. A. JACOBS, Killam Memorial Professor of Science and Director of Institute of Earth and Planetary Physics, University of Alberta, Edmonton, Alberta T6G 2EI, Canada (*currently* Professor of Geophysics, Cambridge University, Cambridge, England)

DR. G. A. L. JOHNSON, Department of Geological Sciences, University Science Laboratories, University of Durham, South Road, Durham DH1 3LE, England

MR. C. B. JONES, School of Physics, University of Newcastle upon Tyne, Newcastle upon Tyne NE1 7RU, England

DR. J. KEENEY, The Superior Oil Company, Geophysical Laboratory, 12401 Westheimer, Houston, Texas, USA

Contributing Authors

DR. ROBERT A. KINZIE III, Department of Zoology, University of Hawaii, Honolulu, Hawaii 96822

MR. ROLAND E. MOHR, Department of Geology and Geophysics, University of Minnesota, Minneapolis, Minn. 55455, USA

MR. LESLIE V. MORRISON, Principal Scientific Officer, Royal Greenwich Observatory, Herstmonceux Castle, Hailsham, Sussex, England

DR. PAUL MULLER, Jet Propulsion Laboratory, Pasadena, California, USA (*currently* at School of Physics, University of Newcastle upon Tyne)

MR. J. R. NUDDS, Department of Geological Sciences, University Science Laboratories, University of Durham, Durham DH1 3LE, England

MR. N. P. J. O'HORA, Royal Greenwich Observatory, Herstmonceux Castle, Hailsham, Sussex BN 27 1RP, England

DR. GIORGIO PANNELLA, Geology Department, University of Puerto Rico, Mayaguez 00708, Puerto Rico

DR. ANGELO POMA, Astronomical Institute of the Cagliari University, Via Ospedale 72-09100, Cagliari, Italy

PROFESSOR EDOARDO PROVERBIO, Director, Astronomical Institute of the Cagliari University, Via Ospedale 72-09100, Cagliari, Italy

DR. GARY D. ROSENBERG, School of Physics, University of Newcastle upon Tyne, Newcastle upon Tyne NE1 7RU, England

PROFESSOR S. K. RUNCORN, FRS, Head, School of Physics, University of Newcastle upon Tyne, Newcastle upon Tyne NE1 7RU, England

DR. F. RICHARD STEPHENSON, School of Physics, University of Newcastle upon Tyne, Newcastle upon Tyne NE1 7RU, England

DR. D. H. TARLING, School of Physics, University of Newcastle upon Tyne, Newcastle upon Tyne NE1 7RU, England

PROFESSEUR GENEVIEVE TERMIER, Maître de recherche CNRS, Faculté des Sciences de Paris VI, 4 place Jussieu 75005, Paris, France

PROFESSEUR HENRI TERMIER, Professeur Honoraire, Faculté des Sciences de Paris VI, 4 place Jussieu 75005, Paris, France

DR. IDA THOMPSON, Department of Geological and Geophysical Sciences, Princeton University, Princeton, New Jersey 08540, USA

DR. J. F. TRUSWELL, Department of Geology, University of the Witwatersrand, Jan Smuts Avenue, Johannesburg, South Africa

DR. D. H. WEINSTEIN, The Superior Oil Co., Geophysical Laboratory, 12401 Westheimer, Houston, Texas 77077, USA

DR. PAUL S. WESSON, St. John's College, Cambridge, England

DR. MARTIN A. WHYTE, Geology Department, University of Hull, Hull HU6 7RX, England

CONTENTS

Introduction xiii

A Comment on Terminology: The Increment and the Series
G. D. Rosenberg 1

Growth Increments in Fossil and Modern Bivalves
W. B. N. Berry and R. M. Barker 9

Carboniferous Coral Geochronometers
G. A. L. Johnson and J. R. Nudds 27

Measured Periodicities of the Biwabik (Precambrian) Stromatolites and their Geophysical Significance
R. E. Mohr 43

Facies and Laminations in the Lower Proterozoic Transvaal Dolomite, South Africa
J. F. Truswell and K. A. Eriksson 57

The Effects of Gravity and the Earth's Rotation on the Growth of Wood
G. T. Creber 75

Sedimentary Behaviour and Skeletal Textures Available in Growth Cycle Analysis
H. Termier and G. Termier 89

Periodic Growth and Biological Rhythms in Experimentally Grown Bivalves
G. R. Clark II 103

Growth and Micromorphology of Two Bivalves Exhibiting Non-Daily Growth Lines
J. W. Evans 119

The Chronometric Reliability of Contemporary Corals
R. W. Buddemeier and R. A. Kinzie III 135

Biological Clocks and Shell Growth in Bivalves
I. Thompson 149

Latitudinal Variation in Shell Growth Patterns of Bivalve Molluscs: Implications and Problems
C. A. Hall, Jr. 163

Time, Tide and the Cockle
M. A. Whyte 177

A Technique for the Extraction of Environmental and Geophysical Information from Growth Records in Invertebrates and Stromatolites
J. Dolman 191

Approaches to Chemical Periodicities in Molluscs and Stromatolites
G. D. Rosenberg and C. B. Jones 223

The Seasonal Perspective of Marine-Orientated Prehistoric Hunter-Gatherers
P. J. F. Coutts 243

Palaeontological Clocks and the History of the Earth's Rotation (Plenary lecture)
G. Pannella 253

Palaeontological and Astronomical Observations on the Rotational History of the Earth and Moon (Plenary lecture)
S. K. Runcorn 285

On a Tentative Correlation between Changes in the Geomagnetic Polarity Bias and Reversal Frequency and the Earth's Rotation through Phanerozoic Time (Plenary lecture)
K. M. Creer 293

Tides and the Rotation of the Earth
R. G. Hipkin 319

The Earth's Interior and the Earth's Rotation (Plenary lecture)
J. A. Jacobs and K. D. Aldridge 337

Gravity and the Earth's Rotation (Plenary lecture)
P. S. Wesson 353

Palaeontology and the Dynamic History of the Sun–Earth–Moon System
D. H. Weinstein and J. Keeney 377

Astronomical Evidence of Change in the Rate of the Earth's Rotation and Continental Motion
E. Proverbio and A. Poma 385

Geological Processes and the Earth's Rotation in the Past (Plenary lecture)
D. H. Tarling 397

Climate, the Earth's Rotation and Solar Variations
J. Gribbin 413

The Detection of Recent Changes in the Earth's Rotation
N. P. J. O'Hora 427

INTRODUCTION

Accretionary skeletons of living and fossil organisms record the history of the Earth's rotation and point to the Moon being closer to the Earth in the distant past. Many other aspects of Earth and life history emerge from growth rhythms preserved in skeletons: possible changes in the Earth's gravitational field; aspects of plate tectonics; expansion of the Earth; the evolution of the Earth's interior and of its magnetic field; the evolution of biological clocks; the history of some civilizations. How Earth and life history can be abstracted from a bivalve shell or a coral skeleton, the degree of uncertainty of our inferences, and the importance of our observations for the physical and biological sciences are the subjects of this book.

Since 1963, when Professor John Wells published 'Coral Growth and Geochronometry', a score of workers have been studying skeletal growth rhythms for geophysical purposes. In recent years it has become increasingly apparent that there is a need to improve the accuracy of the data and to test the geophysical inferences more rigorously.

Wells counted the number of daily growth increments added per annual series of increments in the skeletons of living and fossil corals. He determined a year's growth by noting repeating patterns of the seasonal growth recorded in the skeleton. Living specimens add approximately 365 daily increments per year; Devonian fossils added about 400.

Wells suggested that the difference in number of days per year could be accounted for by a change in the Earth's rotation rate, assuming the period of the Earth's revolution about the Sun has remained constant. Wells also recognized the potential of growth rhythms for absolute age dating. Fossils with 400 daily increments per year's growth would date the formation in which they were found as Devonian.

Following Wells, Scrutton (1965) recognized that some Devonian corals have distinct series of monthly increments. Runcorn (1964, 1966) used the determinations of number of days per year and per month to calculate changes in the Earth's angular momentum and moment of inertia since the Devonian. Runcorn realized that these calculations were important in determining the history of the Earth and Moon.

However, accurate determination of the history of the Earth–Moon system on the basis of palaeontological data requires reducing the uncertainty of results of growth rhythm measurements; Wells' hypothesis requires additional substantiation. Additional palaeontological data have begun to confirm the validity of the early measurements (Berry and Barker, 1968; Mazzullo, 1971; Pannella, MacClintock and Thompson, 1968). Furthermore, it is becoming increasingly clear that some species of organisms living in some habitats are reliable chronometers, thus justifying the measurements on fossil relatives (Clark, 1968; Barnes, 1972; Barker, 1964; House and Farrow, 1968; Pannella and MacClintock, 1968; Rhoads and Pannella, 1970; Farrow, 1971, 1972; Evans, 1972; Evans and Lemessurier, 1972; Rosenberg, 1973). Of the many taxa with accretionary skeletons, the bivalves have received the most study, despite Wells' original work on corals. The geophysical potential of such groups as bryozoa and stromatoporoids has yet to be determined.

The importance of Wells' hypothesis is that it provides the only known direct means of determining the rate of the Earth's rotation in the distant past. Consequently, palaeontological data can be used to test the validity of theoretical calculations based on astronomical and geophysical considerations.

Astronomical observations establish variations in the Earth's rotation within the past 3000 years. Modern determinations are made by comparing time-scales based on rate of the Earth's rotation (Universal Time), the orbital motion of the planets (Ephemeris Time), and the oscillation frequency of a ground state emission of the caesium atom (Atomic Time). Ancient eclipse observations recorded in Chinese, Ugaritic, Babylonian, Assyrian and Medieval texts extend knowledge of the Earth's rotation to about 1500 B.C. The observation of a total eclipse is useful for geophysical purposes if the text indicates when and where the eclipse was observed. This fixes the position of the Moon relative to the Earth at the time the eclipse was observed. Assuming the Earth's rotation is constant and the motion of the Moon is known, one can predict where and when a total eclipse should have been observed by ancient civilizations. The actual observations differ from the theoretical, and the anomaly can be accounted for by a variable rotation rate of the Earth and/or an orbital acceleration of the Moon. Analysis of many total eclipse observations throughout the historical period allows these two effects to be separated.

Modern and ancient astronomical data indicate that during the past 3000 years the length of day has been increasing at the constant rate of approximately $2 \cdot 50 \pm 0 \cdot 3$ msec/cy (Muller and Stephenson, this volume). This value has had a long history of determination and is still subject to dispute (see Scrutton and Hipkin, 1973; Morrison, Muller and Stephenson and O'Hora, this volume).

Theoretical calculations based on geophysical assumptions can extend this value back to the Precambrian. One important assumption is that the dissipation of energy due to tidal friction has slowed the Earth. However, the exact nature of tidal friction is unknown because tides are very complex; there are many kinds of tides induced by the Moon and the Sun in the Earth's oceans, land masses, interior and atmosphere. The frequency and amplitude of the tides vary with changes in sea level, area of the land masses relative to area of the oceans, amount of solar radiation (which in turn depends on the position of the planets Mercury, Jupiter, Venus and Earth relative to the Sun), and configuration of the continents and ocean basins.

Loss of angular momentum of the Earth due to lunar tidal friction must be transferred to the orbital angular momentum of the Moon. Consequently, when the Earth was rotating faster in the distant past, the Moon must have been closer to the Earth. However, based on tidal considerations alone, there is no unique solution to the problem of determining when the Earth–Moon distance was zero (Scrutton and Hipkin, 1973; Hipkin, this volume). It is important to establish when the Earth–Moon distance was minimal, for this may help determine whether the Moon originated as part of the Earth, as a body captured by the Earth's gravitational field, or as the result of an accretion process (Runcorn, 1966). The palaeontological means of directly determining the Earth's angular momentum in the distant past thus become important in determining the origin of the Moon.

Tidal friction may not be the only cause of the Earth's deceleration. The universe may be expanding, the acceleration of gravity may be decreasing and as a consequence the diameter of the Earth may be increasing. Redistribution of mass in the Earth's interior accompanying evolution of the Earth's core, convection currents in the Earth's interior, and topographic, magnetic or viscous core–mantle coupling may also affect the rate of the Earth's rotation by affecting the Earth's moment of inertia. Accurate comparison of palaeontological data with geophysical and astronomical considerations will help determine the importance of these non-tidal factors relative to tidal parameters influencing the rate of the Earth's rotation.

References

Barker, R. M. (1964). Microtextural variation in pelecypod shells. *Malacologia*, **2**, 69–86

Barnes, D. J. (1972). The structure and formation of growth-ridges in scleractinian coral skeletons. *Proc. Roy. Soc. London, B*, **182**, 331–350

Berry, W. B. N. and Barker, R. M. (1968). Fossil bivalve shells indicate longer month and year in Cretaceous than Present. *Nature*, **217**, 938–939

Clark, G. R. II (1968). Mollusk shell: Daily growth lines. *Science*, **161**, 800–802

Evans, J. W. and Lemessurier, M. H. (1972). Functional micromorphology and circadian growth of the rock boring clam *Penitella penita*. *Can. J. Zool.*, **50**, 1251–1258

Evans, J. W. (1972). Tidal growth increments in the cockle *Clinocardium nuttalli*. *Science*, **176**, 416–417

Farrow, G. E. (1971). Periodicity structures in the bivalve shell: experiments to establish growth controls in *Cerastoderma edule* from the Thames Estuary. *Palaeontology*, **14**, 571–588

Farrow, G. E. (1972). Periodicity structures in the bivalve shell: analysis of stunting in *Cerastoderma edule* from the Bury Inlet (South Wales). *Palaeontology*, **15**, 61–72

House, M. R. and Farrow, G. E. (1968). Daily growth banding in the shell of the cockle, *Cardium edule*. *Nature*, **219**, 1384–1386

Mazzullo, S. J. (1971). Length of the year during the Silurian and Devonian Periods: New values. *Geol. Soc. Amer. Bull.*, **82**, 1085–1086.

Pannella, G. and MacClintock, C. (1968). Biological and environmental rhythms reflected in molluscan shell growth. *J. Palaeontol. Mem.*, **42**, 64–80

Pannella, G., MacClintock, C. and Thompson, M. N. (1968). Palaeontologic evidence of variations in length of synodic month since Late Cambrian. *Science*, **162**, 792–796

Rhoads, D. C. and Pannella, G. (1970). The use of molluscan shell growth patterns in ecology and palaeontology. *Lethaia*, **3**, 143–161

Rosenberg, G. D. (1973). Calcium concentration in the bivalve *Chione undatella* Sowerby. *Nature*, **244**, 155–156

Runcorn, S. K. (1964). Changes in the Earth's moment of inertia. *Nature*, **204**, 823–825

Runcorn, S. K. (1966). Corals as paleontological clocks. *Scientific American*, **215**, 26–33

Scrutton, C. T. (1965). Periodicity in Devonian coral growth. *Palaeontology*, **7**, 552–558

Scrutton, C. T. and Hipkin, R. G. (1973). Long-term changes in the rotation rate of the earth. *Earth-Science Reviews*, **9**, 259–274

Wells, J. W. (1963). Coral growth and geochronometry. *Nature*, **197**, 948–950.

A COMMENT ON TERMINOLOGY: THE INCREMENT AND THE SERIES

GARY D. ROSENBERG

School of Physics, University of Newcastle upon Tyne

Students of biological growth rhythms have begun to develop their own terminology, or at least adapt the terminology of other disciplines for their own purposes. The resulting terminology is complex. Growth rings, growth lines, lamellae, laminae, growth layers, crinkles, growth ridges, bumps, constrictions, external lines, daily lines, tidal lines, mesobands, type A and B increments, major bands, organic lines, carbonate layers, biochecks—all of these terms and many more refer to growth structures in organisms with accretionary skeletons. Because growth studies are interdisciplinary, geophysicists on the one extreme and classical taxonomists on the other are applying a variety of names to similar structures in different organisms.

The taxonomist insists on a clearly descriptive terminology distinguishing one species from another. The geophysicist wants to count the number of daily increments per annual series of increments. The biologist and palaeontologist study growth patterns to determine their ontogenetic, physiological and evolutionary significance. The biostratigrapher looks for the environmental and sedimentological implications of biological rhythms. Surely it is possible to establish a flexible terminology consistent with the purposes of this diverse group of workers.

One can begin by classifying the different meanings of the words used in growth increment studies:

(1) Structural
(2) Compositional
(3) Temporal
   (A) Period
      (i) period of the increment
      (ii) period of secretion
   (B) Rate of growth (reciprocal of A)
(4) Environmental
(5) Endogenous, physiological and behavioural

Serious problems of understanding arise when the worker describes the structure and composition of growth increments with terms better suited to the temporal, environmental or physiological significance of the increment.

For example, some workers use the term *tidal increment* to mean a unit of growth deposited with regular semi-daily periodicity. However, the period is regular in only a few populations of the species they are studying. Some workers use the term *growth line* to imply a metabolic change in the character of the secreting tissue, but they do not describe the structural appearance of the 'line' itself nor do they demonstrate, rather than simply infer, the metabolic change.

Thus, it is necessary to examine more closely the different meanings of the words used to describe growth rhythms.

(1) Structural factors. Referring to repetitive textures or construction of features such as crystal forms (laths, prisms, etc.), orientation of crystals, ornamentation on the surface of the organism (ridges, grooves, etc.), 'lines' of accretion appearing in thin section of the skeleton, layers of dense calcium carbonate alternating with layers of porous calcium carbonate, etc.

(2) Compositional factors. Referring to repetitive chemical composition, for example the alternation of layers of calcium carbonate and sheets or layers of organic material (conchiolin), repetitive fluctuation of calcium concentration relative to trace element concentration (e.g. Sr, Mg, S, Fe, depending on the organism), types of amino acids.

(3) Temporal factors:

(A) Period

(i) Period of the increment—time represented between the beginning of a unit of structure or composition, and the beginning of the next, adjacent unit. For example, 24 hr may be represented between the deposition of a growth lamina of conchiolin and the next lamina, 12 months between the beginning of a series of narrow winter increments and the next series of winter increments. There may be no regular period of time between increments, giving rise to pseudo-periodicities of some repetitive structures (see Buddemeier and Kinzie, this volume).

It is important to note that because the period of secretion (below) has rarely been determined, and because the period of the increment is of major importance for geophysical purposes, period terms in the name of the increment refer to the *period of the increment*.

(ii) Period of secretion—actual amount of time the secreting tissue takes to deposit a growth unit. Whereas the period represented between the deposition of one growth increment and the next may be 24 hr, the increment in one species may actually be deposited 'instantaneously' and, in another species, a structurally similar increment may be deposited over a protracted interval within the 24-hr period. In few cases has the duration of secretory activity been demonstrated.

(B) Rate of growth. Reciprocal of 3(A).
  (i) Frequency of the increment—number of increments added per unit time, e.g. 30 daily increments per month, 365 daily increments per year. May be expressed in terms of amount of skeleton (length or weight units) as in 3 mm of growth per year.
  (ii) Rate of secretion—amount of skeleton added during the period of activity of the secretory tissue. For example, within a 14-day period (one tidal cycle) 14 increments may be added, but on the average the secretory tissue is active for say 12 hr per increment (hypothetical). If on the average each increment is 25 μm wide, in this sense there is 25 μm growth per 12 hr.
(4) Environmental factors. Referring to the environmental parameters influencing growth increment production, such as tides, light–dark cycles, temperature changes, oxygen or CO_2 concentration fluctuations, nutrition cycles, storms, substrate, etc.
(5) Endogenous factors expressed in terms of metabolism and behaviour. Periodic growth structures may be deposited on one extreme totally by metabolic rhythms independent of environmental changes; other increments may be produced by both metabolic and environmental influence. Rhythms of carbonate metabolism, feeding and assimilation, respiration, reproduction, pumping (clams), excretion, gaping (perhaps associated with feeding), movement (breeding migrations, escape from enemies, search for food), etc. refer to an internal metabolic or behavioural process which may or may not be cued by the environment.

At present, recognition of repeating features of structure or composition is clearly necessary for subsequent determinations of geophysical or biological implications. Of course, it is certain that genetic, growth or metabolic rhythms exist which are not recorded in the skeleton of the organism, but it is clear that, because such rhythms are not available for geophysical analysis, the structural or compositional record can be considered of first importance. Consequently, the *growth increment* is taken as the geophysical unit of measurement, applicable to an accretionary skeleton with repetitive structures. The increment is a repetitive unit of structure or composition in the skeleton. Because it is a term defining a structural or compositional unit, it must be described so that similar features can be recognized and compared with different populations of the same species and with other species. There are often many different kinds of structural and chemical growth increments within an organism. Repetitive chemical increments, increments at the crystallographic level, and increments of ornamentation among others may be present in the same organism, hence the necessity of clearly describing the chemical or structural unit being considered. A *series* of increments is defined simply as a succession of increments within an organism. Often a series repeats itself; several series of tidal increments, for example, are recognized in bivalves as sequences of narrowing and widening increments produced under tidal influence. Repeating series may be recognized by their *patterns* (Pannella, this volume). Terms such as bands, zones, clusters, orders and biochecks are intended to define types of series but unfortunately the meaning of each term frequently varies from author to author. For example, a band can imply a structurally distinct series of increments or a portion of a daily increment. The structural and environmental significance implicit in the term biocheck as defined by Hall (this volume) is yet to be demonstrated as valid for all species of molluscs. Orders of increments implying a hierarchy are numbered successively but, as Pannella indicates, present problems when an order of increment is discovered between two previously established consecutive increment types, causing renumbering of the entire order system. With this difficulty in mind, it is difficult to propose formalization of the procedure of ordering by number at the present time. At the very least, Pannella's reversing the numbering precedent established by Barker (see Pannella, this volume) is an unnecessary complication. It is suggested that an unambiguous description of the increment types constituting the series will serve to define the series. Inversely, use of the term *series* indicates that the constituent increments have been described either structurally or chemically. The entire series of increments recorded in the skeleton until growth stops is the *life series*.

It is important to note that recognition of a boundary is needed for the complete description of the increment and the series. The *boundary* is the structural or chemical limit of the increment or series. Confusion of boundary description with description of the increment or of the series may account for some of the previous difficulties of terminology. For example, the terms biochecks, clusters and bands imply series with well-defined boundaries, consequently distinguished structurally from the preceding and following series. However, the position of the boundary is not always stated or figured by the workers using these terms. Often this is because the structural or chemical characteristic of the boundary is actually not apparent.

It is important to note that terms such as lines and laminae and taxonomic terms such as dissepiment, diaphragm or ridge have been used to locate the interface between two increments (the boundary) and also to denote the entire increment. It is important to distinguish clearly between the two cases.

The problem of defining the boundary is shown in Figure 1. Spherulitic crystals in a bivalve shell are diagrammed in A. The boundary may be placed at either the apexes of the fans of spherulitic crystals or at the edges of the fans. In B, chemical rhythms are shown. Clearly, the boundaries may be placed at the crests, troughs or along the sides of the waves. In C, increments consist of layers of calcium carbonate alternating with layers of conchiolin. The boundaries of an increment may be placed to the left of each conchiolin layer, or they could be

A Comment on Terminology: The Increment and the Series

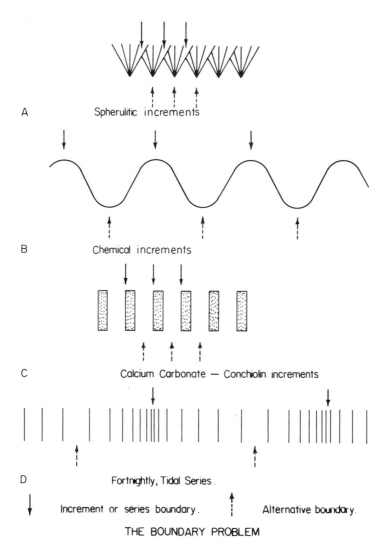

FIGURE 1. The boundary problem. The position of the boundary relative to a structural or compositional character defines the boundary. Direction of growth is to the right in each case. A. Spherulitic increments; B. Chemical increments; C. Calcium carbonate-conchiolin increments; D. Fortnightly, tidal series

placed in either the centre of the conchiolin layer or the centre of the carbonate layer. A similar problem results in placing the boundary of the series (D). Two of many alternative positions of the boundary are shown for this diagrammatic tidal series. The boundaries are placed either at the centre of the widest increment or at the centre of the narrowest.

One cannot at present formalize the positioning of the boundary (e.g. always at the crest of a chemical rhythm, always at the centre of the widest increment). Consider the problem which would arise when structural and chemical increments are superimposed (see Rosenberg and Jones, this volume). Maxima and minima of chemical concentration do not always coincide with spacing of increments. Consider also Pannella's and Evans' (this volume) description of the 'switch-over'. Here the position of the series boundary actually shifts with time. Clearly, proper definition of the boundary means stating the *position* of the boundary relative to the structural or compositional *characteristics* of the increment at that position.

Table 1 classifies some of the words used to describe an increment and a series. The list is not intended to be exhaustive. Usually one or more of these words are selected to refer to the increment or the series. The table proposes to standardize their order if more than one are used: temporal, environmental, endogenous, structural/compositional. Thus, semi-daily, tidal increment refers to periodicity of, and environmental parameter associated with, the increment. An annual series of daily, gaping increments refers to the period of the series and to the period of an endogenous factor associated with the increment. An annual density increment refers to the period associated with the designated structural variations.

Confusion results when temporal, environmental or endogenous terms are applied to the wrong units. Hence, Evans (this volume) writes of semi-daily lines when he means that the increment with the 'line' serving to position the boundary has a semi-daily period; the 'lines' themselves are actually thin layers with a recognizable structure and possibly period (Dolman, Evans, Pannella, this volume), hence may be valid quarter-daily? tidal increments in their own right. Similarly with Clark's use of the term 'daily lines'.

Confusion also results when an increment is named by using a structural or chemical feature which forms only part of the increment. Thus, Evans speaks of width of lines when he means width of increment composed of a layer and a line (actually a thin lamella). A trabecular increment would be one formed only of a trabecula and its boundaries, not a layer bounded by the trabeculae.

Terms which imply width of the increment (e.g. mesoband, microband, layer, lamina, line) should not be used as part of the name, because it is clear that any given increment type has a range of sizes (Pannella, this volume). For example, some increments vary in width between thin laminae and wide layers.

The importance of standardized terminology is emphasized in consideration of the *temporal increment*. The temporal increment is a functional unit of importance to both the geophysicist and the evolutionary biologist. Some

THE INCREMENT and THE SERIES

Temporal	Environmental	Endogenous	Structure or Composition
Hourly, Semi-daily, Sub-daily, Daily, Solar daily, Lunar daily, Diurnal, Circadian, Fortnightly, Monthly, Semi-annual, Annual, 11-yr, 18.6-yr, Sidereal,† Synodical,† Anomalistic,† Equinoctial, Aperiodic, Pseudoperiodic, Seasonal.	Tidal, Erosional, pH, Barometric pressure, Turbidity, Light intensity, Temperature, Storm, Nutrition, Oxygen tension, Sunspot, Gravity, Predator disturbance, Current flow, Salinity, Precipitation, Evaporation.	Physiology: Metabolic, Calcium metabolic, Catabolic, Photosynthetic, Respiratory, Excretory, Assimilative, Genetic, Nuclear division. Behavior: Gaping, Feeding, Breeding, Boring, Pumping, Locomotion, Spawning.	*Crystallographic*: Prismatic, Crossed-lamellar, Foliated, Density, Porosity, Aragonitic, Calcitic, Vateritic. ə *External* (Ornamentation): Ridges, Spines, Constrictions, Bumps, Lines. ∩ *Internal*: Lines, Layers, Bands, Lamellae. See also crystallographic. ɔ *Taxonomic*: Stromatolites: Laminae. Molluscs: Costa, Varix, Suture, Septa, Concentrics. Corals: Trabeculae, Tabulae. Bryozoa: Diaphragm, Dissepiment. Botanical: Xylum, Tracheid, Growth ring. *Sedimentary*: Sorting, Grain size and shape, Crystallinity. Composition: Calcium carbonate, Calcium phosphate, Silica, Cellulose, Conchiolin, Dolomite, Siderite, Hematite, Carbon, Amino acid, Protein.

† Used with daily, monthly, annual

TABLE 1. A classification of words used to describe the increment and the series. Temporal, environmental, endogenous and structural/compositional words in that order are also used to name the increment or the series. The temporal unit is also a functional unit of use to both the geophysicist and the evolutionary biologist. One can denote the precision of the temporal, environmental or endogenous specification with the name. Several terms would increase the precision as in semi-sidereal daily vs semi-daily increment. Quotation marks would indicate approximate as in post-spawning, 'winter' series. A question mark would indicate doubt as in daily? increment

organisms are more suited to geophysical analysis than others, and the next phase of research by students of growth rhythms may be the classification of species on the basis of their chronometric reliability. Stated in terms of the temporal increment, chronometric species can be expected to have:

(1) Readily describable growth increments, and continuous series, each with easily defined boundaries.
(2) Well-preserved growth increments (fossils).
(3) A known period for each increment, possibly associated with an environmental cycle or metabolic rhythm.

The evolutionary biologist can compare temporal increments with similar periods between species as a means to determine the evolution of biological growth rhythms. For example, comparison of the daily increment between species would compel us to note that in some cases the increment consists of a layer of calcium carbonate and a layer of conchiolin and in other cases it consists of a layer of dense calcium carbonate and a layer of porous calcium carbonate. In other species, there may be no daily increment. If tides are found to be the environmental stimulus of the daily increment in one species and day–night cycles the stimulus in the other, we have a means of determining the different selective factors for the different biorhythms. In short, the temporal increment is a functional unit enabling the comparison of (even apparently unrelated) taxonomic characters, and these features combined with the environmental and endogenous information contained in the increment may then be of use in discussing the evolution of biological growth rhythms.

All of this bears on a growth rhythm paradox. Environmental cycles (24-hr light–dark, 14-day tidal, etc.) were of different period long ago when the Earth rotated more rapidly. If, as some people believe, growth rhythms are controlled by internal clocks independent of the environment, one has difficulty in explaining the origin of such clocks and subsequent selection to match changing environmental periods through time. Simplified terminology prompts comparison of growth patterns of different fossil species as one approach to this problem.

Acknowledgments

I thank Colin Scrutton for his review of the manuscript. I am grateful to Marion Turner for typing the paper and to Dorothy Hewett for preparing the figures.

References

Clark, G. R., II (1974). Growth lines in invertebrate skeletons. *Ann. Rev. Earth Planet. Sci.*, **2**, 77–99
Neville, A. C. (1967). Daily growth layers in animals and plants. *Biol. Rev.*, **42**, 421–441

GROWTH INCREMENTS IN FOSSIL AND MODERN BIVALVES

WILLIAM B. N. BERRY
Department of Palaeontology, University of California, Berkeley
AND
RICHARD M. BARKER
Eel Brook, Yarmouth Co., Nova Scotia

Abstract

Growth increments in bivalve shells have attracted attention as possible indicators of tidal activity. Examination of modern bivalve shells indicates that many animals living in the low intertidal–shallow subtidal commonly secrete a shell growth increment in relation to tidal activity during most of the year. The apparently tide-related growth increments appear to be clustered, commonly in groups of about 15, although a group of approximately 29–30 is present in shells of some species. The clusters suggest relationship to fortnightly and monthly tidal rhythms.

Both growth increments and clusters of them similar to those apparently developed in relation to tidal activity in modern bivalve shells may be recognized in shells of certain fossil bivalves. The number of increments in each cluster in fossil shells may be used to suggest the possible number of tides in each fortnight or month in the past. A slightly greater number of apparent growth increments is present in each cluster in fossil shells than may be observed in each cluster in shells of present-day animals. The number of apparent growth increments in each cluster in shells from certain Cenozoic, Mesozoic and Late Palaeozoic strata plotted against geological age of the shells suggest that rate of increase in number of increments in each cluster may have been different in the Late Palaeozoic from what it was in the Mesozoic and Cenozoic. Triassic increments in each cluster are anomalously low compared with those from the Late Palaeozoic and Mesozoic. These data may suggest changes in the Earth's rotation rate and in tidal activity during the past.

Introduction

Interest among geologists and geophysicists in growth increments in modern and fossil organisms grew from Wells' (1963) suggestion that the increments seen on fossil coral epithecae had formed daily and that they appeared to be clustered into what might be concluded to be yearly growth increments. Wells' (1963) suggestion found an appeal among those geophysicists interested in the history of the Earth–Moon system and the effects of tidal forces upon the Earth's rotation in the geological past (Lamar and Merifield, 1966a,b; Runcorn, 1964, 1966). The realization developed that inasmuch as the Moon exerted a primary influence on the tides, and because coral growth increments could be considered to be reflective of tidal activity, then data concerning the influences of tidal activity in the past could be sought by close comparisons of growth phenomena in fossil and modern organisms.

Wells (1963) had little direct experimental data upon which to base his suggestion concerning the daily aspect of the fine growth increments he observed. The essential relationship in favour of his suggestion was the apparently close correspondence between the number of fine growth increments included between major grooves, which were presumed to have formed yearly on certain Devonian coral epithecae, and the number of days in the year in the Devonian obtained by projecting the astronomical data suggestive that the Earth's rotation has been decelerating at the rate of about 2 sec in 100,000 years back into the Devonian. Wells (1963) noted that there should have been about 400 days in the year in the mid-Devonian, based on his examination of growth increments in corals, and he pointed out that this figure obtained from his analysis of the corals was consistent with projecting the astronomical data back in geological time into the Devonian and assuming that the present suggested rate of deceleration of the Earth's rotation had been approximately the same since the Palaeozoic.

Barker (1964) made thin sections of many modern clam shells from localities along the east coast of the United States and recognized a number of sets of growth increments. He pointed out that the finest growth increments were commonly clustered. The clusters of fine growth increments tended to form ridges and grooves that could be observed on the shell surface. Barker indicated that the clusters which formed the ridges and grooves were commonly made up of about 15 or, in some shells, 29–30, fine growth increments. He also drew attention to major constrictions or grooves on the shell surface of some of the clams he examined that included either 12 or 24 of the ridges and grooves. The clusters of fine growth increments that numbered 15 or about 29–30 were suggested to be the result of the influence of fortnightly and monthly tidal phenomena (Barker, 1964). The major grooves were noted to be closely similar to grooves demonstrated to form once a year in response to either no or very slow shell growth during the winter months in the Pismo clam (Weymouth, 1923), certain cockles (Craig and Hallam, 1963; Orton, 1926) and *Tellina tenuis* (Stephen, 1928).

Barker's (1964) observations and suggestions indicated the possibility that clam shells might record the influence of tidal phenomena and that because of this, fossil clam shells could be useful in analyses of the history of the Earth–Moon system as well as of tidal phenomena in the geological past. Consequently, Barker (1970) set out upon a programme of analysis of growth in modern clams in response to tidal as well as other environmental phenomena and comparison of the results of those analyses with the growth increments that could be observed in thin sections of fossil clam shells. Many of the results recorded herein are drawn from Barker's (1970) summary of his studies.

Growth in modern bivalves

Observations of growth in modern clams were made both in the laboratory and in natural habitats. Shell growth in the laboratory specimens that was closely similar to that seen in animals studied in natural habitats was achieved in only one species, *Kellia suborbicularis*.

Of more significance to the relationship between tidal phenomena and shell growth were studies of a number of clam species in staked-out areas in natural settings. Species studied closely include *Chione californiensis*, *Chione undatella*, *Protothaca staminea* and *Mercenaria mercenaria*. Closest attention was given to growth in populations of *Chione undatella* at localities along the shores of the Gulf of California and to a population of *Protothaca staminea* in Bodega Bay on the California coast.

Protothaca staminea

The study of *Protothaca staminea* included staking-out an area in the shallow waters of Bodega Bay and then collecting 23 specimens of *P. staminea*, notching them, and subsequently returning them to their places on the sea floor. The area was visited at 21, 35 and 72 days after the shells had been notched and about 6 individuals were collected at each visit. Little or no shell growth appeared to have taken place in the first 15–20 days after the individuals were notched. Shells collected 35 days after they were notched had 15 fine growth increments. Shells of specimens left undisturbed for 72 days after notching had 52–60 fine growth increments. The fine growth increments in the latter shells were clustered in groups of about 15 to form $3\frac{1}{2}$–4 clusters. The data from the observations of the shell of *P. staminea* were taken to suggest that, when normal shell growth began after notching, a fine growth increment was added at every tide and that the influence of the fortnightly tidal cycle was indicated by the clusters of fine increments.

Chione undatella

A group of 808 individuals of *Chione undatella* was examined in the period March–May, 1967 and in February, 1968 in Cholla Bay (on the southern coast of Adair Bay near Puerto Penasco) on the Gulf of California coast of Sonora, Mexico. The area is near the northern end of the Gulf of California. A sampling grid was established to facilitate analysis of the clams in relationship to the specific area in which they live. *Chione undatella* commonly dwells shallowly buried in tidal flats. It was found in abundance in the tidal flats and along the margins of the main drainage channel in Cholla Bay. Specimens were obtained and marked for study from relatively near-shore stations seaward.

High tides were observed to travel into the marine marsh at the head of the bay, whereas low tides did not flood the marsh. Spring tides have been recorded as much as a mile from the marsh.

Surface water temperatures were recorded daily during April and May, 1967. Temperatures in the main drainage channel at positions near the head of the bay ranged from 24–25·5°C at noon and 15–16°C at 9–10.00 in the evening. Daytime surface water temperatures in the main channel near the bay mouth ranged from 23–24°C.

Chione undatella was found in medium to coarse sands in which large shell fragments were commonly present. They were also found in some silts and fine sands.

The *Chione* specimens examined were observed to be shallow burrowers. Only rarely was any part of the shell exposed except when the tidal flats had been scoured by strong tides related to strong spring tides or to storms.

Observation of specimens whose shell margins had been notched at intervals over a 2-year period indicated that a major groove developed on the shell surface during the winter. The grooves were similar in appearance to those concluded to have formed annually on *Cardium edule* (Orton, 1926; Craig and Hallam, 1963; House and Farrow, 1968; Farrow, 1971, 1972). Within the context of the major grooves that formed yearly, both fine growth increments and clusters of fine growth increments could be identified. Fine growth increments were clustered in groups of approximately 15. Each cluster appeared to comprise a ridge-groove growth component visible on the shell surface.

Disturbance rings were also observed on the shells. In many cases, the disturbance rings were sharply incised and could be easily identified on the shell surface. Some of the disturbance rings appeared similar to the grooves formed in the winter-months interval of slow or no shell growth. Disturbance rings may be related to bottom scouring during storms when the animals are toppled from their normal positions and remain disturbed from their normal life positions for a time.

The number of growth increments between major grooves in the *Chione undatella* specimens examined indicated that little or no shell growth took place for from 3 to as long as $4\frac{1}{2}$ months in the interval November–March. Older animals appeared to have been inactive in terms of shell growth for longer intervals than younger animals.

The fine growth increments added during the summer months in all shells are relatively wider than similar increments added during the spring and autumn. Such growth increments as did develop in some animals during the winter are relatively thin.

When growth increments among those animals living in sandy bottoms were compared with those among animals from silty and muddy bottoms, the growth increments from animals on the coarser grained bottoms were observed to be relatively wider than those in shells of animals living in finer grained sediments. Growth increments were thus wider and more easily seen in animals from the main channel and from positions nearer-to-shore than in animals from the relatively more off-shore positions and areas of finer grained bottoms. Disturbance rings were, however, more numerous in shells of those animals living in nearer-to-shore and main channel positions than in shells of those animals living in relatively deeper areas.

Latitudinal variation in shell growth in Chione undatella

Specimens of *Chione undatella* from Coyote Cove, Concepcion Bay, on the Baja California coast of the Gulf of Mexico, were obtained for comparison of shell growth phenomena with the Cholla Bay specimens. Coyote Cove is approximately 5° latitude south of Cholla Bay. The environments in which the animals are living are similar.

Surface water temperatures at the mouth of Concepcion Bay during February average approximately 8·5°C warmer than similar waters in Cholla Bay in the same month. The waters in Cholla Bay warm to almost the same surface water temperatures in summer as surface waters in Concepcion Bay.

Similar shell growth increments were observed on the Concepcion Bay specimens of *Chione undatella* as recorded in the Cholla Bay specimens. The groove formed during the winter months was not so marked in the Concepcion Bay specimens, however, and several individuals were observed that appear to have added very fine growth increments throughout the winter. Addition of fine growth increments appears to have begun in other specimens 2 to 3 months earlier in the Concepcion Bay specimens than in the Cholla Bay specimens. The longer interval of little or no shell growth in the Cholla Bay specimens appears to be correlated with the colder winter water temperatures in Cholla Bay.

Some conclusions from examination of shell growth in living animals

The observation of shell growth in *Protothaca staminea* and in *Chione undatella* contributes to the data bearing on environmental influences on growth among bivalves. The data developed in study of *Chione undatella* particularly are consistent with observations of growth among cockles recorded by Farrow (1971, 1972), growth in *Chione undatella* noted by Rosenberg (1972), and studies of *Tellina tenuis* (Stephen, 1928), *Mercenaria mercenaria* and *Callocardia morrhuana* (Pratt and Campbell, 1956; Pratt, 1953), *Mya arenaria* (Swan, 1952) and *Cardium edule* (Orton, 1926). The several observations of living clams in natural habitats indicate that certain species that live shallowly buried in the shallow subtidal and in the intertidal secrete a fine growth increment in relation to tidal activity during the growing season. Furthermore, the fine, tide-related, growth increments are clustered into groups of approximately 15 (see also Rhoads and Pannella, 1970). The clusters appear to form in response to the fortnightly tidal rhythms. Not every individual secretes a growth increment with every tide, and some individuals apparently secrete 2 fine growth increments in a day. These observations indicate that large populations of animals from any one area must be examined and counts of fine growth increments in each cluster need to be treated statistically in order to obtain relatively accurate data indicative of tidal activity.

Storm-related tides may disturb the animals such that disturbance rings reflective of no shell growth for one or a few tides may develop. Disturbance rings appear in shells of some but not all members of a clam species' population living in an area over which strong tidal activity has taken place. Disturbance rings may also form in response to relatively long periods of exposure and to frosts (Farrow, 1971, 1972).

Clam species living in temperate waters in subtidal and intertidal environments commonly do not add appreciable amounts of shell material for an interval of time during the colder temperatures of the winter. Young individuals in any population may not cease shell growth for as long a period of time in the winter as older individuals. Latitudinal distribution of a particular species may be reflected in the length of time shell growth may cease during the winter (see also Pratt and Campbell, 1956). The more northerly members of a species may stop shell growth for longer intervals than more southern members of the same species. Bottom conditions also play a role in the appearance of the fine growth increments as does the position of different members of any one population in an area in relation to water depth.

All of these factors suggest that the fine growth increments are commonly deposited either during or in response to tidal activity during an active growing season but that the growing season may be only a few months in duration in older animals in any species, and in those members of a species living in relatively colder waters than other members. Although fine growth increments tend to be relatively easier to distinguish in many bivalves living in coarser sediment than in many of those living in finer, among animals of one population, animals living in coarser bottoms are more easily disturbed by storm activity and hence their shells may not include as complete a record of each tide as those living in finer grained bottoms.

Growth increments in shells of modern bivalves

The studies of shell growth in living animals indicate that certain patterns or clusters of fine growth increments could be sought among clam shells of any age, and that the patterns could be considered indicative of tidal activity. Consequently, a number of shells of modern species obtained from museum collections were examined for the purpose of attempting to identify growth increments similar to those recognized in the shells of living animals. Shells of animals from a number of different habitats and representatives from several different genera and families were examined. Approximately 15 fine growth increments were recognized in individual clusters of shells of 63 species from 19 different families. Bunching of 29–30 fine growth increments which were considered indicative of monthly clustering was identified in 12 species from 9 families. Indistinct or no clusters of fine growth increments were observed in 17 species from 7 families.

Species belonging to genera within the Family Veneridae have the most conspicuous clusters suggestive of fortnightly tidal phenomena. Species in which apparent fortnightly clusters were seen include those from the following venerid genera: *Venus, Mercenaria, Chione, Protothaca, Tivela, Meretrix, Dosinia* and *Macrocallista*. Certain species among the cardiids, crassatellids, mactritids and carditids also had approximately 15 fine growth increments in each primary cluster. Nearly all of the species examined that had fine growth increments clustered to groups of approximately 15 are burrowers into bottoms in the intertidal or live in the relatively shallow subtidal.

Species that lacked distinct bunching of fine growth increments were found among nuculanids known to live in bathyal and abyssal waters as well as relatively shallow subtidal and intertidal dwelling tellinids, anomiids, mytilids, pectinids and oysters.

The survey of shell growth increments seen in shells of representatives from several different bivalve families obtained in museum collections drew attention to at least some taxa the fossil representatives of which might prove useful in an examination of apparent shell growth increments in fossil bivalve shells, to determine the possible influences of tidal activity in the geological past. The results of the survey also pointed out that only certain bivalve taxa could be used in any attempt at deducing the influences of tidal phenomena from bivalve shell growth increments.

Fossil bivalve shells

Observations of bivalve shell growth in response to tidal activity provide a basis for interpretation of growth increments seen in fossil bivalve shells. Fossil bivalve shells were obtained from many parts of the geological record and examined to ascertain if fine growth increments and clusters of them similar to the clusters recognized in living animals could be identified and counted. Thin sections were made of the fossil shells to examine the fine growth increments. The shells were cut approximately normal to the orientation of the growth increments along a midline that passed from the beak through the umbo to the shell margin. Not only fine growth increments but also clusters of them that appeared similar to the clusters seen in shells of living animals were sought.

Preservational factors

Many of the shells examined appeared to have relatively well-preserved growth increments when they were examined superficially, but study of thin sections of shells from numbers of specimens from all parts of the stratigraphic column revealed that many were so highly recrystallized that growth increments were either partly or wholly destroyed or had become so indistinct that reliable counts of fine growth increments could not be made. Shell structure of some bivalves is such that although growth increments may be distinguished on the shell surface,

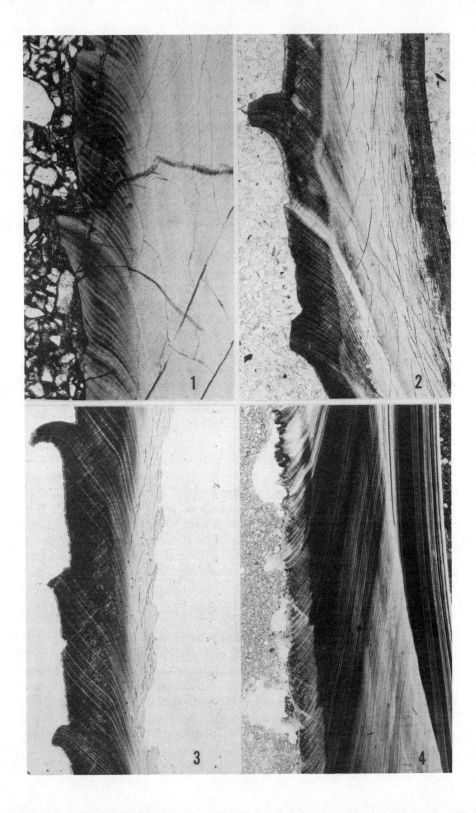

fine growth increments are difficult to identify in thin section. Other bivalve shells, particularly those that include a relatively thick layer composed of needle-like aragonite crystals, are difficult to thin section adequately enough to enable precise counts of the growth increments.

The several preservational aspects of fossil bivalve shells add significantly to the factors that must be considered before bivalve shell growth increments are used as indicators of tidal activity. The preservational factors inherent in fossil bivalve shells as well as the variations in shell growth in response to various aspects of the environment observed in living animals need to be taken into account. The several environmental factors that influence shell growth and the preservational aspects of shells found as fossils indicate that the data obtained from counting shell growth increments in fossil bivalve shells are so imprecise that any conclusions concerning tidal influences in the past which are based on them are highly speculative.

The fossil shell data

Despite the several problems noted in using growth increments in fossil bivalve shells as significant indicators of past tidal activity, growth increments and clusters of them similar to those demonstrated to form in relation to daily and fortnightly tidal rhythms are present in the shells of many fossil bivalves and counts of them may be used for speculations on past tidal influences. Growth increments and clusters of them similar to those in modern bivalve shells have been recognized in fossil bivalve shells from several parts of the stratigraphic record (Barker, 1968; Berry and Barker, 1968; Pannella and MacClintock, 1973; Pannella, MacClintock and Thompson, 1968; House and Farrow, 1968). In addition to the recorded results of counts of growth increments in fossil bivalve shells (Barker, 1968; Berry and Barker, 1968), shell growth increments in shells of some other fossil bivalves have been counted. The species of which the shells are sufficiently well preserved to yield significant numbers of fine growth increments that were clustered, the rock units from which the shells came, and the ages of the rock units are listed in Table 1. The number of fine growth increments in each cluster seen in the shells is summarized in Figure 2. Counts of fine growth increments were made through 2 or more contiguous sets of clusters to minimize the problems inherent in recognizing boundaries between clusters.

FIGURE 1. Thin sections of fossil and recent bivalve shells illustrating fine (daily formed) growth increments which occur in clusters of those that are relatively wide alternating with those that are relatively narrow. The clusters probably developed in response to the fortnightly tidal rhythm.
1. Venerid shell fragment in Meganos Sandstone, Palaeocene, California, ×27·7.
2. *Chione cancellata*, Recent, Bradenton, Florida, ×19·8.
3. *Chione undatella*, Recent, Cholla Bay, Gulf of California, Sonora, Mexico, ×18·6.
4. *Mercenaria mercenaria*, Recent, Bradenton, Florida, ×14·5

TABLE 1. Growth layers counted in fossil bivalves. Fine growth increments per cluster (M is mean number of fine growth increments in each cluster examined closely for all specimens of each species cited)

Geological age, species and locality	Specimen	Number of contiguous clusters	Number of fine growth increments
Pleistocene			
Chione californiensis ($M = 14.73$)	1	16	236
Sandstone near Punta Cholla, Sonora,	2	5	74
Mexico	3	10	148
	4	5	73
	5	7	103
	6	5	74
	7	8	118
	8	6	86
Chione succincta ($M = 14.75$)	1	4	59
Palos Verdes Formation, Newport	2	6	89
Beach, California	3	3	45
	4	4	57
	5	3	45
Chione undatella ($M = 14.78$)	1	11	164
San Pedro Standstone, San Pedro,	2	3	43
California	3	8	119
	4	5	73
	5	2	30
	6	4	59
Pliocene			
Chione cancellata ($M = 14.82$)	1	7	105
Caloosahatchee Formation, Southern	2	3	45
Florida	3	6	89
	4	3	44
	5	3	43
Codakia orbicularis ($M = 14.84$)	1	5	74
Caloosahatchee Formation, Southern	2	5	73
Florida	3	4	60
	4	4	60
Oligocene			
Crassatella lincolnensis ($M = 14.81$)	1	8	119
Sacate-Gaviota Formation, Santa	2	7	103
Inez, California	3	4	60
	4	5	74
	5	8	118
Macrocallista hornii ($M = 14.77$)	1	2	30
Sacate-Gaviota Formation, Santa	2	4	59
Inez, California	3	6	88
	4	3	44
	5	2	30

TABLE 1—*continued*

Geological age, species and locality	Specimen	Number of contiguous clusters	Number of fine growth increments
Eocene			
Meretrix splendida ($M = 14·86$)	1	7	105
Lutetian, Morigny, France	2	5	73
	3	2	30
Palaeocene			
Venerid shell fragments in thin sections	1	4	60
of Meganos Sandstone, Byron Quad-	2	4	60
rangle, California ($M = 14·85$)	3	4	59
	4	3	45
	5	3	44
	6	2	29
Cretaceous			
Crassatella vadosus ($M = 14·86$)	1	7	104
Ripley Formation, Coon Creek,	2	6	89
Tennessee	3	5	74
	4	5	75
	5	2	30
	6	3	44
Idonearca vulgaris ($M = 14·81$)	1	6	88
Ripley Formation, Coon Creek,	2	5	74
Tennessee	3	4	60
	4	3	44
	5	3	45
Glycymeris lacertosa ($M = 14·89$)	1	8	119
Ripley Formation, Coon Creek,	2	5	74
Tennessee	3	4	60
	4	2	30
Lucina subundata ($M = 14·92$)	1	7	104
Cody Shale, Greybull, Wyoming	2	5	75
Jurassic			
Lima gigantea ($M = 14·92$)	1	3	44
Leamington, England	2	2	30
	3	4	61
	4	3	44
Triassic			
Septocardia sp. ($M = 14·88$)	1	5	74
Luning Formation, Pilot Mountains,	2	2	30
Nevada	3	4	60
	4	6	89

TABLE 1—continued

Geological age, species and locality	Specimen	Number of contiguous clusters	Number of fine growth increments
Carboniferous			
Astartella concentrica ($M = 15.0$)	1	2	31
Graham Formation, Texas	2	3	44
	3	3	45
Myalina subquadrata ($M = 15.1$)	1	5	75
Wayland Formation, Brownwood,	2	4	62
Texas	3	3	45
	4	3	44
Conocardium sp. ($M = 15.1$)	1	6	91
Carboniferous, Belgium			
Devonian			
Conocardium sp. ($M = 15.25$)	1	8	122
Alpena Limestone, Michigan			

An effort was made to obtain well-preserved fossil shells of members of the Family Veneridae because shells of most modern venerids examined appeared to record tidal activity most faithfully. As many individuals as it was possible to obtain from each locality for each fossil species were examined to calculate an average number of shell growth increments in shells of the same species from the same locality. Study of growth increments in large numbers of shells should permit an average value for the number of fine growth increments in each cluster to be obtained and the different responses in terms of shell growth of each individual animal to environmental factors may be averaged in the calculations. Commonly, however, only a few shells of a species could be obtained from a locality that were well enough preserved to permit counts of their growth increments. The data obtained from the counts of growth increments may thus be biased toward animals that laid down growth increments more often than every tide or toward animals that did not lay down a shell growth increment in response to every tide. The available data plotted in Figure 2 are thus relatively imprecise and may be biased away from reflecting true tidal activity. Any conclusions suggested by patterns indicated in that plot are only speculations.

The number of fine growth increments recognized in each cluster of such increments in the shells of different geological ages examined are considered suggestive of the daily and fortnightly tidal cycles, by analogy with the observations of shells of living animals. The number of increments in each cluster indicated in Figure 2 is suggested to reflect the number of days in the fortnight at different times in the past. The plot of the number of increments in each

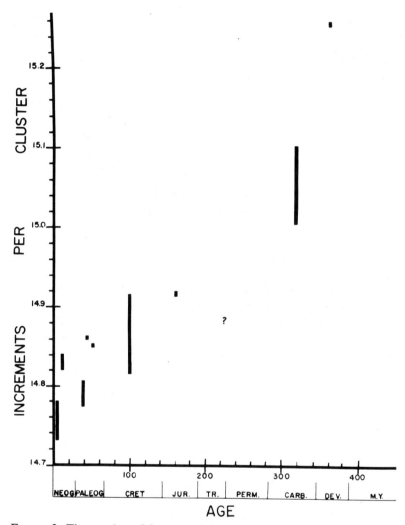

FIGURE 2. The number of fine growth increments recognized in each cluster in fossil bivalve shells is plotted against the ages of the shells. The counts from the Triassic shells are questioned because shell preservation is poor and the fine increments indistinct

cluster includes range in variation in the number of fine increments in each cluster in a single shell, range in variation in increments in each cluster in different individuals of the same species from the same deposit, and range in variation in increments per cluster among shells from 2 or more deposits (if such were available) and 2 or more species (if they were obtained). The width of the bars suggests that shells from different deposits of about the same age (in terms of the geological Periods and Epochs) are probably not precisely coeval. The indicated range in variation in the number of increments in each cluster seen in shells from Triassic and younger age strata is approximately 0·15–0·20 increments per cluster. That range in variation may be considered relatively slight when the factors involved in deposition of the increments by the animal, preservation of the increments through fossilization, and observation of the increments in thin sections are considered. Although the counts are probably as precise as it is possible to achieve from fossil shells by visual observation with the light microscope, they lack the precision requisite for geophysical calculations.

The data plotted for the Triassic and younger shells may be suggestive that a certain amount of deceleration of the Earth's rotation has taken place since the early part of the Mesozoic. Alternatively, however, the data may suggest that tidal activity has been relatively similar since the early part of the Mesozoic, and that the rate of the Earth's rotation has changed little, if at all, since that time.

The few counts of increments in each cluster in shells of Late Palaeozoic age may suggest that a certain amount of deceleration of the Earth's rotation took place from the early part of the Mesozoic into the latter part of the Palaeozoic. The data may also suggest that some event took place during the Late Palaeozoic–Early Mesozoic interval which had an effect on the Earth's rotation rate. The pattern of changes in rate of deceleration of the Earth's rotation may be considered consistent with that suggested by Pannella and others (1968).

If tidal influences on the Earth's rotation changed as palaeogeography changed, then growth increments seen in fossil bivalve shells may reflect such changes. Palaeogeographic reconstructions for the world at intervals through the Palaeozoic and Early Mesozoic (Palmer, 1973; Whittington, 1973; Berry and Boucot, 1968; Boucot, Johnson and Talent, 1969; Ross, 1973; Hill, 1973; Gobbett, 1973; Kummel, 1973) suggest that oceanic areas (particularly shallow shelf seas) were far more widely spread throughout most of the Palaeozoic than during the Late Palaeozoic–Early Mesozoic and that the amount of the Earth's surface occupied by land areas increased markedly in the Late Palaeozoic–Early Mesozoic. If that palaeogeographical change had an effect on the tides and tidal influence on the Earth's rotation, then the change in Earth's rotation suggested by the data from fossil bivalve shells' growth increments is consistent with the evidence from palaeogeographic studies.

The growth increment data from fossil bivalve shells bearing upon changes in the Earth's rotation are meagre and lack the precision needed for geophysical calculations. Nevertheless, the data may be at least suggestive of certain changes in the influences of tidal activity and may be considered with other lines of evidence in reconstructing certain aspects of the Earth's history.

References

Barker, R. M. (1964). Microtextural variations in pelecypod shells. *Malacologia*, 2, 69–86

Barker, R. M. (1968). Fossil shell-growth layering and the periods of the day and month during Late Paleozoic and Mesozoic time, pp. 10–11 in 'Abstracts for 1966'. *Geol. Soc. Am. Spec. Pap.*, 101, 485 pp.

Barker, R. M. (1970). Constituency and origins of cyclic growth layers in pelecypod shells, Space Sciences Laboratory, University of California, Berkeley Report Series 13, Issue 36, 265 pp.

Berry, W. B. N. and Barker, R. M. (1968). Fossil bivalve shells indicate longer month, year in Cretaceous than present. *Nature*, 217, 938–939

Berry, W. B. N. and Boucot, A. J. (1968). Continental development from a Silurian viewpoint. XXIII. *Int. Geol. Congr. Proc.*, 3, 15–23

Boucot, A. J., Johnson, J. G. and Talent, J. A. (1969). Early Devonian brachiopod zoogeography. *Geol. Soc. Am. Spec., Pap.* 119, 113 pp.

Craig, G. Y. and Hallam, A. (1963). Size-frequency and growth ring analysis of *Mytilus edulis* and *Cardium edule*, and their paleoecological significance. *Palaeontology*, 6, 731–750

Farrow, G. E. (1971). Periodicity structures in the bivalve shell: experiments to establish growth controls in *Cerastoderma edule* from the Thames Estuary. *Palaeontology*, 14, 571–588

Farrow, G. E. (1972). Periodicity structures in the bivalve shell: analysis of stunting in *Cerastoderma edule* from the Burry Inlet (South Wales). *Palaeontology*, 15, 61–72

Gobbett, D. J. (1973). Permian Fusulinacea, pp. 151–158 in *Atlas of Palaeobiogeography* (Ed. A. Hallam), Elsevier, Amsterdam, 531 pp.

Hill, D. (1973). Lower carboniferous corals, pp. 133–142 in *Atlas of Palaeobiogeography* (Ed. A. Hallam), Elsevier, Amsterdam, 531 pp.

House, M. R. and Farrow, G. E. (1968). Daily growth banding in the shell of the cockle, *Cardium edule. Nature*, 219, 1384–1386

Kummel, B. (1973). Lower Triassic (Scythian) molluscs, pp. 225–233 in *Atlas of Palaeobiogeography* (Ed. A. Hallam), Elsevier, Amsterdam, 531 pp.

Lamar, D. L. and Merifield, P. M. (1966a). Age and origin of the Earth–Moon system revealed by coral growth lines. *Trans. Am. geophys. Un.*, 47, 486

Lamar, D. L. and Merifield, P. M. (1966b). Length of Devonian day from Scrutton's coral data. *J. geophys. Res.*, 71, 4429–4430

Orton, J. H. (1926). On lunar periodicity in spawning of normally grown Falmouth oysters (*O. edulis*) in 1925, with a comparison of the spawning capacity of normally grown and dumpy oysters. *J. mar. biol. Ass. U.K.*, 14, 195–225

Palmer, A. R. (1973). Cambrian trilobites, pp. 3–11 in *Atlas of Palaeobiogeography* (Ed. A. Hallam), Elsevier, Amsterdam, 531 pp.

Pannella, G. and MacClintock, C. (1968). Biological and environmental rhythms reflected in molluscan shell growth, pp. 64–80 in *Paleobiological aspects of growth and development* (Ed. D. B. Macurda, Jr.), Palaeont. Soc. Mem. 2 (*J. Palaeont.*, 42, no. 5, supp.), 119 pp.

Pannella, G., MacClintock, C. and Thompson, M. N. (1968). Palaeontological evidence of variations in length of synodic month since Late Cambrian. *Science*, 162, 792–796

Pratt, D. M. (1953). Abundance and growth of *Venus mercenaria* and *Callocardia morrhuana* in relation to the character of bottom sediments. *J. mar. Res.*, 12, 60–74

Pratt, D. M. and Campbell, D. A. (1956). Environmental factors affecting growth in *Venus mercenaria. Limnol. Oceanogr.*, 1, 2–17

Rhoads, D. C. and Pannella, G. (1970). The use of molluscan shell growth patterns in ecology and paleoecology. *Lethaia*, **3**, 143–161

Rosenberg, G. D. (1972). Correlation of shell structure and chemistry with life habitat of the bivalve *Chione undatella* Sowerby. *Geol. Soc. Am., Abstracts with Programs*, **4**, 227

Ross, C. A. (1973). Carboniferous foraminiferida, pp. 127–132 in *Atlas of Palaeobiogeography* (Ed. A. Hallam), Elsevier, Amsterdam, 531 pp.

Runcorn, S. K. (1964). Changes in the Earth's moment of inertia. *Nature*, **204**, 823–825

Runcorn, S. K. (1966). Corals as paleontological clocks. *Scient. Am.*, **215**, 26–33

Stephen, A. C. (1928). Notes on the biology of *Tellina tenuis* da Costa. *J. mar. biol. Ass. U.K.*, **15**, 683–702

Swan, E. F. (1952). The growth of the clam *Mya arenaria* as affected by the substratum. *Ecology*, **33**, 530–534

Wells, J. W. (1963). Coral growth and geochronometry. *Nature*, **197**, 948

Weymouth, F. W. (1923). The life history and growth of the Pismo clam (*Tivela stultorum* Mawe). Fish Bulletin No. 7, pp. 1–120, California Fish and Game Commission, Sacramento

Whittington, H. B. (1973). Ordovician trilobites, pp. 13–18 in *Atlas of Palaeobiogeography* (Ed. A. Hallam), Elsevier, Amsterdam, 531 pp.

DISCUSSION

ROSENBERG: When I studied living *Chione* from different areas of a California coastal lagoon, specimens from some habitats showed more prominent tidal series. I think clams from different habitats would give different periodicities. How do you know you aren't getting such effects in the fossil record?

BERRY: We tried to get our fossils from sandstones, or at least we selected for specimens which showed similar patterns to those of living specimens from sandy substrate environments where there were more prominent tidal patterns.

FARROW: There seems a basic difference between *Chione undatella* calcification patterns which deposit a periostracal band when closed at night, and *Cardium* and *Mercenaria* which calcify at night.

BERRY: Our work was not intended to be so precise as to support such a generalization at this time.

CARBONIFEROUS CORAL GEOCHRONOMETERS

G. A. L. JOHNSON AND J. R. NUDDS
University of Durham, England

Abstract

Preservation of the fine ornament of the epitheca of rugose corals is unusual, but it has been found in 6 genera from the marine limestones and shales of Carboniferous age in northern England. These specimens show the daily growth bands with some indication of monthly banding. No specimens have been found that show the detail of the epitheca over the complete length of the corallite and the best specimens show only a limited number of consecutive monthly bands. The best developed epithecal banding has been found in fasciculate *Lithostrotion*. In *L. martini* M. Edwards and Haime epithecal banding gives an average monthly count of 30·2 days, indicating 391·09 days in the Lower Carboniferous year; this figure compares well with the estimate of 393 days based on geophysical methods (Wells, 1963). Growth rate in this species varies between 3 and 5 mm per month and seems to be controlled partly by environmental factors though there is a general trend for the growth rate to quicken up the stratigraphical column. With the increase in growth rate of *L. martini* up the geological column it has been found that there is a gradual wider spacing of the tabulae. Initial calculations have shown that the number of tabulae formed during a month may be constant for this species at 7–8 tabulae per month. If the constancy of tabulae development could be proved in other species of rugose corals the possibilities of using them as geochronometers would be greatly increased.

Introduction

The time-growth significance of parallel concentric markings on the epitheca of well-preserved corals was first recognized by Whitfield (1898) who suggested that they were annual growth increments brought about by seasonal temperature changes. Fossil corals were used as geochronometers by Wells (1963) to supply an independent check to radiometric age dating and compile an absolute geological time-scale based on fossil data. He recognized epithecal banding of 2 orders, annual banding well separated on the corallum and finely spaced ridges caused by daily growth increments. The annual growth banding is thought to be caused by temperature control of lime deposition by the coral polyp. Where seasonal temperature changes are sufficient lime secretion by the polyp would either be slowed down or cease altogether during the colder part of the year, giving the annual epithecal banding. Wells checked annual and daily banding in living corals in which the annual linear growth-rate was fairly well known and found reasonable confirmation of his deductions. The cause of the daily increment ridges is the variation in the rate of lime intake by the tissue of coral polyps between the light conditions of the day and the darkness of night. Despite the difficulty of finding fossil corals of sufficiently fine preservation to show the daily and annual growth increments Wells (1963) obtained specimens of Middle Devonian corals, including *Heliophyllum halli* M. Edwards and Haime, from which it was possible to count 385 to 410 (average 400) fine daily ridges per annual increment. He postulated therefore that the Middle Devonian year contained roughly 400 days. This figure agrees well with the calculated estimate assuming that the Earth's rotation about its polar axis has been slowing down owing to loss of rotational energy by tidal friction of about 2 sec per 100,000 years, as Wells was able to show.

Further important advance in the understanding of epithecal banding was made by Scrutton (1965), again working on Middle Devonian corals. He was able to recognize daily growth ridges and regular bands of much shorter period than the annual growth increment which he was unable to identify. By counting he found that there was an average of some 30·59 daily ridges to each of the bands; this figure was recalculated in 1970 as 30·66 daily ridges per band. Using Wells' (1963) data Scrutton assumed that there were 399 days in the Middle Devonian year. This means that 13·04 bands of daily ridges (later recalculated to 13·01 bands per year) were produced every year and Scrutton concluded that this suggested a lunar monthly effect on the accretionary growth of the coral skeleton. Studies of modern corals give little support to this, but less lime may be deposited at the full Moon when the corals are preoccupied with breeding and a monthly constriction has been recorded in the epitheca in some cases. Alternatively, a direct tidal influence may play an important role, but in either case it is the synodic month that is presumably being recorded. Scrutton and Hipkin (1973) stress that the supposed lunar monthly banding is not supported by satisfactory observations on living corals and the status of the bands is based on a numerical relationship between them and other accretionary banding of the epitheca.

Further analysis of daily growth increments and monthly banding on Silurian and Devonian corals and brachiopods has been completed by Mazzullo (1971). His figures are in general agreement with those of previous workers except that he uses maximum counts rather than the mean or mode of a group of data and his figures are thus proportionately exaggerated. Growth rhythms in bivalves have also been used as a source of growth increment data and cephalopods and stromatolites have also received some attention. This work is ably summarized by Scrutton and Hipkin (1973) who review the whole field of fossil geochronometers and the application of the results obtained from them.

The recognition of regular epithecal banding in corals corresponding to the synodic month allows important calculations to be made in material which does not show annual banding. Thus, as long as the monthly banding is finely preserved, 2 important measurements can be made. First, the number of days in the month can be counted and by simple multiplication by the number of lunar months in a year the number of days in the year can be calculated: the calculation of the number of synodic months in the year is discussed later in the text.

Secondly, the rate of accretionary growth of the epitheca of the corals can be measured per synodic month. Figures derived by this method are only available for the Middle Devonian corals at the present time. To supply an independent check various rugose coral genera from the marine limestones and shales of Lower Carboniferous (Viséan) and Upper Carboniferous (Namurian) age from Northern England have been studied and the results of this work are presented in this paper.

Annual and monthly growth increments

In northern England marine sediments with thick limestones containing rich coral faunas occur in both the Dinantian and the Namurian. An extensive unstable shallow shelf sea covered the region during this period in which bottom conditions were often marked by strong current action. Overturned and displaced coral colonies are common and rolled corals, in which the epitheca is highly abraded, are ubiquitous. Preservation of the fine detail of the epitheca is very rare and is normally associated with corals found in shale and limy shale matrices. In cerioid colonies of *Lithostrotion* and *Lonsdaleia* well-preserved epithecal banding has been found on internal corallites when the partly weathered corallum is split. Thamnastraeoid colonies of *Orionastraea* show fine banding on the holotheca when preservation is in soft calcareous shale. It must be stressed that though well-preserved accretionary growth banding has been found in 6 genera of Carboniferous corals in northern England, preservation of this excellence is rare and the specimens referred to here have been brought together over a period of some 10 years.

Though the corals from northern England show clearly defined daily-growth bands and some indication of monthly banding (Figures 1 and 2), no annual growth increments have been observed. Annual growth increments have been recorded in living corals by Whitfield (1898) and Vaughan (1915) and in Middle Devonian corals by Wells (1963). Seasonal temperature fluctuation is thought to be responsible for the annual banding on the coral epitheca with reduced secretion or non-secretion of lime in the skeleton during the colder winter period. If this mechanism for the production of the annual banding is correct the annual bands would be expected to be strongest away from the equator where the seasonal temperature effects are more marked. Conversely, near the equator, where the seasonal temperature changes are slight, annual banding would be expected to be poorly developed or absent. It is therefore significant that the Viséan and Namurian corals of northern England show no indications of annual banding, particularly as they have been collected within 10° of the position of the Carboniferous equator (Figure 3). In agreement with this hypothesis the Middle Devonian corals which Wells (1963) used to demonstrate annual banding come from seas which must have been almost 40° south of the Devonian equator (Figure 4). The absence of annual banding would seem to indicate coral growth within the broad equatorial belt of the Earth.

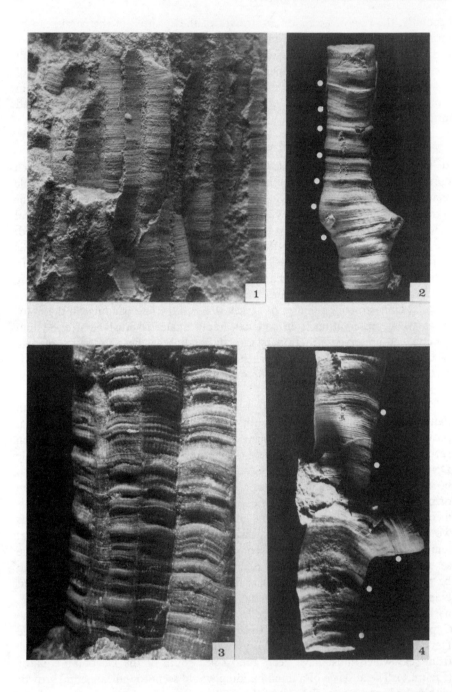

Epithecal banding, believed to be produced by synodic month periodicity, is clearly developed in a few corals from northern England when preservation is in lime-rich sediment. Unfortunately, well-preserved daily banding in specimens preserved in shale does not show satisfactory monthly banding. Up to 5 monthly bands have been found on a single broken corallite of *Lithostrotion martini* and on specimens of this type accurate counts of the days in a month can be made (Figure 1).

Scrutton (1970) states, from studies of Middle Devonian corals, that there are 13·01 synodic months in the Devonian year. At the present day there are 12·53 months in the year, thus by comparison of these figures it appears that the synodic month is increasing in length with time. Assuming that this rate of increase in the length of the month is constant we can calculate that there must have been 12·95 months in the Lower Carboniferous (Viséan) year. Another assumption which has to be made in using epithecal banding data is that the period of the Earth's rotation about the Sun, giving us the year, has been constant during the Phanerozoic.

Number of days in the Viséan year

Rugose corals from northern England used in the present study show daily growth bands and some indications of monthly banding. In no case has annual banding been observed in these corals. By far the best epithecal banding has been found in fasciculate *Lithostrotion* preserved in shale and shaly limestone. The soft nature of the surrounding sediment does seem to play a part in the preservation of fine detail of the ornament of the epitheca though, as Wells (1963) states, epithecal diurnal growth-lines are commonly abraded or corroded in living corals even before the death of the polyp. Still water conditions and relatively quick burial in a protective medium would help to preserve the fine epithecal detail. Corallites of *Lithostrotion martini* M. Edwards and Haime from the lower Viséan C_2-S_1 zone beds of Ashfell Edge, Westmorland, show distinctive synodic monthly banding with well-preserved diurnal banding. The coralla are crushed in the enclosing calcareous shale matrix and no corallites long enough to contain a year of 13 monthly bands have been collected; yearly increments cannot therefore be recognized. The number of diurnal ridges per monthly band

FIGURE 1. 1. *Lithostrotion junceum* (Fleming). Jew Limestone, Orton, Westmorland. D_2 zone, Viséan. Well-developed daily increments, but no apparent monthly banding.
P3581, × 4

2 and 4. *Lithostrotion martini* M. Edwards and Haime. Ashfell Sandstone, Ashfell Edge, Westmorland. C_2-S_1 zone, Viséan. 2: well-developed daily and monthly banding. No. 3Ci, × 2. 4: clear daily ridges run from parent to lateral offset. No. 3Ck, ×3·2.
White dots mark the monthly bands

3. *Lithostrotion minus* (McCoy). Great Scar Limestone, Horton-in-Ribblesdale, Yorks. Daily increments and well-developed monthly banding. P4688, ×2

FIGURE 2. *Lithostrotion martini* M. Edwards and Haime. Ashfell Sandstone, Ashfell Edge, Westmorland. C_2–S_1 zone, Viséan. Composite photograph showing the entire epithecal surface of a cylindrical corallite. Four monthly bands are shown with daily increments, marginal dots mark the monthly boundaries. No. 3Ci, × 6·5

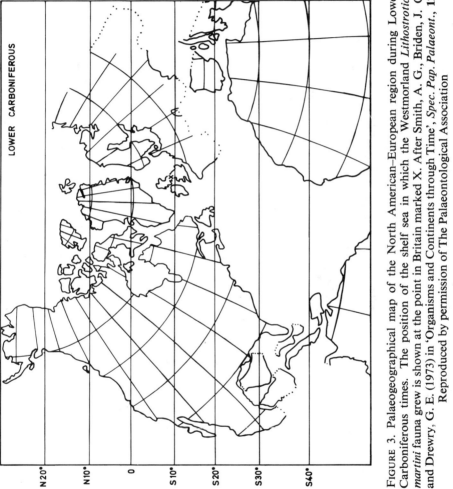

FIGURE 3. Palaeogeographical map of the North American–European region during Lower Carboniferous times. The position of the shelf sea in which the Westmorland *Lithostrotion martini* fauna grew is shown at the point in Britain marked X. After Smith, A. G., Briden, J. C. and Drewry, G. E. (1973) in 'Organisms and Continents through Time', *Spec. Pap. Palaeont.*, **12**. Reproduced by permission of The Palaeontological Association

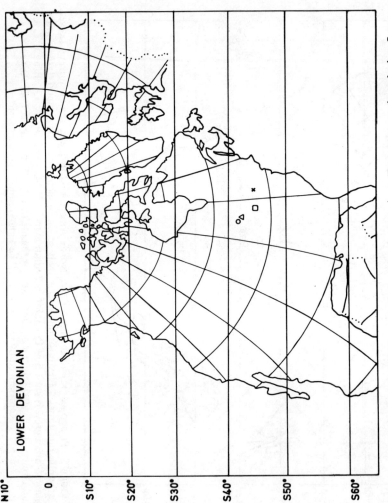

FIGURE 4. Palaeogeographical map of the North American–European region during Lower Devonian times. The position of the shelf seas from which Devonian corals were obtained by Wells and Scrutton is shown by the following symbols: ○, Roger's City, Michigan; △, Alpena City, Michigan; □, London, Ontario; ×, Western New York State. After Smith, A. G., Briden, J. C. and Drewry, G. E. (1973) in 'Organisms and Continents through Time'. *Spec. Pap. Palaeont.* **12**. Reproduced by permission of the Palaeontological Association

Carboniferous Coral Geochronometers

has been calculated using corallites where the epithecal surface is particularly well preserved and these counts give a mean figure of 30·2 days to the synodic month. This figure has been checked with other corallites from the same horizon in which the epithecal surface is less well preserved and counts for the days in the month agree well. If we calculate that there are 12·95 synodic months in the Viséan year we arrive at a figure of 391·09 days in the Viséan year. The deduced Viséan year can now be compared with the Middle Devonian year of 399 days calculated by Wells and Scrutton and we have clear evidence supplied by coral geochronometer evidence alone that the number of days per year is decreasing and that the Earth's rotation about its polar axis must be slowing down.

Slowing down of the Earth's axial rotation

From the epithecal banding data obtained from corals the rate of slowing down of the rotation of the Earth can be calculated and compared with the rates worked out by the geophysicist. Using work on astronomical observations over the past 200 years, Bott (1971) states that the slowing down of the rotation of the Earth about its axis is attributable to 2 factors. Lunar tidal friction causes a slowing down of about 18·1 sec/My, while tidal interaction between the Sun and the Earth contributes a further slowing down of up to 5 sec/My, a total slowing down of some 23 sec/My. The reliability of these figures over periods of the Earth's history longer than 200 years is doubtful as the factors causing deceleration are variable. They depend largely on frictional forces of ocean tides caused by the distribution of land and sea, the shape and number of large tidal estuaries and the relative development of wide shallow shelf seas.

Now only average values for the rate of deceleration of the Earth's rotation about its axis through geological time can be obtained from the estimate in recent times and also from the evidence provided by epithecal banding of Middle Devonian corals. The new figures provided by the counts of epithecal banding of Viséan corals make possible a comparison of the rate of deceleration between different periods of geological time using the following equation:

$$R = \left(\frac{31471200}{D_i} - \frac{31471200}{D_0} \right) \frac{1}{t} \qquad (1)$$

where R = Rate of deceleration
D_i = Days per year of the younger period
D_0 = Days per year in the older period
t = Time in My between D_i and D_0
31471200 = Sec per present day year of 364·25 days.

A calculation of the rate of deceleration of the Earth's rotation can now be made between the present day and the Middle Devonian using Wells' figures. Wells did not calculate the rate of slowing down from his coral data, but he did plot a

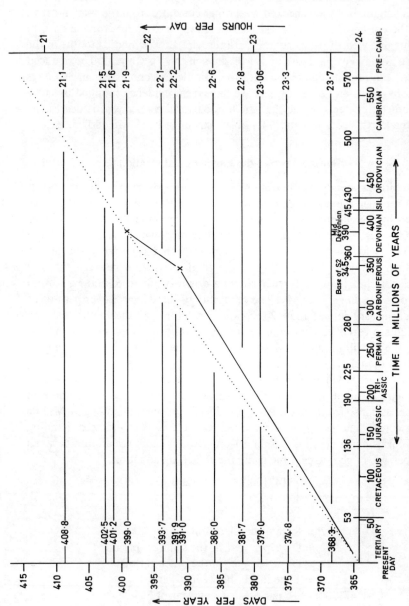

FIGURE 5. Plot of days in the year against geological time showing the changes in the rate of the Earth's axial rotation. The solid line shows changes in the rate of rotation between the Devonian to Carboniferous and Carboniferous to present day. Devonian and Carboniferous counts for days/year based on coral epithecal data. Age date of Carboniferous corals from Fitch *et al.* (1970) and date for Devonian corals from Lambert (1971). The dotted line is the average rotation rate between the present day and the Devonian and projected back to the Cambrian as originally published by Wells (1963)

graph of days in the year against geological time based on his counts for the Middle Devonian year (Wells, 1963, Figure 1). From this graph the rate of deceleration is seen to be 20·8 sec/My. For this graph Wells was using radiometric age data of Kulp (1961) which has since been revised: in particular the base of the Devonian system is now believed to be some 10–20 My older than was previously estimated (Lambert, 1971). This revision has the effect of reducing the rate of deceleration of rotation of the Earth to 19·29 sec/My, i.e. the length of the day increases by 19·29 sec every My.

The overall rate of deceleration of the rotation of the Earth between the present day and the Viséan (Lower Carboniferous) can be calculated using the epithecal counts of *Lithostrotion martini* from Westmorland. This produces a rather slower rate of deceleration of 17·18 sec/My (from equations (1) and (2), where $D_0 = 391·09$ and $D_1 = 364·25$). A comparison of the Devonian and Carboniferous rates of deceleration with the present day suggests that the rate was unusually high between the Devonian and the Carboniferous. This deceleration rate comes out at 35·45 sec/My (from equations (1) and (2), where $D_0 = 399$ and $D_1 = 391·09$). It has already been noted that the cause of deceleration in the rate of rotation of the Earth is largely brought about by frictional forces of oceanic tides that are increased by shallow shelf seas and wide tidal estuaries. Palaeogeography of later Devonian and Lower Carboniferous times involves the widespread marine transgression of mature coastal plains and lowland basins by a shallow shelf sea (Gignoux, 1955; Johnson, 1967). Clearly the palaeogeographical conditions are right for tidal forces to be abnormally high during this period of time and this is supported by the high deceleration rate for the rotation of the Earth at this time, calculated from the epithecal banding data of fossil corals.

The graph plotted by Wells (1963) of days in the year against geological time produced a straight line that assumes a constant rate of deceleration of the rotation of the Earth during the Phanerozoic. We now know from the comparison of the Devonian and Carboniferous data from fossil corals that this rate is variable. A future aim will be to make further epithecal counts on corals from other geological periods and in this way deduce the pattern of change in the rate of the Earth's rotation through as much of the Phanerozoic as possible. The graph can now be redrawn, modified by using the most recent radiometric age dates for the Phanerozoic and including the Carboniferous epithecal counting data (Figure 5). Calculations of the number of days in the year at unknown points in geological time cannot be made directly from the rate of deceleration of the Earth's rotation owing to the errors involved with variations in the deceleration rate. In the graph (Figure 5) the number of days in the year between two known points has been calculated independently of deceleration rate using the following equation:

$$N_x = \left(\frac{D_0 - D_1}{t_1}\right) t_0 + D_1 \qquad (2)$$

where N_x = Number of days at required time x
D_0 = Days per year in older period
D_1 = Days per year in younger period
t_0 = My between required time x and D_1
t_1 = My between D_1 and D_0.

Growth rate in Carboniferous corals

An independent check on the identity and authenticity of the epithecal banding of fossil corals can be made by growth rate comparison with living corals. If the growth rate of fossil corals from the Carboniferous falls within the known range of growth rate of present-day corals we have further evidence that the growth behaviour is similar and that the epithecal banding is directly comparable. According to Wells (1956) growth rate in living scleractinian corals varies in different areas and is greater in places where the average water temperature throughout the year is moderately higher. It also varies according to the structure of the skeleton, being slower in types with dense coralla and faster in types with light porous skeletons. Annual increments in overall height between 5–82 mm are given by Wells (1963) for warm-water reef corals of differing skeletal morphology. Growth rates recorded by Whitfield (1898) and Vaughan (1915) on *Acropora palmata* from the Bahamas was reviewed by Wells (1963) who shows that annual increments of between 40–100 mm per year are reliable estimates. The Bahamas lie at latitude 24° north in warm seas and they are broadly comparable to the conditions during Viséan times in Britain though here the position was only 10° south of the equator. Calculated growth rates from the diurnal and monthly epithecal ridges of 6 genera of Viséan corals give figures which vary from 36–69 mm per year (Table 1). From the table it can be seen that the growth rates of Viséan corals do fall within the known growth rates of present-day corals. Furthermore, those from the Viséan of northern England, living near to the equator (see Figure 4), show faster rates of growth (up to 69 mm per year) than those for the Devonian of New York living 40° south of the equator (20 mm per year).

A factor which has become apparent during the study of *Lithostrotion martini* from Westmorland is that the development of monthly banding on the epitheca may vary widely, particularly in different lithologies of rock matrix. Thus *L. martini* normally has well-developed monthly banding, but specimens collected from shaly beds have no monthly banding developed. If the monthly banding is a lunar breeding periodicity feature expressed on the epitheca of the corals then possibly the muddy environment had an adverse effect on the breeding. A further factor which came out of this study was that the growth rate of *L. martini* quickens up the Viséan stratigraphical column from 3·6 mm per month in the centre of the C_2-S_1 zone to 4·5 mm per month at the top of the zone. This variation is independent of lithology, for at each horizon growth rate is influenced by environmental

TABLE 1. Rates of growth of some Carboniferous rugose corals

	Growth/month in mm									Mean growth/month	Mean growth/year
	1	2	3	4	5	6	7	8	9		
Aulophyllum fungites (Fleming) Horizon uncertain (Viséan)	3·0	3·0	3·87	3·0	3·25					3·22	41·7
Cyathaxonia cornu Michelin Tyne Bottom Lst.	3·9	3·5	3·5	3·6						3·6	46·6
Dibunophyllum bipartitum (McCoy) E₁, Great Lst.	4·5	3·6	4·2							4·1	53·1
Dibunophyllum bipartitum (McCoy) E₂, Botany Lst.	3·0	3·8	3·1	4·0	4·5					3·68	47·7
Lithostrotion minus (McCoy) Great Scar Lst.	4·8	5·0	3·1	3·0	4·0	5·5	6·0	5·5	4·2	4·45	57·7
Lithostrotion martini Edwards and Haime Low C₂-S₁ zone	3·9	4·2	3·0	4·0	4·0	4·0	3·2	3·0	3·3	3·6	46·6
L. martini, Ashfell Sandstone (limestone band)	4·3	5·1	5·4	3·5	4·0	5·0				4·5	58·3
L. martini, Ashfell Sandstone (shaly limestone)	4·4	4·5	4·5	4·5						4·5	58·3
L. martini, Ashfell Sandstone (limey shale)	3·0	2·5	3·0	2·5						2·75	35·6
Lithostrotion junceum (Fleming) Jew Lst.	3·0	3·0	3·0	3·0	3·1	3·6	3·0	3·1	3·0	3·1	40·1
Lonsdaleia floriformis (Martin) Tyne Bottom Lst.	5·0	5·0	7·0	4·0	5·0					5·2	67·3
Orionastraea phillipsi (McCoy) Bankhouses Lst.	6·0	4·4	4·3	6·0	6·0					5·33	69·0

differences. The lithology factor is demonstrated by *L. martini* from limestone and slightly shaly limestone (up to 20% clastic material) with a growth rate of 4·5 per month, whereas in an adjacent limey shale (60% clastics) the growth rate was only 2·75 mm per month; the latter conditions are clearly poor for coral growth.

TABLE 2. Insertion of tabulae in *L. martini* from the Viséan of Westmorland, England

Horizon	Growth/month (mm)	Tabulae/5 mm	Tabulae/month
Ashfell Sandstone (limestone band) top of C_2-S_1 zone	4·55	7·5	6·8
Ashfell Sandstone (shaly limestone band)	4·50	7·5	6·7
Ashfell Sandstone (calcareous shale band)	2·75	13·0	6·5
Thysanophyllum Lst. low C_2-S_1 zone	3·60	10·0	7·2

The variation in the growth rate of *L. martini* seems to be linked to the spacing of the internal horizontal skeletal plates termed tabulae. In the specimens from the C_2-S_1 zone of Westmorland, there is firstly a tendency for the tabulae to become wider spaced, passing up the succession and secondly, closer spacing of tabulae develops in specimens collected from shaly horizons. The growth rate per month has been calculated for four samples from this zone and in each case longitudinal sections have been cut and the number of tabulae per 10 mm counted. From these figures the number of tabulae per month was determined (Table 2). These figures indicate that in *L. martini* there may be a time relationship to the formation of tabulae with 6·5–7·5 tabulae being deposited every lunar month. It is probable that this relationship in *L. martini* is caused by the polyp building up sufficient lime to form a new tabula once every 4·6 days. Further work is required to substantiate the relationship between growth rate and the formation of tabulae, but if a similar constancy of tabula formation could be arrived at in other species and genera a method of estimating growth rate independently of the preservation of the epitheca would be possible. In this case the use of corals as geochronometers might be greatly increased.

References

Bott, M. H. P. (1971). *The Interior of the Earth*, Edward Arnold, London, 316 pp.

Fitch, F. J., Miller, J. A. and Williams, S. C. (1970). Isotopic ages of British Carboniferous Rocks 2: pp. 771–789. In *Sixth Congre. Int. Strat. Geol. Carb. Sheffield 1967*, 4 vols, Ernest Van Aelst, Maastrich (Netherlands)

Gignoux, M. (1955). *Stratigraphic Geology*, English Translation from Fourth French Edition, 1950. Freeman, San Francisco, California, 682 pp.

Johnson, G. A. L. (1967). Basement control of Carboniferous sedimentation in Northern England. *Proc. Yorks. Geol. Soc.*, **36**, 175–194

Kulp, J. L. (1961). Geologic time scale. *Science*, **133**, 1105–1114

Lambert, R. St. J. (1971). The pre-Pleistocene Phanerozoic time scale—A review, pp. 9–31. In Part I, *The Phanerozoic Time Scale—A Supplement, Special Publication of the Geological Society*, no. 5, London

Mazzullo, S. J. (1971). Length of the year during the Silurian and Devonian Periods: New values. *Geol. Soc. Am. Bull.*, **82**, 1085–1086

Scrutton, C. T. (1965). Periodicity in Devonian coral growth. *Palaeontology*, **7**, 552–558

Scrutton, C. T. (1970). Evidence for a monthly periodicity in the growth of some corals, pp. 11–16. In S. K. Runcorn (Ed.), *Palaeogeophysics*, Academic Press, London, 518 pp.

Scrutton, C. T. and Hipkin, R. G. (1973). Long-term changes in the rotation rate of the Earth. *Earth-Science Reviews*, **9**, 259–274

Smith, A. G., Briden, J. C. and Drewry, G. E. (1973). Phanerozoic world maps, pp. 1–42. In Hughes, N. F. (Ed.), *Organisms and Continents through Time: Spec. Pap. Palaeont.*, **12**, vi + 334 pp.

Vaughan, T. W. (1915). The geological significance of the growth rate of the Floridian and Bahamian shoal water corals, *Washington Acad. Sci.*, **5**, 591–600.

Wells, J. W. (1956). Scleractinia, pp. F328–F444. In R. C. Moore (Ed.), *Treatise on Invertebrate Palaeontology. Part F, Coelenterata*. Geol. Soc. Am. and University of Kansas Press, 498 pp.

Wells, J. W. (1963). Coral growth and geochronometry. *Nature*, **197**, 948–950

Whitfield, R. P. (1898). Notice of a remarkable specimen of the West India coral *Madrepora palmata*. *Bull. Amer. Mus. Nat. Hist.*, **10**, 463–464

DISCUSSION

PANNELLA: Were your counts carried out by independent observers, as well as by you?
JOHNSON: Only Nudds and I made our counts, but they were independent of each other, and they were consistent.
SCRUTTON: Were they made by continuous consecutive counting or were they counted between what you first determined were individual constrictions (monthly series)?
JOHNSON: We tried to count consecutively if possible, where not possible we did try to distinguish good constrictions and count between them.
EVANS: Why couldn't you just assume the growth lines were tidal lines, then the constrictions would be semi-monthly?
JOHNSON: Our specimens were not from the tidal zone, their morphology indicated they were subtidal. Assuming the rugose corals are somewhat similar to modern scleractionian species, I infer that reproduction in the fossils occurred by lunar periodicity. Thus, to account for the growth periodicity I said it was monthly.
BUDDEMEIER: You must be careful counting tabulae, for my experience shows tabulae pseudo-periodicities which are not time-dependent. Also your tendency to multiply numbers with 2 significant digits, to get numbers with 4 significant digits (as in your 399·0 day Devonian year), lends an aura of spurious precision to the data.
HIPKIN: It seems that the essential difference between the use of corals and molluscs as geochronometers is the matter of the morphology of the unit. Could you give a unique description of the counting unit, that is, could you recognize the monthly unit without first counting the approximate number of growth lines between apparent ridges?
JOHNSON: Our corals displayed periodic constrictions, a constant distance apart with constant growth, and these periodic constrictions were clearly the monthly unit.
O'HORA: What is the range in monthly counts?
JOHNSON: 28–31 increments per month, and there is no trend with age of the animal.

MEASURED PERIODICITIES OF THE BIWABIK (PRECAMBRIAN) STROMATOLITES AND THEIR GEOPHYSICAL SIGNIFICANCE

ROLAND E. MOHR
University of Minnesota, Minneapolis, USA

Abstract

Periodicities in the growth patterns of 19 mat-like digitate stromatolites have been measured in this study. The method involved measuring the X and Y coordinates of every third to fifth micro-lamination down the entire colony while noting the number of micro-laminations between measurements. At this point in the process no reference was made to any periodicities, either in spacing or in colour of the laminations. Later the specimens were re-examined at lower magnification and the X and Y coordinates of the periodicities noted. At no time does the observer actually count the number of micro-laminations between dark coloured bands. By separating the process of data collecting into two phases, neither of which directly relates to the number of laminations per band, observer objectivity is probably enhanced. Also, because of the digital format, the data may be subjected to numerous statistical tests.

The results suggest a dominant periodicity of about $12·8 \pm 0·3$ (1σ) micro-laminations per band for 262 dark bands. If we assume that the micro-laminations represent daily growth increments, the semi-monthly period seems the best candidate for the dominant periodicity; twice monthly a given tide height and solar zenith angle coincide. The implication is a synodic month of $25·6 \pm 1·2$ (2σ) days during the middle Precambrian, a value less than at present. Although there are not enough data available at present to make an independent estimate of the number of solar days per year, the conservation of angular momentum suggests a value between 800–900.

If the tidal lag angle has been constant throughout geological time the radius of the lunar orbit would be rapidly changing during this phase of its evolution. However, the catastrophic consequences of the heat pulse associated with rapid orbital changes can be avoided if the lag angle was smaller in the past. These data offer a way to calculate such changes from observation. By equating orbital energy loss with frictional heat production, a simple expression can be derived which relates the product of $\mu.Q$ to measurable orbital properties. In this case μ and Q are tidal effective values of rigidity and the specific dissipation function, respectively, for the entire Earth, rather than specific values for any individual earth materials. The results are independent of the specific nature of the dissipation mechanism.

Introduction

Stromatolites are perhaps the oldest known traces of biological organisms (Cloud, 1965). They are the fossilized remains of structures produced by blue-green algae and bacteria. Because they might have preserved within them identifiable traces of periodic environmental perturbations, such as diurnal, tidal and seasonal fluctuations in temperature or light intensity, they offer the possibility of extending our knowledge of these fluctuations into the remote geological past (Pannella *et al.*, 1968). With a knowledge of the frequency distribution of tidal and seasonal cycles with respect to the diurnal cycle, data pertaining to the orbit of the Moon and the rotation rate of the Earth during the Precambrian can be obtained.

The samples

The specimens used in this study were collected at the Mary Ellen mine in the Biwabik Iron-Formation near Gilbert, Minnesota. The age of the rock is of the order of 2×10^9 years (Goldich, 1973). The samples are so far removed from the contact aureole of the Duluth Gabbro that they show no metamorphic characteristics except possibly a pervasive low-grade regional metamorphism (French, 1968). They show some evidence of secondary oxidation. The rocks have a remarkable level of detail preserved within them. Original compaction structure is still visible on millimetre-sized, concentrically banded oöliths and in some specimens submicroscopic structures similar in morphology to modern blue-green algae and bacteria are preserved (Barghorn and Tyler, 1965). The rocks are almost pure silica with small amounts of iron oxide and carbonate. The colonies of the stromatolites are layered with from 5–45 laminae/mm of alternating red-coloured iron-oxide-rich silica and transparent oxide-poor silica. Occasionally a group of laminae will be dark-coloured rather than red. Under visual observation there is a general tendency for the laminae to be more closely spaced where they are darker in colour. The pervasive red colouration is due to very fine-grained dispersed haematite, while the dark colour appears to be due to more coarsely crystalline specular haematite. Colonies of two general morphologies have been examined: the digitate and mat-like stromatolites. The digitate form is *Gruneris biwabika* (Cloud and Semikhatov, 1969). In cross-section the digitate forms appear as columns about 1 cm across and up to 10 cm long, and show branching and a sinuous upward growth. The area between the columns is filled with oöliths. At any given point during their growth, the exposed top surfaces of the stromatolites covered about one-fourth of the surface of the sediment–water interface. The mat-like forms covered nearly the entire (local) sediment–water interface, and there are few oöliths in association with them. The mat-like forms are more finely laminated.

Methods

The method used to extract data on periodicities differs from the techniques of Pannella and others (Pannella, 1972; Wells, 1963; Scrutton, 1964). While they used acetate peels of polished sections, thin sections and surface features, I have measured polished thick sections of the specimens. The sections are in the order of 3 cm thick and about 15 cm on a side. Each section contains 5–20 stromatolite colonies. The specimen is mounted on a milling machine modified to support a binocular microscope. Because the silica matrix of the rock is nearly transparent, it is possible to look down into the specimen. The surface of an individual lamina is viewed more or less edge on, so that its colour contrast is enhanced. Even so, the laminae are diffuse, indistinct and difficult to measure.

The X and Y coordinates of a lamina at one end of a colony are noted. Several laminae (3–5) are visually counted-off and the microscope positioned over the new point whose coordinates are also measured. The entire column is stepped-off in this manner. A typical column will then be described by from 50–200 individual sets of X–Y coordinates and the number of laminae between each coordinate pair. At this point in the process no heed is paid to any periodicities which may be present. Later the specimen is gone over again. This time the X and Y coordinates of any recurring changes in structure or colour, in particular the dark bands of specular haematite, are noted. At this point no attention is paid to the number of laminae between the dark bands. In fact, the second pass is generally made at a lower magnification so that individual laminae are difficult to resolve, but the dark bands stand out more clearly. The information extracted on the two passes over the sample can be combined to give the number of laminae between dark bands, the average number of laminae/mm as a function of the number of laminae, the Fourier power spectrum and numerous other quantitative descriptions of the periodicities in the sample.

An advantage of this method of taking data is that it separates the process of extracting periodicities into two steps, neither of which is complete by itself. At no point does the observer actually count the number of laminae between dark bands even though this is the quantity of paramount importance. The spacing of the dark bands is seldom uniform, and the number of laminae/mm is also variable, so the observer has little idea of the number of laminae between the dark bands. The first time he sees the result is when the computer returns a histogram of the number of laminae between each dark band. Removing from the observer the responsibility of actually counting periodicities probably increases his objectivity by at least a small amount. The results are a more complete description of the sample. If there happens to be an interval where there are 215 laminae between two dark bands, that value will appear in the results. Such information is valuable because it may represent supermultiples of the dominant frequency. On the other hand, the results obtained by coordinate measurement probably have a somewhat lower resolution and are considerably more tedious than simply counting.

The data

Figure 1 shows a histogram of the results obtained from the measurement of 341 dark bands. At first sight it seems rather peculiar, but statistical tests have shown that the observed distribution could be drawn with reasonable probability from a model parent distribution of reasonable characteristics. Consider a situation in which intervals between periodicities are measured for a sample in which a few individual dark bands have either been randomly removed from the record or have escaped detection during measurement. There is considerable

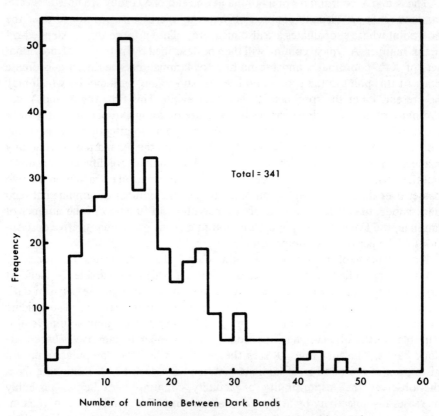

FIGURE 1. Histogram of measured dark band intervals

evidence to suggest that the stromatolites are a parent distribution of this type. The histogram of a sample drawn from such a parent will have a number of peaks. The peak with the lowest mean will be the largest and the amplitude of successive peaks will fall off approximately exponentially. The rate of exponential decay is a measure of the completeness of the record. If the standard deviation of the individual peaks is small compared to the spacing between them, the peaks will occur as isolated individuals. But if the standard deviation of each peak is of the order of the spacing between peaks, there will be a considerable overlap of points belonging to adjacent peaks. This problem complicates the analysis in such cases.

The portion of the distribution in Figure 1 up to 22 laminae per dark band (where noticeable pollution from the second peak becomes apparent) fits a gaussian distribution with a mean of 12·8 and a standard deviation of 5. These parameters correspond to a minimum value of χ^2 (9·9 for 9° of freedom) for all possible combinations of means and standard deviations of the first peak. The probability is roughly 0·5 that a random sample drawn from such a parent distribution will have a χ^2 larger than the observed data. The standard deviation of the mean for the first peak is approximately 0·6 (2σ). The goodness of fit for the entire distribution can be tested by assuming that each successive peak has a standard deviation of 5 with a mean which is a multiple of 12·8 and an amplitude which decays exponentially with the index number of the peak. The index number of the first peak is 1, the second peak is 2, and so on. The minimum value of χ^2 is 29·0 for a decay parameter of 1·3. For 27° of freedom this again yields a probability of about 0·5 that a random sample drawn from the parent distribution will have a χ^2 larger than the observed distribution. The advantage of this method of analysis is that it uses all the data from the entire distribution. Although it is model-dependent, it is probably a more honest technique than confining attention to a single well-developed peak.

The major source of discrepancy between the observed and assumed parent distribution is probably not random fluctuations or uncertainty in the determination of the mean, but the assumption that all data were drawn from distributions with the same standard deviation. The data for both digitate and mat-like

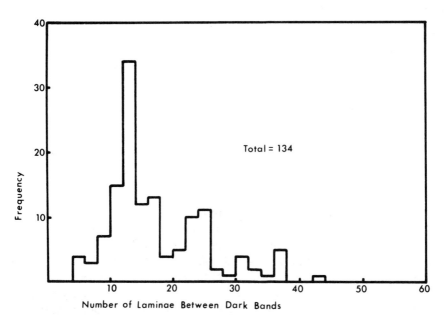

FIGURE 2. Histogram of counted dark band intervals

stromatolites are lumped together in Figure 1, but the mat-like forms probably yield data with a larger standard deviation because they are more finely laminated and harder to measure. When more data have accumulated, this hypothesis can be significantly tested.

Figure 2 shows a histogram for 134 dark bands counted by a technique similar to that of Pannella. The mean of the first peak of the distribution is similar to the measured distribution, but markedly different from Pannella's result for the middle Precambrian (Pannella, 1972). The standard deviation of the counted peaks is smaller so they stand out more clearly as individuals. The counts were made by several skilled observers who were relatively ignorant of the 'right answers', and by the author who, no doubt, had some preconceptions. The data in Figure 2 were counted to see if the two methods were in agreement *after* considerable effort had been expended in measuring coordinates. The results can be taken to mean either that both methods are observationally sound, or that bias has been introduced into (in this case) the method of counting. In view of the differences in sample preparation, perhaps these data and Pannella's vary because different things were measured. The disparity is serious and should be resolved.

The technique of measurement of the samples gives not only the number of laminae between dark bands, but also the average number of laminae/mm for each dark-band interval. If diagenetic or metamorphic alterations had obliterated the fine detail in some portions of the sample, the micro-laminations would probably be the first to be erased. This effect would reduce the number of laminations between dark bands in that portion of the record. We can test this possibility by removing from the sample those dark-band intervals with the fewest average laminae/mm and looking at the resulting change in the mean. A cut-off was established by a 'filter factor' which, when multiplied by the average number of laminae/mm in the whole colony, gave the minimum number of laminae/mm required for an individual dark-band interval to be retained. A filter factor of 0·5, for example, would remove from the results all dark-band intervals with less than 0·5 times the average number of laminae/mm of the colony as a whole. By setting the cut-off relative to the average number of laminae/mm for a given colony we can account for variations in average growth rate for different types of stromatolites, while by applying a single cut-off to data from all colonies we can search for entire colonies which have been altered. A complete statistical study is not available, but preliminary results suggest that, as the filter factor increases from zero to about 2·0, the mean value of the number of laminae per dark band for the remaining intervals does increase slightly. However, because data have been removed, the standard deviation of the mean also increases such that the confidence interval of the mean of the filtered data still encompasses the mean of the unfiltered data. We can tentatively conclude that secondary alteration of the rock specimens has not seriously affected the estimation of the mean value of the periodicity.

Occasionally a bedding surface can be traced from one stromatolite colony to another. If preservation and measurement were perfect, there should be the same number of laminae between 2 such correlative surfaces on 2 adjacent colonies. Measuring such correlations is a stringent test of the entire process. Surfaces which can be confidently traced from one colony to another are rare, but for 10 surfaces traced between 4 colonies in a single rock specimen, the standard deviation of the number of laminae between correlative surfaces was 24% of the mean number. These results seem rather disappointing, but it is a stern test. More work is needed on the collection and interpretation of such data because it is one possible measure of absolute rather than relative accuracy, and the absolute accuracy is what concerns the astronomer or geophysicist.

As Pannella (1972) has pointed out, long-period cycles are more uncertain simply because the data accumulate more slowly. Also, complete records of several periods of long-term fluctuations are seldom encountered. At this point, insufficient data are available to determine the mean value of the longer periods by spectral analysis of the number of laminae/mm as a function of the cumulative number of laminae.

In order for measured periodicities to be of real value, they must be convincingly related to the proper causal agent. The data obtained from Precambrian stromatolites are especially difficult to interpret because blue-green algae under present environmental conditions do not form stromatolite colonies which are completely analogous in structure or composition to those found in the middle Precambrian. Today, the surface environment is oxidizing, and blue-green algae precipitate calcareous stromatolites in all except a few aberrant environments (Walter *et al.*, 1972). During the middle Precambrian and earlier, the environment was reducing and the algae frequently precipitated silica stromatolies. Thus, we are deprived of an important empirical test: we cannot necessarily expect that when we apply the same techniques of analysis to modern stromatolites that we have applied to the fossil specimens, we will obtain analogous results. It would be gratifying if such observations were in agreement with the present-known periodicities of tides and seasons, but it would not be a compelling reason to assume a similar relation must have held true in the middle Precambrian. For this reason, we must apply the techniques and the methodologies of many disciplines in an attempt to deduce a reasonable and internally consistent model for the formation of stromatolites under the conditions presumed to have existed during the Precambrian. A study of the sedimentary geochemistry and geology of iron-bearing stromatolites with special attention to the effect of cyclical changes in temperature and oxygen fugacity would be especially interesting. Unfortunately, there is no unanimity of opinion on the mode of formation of the Precambrian iron formations (James and Sims, 1973), and the petrology of the beds containing stromatolites has been little studied.

If the laminations represent daily growth increments, the rate of accumulation of silica at the time of deposition was of the order of 10 cm/year for some of the digitate forms and half that value for mat-like colonies. Many minor unconformities exist in the specimens, so the net deposition rate over 1000 years might be only a small fraction of these values. Over time-scales on the order of years, however, these are very high deposition rates, especially considering that the linear extent of the Biwabik shoreline was of the order of 1000 km. The stromatolites of the Biwabik occur not in peculiar isolated pockets, but as a pervasive feature of the environment. We must ultimately provide a plausible mechanism for the transportation and deposition of large amounts of silica before the laminations can be confidently ascribed to daily environmental fluctuations. The problem is not an easy one because of the limitations of our knowledge of middle Precambrian atmospheric and hydrospheric conditions.

Blue-green algae living today are photosynthetic oxygen producers. If Precambrian algae also produced oxygen, their presence could have had important local effects on the geochemical balance of sea water (Cloud, 1965). If the atmosphere as a whole was reducing, iron could have been transported in solution in the ferrous state. An influx of ferrous-rich waters over the colonies of stromatolites might cause precipitation of insoluble ferric iron during times when the algae were producing oxygen. In this way a daily cycle of relatively iron-rich silica during daylight hours alternating with iron-poor silica during the night might have been deposited.

The oöliths found in association with the stromatolites, the presence of many minor unconformities, and the strong likelihood that the algae required sunlight for their existence all suggest that the stromatolites were shallow-water deposits formed in a relatively high-energy environment. This in turn implies that tidal modulation of water depth could have had a considerable relative effect on the local environment. Of all the various tidal periodicities the dominant one under conditions where light intensity is an important variable would be the semi-synodic period (given an approximately circular lunar orbit of moderate inclination to the ecliptic and a tropical or temperate latitude on an Earth with fairly low obliquity of axis). This is true because twice monthly a given tidal stage occurs with a given hour angle of the Sun. For example, twice monthly a rising tide occurs at noon. If the rising tide brings with it a fresh supply of ferrous-rich water, the algae, under the influence of high ambient light levels, might deposit larger amounts of ferric iron. This is only one of many possible schemes to account for the deposition of periodic dark bands, but any scheme which depends on both the tidal stage and light level will be dominated by the semi-monthly period.

Implied orbital changes

If this interpretation is correct, the number of days/synodic month was $25·6 \pm 1·2$ (2σ) during Biwabik times, a value less than at present. A measurement of the number of days/year is unavailable, but an estimate can be made by a method similar to that of Lamar and Merifield (1966). Curve A in Figure 3 shows the variation in the number of days/synodic month as a function of days/year required by the conservation of angular momentum of the Earth–Moon system, if both the total angular momentum and the Earth's moment of inertia have been constant at their present values. However, solar-produced tides are removing angular momentum from the system so that in the past the total was

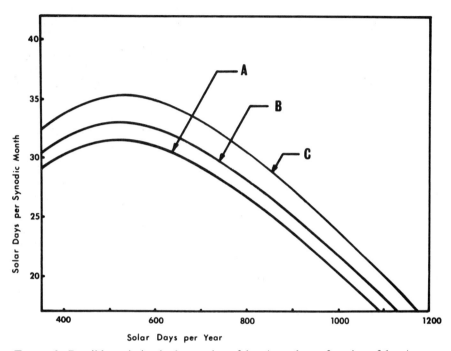

FIGURE 3. Possible variation in the number of days/month as a function of days/year. Curve A is drawn for present value of total angular momentum of Earth plus Moon. Curves B and C are drawn for 2 and 3 billion years B.P., respectively, assuming a constant solar torque of -1×10^{23} dyne cm

larger. Curve B is drawn for the total angular momentum of the Earth–Moon system during Biwabik times and Curve C for the total during Bulawayan times (about 3×10^9 years B.P.). Both curves assume a constant solar torque at one-fourth the present value of the lunar torque, with a constant moment of inertia for the Earth. If the Earth's moment of inertia were larger in the past, the curves would be displaced downward. For example, an Earth with a core and a homogeneous mantle in a system with a total angular momentum appropriate to Biwabik times would evolve along a path approximately coincident with Curve A. In all the curves the effects of any reasonable variations in the Moon's rotation rate are negligible. The band between Curves A and C defines the possible combinations of days/synodic month and days/year. The system occupies a point on Curve A at present and would have been somewhere along Curve B during Biwabik times. The question is, where along the curve? The data presented here suggest that the number of days/month was 25·6 therefore, reading Curve B, the corresponding number of days/year was about 880. A discussion of the consequences of these values will be taken up later.

Except at times of rapid change in the Earth's moment of inertia, the path of evolution of the Earth–Moon system for the last 3 billion years should be a smooth curve on Figure 3 lying between Curves A and C. Pannella's interpretation of his 1972 data obtained from the approximately 3-billion-year-old Bulawayan stromatolites was a month of about 38 days. From Figure 3 we see that Curve C never reaches a value of 38 days/month. One could argue that the magnitude of the solar torque was larger in the past and therefore Curve C should be drawn higher. But if the solar torque was larger, the lunar torque would have been larger also, so that changes in radius of the Moon's orbit would have been more rapid. This, coupled with the fact that the first and largest peak in Pannella's data has a value of approximately 10 laminae/group, suggests that a month of about 20 days rather than 40 might be a reasonable interpretation. It is the only way Pannella's Bulawayan data can be reconciled with the data for the Biwabik presented here. It should be mentioned that Pannella's mean value of 32 days/month and maximum value of 448 days/year during the Biwabik plot in a perfectly reasonable position on Figure 3.

The radius of the lunar orbit and the Earth's rotation rate as a function of time are shown in Figure 4. The error bars correspond to ±50 days/year, a value chosen simply to gauge the effect of our uncertainty in the true value. For com-

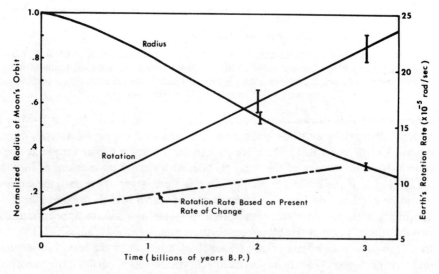

FIGURE 4. Orbital properties based on the data presented in this paper

parison an extrapolation of the present time rate of change of the Earth's rotation rate is also shown. If the interpretation offered here is correct, the average rate of change for the past 3 billion years has been nearly double the present value.

Consequences

MacDonald and others (MacDonald, 1964; Lamar et al., 1970) have dealt with the probable consequences of the rapid evolution of the lunar orbit associated with the period of time when the number of days/month was increasing. Although by the interpretation presented here the number of days/month was increasing throughout the interval of time between the deposition of Bulawayan and Biwabik stromatolites, a glance at Figure 4 shows no intervals of rapid change in the rotation rate of the Earth or the orbital radius of the Moon. Rapid change in orbital radius and a large, but short-term, frictionally produced heat flow pulse are necessary products of the early evolution of an Earth–Moon system with constant rheological properties. The curves in Figure 4, on the other hand, imply that the rheological response of the Earth to tides produced by the Moon has changed substantially over the last 2–3 billion years. An estimate of the amount of this change can be calculated from the data by equating the energy loss from the Moon's orbit and the Earth's rotation with the energy dissipated by frictional heating of the Earth. The energy loss from the orbit and from changes in the Earth's rotation is given by

$$\frac{dE}{dt} = C(\omega - \omega_L)\frac{d\omega}{dt}$$

where C is the Earth's moment of inertia, ω and ω_L are the rotation rate of the Earth and rate of revolution of the Moon, respectively, and $d\omega/dt$ is the time rate of change of the Earth's rotation. For a nearly perfectly elastic body subjected to a cyclical stress of frequency $2(\omega - \omega_L)$, the rate of heat production per unit volume is given by

$$\frac{dE}{dt} = \frac{\sigma_m^2(\omega - \omega_L)}{\mu Q}$$

where σ_m is the maximum stress during a cycle, μ is the rigidity and Q is the specific dissipation function. By equating the two expressions for dE/dt we obtain an expression for the product μQ in terms of measurable quantities, namely

$$\mu Q = \frac{V\sigma_m^2}{C\left(\dfrac{d\omega}{dt}\right)}$$

where V is the volume of the Earth subjected to stress. Strictly speaking, $d\omega/dt$ is that portion of the deceleration of the Earth due to stresses of frequency $2(\omega - \omega_L)$. Both μ and Q are the tidal effective values for the entire Earth and not the value for any particular Earth materials. Although the expression is independent of the specific nature of the energy dissipation mechanism, so long as the loss per cycle is small compared to the maximum elastic energy stored per cycle, Q itself may be a complicated function containing powers of the tidal frequency. The maximum stress of the lunar tides is proportional to the inverse cube of r, the Earth–Moon distance normalized to the present value. If we assume, for illustrative purposes, that C has not changed, then the ratio of μQ in the past to μQ now is given by

$$\frac{(\mu Q)_p}{(\mu Q)_N} = \frac{V_p (d\omega/dt)_N}{V_N (d\omega/dt)_p r^6}$$

where p denotes past and N denotes the value now.

Assuming that the volume of the Earth yielding appreciably to the applied stress in the past was roughly comparable to the present volume, we obtain numerical values of about 20 in the middle Precambrian and 120 during Bulawayan times for $(\mu Q)_p/(\mu Q)_N$.

At first sight these values seem impossibly high, but they are average values for the whole Earth, therefore the size and depth of the oceans affect the value. It is also possible that the volume of the Earth yielding to stress has increased through time. The expression can be regarded as a measure of the change in tidal lag angle required to give the observed orbital properties during some past epoch.

Conclusions

The method of coordinate measurement makes possible a number of tests of internal consistency and probably increases objectivity over the method of directly counting periodicities. However, a number of problems still remain to be solved before the results can be accepted without reservation. Data on long period fluctuations would offer a test of absolute accuracy, because even if the fine details have been uniformly degraded in such a way as to escape detection in the measurements of the number of days/year or days/month separately, the ratio of days/month to days/year will show a disparity in Figure 3. Another potential source of error which is not reduced by any number of counts or measurements is the possibility of misidentifying the cause of a periodicity preserved in the sample. This problem is best attacked by a careful consideration of the effects of cyclical environmental perturbations on the depositional regime. This is especially important in the case of Precambrian stromatolites, for which modern analogues living under conditions similar to those of the past are not available for study.

Although the results presented here are preliminary and additional data might modify the exact numerical values quoted, most of the fundamental ideas depend only on the correct identification of a semi-monthly period determined to be less than the present value. If future work should confirm this, we will gain insight into both the evolution of the lunar orbit, and the physical state of materials within the Earth in the remote geological past.

References

Barghorn, E. and Tyler, S. (1965). Microorganisms from the Gunflint Chert. *Science*, **147**, 563–577

Cloud, P., Jr. (1965). The Significance of the Gunflint (Precambrian) Microflora. *Science*, **148**, 27–35

Cloud, P., Jr. and Semikhatov, M. (1969). Proterozoic Stromatolite Zonation. *Amer. J. Sci.*, **267**, 1017–1061

French, B. (1968). Progressive Metamorphism of the Biwabik Iron-Formation, Mesabi Range, Minnesota. *Minn. Geol. Survey Bull.*, **45**, 103 pp.

Goldich, S. (1973). Ages of Precambrian Banded Iron-Formations. *Econ. Geol.*, **68**, 1126–1134

James, H. and Sims, P. (Ed.) (1973). Precambrian Iron-Formations of the World. *Econ. Geol.*, **68**, 913–914

Lamar, D., McGann-Lamar, J. and Merifield, P. (1970). Age and Origin of Earth–Moon System, pp. 41–52. In *Palaeogeophysics* (Ed. S. K. Runcorn), Academic Press, London, 518 pp.

Lamar, D. and Merifield, P. (1966). Length of Devonian Day from Scrutton's Coral Data. *J. Geophys. Res.*, **71**, 4429–4430

MacDonald, G. (1964). Tidal Friction. *Revs. Geophys.*, **2**, 467–541

Pannella, G. (1972). Paleontological Evidence on the Earth's Rotational History Since the Early Precambrian. *Astrophys. Space Sci.*, **16**, 212–237

Pannella, G., MacClintock, C. and Thompson, M. (1968). Paleontological Evidence of Variations in Length of Synodic Month Since Late Cambrian. *Science*, **162**, 792–796

Scrutton, C. (1964). Periodicity in Devonian Coral Growth. *Palaeontology*, **7**, 552–558

Walter, M., Bauld, J. and Brock, T. (1972). Siliceous Algal and Bacterial Stromatolites in Hot Spring and Geyser Effluents of Yellowstone National Park. *Science*, **178**, 402–405

Wells, J. (1963). Coral Growth and Geochronometry. *Nature*, **197**, 948–950

DISCUSSION

PANNELLA: I believe your work supports mine. The peaks in your data around 9, 12–13 and near 20 days are important; stromatolites always give a number (of days/month) which are equal or less than the (believed) actual number. I think the Mink material is highly reliable because it has strong seasonal bands, hence good control on the number of days/year. There probably were 25–26 fortnights/year recorded by Mink data.

MOHR: The value of number of days/year given by stromatolites depends on how they were counted.

PANNELLA: I counted almost 50 years consecutively.

MOHR: If you had actually missed a few tidal cycles, your value for number of days/tidal cycle would actually be too high, rather than too low.

EVANS: What is the periodicity of the stromatolite increment?

MOHR: The laminae are probably solar daily, as the growth of the algae was probably light-controlled. However, geochemical and petrological models are difficult to deal with because banded iron formations are not being formed today. If tides were an important influence on growth as well as light, one would see a semi-synodic period as dominant. Twice a month, a tidal bulge will coincide with a given hour-angle of the Sun.

BUDDEMEIER: Have you tried x-ray techniques to determine changes in the iron content?

MOHR: So little iron was present, it probably wouldn't be detected by x-ray. Some bands could contain as much as 10% iron, so perhaps, eventually, x-ray techniques could be used successfully.

DOLMAN: Did you find a patterned oscillation to the number of counts per fortnight, as I wonder whether the stromatolite might not have remained uncovered by the tides for 3–4 days/month.

MOHR: I assumed tides were of equal amplitude. Any pattern of varying number of counts per fortnight was lost in the noise.

FARROW: It would therefore be important to check the field association of the stromatolites and to do lateral variation studies.

PANNELLA: Whereas the Biwabik is oölitic (implying a high energy environment), the Mink is not (implying subtidal). Where the Mink is oölitic, one finds the same banding as the Biwabik. Because the Biwabik is higher with respect to the tides, the Biwabik record was probably less complete.

MOHR: The number of interruptions to growth are important, especially if they are periodic, but I believe my digitate stromatolites were seldom exposed at low tides.

FACIES AND LAMINATIONS IN THE LOWER PROTEROZOIC TRANSVAAL DOLOMITE, SOUTH AFRICA

J. F. TRUSWELL AND K. A. ERIKSSON
University of the Witwatersrand, Johannesburg, South Africa

Abstract

The Transvaal Dolomite is approximately 2250 My old. Other stromatolitic units older than this are of local extent, related to vulcanicity, and unlikely to present evidence that could relate them to tidal concepts.

Stromatolites are commonly considered to be peritidal phenomena but the use of grouped stromatolitic laminations to suggest a lunar, seasonal or annual cycle has several prerequisites. These include the recognition of precise criteria actually indicative of tidal sedimentation, the distinction of specific environments within the framework of this, and means of establishing whether or not the lamination is diurnal.

In the Transvaal Dolomite a facies that is analogous to that developed in contemporary embayments, together with eroded columnar stromatolites and interference ripple-marks, are among the lines of evidence taken to suggest a tidal regime. Lamination is developed within an upper intertidal to marginal supratidal setting in an embayment facies; in lower intertidal domes in a semi-protected environment; within large elongate subtidal mounds under more exposed conditions; and in a lagoonal facies.

The ephemeral embayment lamination contains evidence suggestive of diurnal deposition and of seasonal banding although the latter is characteristically truncated by desiccation. Other intertidal lamination was effectively distorted during recrystallization and the formation of secondary chert. The colour banding developed in discrete and linked intertidal columns is the result of recrystallization. Subtidal lamination is vertically persistent, but the different orders of laminae present on weathered surfaces are not recognizable in thin section. Lagoonal lamination is vague; banded iron formations accumulate under similar restricted lacustrine conditions and require further study, for the finest banding in them may be diurnal, and coarser mesobands seasonal.

Introduction

Stromatolites are now generally considered to be analogous to contemporary organo-sedimentary mats. Many stromatolitic carbonates are finely laminated, and the suggestion has been made that certain of the laminations may be diurnal, and that if this is so it might be possible to recognize groupings of laminations, representing lunar, seasonal or annual cycles; further, that these might provide evidence on the length of day through the geological record (Runcorn, 1966). This idea has been developed further (McGregor, 1967; Pannella *et al.*, 1968; Pannella, 1972).

TABLE 1. Stromatolitic units in southern Africa

	Local unit	Age ($\times 10^6$)	Stromatolite reference(s)
Proterozoic	Nama, SWA, SA	>550, probably <700 Welke et al. (1973)	Germs (1972)
	Damara, SWA	>550 ± 100 <1200 ± 100 Clifford et al. (1969)	Kruger (1969)
	Katanga, Z	>620 < 1300 Cahen (1970)	Malan (1964)
	Umkondo, R	>1785 Vail and Dodson (1969)	Tyrwhitt (1966)
	Lomagundi, R	>1940 ± 70 Vail and Dodson (1969)	Winnall (1971)
Transvaal, SA	Pretoria	2224 ± 21 Crampton	Button (1972)
	Transvaal Dolomite	>2224 ± 21 Crampton	Young (1934), Truswell and Eriksson (1972, 1973), Eriksson and Truswell (In press)
	Wolkberg	pre-Transvaal Dolomite	Button (1973)
Archaean	Ventersdorp, SA	2300 ± 100 van Niekerk and Burger (1964)	Winter (1963)
	Bulawayan, R	>3000 Bond (1972)	McGregor (1940), Winnall (1971), Schopf et al. (1971)

SA = South Africa, SWA = South West Africa, R = Rhodesia, Z = Zambia.

The stratigraphical units that include stromatolitic rocks in southern Africa are shown, with their indicated ages, in Table 1. Button (1973) has stressed the distinction between the 3 oldest of them and the remainder. In the Bulawayan, Ventersdorp and Wolkberg (see Table 1 for references) the stromatolitic rocks are restricted in vertical and lateral extent, and are closely related to vulcanicity. Such local lake-type deposits contrast with the succeeding extensive shallow-water shelf carbonate sequences. As a result we consider it unlikely that a tidal influence will be recognized in the older group, nor indeed has it actually been established whether any of them actually contain *algal* stromatolites. The oldest of the younger group is the Transvaal Dolomite, now regarded as older than 2224 ± 21 My (Crampton, personal communication). This unit is one of a number that formed within intracratonic basins on the early Archaean Kaapvaal craton between 3000–1800 My ago (Anhauesser, 1973), most of which are striking for their lack of structural disturbance and metamorphic effect.

Algal mats accumulating today form largely in an intertidal setting, but may be represented in the supratidal and subtidal zones as well. On uniformitarian grounds a general assumption follows that stromatolites are peritidal and subtidal, but the point of whether or not this really is the case requires careful environmental consideration. It appears to us particularly relevant to assess this in the case of the Transvaal Dolomite in view of the antiquity of this unit. Furthermore, an appreciation of where the laminated carbonate may have formed is surely essential before any consideration of the lamination itself: subtidal lamination will be influenced by different factors from, say, that which accumulates in an upper intertidal situation. In addition to the tidal concept outlined above, an understanding of the conditions under which stromatolites accumulate relies heavily on contemporary environments, notably parts of Shark Bay, Western Australia, where different forms have been related to tidal embayments, headlands and semi-protected bights (Logan *et al.*, in press).

Tidal sedimentation in the Transvaal Dolomite

The Transvaal Dolomite has been studied in a basal section near Boetsap in the northern Cape Province (Truswell and Eriksson, 1973); and at Zwartkops north-west of Johannesburg in the Transvaal, where a 1400-m succession has been examined (Eriksson and Truswell, in press, a, b) (Figure 1).

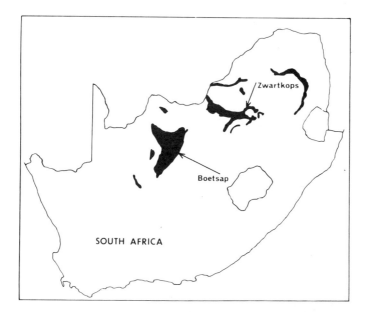

FIGURE 1. Locality map showing distribution of the Transvaal Dolomite. Reproduced with permission from I. W. MacGregor, J. F. Truswell and K. A. Eriksson, *Nature*, **247**, 538 (1974).

At Boetsap a modification of Irwin's (1965) limestone shelf model has been developed, and the concept that domes increase in size outwards in the intertidal (Logan, 1961; Monty, 1972) has been extended out into the subtidal regime. In the resulting model, which is dominated by domes, accumulation is considered to have taken place through a telescoped range of environments from the intertidal through a high-energy agitated zone and into the subtidal. The agitated zone material is dominated by breccias, ripple-marks, oölites and oncolites and contains little lamination. The largest features in the subtidal zone are elongate mounds up to 40 m long and 10 m wide, with inheritances through heights of up to 13 m. These mounds contain a characteristic delicate crinkled lamination as the ubiquitous minor structure. Columnar stromatolites are developed in both the shallow subtidal and the intertidal zones, and contain discontinuous colour-banded laminations. Certain of the intertidal columns are eroded, and their scouring is taken as direct evidence of direct exposure in an intertidal zone.

At Zwartkops, localities have been recorded in the uppermost part of the succession, or Hennops River Formation, at which finely laminated material may display flat, small domical, crinkled or pustular surfaces. The lamination may be overfolded, lens-like flat pebble breccias are common, and features taken to be vertical algal moulds (palisade structures) are also present. This facies is analogous to contemporary tidal flat sedimentation, notably that at Abu Dhabi on the Trucial coast (Kendall and Skipwith, 1968), even more so that of the embayments at Shark Bay (Davies, 1970). As a result an upper intertidal–marginal supratidal setting is indicated for these specific horizons. The structures with which this restricted facies is associated in the field include larger domes, coarser bedding, oölites and ripple-marks; by inference many of these structures are placed in the mid-to-lower intertidal zone in semi-protected settings. The ripple marks, particularly the interference variety, provide diagnostic evidence of an intertidal setting. There are a number of marker horizons in this inferred setting which, although varying in nature from one to another, are dominated by elongate domical structures with a relief of 1–2 m. These marker horizons contain distinctive lamination. Elongate mounds of similar dimensions to those of the subtidal facies at Boetsap are present in the Zwartkops succession. They are not, however, linked to the intertidal regime through an agitated zone but rather by the presence of talus breccias considered to have developed on steeper-than-normal slopes. A lagoonal facies which developed behind a subaqueous oölite bar has also been recognized at Zwartkops.

Lamination types and facies

Most of the rock types in the Transvaal Dolomite are laminated. Figure 2 indicates the relationship between distinctive sedimentary facies and forms of lamination developed in them. The lamination of the marginal supratidal to upper intertidal embayment facies contains lighter and darker carbon-rich

Facies and Laminations in the Lower Proterozoic Transvaal Dolomite 61

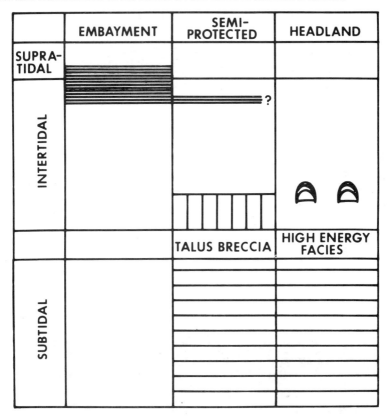

FIGURE 2. The relationship of lamination types to distinctive depositional environments

layers (Figure 3). Palisade structures are present in certain of the light layers, and in thin section extend vertically through thicknesses of up to 5 mm as elongate aligned dolomite crystals (Figure 4). Apparent greater thicknesses in hand specimens of up to 1 cm were in fact composite (Figure 3). Bipartite laminae were recorded in the darker layers, consisting of a thin carbonaceous parting 0·01 mm in thickness, and a thicker dolomitic layer ±0·07 mm in thickness. The preservation of laminae of this sort is, however, rare since in most instances finer carbonate of average grain size 0·015 mm is recrystallized to a coarser grain size of up to 0·10 mm. Palisade structures have only been recorded from contemporary supratidal settings where they represent seasonal growths (Monty, 1967; Shinn et al., 1969). If the palisade structures observed represent either a part or the whole of a seasonal growth it appears reasonable to assume that the thicker but uniform dark layer of Figure 3 is also seasonal. This, in turn, is desiccated such that the 1-cm thickness represents a minimum seasonal growth increment.

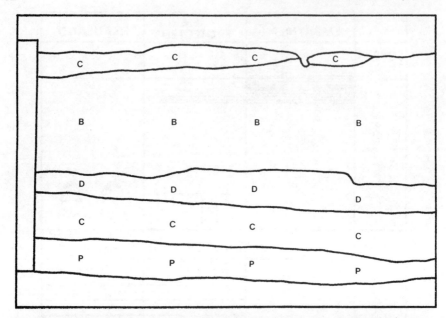

FIGURE 3. An association of upper intertidal–marginal supratidal structures. P: palisade structures; C, D: carbon-rich flat lamination; B: breccia

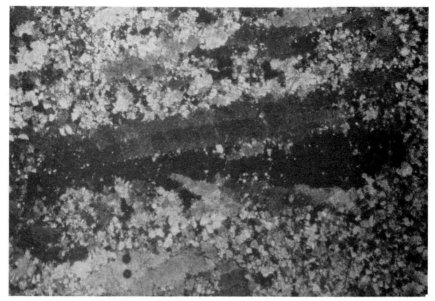

FIGURE 4. Photomicrograph of palisade structures (×20)

FIGURE 5. Lamination from lower intertidal marker domes indicating the transgressive and replacement nature of chert. (Specimen cut and etched)

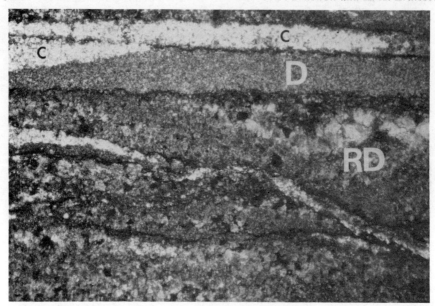

FIGURE 6. Photomicrograph of lamination from lower intertidal marker domes. R_xD_x: recrystallized dolomite; D: primary fine-grained dolomite; C: chert. (×20)

FIGURE 7. Colour-banded dolomite from intertidal columnar stromatolites

Other intertidal lamination which is discussed occurs either within domical or columnar stromatolites. That within the lower intertidal marker domes has an average thickness of 0·4 mm, again consisting of a bipartite carbonaceous and dolomitic grouping. No higher order banding is recognizable. A primary grain size (0·01 mm) is distinguished from a recrystallized secondary grain size (0·24 mm). This recrystallization, together with the pervasive chertification, often results in partial-to-complete destruction of the lamination (Figures 5, 6). Micro-unconformities also restrict the lateral continuity of laminae.

FIGURE 8. Subtidal stromatolitic mounds showing flaggy bedding and some finer lamination. The length of the hat is 35 cm

Intertidal columnar stromatolites are characterized in hand specimen by a well-defined colour-banding (Figure 7). In thin section, however, it is clear that this banding is secondary with the lighter bands composed of coarse patches of dolomite, with average grain sizes of 0·2 mm, recrystallized from a primary grain size of 0·01 mm which constitutes the darker laminae. No lamination was distinguishable within the darker bands. The poor preservation, presence of micro-unconformities and limited lateral persistence render this lamination unsuitable for further study.

Subtidal lamination is present in the large elongate domes at Boetsap and Zwartkops. At Boetsap the crinkled stratification appears hierarchical: the outcrop is dominated by 6–15-cm thick flaggy bedding (Figure 8). Cut surfaces show an alternation of subtly different colour-bands 5–15-mm thick within this

(Figure 9). A third and fourth order of bedding are present on the weathered surfaces, both as a subdivision of the second order which may be developed, and as fine lamination in it (Figure 10). Microscopic work is, however, uninstructive and serves only to distinguish a very fine, darker micritic component with a grain size of 0·012 mm, from a coarser light-coloured recrystallized dolomite with a grain size of 0·040 mm. At Zwartkops the minor structure within the large elongate domes is often flat lamination. In this there is an alternation of

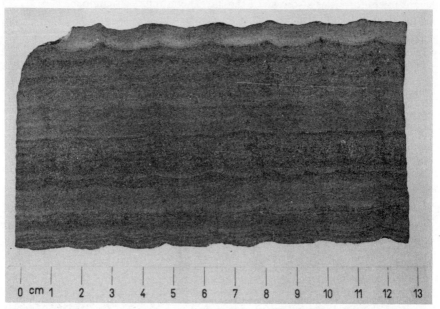

FIGURE 9. Subtidal lamination with lighter bands containing more recrystallized dolomite. (Cut and etched surface)

slightly finer and coarser units 5–15-mm thick. The darker material is better laminated than the lighter, but as at Boetsap the distinction reflects recrystallization which may or may not be a function of a primary fabric. Nor does thin-section work provide any evidence of lamination thickness. The vertical persistence of these subtidal features is considerable: at Boetsap, for example, individual structures extend upwards for 13 m.

Lagoonal dolomites consist of fine-scale, flat lamination in hand specimen but in thin section fine-grained carbonaceous dolomite lacks any recognizable lamination. Banded iron formations in sedimentary settings in contrast to those directly associated with vulcanism, are considered by some investigators (see, for example, Govett, 1966; Eugster, 1969) to form in alkaline lakes which can be most readily equated with extensive lagoons. Above the Transvaal Dolomite a thick succession exists, as in Australia and North America, which consists of alternating siliceous and ferruginous layers. Where studied in detail, banded iron formations contain mesobands varying between 0·25–2 in in thickness (Trendall, 1968) with internal microbands which average 0·15 mm. Preliminary examination of banded iron formations above the Transvaal Dolomite has revealed a well-defined mesobanding varying between 2·50–6·25 mm, consisting of alternating silica-rich and iron-rich layers. Microbands consist of siliceous (±0·006 mm) and ferruginous (±0·012 mm) stringers within the iron-rich and silica-rich mesobands respectively. Banding in iron formations clearly represents

FIGURE 10. Weathered surface of subtidal lamination

a response to some cyclical variation in those processes controlling precipitation. It has been suggested that a banding related to pH differences in humid and arid seasons is recognizable in banded iron formations (Trendall, 1968; Eugster, 1969). The coarser banding discussed above may be of this type. Walter (1972) considered that a diurnal banding was also recognizable but in the banded iron formations examined this was not obvious. Further examination of the microbands is, however, warranted in this regard.

Significance of lamination

The use of any grouping of stromatolitic laminae to suggest lunar, seasonal or annual cycles can only retain relevance where the basic lamination is in fact of diurnal nature; and the point can only be proven in the contemporary setting. In the Bahamas, Monty (1967) has reported that no laminae may be added for several days as intertidal stromatolites are often buried under heavy sediment influx. We would feel that this is particularly the case immediately landwards and seawards of an agitated zone which may be recognizable in the geological record. It is for this reason that the lamination developed in the lower intertidal marker domes as well as that in the columnar stromatolites probably does not contain any ordered cyclical groupings. This suggestion is supported by the abundant development of micro-unconformities in these structures. The potential complexity of intertidal lamination was further illustrated when Gebelein and Hoffman (1968) found that two stromatolitic laminations develop each day in South Florida, while in contrast only a few laminae are added per year to the Shark Bay intertidal columnar stromatolites (Hoffman, 1973). The carbonaceous horizon in Figure 3 probably provides the most absolute information on the nature of tidal flat lamination in the Transvaal Dolomites although in this instance the record is incomplete. The average lamination thickness of 0·07 mm is divisible into the preserved thickness of the dark horizon with some sensible result to suggest that the individual laminae are diurnal instead of tidal.

The remarkable vertical persistence of subtidal lamination and the absence of erosional features suggests that the elongate mounds contain the most complete record of any facies in the Transvaal Dolomite. Different orders of laminae are recognizable in these structures. Those seen on etched cut surfaces reflect recrystallization. It is possible that this recrystallization might reflect a seasonal influence, and if it is so then the higher-order flaggy bedding would represent accumulation during a period of several years. The fine lamination present on weathered surfaces was not seen in thin section. Caution should, however, be exercised in the interpretation of subtidal lamination in that physical tidal forces are absent. Monty (1967) and Gebelein (1969) illustrated the diurnal nature of subtidal lamination at Bermuda resulting from sediment trapping during the day followed by nocturnal algal binding of the particles. It is questioned, however, whether subtidal lamination will record any lunar influences. Rather, any higher-order cycles may be of a seasonal or annual nature.

Summary

There is clear field evidence that the Transvaal Dolomite is of tidal origin. It is of Lower Proterozoic age and doubt can be expressed whether a tidal concept can be applied to any carbonate rocks that are significantly older than this.

Study of lamination from different Transvaal Dolomite facies has not, however, led to the recognition of any tidal cycles in it. The most precise information on the lamination comes from a marginal supratidal to upper intertidal facies, in which a suggested diurnal lamination and evidence of interrupted seasonal banding is developed. Lower in the intertidal, columns contain numerous micro-unconformities and the colour-banding in them is not primary; while the lamination associated with marker horizons is largely destroyed by recrystallization

TABLE 2. Lamination data

Facies	Upper intertidal marginal supratidal	Lower intertidal domes	Intertidal columns	Subtidal	Banded ironstones
Primary grain size	±0·012 mm	±0·012 mm	±0·012 mm	<0·012 mm	Not recognizable
Recrystallized grain size	±0·100 mm	±0·240 mm	±0·200 mm	±0·040 mm	Not recognizable
Lamination thickness	Bipartite ±0·070 mm Palisade structures ±5·0 mm	Bipartite ±0·400 mm	Not recognizable	<1 mm[a] Bipartite coarse and fine carbonate layers ±2 cm[a]	Chert ±0·006 mm Magnetite ±0·012 mm Magnetite layer with chert laminae ±2·50 mm
Lamination groupings	Carbon-rich dolomite >1·0 cm	Not recognizable	Not recognizable	Flaggy bedding 6–15 cm[b]	Chert layer with magnetite laminae ±6·25 mm

[a] Hand specimen observation.

and chertification. Subtidal lamination has considerable vertical persistence but the significance of this is as yet not understood. The lamination data obtained are summarized in Table 2.

Acknowledgments

We are indebted to the Council of the University of the Witwatersrand and the C.S.I.R. for financial support. We have benefited from discussion on banded iron formations with N. J. Beukes. We thank Mark Hudson for photographic assistance, and F. T. Magadla and J. S. Matlala for technical assistance.

References

Anhauesser, C. R. (1973). The evolution of the early Precambrian crust of Southern Africa. *Phil. Trans. Roy. Soc., London, A*, **273**, 359–388

Bond, G. (1972). Milestones in Rhodesian Palaeontology. *Trans. geol. Soc. S. Afr.*, **75**, 149–158

Button, A. (1972). Early Proterozoic algal stromatolites of the Pretoria Group, Transvaal Sequence. *Trans. geol. Soc. S. Afr.*, **75**, 201–210

Button, A. (1973). Algal stromatolites of the Early Proterozoic Wolkberg Group, Transvaal Sequence. *J. Sediment. Petrol.*, **43**, 160–167

Cahen, L. (1970). Igneous activity and mineralisation episodes in the evolution of the Kibaride and Katangide orogenic belts of Central Africa, pp. 99–117, in Clifford, T. N. and Gass, I. C. (Eds.), *African Magmatism and Tectonics*, Oliver and Boyd, Edinburgh, 461 pp.

Clifford, T. N., Rooke, J. M. and Allsopp, H. L. (1969). Petrochemistry and age of the Franzfontein granitic rocks of northern South-West Africa. *Geochim. et Cosmochim. Acta*, **33**, 973–986

Davies, G. R. (1970). Algal-laminated sediments Gladstone Embayment, Shark Bay, Western Australia, pp. 169–205, in Logan, B. W., Davies, G. R., Read, J. F. and Cebulski, D. E. (Eds.), *Carbonate sediments and environments, Shark Bay, Western Australia*, Amer. Assoc. Petrol. Geologists, Memoir 13, 223 pp.

Eriksson, K. A. and Truswell, J. F. (In press, a). Tidal flat associations from a Lower Proterozoic carbonate sequence in South Africa. *Sedimentology*

Eriksson, K. A. and Truswell, J. F. (In press, b). Stratotypes from the Malmani Group north-west of Johannesburg, South Africa. *Trans. geol. Soc. S. Afr.*

Eugster, H. P. (1969). Inorganic bedded cherts from the Magadi area, Kenya. *Contr. Mineral. and Petrol.*, **22**, 1–31

Gebelein, C. D. (1969). Distribution, morphology, and concretion rate of recent subtidal algal stromatolites. *J. Sediment. Pet.*, **39**, 49–69

Gebelein, C. D. and Hoffman, P. (1968). Intertidal stromatolites and associated facies from Lake Ingraham, Cape Sable, Florida. *Geol Soc Amer., Ann. Meet., Mexico City, Abstr.*, 28–29

Germs, G. J. B. (1972). The stratigraphy and palaeontology of the Lower Nama Group, South West Africa, Bull. 12 Precambrian Res. Unit., Univ. Cape Town, p. 250

Govett, G. J. S. (1966). Origin of banded iron formations. *Geol. Soc. Amer. Bull.*, **77**, 1191–1212

Hoffman, P. (1973). Recent and ancient algal stromatolites: Seventy years of pedagogic cross-pollination, pp. 178–191, in Ginsburg, R. N. and Potter, P. E. (Eds.), *Evolving Concepts in Sedimentology*, Johns Hopkins, Baltimore

Irwin, M. L. (1965). General theory of epeiric clear water sedimentation. *Amer. Assoc. Petrol. Geologists*, **49**, 445–460

Kendall, C. G. St. C. and Skipwith, P. A. d'E. (1968). Recent algal mats of a Persian Gulf Lagoon. *J. Sediment. Petrol.*, **30**, 1040–1058

Kruger, L. (1969). Stromatolites and oncolites in the Otari Series, S.W.A. *J. Sediment. Petrol.*, **36**, 1046–1056

Logan, B. W. (1961). Cryptozoan and associated stromatolites from the Recent, Shark Bay, Western Australia. *J. Geol.*, **69**, 517–533

Logan, B. W., Hoffman, P. and Gebelein, C. D. (In press). Algal mats, cryptalgal fabrics and structures, Hamelin Pool, Western Australia. In *Carbonate Sedimentation and Diagenesis, Shark Bay, Western Australia*, Amer. Assoc. Petrol. Geol., Memoir Series

Malan, S. P. (1964). Stromatolites and other algal structures at Mufulira, Northern Rhodesia. *Econ. Geol.*, **59**, 397–415

MacGregor, A. M. (1940). A pre-Cambrian algal limestone in Southern Rhodesia. *Trans. geol. Soc. S. Afr.*, **43**, 9–16

McGregor, A. (1967). Possible use of algal stromatolite rhythms in geochronology. *Geol. Soc. Amer., Ann. Meet., New Orleans, Abstr.*, p. 145

Monty, C. L. V. (1967). Distribution and structure of recent stromatolitic algal mats, Eastern Andros Island, Bahamas. *Annals. Soc. géol. Belgique*, **90**, 55–102

Monty, C. L. V. (1972). Recent algal stromatolitic deposits, Andros Island, Bahamas. *Preliminary Report, Geol. Rundschau*, **61**, 742–783

Pannella, G. (1972). Precambrian stromatolites as palaeontological clocks. *24th Int. geol. Congr.*, **1**, 50–57

Pannella, G., MacClintock, C. and Thompson, M. N. (1968). Palaeontological evidence of variations in length of synodic month since Late Cambrian. *Science*, **162**, 792–796

Runcorn, S. K. (1966). Corals as palaeontological clocks. *Sci. Amer.*, **215**, 26–33

Schopf, J. W., Oehler, D. E. I., Horodyski, R. S. and Kvenfolden, K. A. (1971). Biogenicity and significance of the oldest known stromatolites. *J. Palaeont.*, **45**, 477–485

Shinn, E., Lloyd, R. M. and Ginsburg, R. N. (1969). Anatomy of a modern carbonate tidal-flat. *J. Sediment. Petrol.*, **39**, 1202–1228

Trendall, A. F. (1968). Three great basins of Precambrian banded iron formation deposition—a systematic comparison. *Bull. geol. Soc. Amer., Bull.*, **79**, 1527–1544

Truswell, J. F. and Eriksson, K. A. (1972). The morphology of stromatolites from the Transvaal Dolomite north-west of Johannesburg, South Africa. *Trans. geol. Soc. S. Afr.*, **75**, 99–110

Truswell, J. F. and Eriksson, K. A. (1973). Stromatolitic associations and their palaeoenvironmental significance—a reappraisal of a Lower Proterozoic locality from the Northern Cape Province, South Africa. *Sediment. Geol.*, **10**, 1–23

Tyrwhitt, D. S. (1966). Petrology, mineralogy and genesis of some copper deposits in the Middle Sabi Valley, Rhodesia. Unpubl. Ph.D. thesis, Univ. Leeds

Vail, J. R. and Dodson, M. H. (1969). Geochronology of Rhodesia. *Trans. geol. Soc. S. Afr.*, **72**, 79–113

Walter, M. R. (1972). A hot spring analog for the depositional environment of Precambrian iron formations of the Lake Superior region. *Econ. Geol.*, **67**, 965–980

Welke, H. J., Allsopp, H. L., Köstlin, E., Corner, B. and Kröner, A. (1973). Review of age measurements from the Richtersveld Complex and its environs. *Geol. Soc. S. Afr., 15th Ann. Cong., Bloemfontein, Abstr.*, 92

Winnall, N. J. (1971). Some occurrences and aspects of stromatolites in Rhodesia. Unpubl. B.Sc. Hons. project, Univ. Rhodesia

Winter, H. de la R. (1963). Algal structures in the sediments of the Ventersdorp System. *Trans. geol. Soc. S. Afr.*, **66**, 115–121

Young, R. B. (1934). A comparison of certain stromatolitic rocks in the Dolomite Series of South Africa with marine algal sediments in the Bahamas. *Trans. geol. Soc. S. Afr.*, **37**, 153–162

DISCUSSION

TARLING: Can you make any estimates of tidal amplitude from the height of stromatolite laminae?

TRUSWELL: We came up with a figure of $\frac{1}{2}$–1 m based on the development of columnar forms. This was one of the reasons for believing there was a tidal influence in growth, and we believe our stromatolites were growing in a tidal area, although there was no evidence of tidal lamination.

JOHNSON: I suggest a comparative study with stromatolites from areas such as Pyramid Lake, Nevada. It would be interesting to study stromatolite periodicities from a modern salty sea and compare them with Precambrian data.

THE EFFECTS OF GRAVITY AND THE EARTH'S ROTATION ON THE GROWTH OF WOOD

G. T. CREBER

Department of Botany, Birkbeck College, London, UK

Abstract

The trunk and branches of a tree consist largely of wood; this is secondary xylem tissue formed by the activity of a single layer of cells called the vascular cambium. Active cell division of this layer produces cells which expand and displace the cambium outwards. The extent to which the newly formed cells expand is influenced by a number of factors both internal and external, e.g. seasonal changes in temperate latitudes cause the well-known phenomenon of growth rings. In the humid tropics the absence of such rings results from the constancy of the environment.

Wood is abundant in the fossil record, the earliest being found in the Devonian (ca. 360×10^6 years). Its characteristics are a reflection of the environment in which it grew; many specimens from the Carboniferous are devoid of growth rings but are found in rocks now in temperate latitudes, providing evidence of continental movement or changes in the Earth's axis of rotation.

Wood that is formed under the effects of a transverse or oblique gravitational stimulus is known as *reaction wood*. This is of normal occurrence in lateral branches which grow more or less at right-angles to the line of action of the force of gravity; it occurs also in trunks which have become displaced from the vertical. The *reaction wood* of angiosperm (flowering plant) trees, known as *tension wood*, forms along the upper sides of trunks and branches while in gymnosperms (e.g. conifers) it is called *compression wood* and forms on the lower sides of trunks and branches. The formation of *reaction wood* results from changes in the distribution of hormone (auxin) in the affected parts.

Wood is the secondary xylem tissue formed by the activity of a layer of cells known as the vascular cambium. Division of these cells in a tangential plane leads to the production internally of xylem initials. Some or all of these may divide again before finally differentiating into mature xylem cells.

In gymnosperms (e.g. conifers such as *Pinus*) the xylem is a homogeneous tissue composed largely of tracheids (Figure 1). These are elongate cells which may attain 2-3 mm in length and are very regularly arranged in rows which follow closely the cambial cell from which they were originally derived (Figure 2). The extent to which the xylem initials enlarge depends, in temperate latitudes, upon the time of year. In early summer the tracheids achieve a greater radial diameter and have thinner cell walls than those differentiated later in the growing season. A marked boundary is seen (Figure 2) between the last-formed cells of one season and the first of the next. The growth increment between two such boundaries is known as a growth ring.

These basic features which are exhibited by wood result from the effects of a

FIGURE 1. A transverse section of *Xenoxylon latiporosum*, fossil gymnosperm wood from the Upper Jurassic of Franz Joseph Land, showing the homogeneity of this type of wood. This portion represents about seven years growth. (× 4·9)

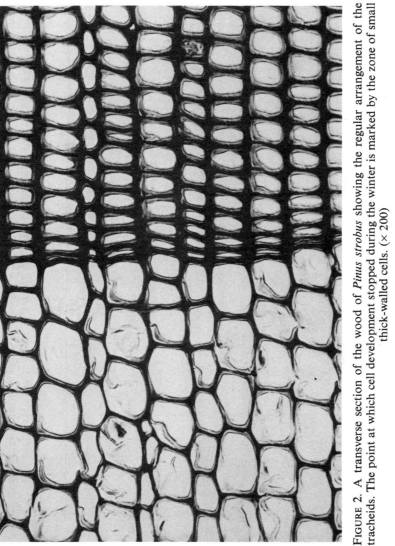

FIGURE 2. A transverse section of the wood of *Pinus strobus* showing the regular arrangement of the tracheids. The point at which cell development stopped during the winter is marked by the zone of small thick-walled cells. (× 200)

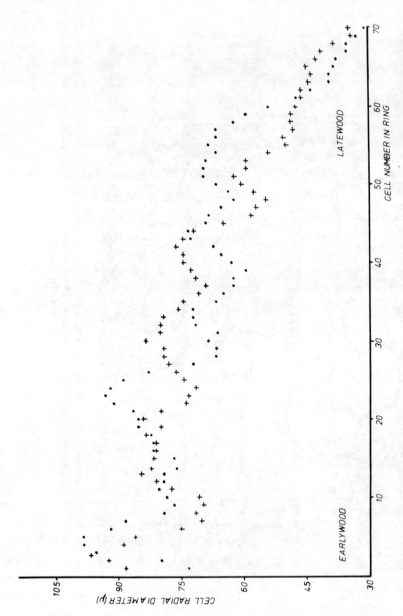

FIGURE 3. A graph of data from a late Jurassic gymnosperm to show the diminution in cell radial diameter through one growth ring. The measurements of one row of cells are shown (dots) superimposed upon the readings from the next row but one (crosses). Out-of-phase relationships can be seen at various points along the curves. The data have been smoothed with a five-point moving average

number of environmental factors. The growth rings owe their existence, through seasonality, to the inclination of the Earth's axis of rotation. The intensity of the growth rings will be influenced by the latitude of the place of growth. The force of gravity will cause asymmetries in the growth rings. Comparison of successive growth rings reflects some aspects of the totality of each growing season, as far as that particular tree is concerned; whereas the pattern of cell size within each ring is to some extent a function of environmental conditions through the duration of that particular growing season. In temperate latitudes, an exceptionally cold or exceptionally dry summer will generally produce a smaller growth increment than a milder one with adequate rainfall. Fossil wood obviously offers the potential of yielding information bearing on past climate and climatic change (Chaloner and Creber, 1973). In order to attempt an interpretation of the environmental significance of growth ring structure, the present author started a study of cell size within individual growth rings. It soon became evident that within the simple seasonal periodicity referred to, there exists a more complicated 'internal' rhythm of cell size increase.

Although there is, in general, a decline in radial diameter from the first cells to the last in a growth ring, there is not by any means a continuous diminution in size. Plotting a graph of the cell radial diameters reveals a cyclic pattern superimposed upon the slope (Figure 3). The cause of the cyclicity is difficult to determine; for the moment it is assumed that it results from the interaction of the xylem initials, each competing with its neighbours for space as expansion proceeds. That the cyclicity is a purely internal matter, and not one determined externally, can be shown by superimposing the measurements of one row upon another (Figure 3). It is evident that while the cells of one row are passing through an ascending phase, in the other row they are descending. Fourier analysis, too, has shown that there is little or no similarity in the periodicities of adjacent rows. Both the cyclicity and the out-of-phase relationships have so far been observed in gymnosperm woods from a wide range of geological time from the Upper Jurassic (Creber, 1972) to the present day. Furthermore, the effects are observable in genera that are not closely related. In the absence, as yet, of observations to the contrary it is assumed that these phenomena are fundamental to the growth of gymnosperm secondary xylem.

The wood structure of angiosperms (flowering plants) is generally not as homogeneous as that of the gymnosperms. The presence of vessels (Figure 4) distorts the pattern of cells but, nevertheless, wood that has developed in temperate latitudes will show growth rings in just as marked a fashion as in gymnosperms (Figure 4).

Hormonal control of xylem development

The growth in diameter of developing xylem cells is controlled by hormones diffusing down from the foliar crown of the tree. This hormone flow in the early

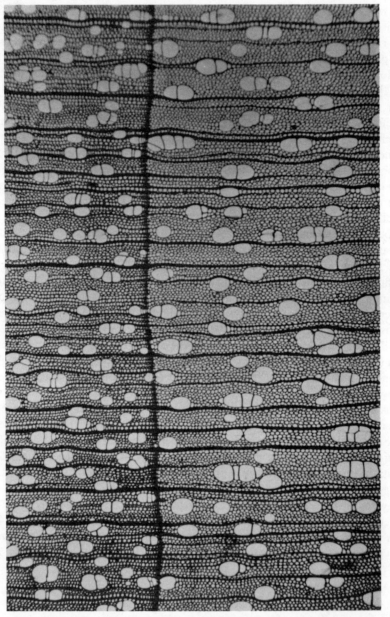

FIGURE 4. A transverse section of the wood of *Populus alba* showing the disturbance of the other elements of the wood caused by the expansion of the large vessels. The narrow zone of small cells across the photograph marks a growth ring boundary where the development of cells ceased during the winter. (× 42)

summer (in temperate latitudes) results in the formation of large cells in the early part of the growth ring. Although it has been shown (Wareing *et al.*, 1964) that indolyl acetic acid and gibberellic acid are the principal growth substances that control xylem formation, it has also been pointed out that not too much emphasis should be placed on these two substances (Robards *et al.*, 1969). They have shown that an array of substances is probably involved.

Although controlled by hormones, the actual onset of new wood formation in temperate latitudes is, however, determined by a phenomenon known as photoperiodism. This involves the perception by the plant of changes in the length of the daylight hours (day-length). As this increases during spring, the buds burst and leaves develop; the latter activity then releases the hormones which stimulate the cambium. It would seem that the evolution of photoperiodic control reflects the biological success of this method rather than one based on the seasonal rise in temperature which is often rather haphazard. Much work has been done on this control mechanism; for instance Larson (1964) has shown that a sequence of artificial growth rings may be induced by subjecting trees to a succession of varied photoperiods. Although most of the work on photoperiodic control has been carried out on temperate species some interesting work by Njoku (1963, 1964) in Nigeria has established that it may also operate in tropical conditions where the day-length variation may be only between $11\frac{1}{4}$–$12\frac{1}{2}$ hr.

It is exceedingly doubtful whether there is any evidence to show exactly when photoperiodic control of wood growth evolved in geological time. However, the existence of growth rings in wood since the Upper Devonian and the evolution of a photoperiodic mechanism involved in their formation both provide supporting evidence for the inclination of the Earth's axis of rotation for a considerable period of time.

The occurrence of considerable growth rings in the wood of trees which grew apparently very close to the contemporaneous pole raises some major problems. In particular, Schopf (1961) described growth rings of substantial proportions in some Permian woods which at the time of growth must have been within a few degrees of the Permian pole. This represents a biological paradox which is not overcome by invoking drift or polar movement.

Gravitational effects

Much of the trunk wood of trees develops with the line of action of the force of gravity passing vertically through the axis of the trunk cylinder. Provided that the trunk does not lean, the axis of symmetry is vertical and the force of gravity does not have any asymmetric effect on it. However, in the case of leaning trunks, and also, of course, of branches, gravity has a marked asymmetric effect and leads to the production of what is known as *reaction wood*. Angiosperms and gymnosperms differ as to the position of the reaction wood in the affected parts. In angiosperms it is known as *tension wood* and it develops on the upper sides of branches and leaning trunks. Conversely, the *compression wood* of gymnosperms forms on the lower sides of branches and leaning trunks. It must be recorded, however, that there are exceptions to these general rules (White, 1962; Robards, 1965).

Reaction wood differs both macroscopically and microscopically from normal wood. To the unaided eye it is obvious that the pith is not central and the growth rings are elliptical (Figure 5). With the microscope it can be seen that the cells of reaction wood are atypical. In tension wood in angiosperms the vessels are smaller and there are fewer of them per unit volume of the wood (White, 1965). It is, however, the presence of so-called gelatinous fibres that really characterizes tension wood. These cells can be recognized because they are thicker walled than normal fibres and the innermost parts of their walls remain unlignified. That is to say, part of the original cellulose wall does not become impregnated with the

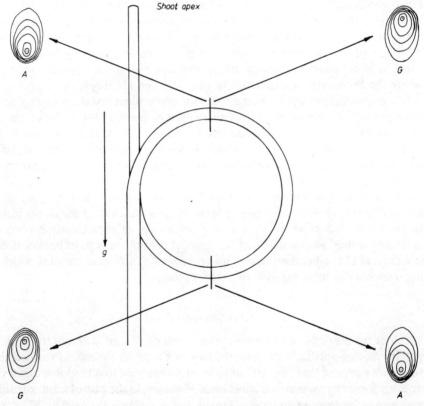

FIGURE 5. A diagrammatic representation of the distribution of reaction wood in a loop made in a stem. The asymmetry of the growth rings in angiosperms (*A*) is opposite in direction to that of gymnosperms (*G*). *g*: the line of action of the force of gravity

hardening substance, lignin, as would normally occur. Indeed, the hardness of wood is due to the fact that in general the majority of the cells become lignified. These partially lignified cells in tension wood are therefore particularly noticeable and can be shown up by various staining techniques. Tension wood has a high tensile strength on account of the large amount of cellulose present. Although this is easily demonstrated in living wood, it would probably be difficult to identify gelatinous fibres in fossil wood, owing to the changes that inevitably take place during the process of fossilization.

In gymnosperms, compression wood is also recognizable by the excentric pith and elliptical growth rings; with the further point that it is often tinged with a red tone; the German authors use the term 'rotholz'. Microscopically its special feature is that the tracheids of which it is largely composed are round in transverse section and there are consequently large intercellular spaces between them. They are also usually much thicker walled and more heavily lignified than in normal wood, making it much heavier in bulk. Timber for commercial use may be severely reduced in value if it contains compression wood on account of its considerable longitudinal shrinkage on drying. Planks may be distorted or even split as a result and the high ratio of lignin to cellulose causes brittleness. It would be unwise, however, to assume that compression wood functions by means of these distortions in the living tree.

Much experimental work has been carried out over the last 70 years since the pioneer studies of Hartig (1901). This work has shown that gravity is the primary stimulus for the production of reaction wood and that this stimulus results in changes in the distribution of growth substances diffusing down from the apex to points lower in the stem. In testing the hypothesis that the compression wood on the under sides of the branches in gymnosperms develops in response to high concentrations of indolyl acetic acid in those regions, Wershing and Bailey (1942) were able to induce compression wood formation in Pine seedlings by applying high concentrations of this auxin. Nečesaný (1958) has shown that in *Populus alba* auxin also accumulates on the lower sides of horizontal angiosperm branches. Here, however, it suppresses the formation of tension wood on the lower sides. Furthermore, he found that application of high concentrations of auxin to the upper sides of horizontal angiosperm branches suppressed tension wood formation. Reaction wood in angiosperms forms therefore in the presence of lower than normal concentrations of auxin whilst in gymnosperms it is the higher concentrations that induce its formation.

The opposite behaviour of gymnosperms and angiosperms to the same concentration of auxin does not in any way cast doubt on the validity of the auxin theory, since a similar situation arises in the response to gravity of the roots and shoots of herbaceous plants. Here, gravity brings about a redistribution of auxin in both root and shoot but each behaves in an opposite way to the same concentration of the substance. High concentrations promote cell elongation in stems but suppress it in roots. Additional work to demonstrate gravity as the primary

FIGURE 6. A transverse section of a piece of gymnosperm wood from the Upper Jurassic of Norfolk showing asymmetric growth rings. (× 4·2)

cause of reaction wood formation is illustrated in Figure 5. Such work has been carried out by Ewart and Mason-Jones (1906) on gymnosperms and by Onaka (1949) on angiosperms; more recently further work has been done on angiosperms by Robards (1965). It has been shown conclusively that the formation of reaction wood is not related to one particular side of the shoot experimented upon but is connected solely with the line of action of the force of gravity. By bending the shoot into a complete loop it was possible to show that in the upper part the reaction wood formed on the morphologically lower side; in the lower part of the loop the reaction wood formed on what was actually the opposite side to that higher up. Thus the reaction wood formed where the shoot was traversed by the force of gravity; parts of the shoot that experienced the force longitudinally showed no development of reaction wood.

Previous workers had suggested that the quantity of reaction wood in a leaning trunk is proportional to the sine of the displacement angle; this proportionality became known as the sine rule. Robards (1965) has, however, clearly demonstrated that this rule must now be modified. By growing a large number of willow saplings inclined at various angles to the vertical (5°–180°) he showed that the maximum amount of tension wood was produced at 120° to the vertical. It is interesting to note that this is in close agreement with Audus' work (1964) on geotropic responses in herbaceous plants. It also became clear from Robards' work that deviations of as little as 5° from the vertical were quite sufficient to cause the production of tension wood. He is of the opinion that the site of graviperception is in, or very close to, the vascular cambium.

The extent to which reaction wood develops in a trunk or branch will therefore depend on three factors; the first is g itself, the second is the load that the part is carrying and the third is the angle of deviation from the vertical. That is to say, there will be a considerable development of reaction wood in a trunk with a marked deviation from the vertical. Similarly there will be much reaction wood in a branch which is long and carries many smaller branches. It is likely therefore to be a very difficult task to assess, in a piece of fossil wood, the individual values of g, the load and the angle of deviation. As regards the occurrence of reaction wood in fossil specimens, Figure 6 shows what are apparently asymmetric growth rings in a late Jurassic gymnosperm. Such growth rings have also been observed in the lycopod genus *Lepidodendron* from the Carboniferous but in general there is little or no reference to reaction wood in fossil specimens. A re-examination of fossil material specifically to identify reaction wood might well prove to be a most useful exercise, especially if it were possible to quantify the extent of its development through a wide range of geological time.

References

Audus, L. J. (1964). Geotropism and the modified sine rule; an interpretation based on the amyloplast statolith theory. *Physiologia Pl.*, **17**, 737–745

Chaloner, W. G. and Creber, G. T. (1973). Growth rings in fossil woods as evidence of past climates, pp. 425–437 in *Implications of Continental Drift to the Earth Sciences* (Ed. D. H. Tarling and S. K. Runcorn), Academic Press, London

Creber, G. T. (1972). Gymnospermous wood from the Kimmeridgian of East Sutherland and from the Sandringham Sands of Norfolk. *Palaeontology*, **15**, 655–661

Ewart, A. J. and Mason-Jones, A. J. (1906). The formation of red wood in conifers. *Ann. Bot.*, **20**, 201–203

Hartig, L. (1901). *Holzuntersuchungen. Altes und Neues*, Springer-Verlag, Berlin, 99 pp.

Larson, P. R. (1964). Some indirect effects of environment on wood formation, pp. 345–365 in *The Formation of Wood in Forest Trees* (Ed. M. H. Zimmermann), Academic Press, New York

Nečesaný, V. (1958). Effect of B-indole-acetic acid on the formation of reaction wood. *Phyton, B. Aires*, **11**, 117–127

Njoku, E. (1963). Seasonal periodicity in the growth and development of some forest trees in Nigeria. I. Observations on mature trees. *J. Ecol.*, **51**, 617–624

Njoku, E. (1964). *Ibid*. II. Observations on seedlings. *J. Ecol.*, **52**, 19–26

Onaka, F. (1949). Studies on compression and tension wood. *Bull. Wood Res. Inst. Kyoto*, **1**, 1–83

Robards, A. W. (1965). Tension wood and eccentric growth in crack willow (*Salix fragilis* L.). *Ann. Bot.*, **29**, 419–431

Robards, A. W. (1966). The application of the modified sine rule to tension wood production and eccentric growth in the stem of crack willow (*Salix fragilis* L.). *Ann. Bot.*, **30**, 513–523

Robards, A. W. (1969). The effect of gravity on the formation of wood. *Sci. Prog., Oxf.*, **57**, 513–532

Robards, A. W., Davidson, E. and Kidwai, P. (1969). Short-term effects of some chemicals on Cambial activity. *J. exp. Bot.*, **20**, 912–920

Schopf, J. M. (1961). A preliminary report on plant remains and coal of the sedimentary section in the central range of the Horlick Mountains, Antarctica. *Inst. Polar Studies*, Rept. 2, 61 pp.

Wareing, P. F., Hanney, C. E. A. and Digby, J. (1964). The role of endogenous hormones in Cambial activity and xylem differentiation, pp. 323–344 in *The Formation of Wood in Forest Trees* (Ed. M. H. Zimmermann), Academic Press, New York

Wershing, H. F. and Bailey, I. W. (1942). Seedlings as experimental material in the study of 'red-wood' in conifers. *J. For.*, **40**, 411–414

White D. J. B. (1962). Tension wood in a branch of sassafras. *J. Inst. Wood Sci.*, **10**, 74–80

White, D. J. B. (1965). The anatomy of reaction tissues in plants, pp. 54–82 in *Viewpoints in Biology*, 4 (Ed. J. D. Carthy and C. L. Duddington), Butterworth, London

DISCUSSION

TARLING: If you wish to see a change in g reflected in the growth of trees, you should see this in trees growing at different elevations.

CREBER: Yes, and the effect in a single tree would be maximum in the giant Sequoias. It would be a different matter comparing these observations with fossil trees whose life orientation and elevation one was uncertain of.

BUDDEMEIER: Some of the more optimistic researchers in the field might even be able to correlate the rhythm in cell size of trees with tidal cycles.

EVANS: Is it possible for fluctuations in size of cells to reflect variations in cell cross-section through different parts of the tree?

CREBER: The pattern of tapering of the cells and the position of the cross-section through them might give apparent periodicities, but much more research needed to be done in this regard.

HIPKIN: How long does it take for the tree to add one growth increment?

CREBER: It varies with locality. I believe May–October was a reasonable growing season for British plants.

BUDDEMEIER: Photoperiodic growth rhythms demonstrated to be important in plants should be considered more seriously by marine biolgists in relation to their organisms.

ROSENBERG: I recommend researchers distinguish between the variations in seasonal length of daylight (l.o.dl.) and the variable geophysical parameter, length of day (l.o.d.). Both vary with the changing rate of the Earth's rotation, and the distinction would seem especially significant to plant evolution in the distant past.

SEDIMENTARY BEHAVIOUR AND SKELETAL TEXTURES AVAILABLE IN GROWTH CYCLE ANALYSIS

H. TERMIER AND G. TERMIER
University of Paris, France

Abstract

Cosmic rhythms control all biological and physiological functions. Biological mineralized secretions are really sedimentary as well as temperature and sexually controlled concretions. Among all mineralized, algal and animal, secretions and concretions, laminated crusts display more rhythmic growing than branching tests. For this reason, the prismatic outer layers of molluscs as well as coral epithecae and stromatolitic sedimentary concretions are the best material for growth cycle analysis.

From that viewpoint we shall consider encrusting laminated examples as different as sedimentary stromatolitic algae, *Lithothamnia*-thallus, Stromatoporoidea, Tabulata, Chaetitida and massive colonies of Ectoprocta (fistuliporids). Some common features of the cyclic growth (tidal and sexual markings) of these attached tests are pointed out. A comparison shows that the ramose forms of the same diversified sorts of organisms do not give the same responses in matter of growing (branching timing, for example). Benthonic free-living invertebrates appear to be synchronized to seasonal rhythms, pelagic ones far less. Besides the most important tidal rhythms, it is suggested that a reconstruction of climatic geological variations would be fairly deduced from the cycles displayed by mineralized biological products.

As a matter of fact, it seems important to extend the palaeobiological study of cycles to the minute and identifiable test fragments included in sedimentary rocks that are not uncommonly better preserved by quick sedimentation processes.

1. Biological and non-biological concretions

Among the different cosmic rhythms which affect all living creatures, the nychthemeral-circadian and lunar-synodic ones control the formation of biological and non-biological concretions, as also do the tidal and monthly rhythms. Examining the different kinds of skeletal concretions, the best material for such a study is to be found in prismatic and sphaerolithic concretions.

Algae

Stromatolitic or travertinous concretions ensure a perfect transition between purely physicochemical deposits as perhaps laminites (Figure 1) and biochemical schizophytic algae deposits (Figure 2). One can easily understand how daily, seasonal or annual rhythms affect these structures, each of them reacting differently to supratidal seasonal drying, tidal drying during low tides, lagoonal

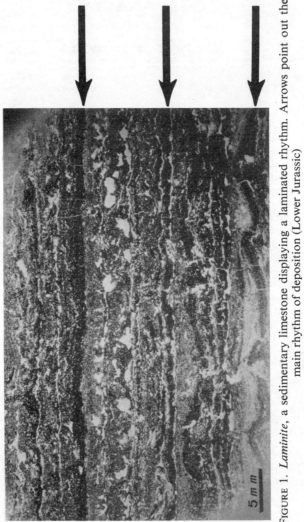

FIGURE 1. *Laminite*, a sedimentary limestone displaying a laminated rhythm. Arrows point out the main rhythm of deposition (Lower Jurassic)

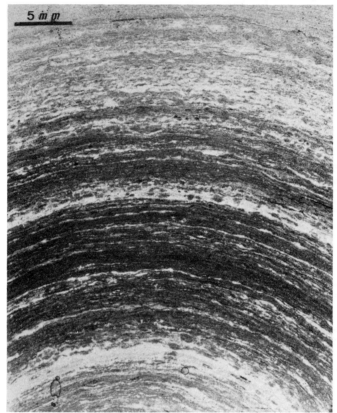

FIGURE 2. A transverse thin section through a columnar stromatolite (*Conophyton*) from Proterozoic

conditions, etc. A cross-section through a stromatolitic concretion shows characteristic breaks with microsedimentary rhythms, while some of them offer a fortnightly rhythm (Pannella, MacClintock and Thompson, 1968) probably corresponding to the sexual rhythms of algae. But drying or burying also mark the stromatolitic crusts, with breaks seemingly corresponding to death, followed by rejuvenation. These drastic events may be due either to synodic or to annual rhythms (Figures 3–5).

The regular 'circadian' zonation, frequent in all categories of algae, seems to result from the multi-axial coordinated growth of branching filaments, at the cellular level (Wardlaw, 1955; Dahl, 1971). Superimposed onto this zonation are environmental periodic patterns such as nychthemeral and seasonal changes in lighting and tidal rhythms. In the Phaeophyta *Dictyota dichotoma*, reproductive structures develop once a month during the full Moon spring tides as bands of sori crossing throughout the thallus, while the release of sexual products and vegetative thallus corresponds to the new Moon spring tides (Umamaheswarara

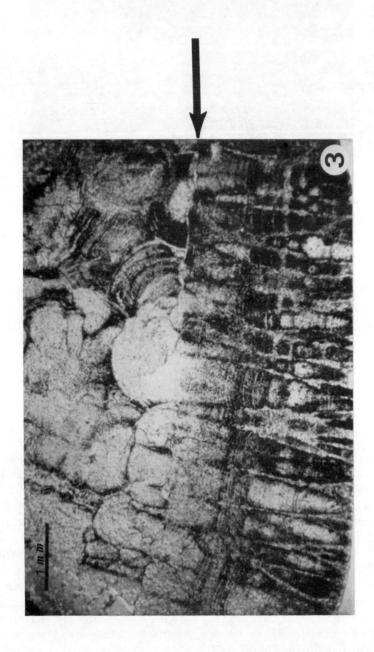

FIGURES 3, 4, 5. A thin section through a non-columnar stromatolite (*Collenia*) from Lower Carboniferous. Figure 3 is a general view displaying two different parts separated by a (seasonal?) break line (indicated by the arrow). Figure 4 is a detail of the upper part of the section displaying regular circadian striation, gathered in from secondary (tidal?) cycles (indicated by the three points). Figure 5 is a detail of the lower part of the section displaying fasciculated fibres constructed with the same circadian striation and tidal cycles as the upper part

and Sreeramulut, 1970): a synodic rhythm (Liddle, 1968) associated with a moonlight synchronization (Müller, 1962; Vielhaben, 1963). Probably, in the tidal zone, some variability is introduced by photoperiod and seasonal temperature fluctuations.

Tidal conditions involve phases of burying followed by phases of exposure to full sunlight, resulting in a semblance of death. Yet some Dictyotales, buried (sometimes in anaerobios) for over 6 months, are still able to regenerate (Dahl, 1971). The thallus then shows an obvious dark zone.

In other forms, the gamete maturation also lasts 12–14 days. It is the same kind of monthly periodicity, divided into two different fortnightly periods, that seems to have influenced the calcareous secretion of some stromatolites (see above).

Archaeolithothamnium (Cretaceous-Eocene), a calcareous encrusting Lithothamnial Rhodophyta, displays a good example of cyclic growth with filamentous parts following a circadian tempo, separated by fertile bands due to a synodic rhythm (Figures 6 and 7).

Recent Rhodophyta rhythms have been studied in a laboratory. The photoregulation is obvious, especially in Bangiophyceae (Dixon and Richardson, 1969). Here, the alternation has been observed of a leafy vegetative phase (*Bangia*), associated with a photoregime of less than 12 hr/day, and of a filamentous (sometimes shell-boring) fertile phase (*Conchocelis*), associated with a photoregime of more than 12 hr/day. *Conchocelis* fertile cells are monospores, while the light range of *Bangia* stability produces fertile cell rows. In the tidal zone, crusts of 'stromatolitic' appearance, including blue-green algae, bacteria, diatoms and sand grains, bear vegetative leaves of *Bangia* that die during the long daylight season. Rejuvenation of uniseriate filaments of *Conchocelis* or of leafy *Bangia* depends on the photoregime.

Invertebrate animals

Mineral concretions of animal origin appear at first glance to result from fewer sedimentary processes than the algal ones. However, very similar formations are found in the external prismatic cuticular layers of mollusc shells, caused by contact with an external fluid environment (sea, brackish or fresh water).

Actually, the best material for an evolutive study of periodical rhythms is provided by cnidarian epithecae (Wells, 1963; Scrutton, 1965) and molluscan prismatic layers, i.e. the more external skeletal parts. It does not seem to make any difference whether the mineralogical material is aragonitic or calcitic or, more frequently, a mixture of both. Among other animal tests, brachiopod shells appear to be promising: the phosphatic *Lingula* shell develops by piling of mineralized layers with partitions of organic matter; the prismatic 'primary layer' of Spiriferids shows external growth increments, and the spiny bands of Productid *Pustula* or *Echinoconchus*, for example, also appear to have registered a growth rhythm.

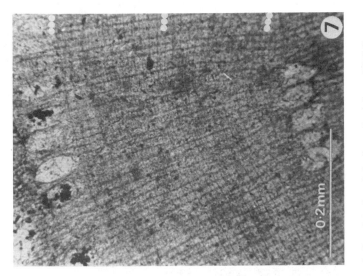
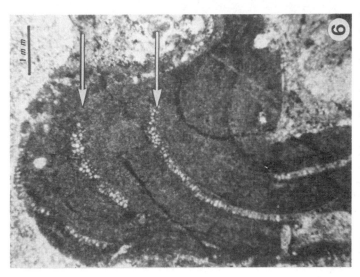

FIGURES 6, 7. A thin section through a Rhodophyte thallus (*Archaeolithothmanium*) from Middle–Upper Cretaceous. Figure 6 is a general view displaying the dark vegetative parts cut by clear alignment of fertile cells (indicated by arrows). Figure 7 is a detail of the same, between two fertile alignments, showing the fibrous vegetative thallus with a circadian striation of about 29–30 striae, corresponding to the synodic rhythm

Further studies on these problems have suggested that the prismatic layer of any mollusc is a special case of sphaerolithic structures: for example, Taylor (1973) quotes the bivalve *Thracia* in which the late prismatic shell layer is replaced by true isolated relict sphaeroliths. The cnidarian polypary epithecae appear to be similarly sphaerolithic (Von Koch, 1882). Such an homology opens up a large field of investigation, since sphaeroliths are widely distributed among invertebrate skeletal concretions: for example, the 'building stones' of Pharetronid sponges and sclerosponges (*Ischyrospongia*) (Termier and Termier, 1972), *Hydrocorallia*, *Scleractinia* (Bryan and Hill, 1941) and so on. Sphaerolithic skeletons are potentially more lacy. Among them, those having an encrusting habitus can respond to cosmic rhythms in the same way as stromatolitic and encrusting Algae.

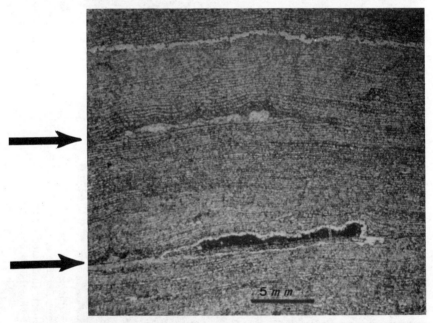

FIGURE 8. A thin section through a Stromatoporoid Sponge (*Actinostroma*) from Lower Silurian. Arrows point out the erosional or depositional breaks between *latilaminae* comprising *laminae* in a circadian rhythm. The number of these *laminae* generally exceeds the synodic one and seems to be bimestrial

Fossil Stromatoporoidea are generally encrusting, with a laminated nature. Basic laminae are clustered and form latilaminae, probably of a synodic kind. In some specimens of the genus *Actinostroma* (Figure 8), found respectively in Lower Silurian, Middle and Upper Devonian, the upper surface of a latilamina often appears to be sharp, maybe corresponding to a micro-hardground due to a short erosive phase and individual death, immediately preceding a short sedimentary phase, before the beginning of a rejuvenation. During diagenesis, these break-surfaces provide a basis for the usual silification process.

In the Sclerosponge theory of the Stromatoporoid evolution (Hartmann and Goreau, 1970), the fleshy cortex is considered as having been supported during life by some siliceous sclerites. Subsequently, the growth stoppage of the calcareous choanosomal skeleton might have allowed a free deposit of such siliceous sclerites on the temporarily dead surface, preceding dissolution and silification. Therefore the synodic rhythms in encrusting Stromatoporoidea make it possible to reconstruct part of their behaviour. Since they were the main reef-building organisms during Upper Ordovician–Upper Devonian times, they presumably lived in warm tropical seas, and the breaks between latilaminae were not due to low temperature but almost certainly were caused by the lowest tides, which perhaps created dry phases, as well as by sexual cycles, in the same way as the rhythms shown by algae.

Among Palaeozoic Ectoprocta, a number of fistuliporids display encrusting zoaria consisting of several growth layers separated by distinct planes. The phenomenon is of the same kind as found in encrusting stromatolites and encrusting Stromatoporoidea. It seems to be the *rejuvenation* of an old or almost dead organism or consortium, for which affinities are unimportant. We think that the modalities of concretion, probably following the same tidal–sexual rhythm, are predominant because of anatomical and physiological peculiarities. An advantage of using warm sea animal concretions for rhythm research therefore

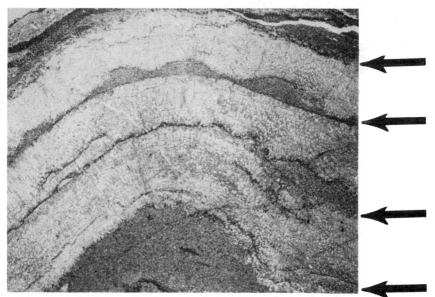

FIGURE 9. A thin section through a *Chaetetes* from Lower Namurian showing three successive crusts with erosional lines between them (indicated by arrows)

lies in the lack of any cold season, since winter often induces great disturbance in skeletal concretions, bringing a break associated with a growth stoppage of the shell. Such winter-constricted growth breaks may be observed on *Pecten maximus* shells, found today in the Brest Bay (Lucas, 1973): similar marks are found on Middle Devonian Pectinoid clams from Morocco (*Ateyapecten*, Termier and Termier, 1973) and on their neighbour *Lyriopecten*, from the Upper Devonian of Catskill Delta, both regions being much closer climatically to each other during Devonian times (before Atlantic opening) than nowadays.

FIGURE 10. A thin section through a massive Ectoproct (*Prasopora*) from Upper Ordovician, showing the zoning of successive crusts. ($\times$ 5)

For fresh-water molluscs, winter brings drastic disturbance, leading to the stoppage of sexual development and the renewal of new shell structures.

Tidal or seasonal breaks have sharp bathymetric and behavioural limits: away from the tidal zone and atmospheric influences there are no effects from drying out or cold weather. For the recent littoral *Cerastoderma edule*, Farrow (1972) has shown major breaks due to synodic (29 day) tidal drying rhythms and also to winter 8-day breaks, explaining the deep marks on the shell. For deep-sea molluscs, the only lamellae are those of circadian growth and therefore they are very regular (Evans, 1972). Similarly, pelagic molluscs are also exempted from growth irregularities.

2. Internal structures

The internal tests, being progressively isolated from ambient conditions, are more and more bound to the individual physiology. The 'building stones' are often sphaerolithic as in the external structures, but their significance is different. Circadian rhythm is always subjacent but not dominant. Sexual cycles may stop growth, sometimes several times a year (twice for a *Littorina* from temperate countries), giving rise to shell pads.

In some invertebrates, the conic-tubular test is marked out by secreted horizontal diaphragm tabulae or septa (Sclerospongia such as *Merlia*; Cnidaria; Cephalopoda.) It is commonly believed that growth is practically infinite, but in fact it is interrupted not only by every dry or cold season, but also by the coming of the adult stage and finally by senescence. Several Ammonoid shells display a slowing down, ending in complete stoppage of shell growth. At adult stage, while septal secretion continues (for example *Agathiceras suessi* Gemmellaro, in Afghan Permian), the spaces between the septa decrease, until they finally join. Therefore, for some time, while attaining the full adult stage, the shell keeps on registering the rhythm of the discontinuous secretion of septa (Termier, Termier, Desparmet and Montenat, 1972).

3. Practical suggestions

The growth increments of the best fossil specimens available for the study of their faithful record of cosmic cycles call for microscopic inspection. We think that application of the same method to fragments of concretions or tests of fossil marine organisms would be equally valuable. For example, in a microfacies slide, it is sometimes possible to find one or two pieces of shell or thallus preserved by a sudden fossilization (see Figures 3 and 7), more easily than in complete organisms. The method offers the further advantage of providing valuable data concerning ecology and environment.

Conclusion

Rhythmic phenomena recorded by mineralized organisms appear to be mainly cosmic patterns influencing every terrestrial fact, of organic or non-organic nature: sunlight (circadian rhythm); Moon-tide (synodic and fortnightly rhythms). The neritic zone, well-lit and subject to tides, is the best field for these investigations. Sessile encrusting organisms deliver the most reliable record since they behave as recording tapes. Consortial or monospecific algae, displaying only cellular processes, give the clearest patterns, since the only living phenomena in them are of a sexual kind and clearly synodically controlled. Algae and animals such as sponges and ectoprocts respond similarly to destructive processes such as dryness, burying or overlighting: sublethal phases occur, followed by reviviscence or rejuvenation. Mineralized encrusting algae and invertebrate animals appear to be the best recording material because of their behavioural stability throughout the ages.

We must compare like to like, reconstructing the ecological place in the tidal zonation, the climatic status, and the category of the fossil. This is also true of stromatolites. It is necessary to know the pattern of growth and its physiological relations, generally sexual (often of lunar-tidal dependence).

References

Barker, R. (1966). Fossil shell-growth layering and the periods of the day and month during Late Paleozoic and Mesozoic Time. *Ann. Meet. Geol. Soc. Amer.*, 10–11

Bryan, W. H. and Hill, D. (1941). Spherulitic crystallization as a mechanism of skeletal growth in the Hexacorals. *Proc. Roy. Soc. Queensland.* LII (9), 78–91

Dahl, A. L. (1971). Development, form and environment in the Brown Algae *Zonaria farlowii* (Dictyotales). *Botanica marina*, XIV, 76–112

Dixon, P. S. and Richardson, W. N. (1968–69). The life histories of *Bangia* and *Porphyra* and the photoperiodic control of spore production. *Proc. Int. Seaweed Symp.*, 6, 133–139

Evans, J. W. (1972). Tidal growth increments in the cockle *Clinocardium nuttalli*. *Science*, 176, 416–417

Farrow, G. E. (1971). Periodicity structures in the bivalve shell: experiments to establish growth controls in *Cerastoderma edule* from the Thames estuary. *Palaeontology*, 14, (4), 571–588

Farrow, G. E. (1972). Periodicity structures in the bivalve shell: analysis of stunting in *Cerastoderma edule* from the Burry Inlet (South Wales). *Palaeontology*, 15 (1), 61–72

Foster, M., Neushul, M. and Chi, E. Y. (1972). Growth and reproduction of *Dictyota binghamiae* J. G. Agardh. *Botanica marina*, XV, 96–101

Hartmann, W. D. and Goreau, T. F. (1970). Jamaican coralline sponges: their morphology, ecology and fossil relatives. *Symp. zool. Soc. London*, 25, 205–243

Koch, G. von (1882). Uber sie Entwicklung des Kalkskeletes von *Asteroides calycularis* und dessen morphologischer Bedeutung. *Mitt. Zool. Stat. Neapel*, III, 284–290

Liddle, L. B. (1968). Reproduction in *Zonaria farlowii*. I. Gametogenesis, Sporogenesis and Embryology. *J. Phycol.*, 4, 298–305

Lucas, A. (1973). La croissance de *Pecten maximus* en Rade de Brest, dans les conditions naturelles et en vivier. Le Congr. Soc. Franç. Malacologie, Lyon, *Haliotis, S.F.M.*

Miller, R. H. (1969). Synodic month: Variations in the geologic past. *Science*, 164, 67–68

Müller, D. (1962). Uber jahres- und lunarperiodische Erscheinungen bei einigen Braunalgen. *Bot. Mar.* (4), 140–155

Müller, D. (1962). Untersuchungen über die lunarperiodische Entleerung der Geschlechtsorgan von *Dictyota dichotoma. Deut. Bot. Ges.*, N.F. 1, 173–177

Pannella, G., MacClintock, C. and Thompson, M. N. (1968). Paleontological evidence of variations in length of synodic month since Late Cambrian. *Science*, 162, 792–796

Reinberg, A. (1970). La chronobiologie, une nouvelle étape de l'étude des rythmes biologiques. *Science*, I (4), 181–197

Rhoads, D. C. and Pannella, G. (1970). The use of molluscan shell growth patterns in ecology and palaeoecology. *Lethaia*, 3, 143–161

Scrutton, C. T. (1965). Periodicity in Devonian coral growth. *Palaeontology*, 7 (4) 552–558

Taylor, J. D. (1973). The structural evolution of the bivalve shell. *Palaeontology*, 16 (3), 519–534

Termier, H. and Termier, G. (1972). Stromatopores et Eponges coralliens, Scléro-sponges et Pharétrones. *Jubilé Marcel Solignac*, Tunis (in press).

Termier, H., Termier, G., Desparmet, R. and Montenat, Ch. (1972). Les Ammonoïdes du Permien (Kubergandien) de Tezak (Afghanistan central). *Ann. Soc. Géol. Nord*, **XCII**, 105–115

Termier, H. and Termier, G. (1973). Un biotope à Ptérinopectinidés (*Ateyapecten scougouicus* nov. gen. nov. sp.) dans le Dévonien moyen de Dohar Ait Abdallah (Mazoc Central), *Haliotis, S.F.M.* (in press)

Umamaheswarara, M. and Sreeramulut, (1970). The fruiting behaviour of some marine Algae at Visakhapatnam. *Bot. Marina*, **XIII**, 47–49

Vielhaben, V. (1963). Zur Deutung des semilunaren Fortpflanzungszyklus von *Dictyota dichotoma. Z. Bot.*, **51**, 156–173

Wardlaw, C. W. (1955). *Embryogenesis in Plants*. Methuen, London, 381 pp.

Wells, J. W. (1963). Coral growth and geochronometry. *Nature*, **197**, 948–950

DISCUSSION

BUDDEMEIER: I studied sclerosponges by x-ray and found that they have interesting growth patterns, and those from the Marshall Islands do have growth increments.

TERMIER: The growth behaviour and life environment of sclerosponges would be different from stromatolites. Sclerosponges are found in caves, stromatolites in reefs.

CLARK: Some caves have living green algae in them, so there must be sunlight in them, hence the sclerosponge habitat may be more similar to stromatolite habitats than is first apparent.

BUDDEMEIER: Sclerosponges may now be found in caves because they can't compete with corals.

PERIODIC GROWTH AND BIOLOGICAL RHYTHMS IN EXPERIMENTALLY GROWN BIVALVES

GEORGE R. CLARK II
University of New Mexico, Albuquerque, New Mexico, USA

Abstract

The fine concentric ridges found on many pectinids have been found to form with a daily periodicity on several species studied in laboratory and field experiments. In some cases experiments with artificial lighting regimes have demonstrated an underlying biological rhythm, and thus confirmed the natural period as a solar (24-hr) day. Observations by time lapse photomicrography show that the ridges do not mark halts in calcification, but merely changes in direction by an otherwise continuous process.

Despite these advances in our understanding of the nature of these daily growth lines, there are still difficulties in their interpretation. Organisms living in variable environments, such as shallow water, are subject to disturbances which can limit or preclude growth for periods of days or weeks, resulting in gaps in the growth line record. In many cases such growth checks can be recognized, and in some situations their duration can be determined. A more challenging problem is the observation that under some conditions the periodicity may be altered; in one such experiment the periodicity varied from daily to semi-daily.

Experiments with living organisms, originally envisioned to confirm or deny the periodicity of growth lines, have instead shown an intermediate situation where periodicity exists but is preserved only under the best of circumstances. Much more experimentation will be required before the limits of these circumstances can be determined and the accuracy of growth line records in fossils be established.

Introduction

In 1963 John Wells took a number of well-established but apparently unrelated hypotheses and assembled them into a remarkable and elegant speculation. Stated briefly, this speculation was that the growth lines found on many fossil organisms preserve a useful record of changes in astronomical periodicities.

Wells' speculation has stimulated a great deal of discussion and investigation, but appears unlikely to become widely applied until one of its basic assumptions, the hypothesis that growth lines are faithful records of environmental periodicities, is more fully examined. Such examination is not a simple process; as discussed in an earlier paper (Clark, 1974), there are several periodic and random environmental variables which may influence the formation of growth lines, singly and in combination. There are also great differences in the response of various organisms to these variables, so that data acquired from one group of organisms cannot necessarily be applied to another, no matter how closely related.

This is a report of investigations on one type of periodic line in one group of organisms: daily growth lines in the bivalve family Pectinidae. The studies reported here have been selected for the variety of methods and approaches used as well as for the results obtained. Two of the four studies reported in detail were aimed at establishing the periodicity of the growth lines, the third was designed to determine the environmental stimulus behind their formation, and the fourth sought to look at the growth processes involved in growth line formation. A short summary is also provided of the results obtained from other experiments.

The growth lines found in the family Pectinidae are external growth (see Clark, 1974). In the 'supergenera' *Pecten* and *Argopecten* the most common growth lines are very thin frills or ridges best developed between the ribs; in the genus *Chlamys* they consist of rows of spines on the ribs. Other pectinids have other types of growth lines, and in some cases more than one kind of growth line may be present.

Growth line experiments

The simplest kind of experiment to determine the periodicity of growth lines involves tagging a group of the organisms in question and permitting them to grow under approximately normal conditions for a length of time many times greater than the period between growth lines. The new growth lines formed during this time are counted and related, if possible, to periodic variations in the environment. This was the basic approach used in preliminary experiments, although several variations were used.

In one such experiment, discussed in more detail in an earlier report (Clark, 1968), 12 specimens of *Pecten diegensis* were grown in running seawater aquaria for a period of 51 days. Despite the departure from normal conditions, 3 of the 12 formed 50–51 growth lines during this period (see Figure 1a). The remaining 9 specimens formed fewer than 50 growth lines (see Figure 3a), ranging from 34–49; none formed more than 51 growth lines. This distribution suggested that *Pecten diegensis* does tend to form one growth line each day, and that circumstances which cause it to form fewer lines than days are much more important than circumstances which produce more lines than days. This in turn suggests that these circumstances might be entirely unrelated, permitting the hypothesis that the low numbers might be due to occasional halts in growth rather than a slower periodicity. Halts in growth are in fact a common response to environmental stress, and often leave evidence of their occurrence in the form of disturbance lines. Many of the specimens in this experiment have disturbance lines (arrows in Figure 1a), and in general the specimens which produced the fewest growth lines have the most prominent disturbance lines.

Efforts to determine whether the low growth-line counts were in fact because of 'missing' lines due to halts in growth led to the proposal of a new hypothesis.

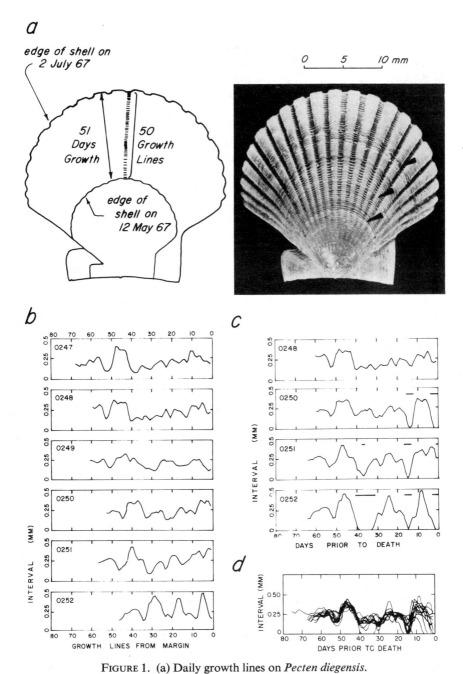

FIGURE 1. (a) Daily growth lines on *Pecten diegensis*.
(b) Growth-rate curves for six experimentally grown *Pecten diegensis*.
(c) Four of six growth-rate curves shown in (b), adjusted for missing lines (bars).
(d) Overlay of all twelve adjusted growth-rate curves, illustrating common pattern of growth-rate variations

This was that the growth between lines should vary with environmental conditions, and that *the pattern of variation should be the same for all specimens growing in the same environment*. Of course this could happen only if the lines form with a very regular periodicity, therefore finding correlations in the growth patterns of individuals is in itself a strong argument for some sort of periodicity. Growth-rate curves were constructed for the 12 specimens; 6 of these curves are shown in Figure 1b. Note the very strong similarity between the curves for specimens 0247 and 0248, both of which formed 51 lines in 51 days; this seems to establish the hypothesis. The other growth-rate curves, which initially showed little similarity to the two 'complete' curves, could then be broken at points corresponding to suspected halts in growth (i.e. disturbance lines), and the resulting fragments separated by gaps equivalent to the number of lines presumed 'missing' in each case; this resulted in a high degree of correlation (see Figure 1c). The overlay of all 12 adjusted growth-rate curves (Figure 1d) gives a very strong impression of a common response to environmental conditions, and strongly supports the hypothesis.

Together, these results provide strong evidence that *Pecten diegensis* tends to form one growth line each day, with a high degree of regularity. Under environmental stress, such as the conditions in laboratory aquaria, growth may stop for intervals long enough for growth lines to be missed. However, if the environment is variable enough to cause growth halts it is likely to cause variations in growth rates as well, and if some specimens have grown continuously the missing lines can be detected in the rest by comparing growth-rate curves.

A second experiment of some interest involved the species *Pecten vogdesi*. In this species the two valves are very different. The right valve is deeply convex, the left flat or slightly concave. The right valve is light in colour, the left richly pigmented. The right valve has smooth rounded ribs and no concentric sculpture aside from disturbance lines. The left valve has flattened ribs with fine concentric growth ridges, very similar to those of *Pecten diegensis*. These features can be seen in Figure 2a.

Eight juvenile specimens of *P. vogdesi* were exposed to a tetracycline solution to form a fluorescent band at the growing margin of their shells; this was to serve as an initial marker (Nakahara, 1961; Bevelander, 1963). The specimens were put in a sand-bottom aquarium, supplied with running seawater, for 31 days. Twice during this period they were removed briefly for measurement.

At the termination of the growth period the specimens were examined for fluorescence. None could be seen through the pigmentation on the left valves, but there was a distinctive line on the right valves of 7 of the 8 specimens. Moreover, the fluorescence was in each case associated with a strong disturbance line, and similar disturbance lines occurred in a corresponding position on the other shells. The ridges were counted between these disturbance lines and the shell margins on the left valves of all but one of the 8 specimens; this one had ridges too indistinct for accurate counting. The results can be seen in Figure 3b; in 3

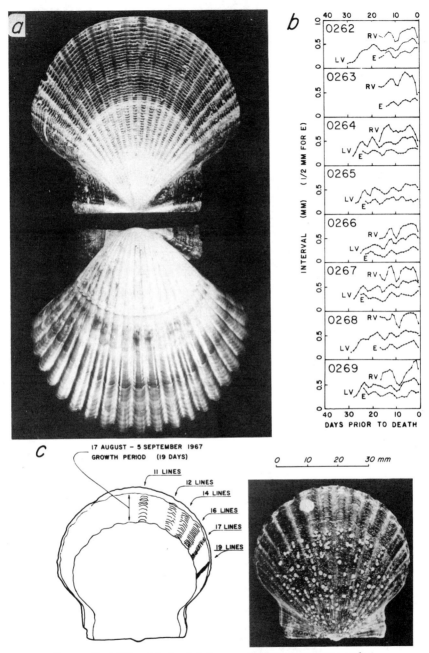

FIGURE 2. (a) Two kinds of daily growth lines on *Pecten vogdesi*.
(b) Growth-rate curves for right valves (RV), left valves (LV), and anterior ears of right valves (E) for experimentally grown *Pecten vogdesi*.
(c) Daily growth lines on *Argopecten irradians*, showing incomplete development on ventral margin

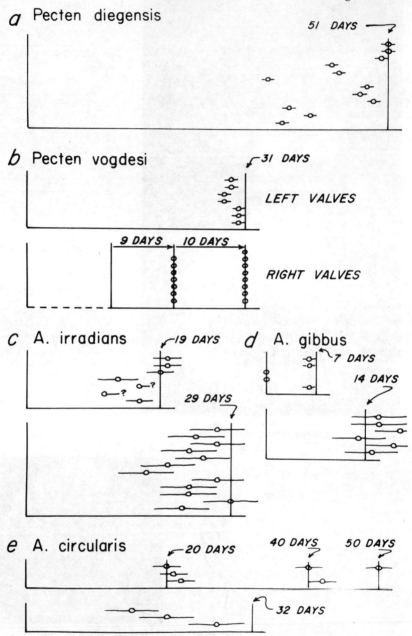

FIGURE 3. Growth lines produced (circles) by experimentally grown species of pectinids, compared to the number of days grown (vertical bars). Error bars, when present, indicate probable limits of error in counting lines between recorded measurements; limits are greatest when lines are closely spaced

cases the growth lines counted were only 1 short of 31, and the lowest count recorded was 28. This is suggestive of daily periodicity and, like the data in the *Pecten diegensis* experiment, suggestive of missing lines. The most likely site of a halt in growth is the strong disturbance line associated with the tetracycline exposure, and indeed tetracycline has been known to inhibit growth, so this would not be unlikely.

There are, however, two other well-defined disturbance lines on the right valves of all specimens, dividing the new growth into 3 subequal zones. They are not apparent on the left valves, possibly due to the strong sculpture. Further examination of the right valves yielded the additional information that much of the new growth was marked with fine concentric pigmented lines, particularly prominent near the margin (see Figure 2a). These lines could not be counted across the entire area of the new growth, but could be counted within the two outer zones in most cases. There were 10 lines in the outer zone in all 8 specimens, and 9 lines in the middle zone in the 7 specimens which could be counted (Figure 3b). Although the lines were too faint to be counted in the inner zone, extrapolation from the outer 2 zones suggests about 30 lines in all. As this implies daily periodicity, the possibility was examined that the disturbance lines could be related to particular environmental stresses 10 and 19 days prior to the end of the experiment; in fact these dates were found to be those on which the animals were measured, and those measurements agree with the positions of the disturbance lines. This suggests that *P. vogdesi* has a remarkably precise growth periodicity.

As a further test of the daily nature of both types of growth lines, they were compared by means of growth-rate curves. Figure 2b shows that the growth-rate curves constructed from right and left valves are in good agreement, strengthening the individual arguments for daily periodicity. This figure also shows considerable similarities between the growth-rate curves of the different specimens, further strengthening the case for periodicity and also supporting the hypothesis of a common response to environmental conditions. Final features of interest are the growth-rate curves constructed from the ridges present on the ears of the right valves; these also correlate well with the other curves, implying that essentially the same curve can be constructed from any part of the shell.

This investigation showed that *Pecten vogdesi* forms two distinct varieties of growth line, one similar to that of *P. diegensis* and one very different. Independent and interdependent lines of evidence show that both types of growth line form with a daily periodicity.

A number of other species of pectinids have been examined in less detail in a continuing survey of growth-line periodicity. In general, the examination has consisted of simple growth experiments with tagged specimens in running seawater aquaria, although some were conducted in the natural environment.

Argopecten irradians. This species is unusual in that the complete series of growth lines commonly appears only at the anterior and posterior margins and on the ears; the ventral margin usually shows only about 70% or less of the full number present (Figure 2c). Two growth experiments were conducted, one for 19 days in the natural environment and one for 29 days in the laboratory. A comparison of the growth lines produced with the days grown (Figure 3c) indicates that these lines may be daily.

Argopecten gibbus. Two experiments were also conducted with this species, one lasting 7 days in the natural environment and one lasting 14 days in the laboratory. The data, shown in Figure 3d, are suggestive of daily periodicity. Experiments now in progress, however, suggest that under some conditions specimens can form two lines per day.

Argopecten circularis. Very few specimens of this species could be obtained, but two experiments, one lasting 50 days for one specimen, were conducted. The results (Figure 3e) suggest daily periodicity.

Three species, *Chlamys hastata hastata*, *Chlamys hastata herica* and *Hinnites multirugosus*, form a predominantly spinose concentric sculpture. Specimens of these species were grown in the laboratory but no relationship was found between the numbers of rows of spines formed and the days grown. Some specimens appeared to have concentric ridges in the early growth stages, but none formed these ridges on increments grown in the laboratory.

Although it is a considerable advantage to know that a particular set of growth lines forms with a daily periodicity, other questions arise. One such is just what environmental variables act as stimuli in the formation of the growth lines.

There are quite a number of potential stimuli. First, there are the tides; these usually occur twice daily (actually, twice in 24·84 hr) but some of their effects could register once a day, and in some areas they occur only once a day for at least part of the month. Next, there are the indirect solar effects, such as temperature variation (in shallow waters), variation in food supply (with vertical migration of plankton), variation in dissolved oxygen (with photosynthesis), and more subtle chemical changes. Finally, there is solar illumination itself. This might seem an unlikely stimulus, but pectinids have eyes and exhibit a characteristic closing reaction at the passage of a shadow.

It might ordinarily be no simple task to devise an experiment to separate these factors, but by a fortuitous set of circumstances most of these factors can be eliminated from serious consideration.

Kerckhoff Laboratory, where these experiments were conducted, is located on the inlet to Newport Bay, a sizeable estuary. When the tide is falling, the water in the inlet is coming from the estuary; when the tide is rising, the water is coming from the open ocean; and the seawater in the laboratory passes through the system rapidly enough to approximate closely to the seawater in the inlet at all times. Thus the characteristics of the seawater in the laboratory should vary with the tides. Figure 4a shows this to be so for the water temperature; other characteristics, such as food content and chemistry, are likely to be just as different. As Figure 4a also shows, the tides at Corona del Mar are semi-daily;

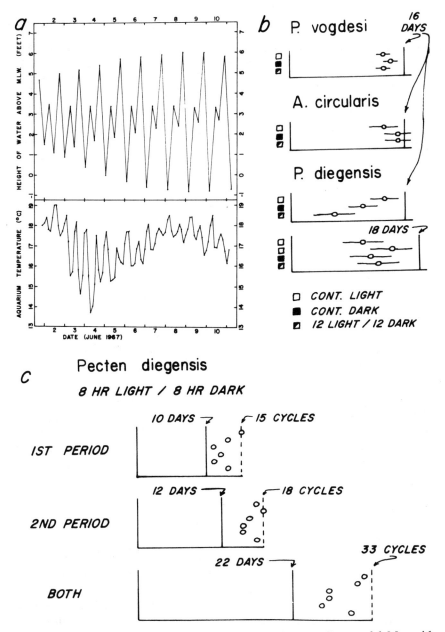

FIGURE 4. (a) Comparison of predicted tidal variations at Corona del Mar with seawater temperatures in laboratory aquaria.
(b) Growth lines produced (circles) by subjects of artificial illumination growth experiments compared to number of days grown (vertical bars). Error bars as in Figure 3.
(c) Growth lines produced (circles) by subjects grown in 16-hr light/dark cycles, compared with days grown (vertical bars) and cycles experienced (vertical dashed lines)

therefore nearly every environmental variable experienced by the experimental animals, even those indirectly related to solar illumination, would have strong semi-daily variations. The only exception appears to be direct solar illumination; the animals were kept in transparent aquaria near a large window, and no artificial illumination was used at night.

It would seem from these observations that as the only daily stimulus available was variation in illumination, this must be the stimulus involved. To investigate this further, some experiments were conducted with artificially controlled illumination.

In the initial experiment, scallops of several species, including *Pecten diegensis*, *Pecten vogdesi* and *Argopecten circularis*, were put into 4 aquaria and subjected to the following light regimes: continuous light, continuous dark, 12 hr light/12 hr dark, and 8 hr light/8 hr dark. These tanks developed particularly severe problems with their flow of seawater, and on a number of occasions the flow stopped completely for several hours at a time in one or more tanks. At the termination of this experiment it was found that the specimens in the 8-hr aquarium had all died and that growth was poor in the other aquaria. This was especially true for the specimens of *P. diegensis*, which are used to much cooler water; a second experiment was run for this species.

The results of these experiments (Figure 4b) were rather surprising; no significant differences in growth-line formation were noted between organisms maintained at constant illumination and those exposed to light and dark at 12-hr intervals. This initially seemed to argue against the influence of illumination upon growth-line formation.

The 8-hr interval experiment was run again using a constant-flow regulator and more precise techniques of marking. The results, shown in Figure 4c, show that the number of growth lines exceeded the number of days and approached the number of light/dark cycles to which they were exposed. Inasmuch as no previous examples of extra lines had been encountered and missing lines are a fairly common event, it would seem likely that the growth lines formed in direct response to the light/dark cycles; yet how is this to be reconciled with the results of the earlier experiment?

There is one phenomenon which can explain these results; whereas some rhythmic behaviour seems to be exogenous, or in direct reponse to environmental stimuli, much of it is now known to be endogenous, or under the immediate control of the organism. Endogenous rhythms, also called biological rhythms or biological clocks, follow a few simple rules. If the organism is in normal surroundings with a 24-hr periodic variation in the environment, the rhythms will continue indefinitely with a precise 24-hr periodicity. If the organism is then placed in constant surroundings the rhythm will continue for many days, but may shift from a 24-hr periodicity to a longer or shorter period and may eventually fade out. If the organism is raised under constant conditions it may show no rhythms at all, but any sudden change in the environment, such as a flash of light in dark conditions, may initiate a rhythm. The phase of the rhythm may be shifted by altering the environmental conditions, such as gradually inverting the light/dark regime. The period of the rhythm may also be shifted, so that under conditions of 10 hr light and 10 hr darkness the rhythm may adjust to a 20-hr period. However, most rhythms will revert to the sort of behaviour exhibited under constant conditions if the artificial cycles differ by more than 6–8 hours from the normal 24-hr cycle.

The results of these controlled illumination growth-line experiments seem to follow the same rules as biological rhythm phenomena. The growth lines are normally formed daily, indicating a 24-hr periodicity. They continue to be formed, approximately one a day, under conditions of either constant illumination or constant darkness. In addition, they can be shifted to a shorter period by a shortened cycle in an artificial environment. The identification of a biological rhythm in the formation of daily growth lines permits the resolution of the apparently conflicting evidence in these experiments, and demonstrates that the cycle of light and darkness is the major stimulus in the formation of growth lines in these species.

If further experiments support this interpretation, daily growth lines may have applications in the study of biological rhythms. Experiments in most rhythmic phenomena require continual sampling or monitoring, either by the investigator or by special apparatus. In the mollusc shell the rhythms form their own record, in the form of growth lines, with no need for external monitors. Experiments need not be limited to highly artificial laboratory situations, for the animals record the periodicity in their natural environment, free from disturbance by the investigator. Moreover, these animals have been responding to similar periodicities for millions of years, preserving a record of rhythms and periodicities over a span of time in which the periodicities themselves have significantly changed.

The last experiment discussed here was designed to look in more detail at the formation of daily growth lines. A special aquarium was designed to hold a specimen of *Pecten diegensis* rigidly under a microscope while maintaining a running seawater environment. The margin of the shell could then be kept in focus under a microscope for periods of several days to permit time-lapse photography of the growth process. The resulting film was viewed first as a motion picture, in which the growth could be observed at an apparent rate more than 1000 times normal, and then as individual frames, which could be examined more closely to resolve details. The information obtained in this manner was supplemented by scanning electron microscopy (SEM) on the structure of the growth ridges and the outer shell layers.

The most significant result of this study was the observation that the growth ridges of *Pecten diegensis* do not form in connection with any halts in the calcification process. Instead, there is a gradual shift in the orientation of the mantle resulting in an upward growth of the very thin margin of the shell (Figure 5a). This upward growth may continue for $\frac{1}{2}$ mm or more and occupy as much as half the time involved in the formation of the entire increment. Eventually, however, the mantle edge retreats and begins building a new shell margin parallel to the principal direction of shell growth; this leaves the upward growing margin isolated as a growth ridge (Figure 5). SEM observations of this region of the shell (Figure 5c) shows that there is no structural break in the transition from the zone which becomes the true shell surface (the interspace) to the zone which becomes the growth ridge.

This has two important implications. Inasmuch as it is the interspace which is measured as the distance between growth lines, and therefore the variations in interspace which are being correlated between shells, the period involved is not the entire day but only one part of it. This period must be quite regular, for the growth-rate curves would not correlate otherwise. This suggests that two environmental stimuli are involved, one to initiate the upward growth and one to terminate it. The second implication is that because the growth ridges are easily broken and become badly eroded on older shells, the growth record best preserved in fossils involves effective, if not actual, halts in calcification. Any studies of subdaily lines or periods must take this into account.

It was hoped that this work would yield other information about the formation of these ridges, such as their relationship to the time of day. Unfortunately, the specimens used in the first experiments did not follow a consistent rhythm; the one which grew the most produced 5 lines in 7 days. At least part of the poor growth appears to be due to the continuous and intense illumination used for the photography; this should no longer be a problem as the equipment is now using stroboscopic illumination, which does not seem to disturb the organisms. Aside from this drifting from the natural rhythm the growth lines produced during the time-lapse study appear no different from those produced under natural conditions, and future work along these lines should be very productive.

Summary

It seems well established that at least two pectinid species form growth lines with daily periodicity, and for one of these it has been shown that the process involves a biological rhythm with solar illumination as the stimulus.

FIGURE 5. (a) Interpretation of growth ridge formation at the margin of *Pecten diegensis* based upon time-lapse microphotography and scanning electron microscopy. The growth ridges are approximately $\frac{1}{4}$ mm apart.

(b) Selected pairs of frames from time-lapse film, showing stages in growth ridge formation corresponding to intermediate stages diagrammed in part (a). The direction of growth is toward the top; the film was photographed from above the shell surface. Each pair of frames includes a view with the mantle in normal position (upper) and a view with the mantle retracted showing the extent of calcification (lower). The growth ridges are approximately $\frac{1}{4}$ mm apart.

(c) Scanning electron micrograph of two growth ridges in *Pecten diegensis*

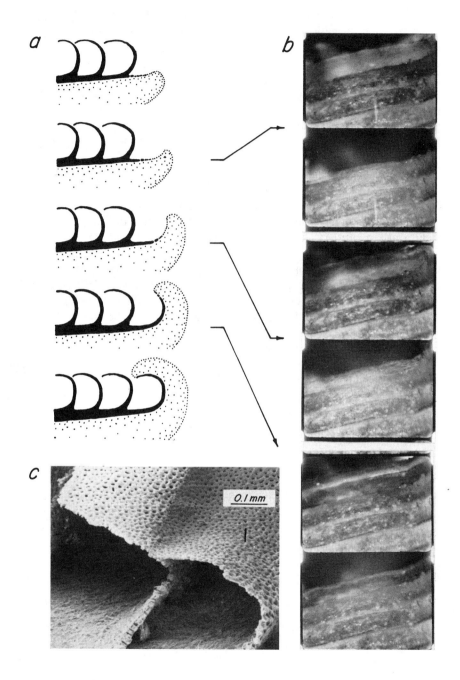

For a number of other species, the evidence points towards a daily periodicity in growth-line formation, although some abnormal observations await further investigations. In a third group, a different sort of growth line appears to be unrelated to daily periodicities; again, the evidence could be improved by further experimentation.

The real achievement of this research goes beyond these few facts to an awareness of the complexities involved in growth-line formation. Growth lines seem to be different things in different circumstances, and the reasons for this must be better understood before we can begin to use fossil growth lines to full advantage.

Acknowledgments

Most of this research was conducted at Kerckhoff Marine Laboratory, Corona del Mar, California, through the courtesy of the Division of Biological Sciences, California Institute of Technology. Thanks are due also to the Bermuda Biological Station, The Friday Harbor Laboratories, and the U.S. Bureau of Commercial Fisheries Biological Laboratory at Woods Hole, for the occasional use of their facilities. The specimens of *Pecten diegensis* were dredged from the R. V. Velero IV through the courtesy of Donn S. Gorsline of the University of Southern California. The *Pecten vogdesi* were collected in the Gulf of California with the permission of the Departamento de Estudios Biologicos Pesqueros. The SEM work was accomplished at the University of California, Los Angeles, with the aid of J. William Schopf. This research was initiated in the course of graduate study at the California Institute of Technology and profited greatly from the advice and inspiration of Heinz A. Lowenstam. Financial support was provided by National Science Foundation grants GB-6275 and GB-20692 and by the Alfred P. Sloan Foundation.

References

Bevelander, G. (1963). Effect of tetracycline on crystal growth. *Nature*, **198**, 1103

Clark, G. R., II (1968). Mollusk shell: daily growth lines. *Science*, **161**, 800–802

Clark, G. R., II (1974). Growth lines in invertebrate skeletons. *Ann. Rev. Earth Plant. Sci.*, **2**, (In press)

Nakahara, H. (1961). Determination of growth rates of nacreous layer by the administration of tetracycline. *Bull. Natl. Pearl. Res. Lab.*, **6**, 607–614

Wells, J. W. (1963). Coral growth and geochronometry. *Nature*, **197**, 948–950

DISCUSSION

PANNELLA: I don't think you can safely extrapolate from laboratory experiments to natural conditions. In addition, I believe Pectens are palaeontologically unfavourable because the group only has external lines difficult to count on a thin shell (subject to rapid diagenesis).

CLARK: These problems are not insurmountable; for example, I can count growth lines on the external moulds of many fossils. In some species there is even an internal expression of lines, despite the coarsely crystalline calcite shell structure.

FARROW: Have you looked at deep-water scallops?

CLARK: I have dredged scallops from 1200 m and they do have growth lines, but they didn't live long enough for me to conduct experiments on them.

GROWTH AND MICROMORPHOLOGY OF TWO BIVALVES EXHIBITING NON-DAILY GROWTH LINES

JOHN W. EVANS

*Biology Department, Memorial University of Newfoundland,
St. John's, Newfoundland, Canada*

Abstract

A number of workers have presented evidence that skeletons of recent invertebrates are deposited in a daily rhythm. This rhythm is recorded in the microstructure of the shell as daily lines. These observations on individual species have often been generalized to invertebrate skeletons as a whole. Some palaeontologists have accepted these generalizations and used the microlines of fossils to determine the duration, in number of days, of past months and years. It is the purpose of this paper to caution palaeontologists against uncritical acceptance of this method by describing the growth patterns of two recent bivalves whose skeletal lines are not deposited on a daily basis.

The microstructures of *Penitella penita*, a rock-boring clam, and *Clinocardium nuttalli*, a sand-dwelling cockle, were studied in detail. Although both of these animals were collected from adjacent sites in the midtidal zone on the Oregon Coast, and both were subjected to similar tidal and climatic experiences, their growth patterns are quite different.

The growth pattern of *Clinocardium nuttalli* appears to be wholly entrained by the rhythm of tidal exposure. When the cockle, which lives close to the surface of the sand, is covered by the sea, the valves gape and calcium carbonate is deposited on the leading edges. When the tide recedes below the cockle's habitat, the valves shut and a layer of conchiolin is secreted. This leaves a near-perfect record of the animal's tidal experiences. This tidal experience is peculiar in that the area is subject to a mixed semi-diurnal pattern of tides. During spring tides conchiolin lines with a periodicity of 24 hr 50 min are produced, while lines with a periodicity of approximately 12 hr 25 min are deposited during neap tides. There is no evidence of daily lines.

The microstructure of *Penitella penita* shows no evidence of the precise tidal pattern which is seen in the cockle. Rather, a unique pattern of lines is seen which is repeated on an approximately fortnightly basis. A group of 5–9, wide, poorly defined lines (no conchiolin separations) is followed by another group of 4–8, well-defined, narrow, conchiolin-rich lines. This pattern appears to bear no relationship to the external environment, i.e. the narrow lines are not produced at any particular phase of the Moon. Different individuals from the same habitat can vary widely in duration of the cycle. This pattern appears to reflect a behavioural cycle of events peculiar to the rock-boring clams. The narrow lines are deposited during a period of intense boring activity when the animal is enlarging its burrow, while the wide lines are related to a relatively quiescent period of growth within the enlarged burrow.

In conclusion, it would appear that micro-growth lines in the skeletons of invertebrates may be deposited under the influence of a number of cyclical factors, other than that of the solar day. Therefore, before palaeontologists draw conclusions about the duration of the month or year in ancient times, using as evidence the counts of microlines, they should be prepared to prove conclusively that these lines are daily in nature.

Anybody looking back over the brief 10-year history of the use of biological growth rings as geochronometers is provided with a striking example of the strengths and weaknesses of a beautiful theory. Wells (1963) first propounded the theory that the duration, in terms of the number of days, of ancient months and years could be measured by counting the daily growth lines on the skeletons of fossil organisms. This theory did what any beautiful theory should do, it inspired and directed the work of many scientists. The problem with a beautiful theory is that it tends to prejudice the objectivity of the scientist. In this case it leads him to assume that the microgrowth lines represent solar daily intervals, for only by doing this can he get on with the business of counting the number of days in a Devonian year. Pannella and MacClintock (1968) on the basis of their experiments with living *Mercenaria mercenaria* concluded that, of over 40 species of living and fossil bivalves which they had examined, all showed daily growth lines. I think that it is an error to make generalizations about invertebrate growth patterns or, more specifically, bivalve growth patterns. I believe that when each species is looked at individually it will be found that many different growth patterns will be demonstrated. Some species will conform to a solar daily cycle, some to a lunar daily cycle, some to an internal cycle with no apparent external control. When the growth patterns of recent species are known, then the palaeontologist will be able to choose with confidence the fossil species which will best answer the questions he is asking.

When the durations of two different external rhythms closely approximate each other, such as solar and lunar daily rhythms, it is often difficult to establish definitely which is the controlling influence. Pannella and MacClintock's (1968) experimental study of the growth lines of *Mercenaria mercenaria* seemed to establish without doubt that the micro-growth lines were of solar daily duration, yet Pannella (1972) claims that they are of lunar daily duration even though his original counts do not bear out this contention. The trouble is that counting growth lines is not easy, as anyone who has tried can testify. It is constantly necessary to make subjective decisions about whether a line is really a line or where an annual or monthly series begins or ends. As a result it is not surprising to find that counts often come out close to the hypothesized values. Clark (1974) provides a good discussion of these problems in his review of 'Growth lines in invertebrate skeletons'.

I have examined in detail the microstructure and growth lines of two recent bivalves. The growth pattern of each is entirely different and neither follows a strict solar daily rhythm (Evans, 1972; Evans and LeMessurier, 1972).

Penitella penita (Fam. Pholadidae), a rock-boring clam, and *Clinocardium nuttalli* (Fam. Cardiidae), the basket cockle, are commonly occurring Eastern Pacific intertidal bivalves. Both have a broad range from Baja California to Alaska. They are abundant in Coos Bay, Oregon, where most of the specimens for this study were collected.

One of the chief characteristics of the tide along the west coast of North America is diurnal inequality, this is also known as mixed semi-diurnal tides (Figure 1a). The effect of this tidal pattern in the Empire region of Coos Bay is that some parts of the shore (between 0·2 m and 0·7 m above datum) are exposed once every lunar day during spring tides and twice a day during neap tides. Lower parts of the shore (from about 0·1 m down to spring low-tide level) are exposed only once a day during spring tides.

Most of the *Clinocardium nuttalli* were collected from protected intertidal sand flats. I collected specimens in Coos Bay, Oregon and collaborators collected others from Humbolt Bay, California and from a sand flat near Amphitrite Point, Vancouver Island. Two other specimens were collected from the middle intertidal zone of a moderately exposed beach in Patricia Bay, Saanich, B.C.

Penitella penita were collected by the author from two locations on the Oregon coast. One was an exposed rocky shore, the Jetty on the south side of the entrance to Coos Bay. The other location, Fossil Point, was a protected area within Coos Bay. All specimens were collected between the −0·3 m tide level and the +0·3 m tide level. The relative hardness of the rock in which the clams were boring was estimated as 2 and 1 or medium and soft, respectively (Evans, 1968a).

The microstructure and growth lines of both species were examined using the acetate peel technique, thin sections, and Scanning Electron Microscopy of ground and etched shell fragments.

Observations on *Clinocardium nuttalli*

There are two kinds of growth structures seen on basket cockle shells. Externally there are fairly evenly spaced concentric ridges located on the top of the radial ribs. Internally (Figure 3a) there are micro-growth lines.

First let us examine the micro-growth lines. Rather than the continuous series of 'daily' lines as seen in *Mercenaria mercenaria* (Pannella and MacClintock, 1968) or the continuous series of external daily ridges as seen in *Pecten diegensis* (Clark, 1968), there are in *C. nuttalli* a whole series of groups of evenly spaced lines which partially overlap each other. Each group usually contains between 18–21 lines. The amount of overlap varies considerably, usually the outer 3–6 lines of neighbouring groups overlap (Figure 2a) but this may go as high as 9 and very occasionally as low as zero (Figure 2c). This overlapping results in a distinctive pattern, groups of widely and narrowly spaced lines follow each other in a continuous series.

There is only one environmental factor that varies in a way that matches the cockle's growth lines and that is tidal exposure. The assumption is made that when the animal is submerged its valves gape for feeding purposes. When the valves gape the mantle can extend out around the periphery of the valve and conditions are right for deposition of $CaCO_3$ on the leading edge. When the

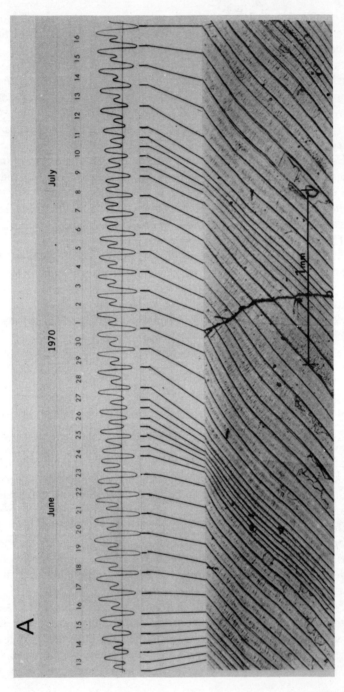

FIGURE 1A. Internal growth lines in *Clinocardium nuttalli* collected at Charleston, Oregon, compared with tidal predictions for the same period. The horizontal line drawn through the tidal curves marks the intertidal position (+0·6 m) at which the specimen was found. Lines are formed when the water level falls below the +0·6 m tide level. Reproduced with permission from Evans, J. W., *Science*, **176**, 416–417 (1972). Copyright 1972 by the American Association for the Advancement of Science

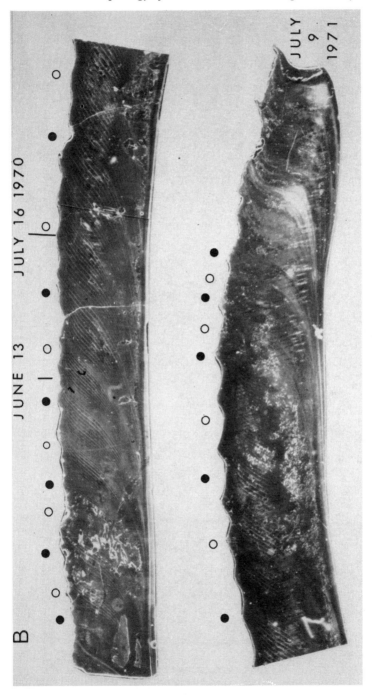

FIGURE 1B. Acetate peel of *C. nuttalli* collection date 9 July 1971. Note location of month-long period of growth illustrated in Figure 1A. Evenly spaced external ridges can be seen in profile. New ● and full ○ Moon symbols are located over appropriate sections of the shell. Note lack of relationship between lunar cycle and external ridges

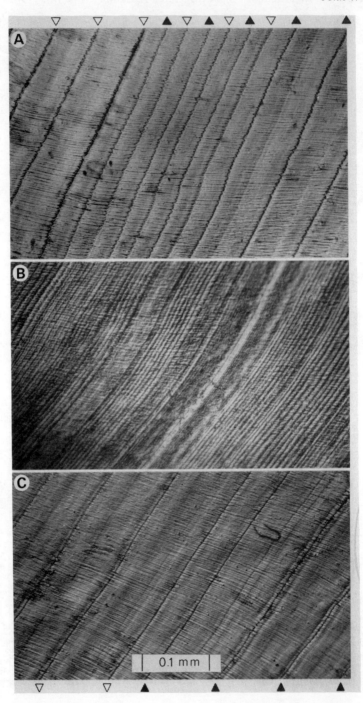

animal is exposed at low tide the valves close, this temporarily terminates shell growth and the developing columns of $CaCO_3$ are capped with a thin dome-like layer of conchiolin; this process forms the growth line.

Because the collection date is known it is theoretically possible to count backwards from the leading edge and establish the exact time and date that a line was deposited. Because the winter growth is difficult to read it was only possible to describe in detail a full year's growth in a few specimens from Coos Bay. A 15-month growth record is illustrated (Figure 4). It is useful to be able to date the deposition of a line because then the height of the predicted tide and the line pattern can be compared (Figure 1a).

The Coos Bay *C. nuttalli* were collected from the midtide zone of a sand flat. No attempt was made to establish the precise height in the intertidal zone. By comparing the tidal predictions with the occurrence or non-occurrence of tidal growth lines it was found that the best correspondence was obtained when it was assumed that the cockles were collected at the +0·6 m intertidal level. In other words, when the tide falls below the +0·6 m level a line is deposited. During the spring tides only the low low tide is represented by a line, the high low tide is not low enough to expose the cockles. During neap tides the diurnal inequality fades and then disappears so that a cockle at the +0·6 m level is exposed twice in a lunar day. The cycle is repeated with the next series of spring tides with the important difference that the pattern of exposure is 12 hr 25 min out of phase with the previous series. This is because the high low tides of one series become the low low tides of the next and *vice versa*.

Specimens of *C. nuttalli* that were collected from the 3 sand flat habitats all exhibited this kind of a growth pattern. However, the specimens collected from the moderately exposed beach near Saanich, B.C. showed a modified pattern. The widely spaced spring tide lines are readily apparent but the narrowly spaced neap tide lines are not. Where they should be is a homogeneous area of growth. It is likely that the cockles were fully exposed in their midtide location during spring tides but were in the swash zone during the neap tide (Figure 3b).

The width and clarity of definition of the growth lines vary considerably; this is the seasonal influence, winter growth lines (Figure 2b) being very narrow and difficult to interpret as compared with summer growth lines (Figure 2a). *C. nuttalli* grows as much as 25 times faster during the summer than it does during the winter. Regardless of the rate of growth, the pattern of growth so clearly evident in the summer shell is also present in the winter shell. Just what

FIGURE 2. Acetate peels of *C. nuttalli*, three different sections of same shell. (a) Summer growth with normal overlap of neighbouring groups of lines. ▽ indicates first group. ▲ indicates second group. (b) Winter growth, overlapping groups cannot be clearly seen because of close spacing of the lines. (c) Summer growth showing atypical non-overlap of neighbouring groups of lines. Leading edge of shell is towards the right. Magnification uniform

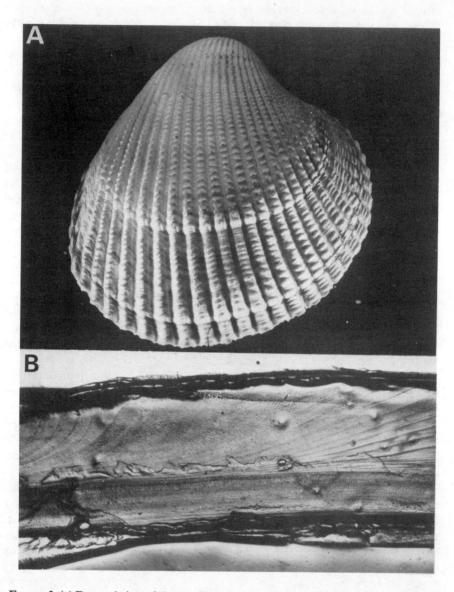

FIGURE 3. (a) External view of *C. nuttalli* shows external growth ridges on top of the ribs. (b) Acetate peel of shell from moderately exposed beach, Patricia Bay, B.C. Spring tide lines show on the left and right sides. The central homogeneous area represents neap tide growth

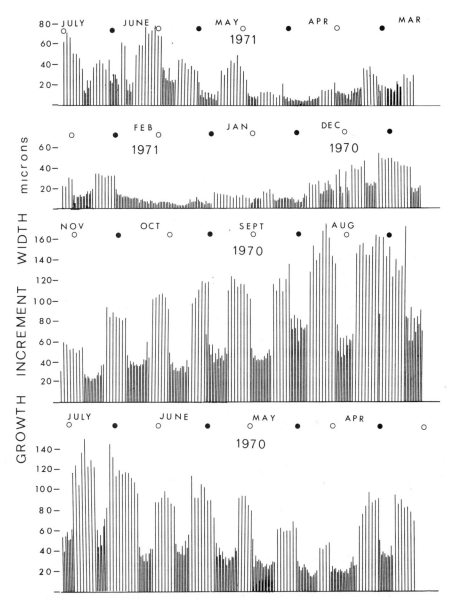

FIGURE 4. Fifteen-month record of growth of specimen illustrated in Figure 1B. Each growth line is represented by a bar. The widely spread bars represent lines deposited every 24 hr 50 min during spring tide periods; narrowly spaced bars represent lines deposited every 12 hr 25 min during neap tides. The length of the bar indicates the width of the growth increment. New ● and full ○ Moons are placed over appropriate areas of growth record. Note that new and full Moons do not lie directly over widely spaced spring tide lines. This is because in Oregon there are two more spring tides per year (27) than there are new and full Moons (25). The cockles' growth therefore reflects the tidal cycle, not the lunar cycle

the cockle is sensitive to is not known. Water temperatures in the bay vary little during the year (max 13·3°C, min 4·4°C) and air temperatures are far from extreme, averaging about 18°C in summer and 4·5°C in winter. By comparison the seasonal differences of growth of *Penitella penita*, living only half a mile away under similar conditions, are barely recognizable.

The fidelity with which *C. nuttalli* records its tidal experiences and the sensitivity of the animal to seasonal change is, I believe, related to its habit of lying just below the surface of the sand. The little neck clam *Protothaca staminea* (Fam. Veneridae) is found in the same area as *C. nuttalli* but lives buried about 15 cm in the sand. A pattern of growth lines similar to that of *C. nuttalli* can just barely be distinguished.

The external concentric ridges (Figure 1b) can be seen in profile on the same section of shell that was used to obtain the 15-month record (Figure 4). The question is, do these represent similar time intervals? Because all the lines can be accurately dated it is possible to place full and new Moon symbols over that part of the shell deposited under their influence. In all cases these correspond to areas of wide growth lines. It can be seen that the external ridges do not correspond to any particular period of the lunar cycle and therefore do not represent similar time intervals. It seems likely that their secretion is more dependent on geometrical spacing. Similar external shell ornamentation has been described in other bivalves by Stanley (1969) who determined that they were useful for gripping the sediment during burrowing movements.

Observations on *Penitella penita*

Externally the shell of *P. penita* can be subdivided into 3 zones, a posterior slope, an anterior slope and an intermediate umbonal ventral sulcus (Figure 5a,c). The surface of the posterior slope is relatively smooth and the margin is sharp and unreflected. The posterior slope appears to function primarily as a protective surface for the internal organs. Microstructural examination revealed no growth lines. The surface of the anterior slope is strongly ridged with quite evenly spaced concentric 'growth bands'. The margin is thick and strongly reflected. This is the portion of the shell that is used as a cutting tool in rock boring. Microgrowth lines are at best very poorly defined (Figure 5b,d).

The umbonal ventral sulcus is a narrow strip of shell which extends from the umbo to the ventral condyle. The ventral condyles are medially located pillow-like protrusions on the ventral edge of the valves. The shell deposited on the condyle accumulates to form the umbonal ventral sulcus. When viewed externally it can be seen that the growth bands of the anterior slope continue onto the floor of the sulcus. In radial section these bands are reflected perpendicular to the direction of growth and form the thick outer shell layer. Within each band the finer growth increment lines can be seen. The lines in the proximal 80% of the band are relatively straight and parallel while those in the distal 20% form a

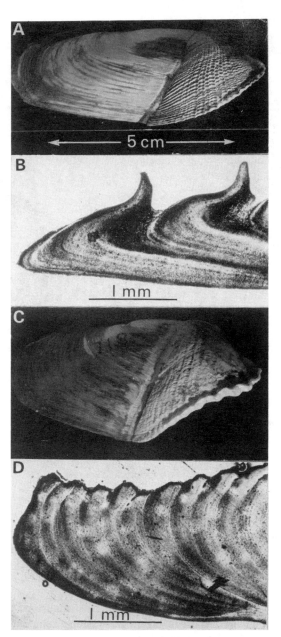

FIGURE 5a, c. External views of *Penitella penita* from soft (a) and medium (c) rock. Posterior end at left.

b, d. Acetate peels of sections through anterior slope of shell of animals from soft (b) and medium (d) rock. Figures 5a–d reproduced by permission of the National Research Council of Canada from the *Canadian Journal of Zoology*, Vol. 50, pp. 1251–1258, 1972

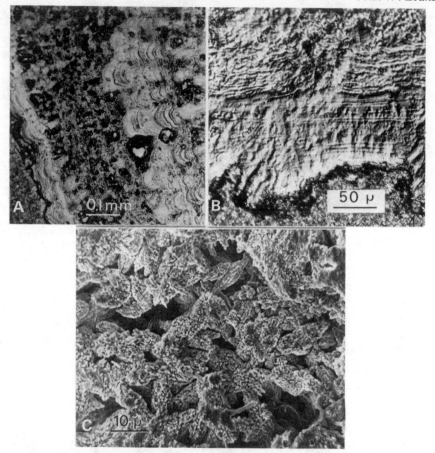

FIGURE 6a, b. Acetate peels of umbonal ventral sulcus of *P. penita* from medium rock. (a) Distal zones show distinct scalloped lines, (b) Nomarski interference phase shows straight lines in the proximal zone.
c. Scanning Electron Micrograph of proximal zone of *P. penita* from soft rock. Note loose arrangement of composite bipyramidal aragonite crystals. Figures 6a–c reproduced by permission of the National Research Council of Canada from the *Canadian Journal of Zoology*, Vol. 50, pp. 1251–1258, 1972

distinct scalloped pattern (Figure 6a). The microstructure of these 2 zones is entirely different. The lines in the distal zone are clearly delineated by thick wavy layers of conchiolin, $CaCO_3$ development between these layers is minimal and sometimes altogether absent. The proximal zone is almost entirely composed of bipyramidal composite crystals of aragonite, conchiolin is virtually absent from this area. The structure of the proximal zone varies with the hardness of the substrate. In animals from soft rock the crystals are loosely organized (Figure 6c). Growth increments are indistinct and are mainly seen as rows of cavities. The proximal zone in animals from medium and hard rock is compact and well-organized. Growth increment lines stand out as low parallel ridges when the acetate peel is examined under Nomarski phase (Figure 6b). In good preparations these lines may occur in bewildering abundance; usually the zone appears to be subdivided into 6–12 major increments. Within the major increments 2–12 sublines can be seen. It is sometimes difficult to decide where the major lines are located in the mass of minor lines. The lines in the proximal zone are unlike lines from other bivalves in that they are not composed of alternating layers of $CaCO_3$ and conchiolin. Rather they appear to be built into the aragonite crystal structure.

Evans and LeMessurier (1972) demonstrated that the conchiolin-rich distal zone of the band is deposited during a period of active boring when the animal is enlarging its burrow. The proximal zone is deposited during a period of quiescence when the animal is growing to fill its newly enlarged burrow. Pannella and MacClintock (1968) claim that in bivalves organic material is deposited when the valves are shut and $CaCO_3$ when they are open. When *P. penita* is actively boring the 2 ventral condyles are rotating upon each other with considerable force, thus giving proper conditions for the deposition of conchiolin. The lack of conchiolin in the proximal zone supports the hypothesis that the animal is quiescent when this zone is deposited.

Duration of growth intervals

A series of shells were available which had previously been used for a growth experiment (Evans, 1968b). These animals had been collected, measured and replaced in artificial burrows in the field for 341 days; then they were recollected and measured for growth. Experimental animals were collected from both medium and soft rock and were replaced in either their own or alternate substrates (Table 1).

Initially the assumption was made that the distinct scalloped lines in the distal zone of the band and the major increments in the proximal zone were daily increments. In order to test this assumption the total number of lines that were deposited, during the 341-day replant period, were counted (Table 1).

The results demonstrate that the increment lines are only approximately daily in duration. Total counts of lines produced during the replant period show that some fit closely the expected value of 341 lines while others are as much as 125 lines above or below this value. The amount of variation from the expected value indicates that the lines cannot be assigned a rigid temporal value. It is only possible to say that they are deposited on a roughly circadian basis with considerable room for individual variation. It has not been possible to relate these variations to any known environmental influences.

Initially it seemed possible that the tidal cycle might control the cyclic deposition of the growth bands. The mean number of circadian lines per band (15·6)

TABLE 1. Summary of line counts on *Penitella penita* replant animals, growing over a period of 341 days and 24 tidal cycles (10 August 1964 to 15 July 1965). If growth lines were deposited daily, and each growth band was deposited in a tidal period, we would expect each specimen to have 341 growth lines and 24 growth bands. $\bar{X}^*$ indicates mean value of four independent counts

Specimen number	Substrate		Number of growth bands	$\bar{X}^*$ total number of growth lines deposited during replant period	Lines per growth band, only complete bands counted	
	Original	Final			$\bar{X}$	Max–Min
1	Medium	Soft	22	360·25	16·8	22–10
2	Soft	Medium	32	464·00	15·1	20–10
3	Soft	Medium	21	273·75	13·3	23–7
4	Medium	Soft	15	217·75	15·1	20–9
5	Medium	Soft	15	269·50	18·3	22–15
6	Soft	Soft	22	354·25	16·2	20–9
7	Medium	Medium	20	330·75	16·7	19–14
8	Medium	Medium	25	384·00	15·9	20–13
9	Medium	Medium	18	236·00	13·3	19–7
$\bar{X}$			21·1	321·1	15·6	
Expected			24	341	14·5	

supported this idea but, again, variations about the mean (see Table 1) are so great that tidal control must play only a minor role in determining the duration of the growth-band interval. There are usually fewer bands ($\bar{x} = 21\cdot1$) than tidal cycles experienced (24).

Conclusion

It is dangerous to assume that all micro-growth lines in bivalves represent the same kind of periodicity. The preceding examination of 2 species of eastern Pacific bivalves demonstrates how variable growth patterns can be.

In the single shell of *Penitella penita* there are three different shell types which relate to three different functions. Only the umbonal ventral sulcus shows a clear pattern of growth lines. Even in this area two different types of lines are deposited which represent a cyclic pattern of behaviour. Wide, straight increments are deposited in the proximal zone of the band during passive growth and narrow, scalloped increments, rich in conchiolin, are deposited in the distal zone during active boring. The microstructure of these lines can provide much useful information to the palaeoecologist but, because of their relative independence of environmental cycles, this species is quite useless as a geochronometer.

Clinocardium nuttalli from the sandflats in Coos Bay, Oregon, grow in a completely different manner from *Penitella penita* which lives near by. Both species are subjected to essentially the same environmental conditions with two important exceptions which help to explain their different growth patterns. The unyielding rocky burrow of *P. penita* forces it to grow in the cyclic manner just described. The cockle, which like most other bivalves lives in a semi-fluid medium, can grow unimpeded by the substrate and thus lacks the cycles of rapid and slow growth. Also, *C. nuttali's* habit of living just beneath the sand surface makes it much more sensitive to tidal exposure than *P. penita*, which shows no clear tidal control of its growth. The cockle's growth is a faithful reflection of its tidal experience, each line representing exposure at low tide. Semi-lunar daily lines are deposited during neap tides, and lunar daily lines during spring tides.

Two complete cycles of spring and neap tide lines represent a period about two days short of a synodic month. This is because in this area there are 27 spring and neap tide cycles per year but only 25 new and full Moons. *C. nuttalli* reflects the tidal not the lunar cycle. The extreme sensitivity of *C. nuttalli* to seasonal changes is also probably related to its location in the substrate.

A fossil which shows a growth pattern similar to *C. nuttalli* can tell a palaeontologist many useful things. He can tell that his animal lived in the middle intertidal zone of a sand flat protected from wave action and experienced mixed semi-diurnal tides. He can know without question that the lines represent semi-lunar and lunar days. He should be able to count the number of spring and neap tide periods in a year and also the number of lunar days in a year. Thus it seems quite likely that cockles are potentially one of the most useful bivalves for future geochronometric studies.

References

Clark, G. R., II (1968). Mollusk shell: daily growth lines. *Science*, **161**, 800–802

Clark, G. R., II (1974). Growth lines in invertebrate skeletons. *Ann. Rev. Earth Planetary Sci.*, **2**, 77–99

Evans, J. W. (1968a). The effect of rock-hardness and other factors on the shape of the burrow of the rock boring clam, *Penitella penita*. *Palaeogeogr. Palaeoclimatol. Palaeoecol.*, **4**, 271–278

Evans, J. W. (1968b). Growth rate of the rock boring clam *Penitella penita* (Conrad 1837) in relation to hardness of rock and other factors. *Ecology*, **49** (4), 619–628

Evans, J. W. (1972). Tidal growth increments in the cockle *Clinocardium nuttalli*. *Science*, **176**, 416–417

Evans, J. W. and LeMessurier, M. H. (1972). Functional micromorphology and circadian growth of the rock boring clam *Penitella penita*. *Can. J. Zool.*, **50**, 1251–1258

Pannella, G. (1972). Palaeontological evidence on the Earth's rotational history since early Precambrian. *Astrophys. Space Sci.*, **16**, 212–237

Pannella, G. and MacClintock, C. (1968). Biological and environmental rhythms reflected in molluscan shell growth. *J. Palaeontal.*, **42** (5) Suppl., 64–80

Stanley, S. M. (1969). Bivalve mollusk burrowing aided by discordant shell ornamentation. *Science*, **166**, 634–635

Wells, J. W. (1963). Coral growth and geochronometry. *Nature*, **197**, 948–950

DISCUSSION

DOLMAN: I think it is necessary to use tidal predictions and tidal characteristics for the area under study, not for an area of the coast miles away. Tides are distorted by the bottom in shallow water, hence are affected by configuration of the coast.

EVANS: The tide predictions I used were corrected for Empire Oregon, 6–7 miles away from my collecting site.

CLARK: Tide predictions are true for a whole coast, and you're not going to find a tide occurring a day or two later because it is a slightly different place.

ROSENBERG: But the tides in coastal lagoons separated from the open sea can be quite different from those predicted for the coast, because the tidal inlets into the lagoons can modulate tidal exchange.

HIPKIN: I think it is important to note that if you had assumed each line was deposited at daily intervals rather than, as demonstrated, semi-daily, your statistical analysis would have lost all true tidal information. But it is important to note that the period of the tides can vary with locality.

PANNELLA: Do you agree that valve movement is important in the production of a line?

EVANS: In some species, for example note my demonstration of boring lines.

PANNELLA: In *Mercenaria* I found the more tightly the valves were shut, the more pronounced the line.

FARROW: *Tridacna* deposits lines even when the valves are open. They do have zooxanthellae and that is important.

THE CHRONOMETRIC RELIABILITY OF CONTEMPORARY CORALS

ROBERT W. BUDDEMEIER AND ROBERT A. KINZIE III

University of Hawaii, Honolulu, Hawaii 96822

Abstract

Skeletal accretion of the hermatypic coral *Porites lobata* has been studied by periodic alizarin staining in the field, followed by x-radiography of slices through the specimens. Short-term skeletal density variations occur with lunar periodicity, with low-density layers apparently associated with the full Moon. Linear growth rates show substantial seasonal variations, with most of the growth occurring as low-density skeletal deposition during the summer months. The lunar fine structure is not observed in the narrow, denser band of winter skeletal accretion, and is apparently recorded only when growth rates are relatively rapid. The variations in bulk skeletal density are related to variations in trabeculae thickness. The spacing of the trabeculae is relatively constant, and their deposition does not appear to have any recognizable periodicity. The nature of growth responses in this species to annual and monthly cycles in the environment appear to make the skeletal record more suitable for evaluating the quality of the growth environment than for retrieving information about solar and lunar periodicities.

1. Introduction

Since the initial report by Wells (1963) and the subsequent work by Wells (1966, 1968) and Scrutton (1965, 1970) on daily, monthly and annual growth patterns in coral epithecae, a significant amount of work has been directed towards the identification and explanation of comparable patterns in the exoskeletons of contemporary hermatypic corals.

Investigation of contemporary coral growth patterns is relevant to interpretation of the fossil records for several reasons. First, and most simply, an increased understanding of growth pattern deposition and its significance in any organism, and especially those whose 'relatives' provide part of the fossil record, will provide a broader base of knowledge from which to view the 'paleontological clocks' (Runcorn, 1966). Second, hermatypic corals come from relatively shallow depths in tropical and subtropical oceans, an environment probably more similar to the depositional environments of many of the fossil species than is the temperate zone coastal environment which provides much of the data on contemporary molluscan growth patterns (Pannella, 1972). Finally, the calcification of hermatypic corals undoubtedly reflects the combined response to environmental stimuli of both the coral animal and its symbiotic algae. Although it seems generally assumed that the species used in fossil growth pattern studies did not contain algal symbionts, this is not conclusively known; further, the use of stromatolites (Pannella, 1972) to extend the fossil record implies intercomparability between marine plant and animal calcification records, making the study of contemporary corals particularly relevant.

Barnes (1970, 1972) has published an exhaustive study of calcification and epithecal growth ridges in contemporary corals. His conclusion that the growth increments are basically a daily phenomenon but have a rather low chronometric reliability suggests caution in applying, but does not discredit, the interpretation of fossil growth records.

The discovery of alternating high- and low-density growth bands comprising annual variations in the bulk density of hermatypic coral skeletons (Knutson, Buddemeier and Smith, 1972) has led to numerous investigations and applications of x-radiography to coral growth studies (Buddemeier, Maragos and Knutson, 1974; Knutson and Buddemeier, 1973; MacIntyre and Smith, 1974; Moore and Krishnaswami, 1974; Glynn, 1974; Dodge, Aller and Thomson, 1974). Of

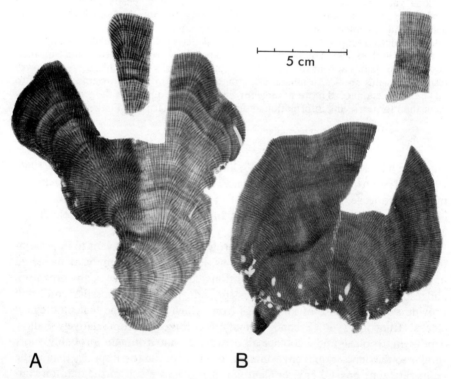

FIGURE 1. x-Ray positives (dark image represents high sample density) of *Porites lobata* specimens showing fine structure in the density banding. Large pieces are 3–4 mm thick; smaller sections are pieces of the same slab ground to 1–2 mm. A. Sample F-16/2, from Fanning Atoll. Note that fine structure is prominent, but seasonal density variations are also evident. B. Sample C-5, from Christmas Island. Fine structure is strong, seasonal structure less evident

particular relevance to the relationship between fossil and contemporary coral growth patterns is the observation (Buddemeier, 1974) of 'fine structure' within the annual density bands in certain species of the genus *Porites*. Figure 1 shows radiographs of samples in which the fine structure density band pairs are particularly evident (it should be noted that these particular samples do not necessarily have the most clearly defined annual density band variations).

The fine structure in the density patterns was observed to have a maximum periodicity (10–14 fine band pairs per 'annual' density band pair) appropriate to a lunar cycle. This paper is a preliminary report on the results of experiments undertaken to determine the cause, significance and physical nature of these apparently monthly variations in coral skeletal density.

2. Methods

The field study site chosen is off Kahe Point on the northwest (leeward) coast of the island of Oahu. Water depth is 3–4 m; tides are mixed semi-diurnal, with a maximum amplitude of about 0·7 m. The bottom is dominated by massive heads of *Porites lobata* interspersed with patches of loose sand. Although the site is approximately 150 m offshore, it is subject to some bottom surge and sediment resuspension during periods of high-to-moderate surf.

A number of colonies of *Porites lobata* were selected for long-term staining experiments; staining was accomplished by bagging the corals in transparent plastic bags containing alizarin red for a period of approximately 24 hr (Barnes, 1972). Beginning in June 1973, the corals have been stained at the time of each full Moon (except during August, when the dive was aborted because of shark warnings in the area); the specimens reported on in this paper are portions of 5 of the colonies collected in early November, approximately 3 weeks after the previous staining episode. Staining experiments are continuing with the rest of the colonies.

Slices 1–3 mm thick were cut along a plane of growth through the approximate centre of each section collected. These slices were bleached, dried, and x-rayed with a Faxitron 805 x-ray unit (typical exposure conditions: 30 KVp, 3 ma, source-to-film distance 1·3 m, Kodak AA film, exposure time 2–5 min, depending on sample thickness. See references cited above for more detailed discussion of technique). Examples are shown in Figure 2.

Both the samples and the x-ray negatives were examined visually and under a low-power microscope. Positions of the alizarin stains were noted and compared with the x-ray density patterns. Density patterns were examined visually and trabeculae separations measured in both high- and low-density regions. Growth rates were calculated both from the monthly stains and from the annual density patterns. The growth rates used were the linear growth rates along the axes of maximum growth. While this measurement is not quantitatively representative for colonies of limited symmetry, it is unambiguously identifiable and

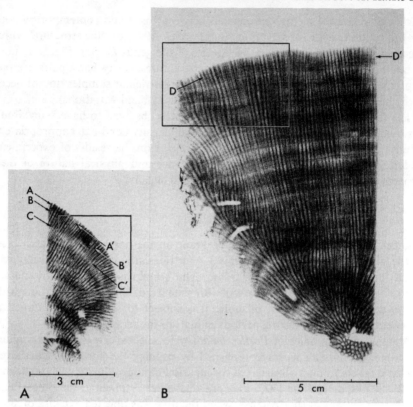

FIGURE 2. x-Ray positives of samples KaS-3b (A) and KaS-1 (B). Arrows A–A', B–B', C–C', and D–D' indicate positions and approximate orientations of alizarin stain lines in the coral skeletons. Boxed portions of coral images are shown on an expanded scale in Figure 3. Note density fine structure in radiographs; see also Table 1. The alizarin stain in KaS-1 is unusually broad and probably represents two or more sequential stains too close to be resolved

is relatively immune to morphological variations induced by micro-environmental conditions.

Also examined for comparison purposes were samples of the same species from Fanning and Christmas Islands (Figure 1), a specimen collected from the area of the study site in early June 1973, and a fossil specimen taken from the Pleistocene raised reef along the shoreline of Kahe Point.

3. Results

Some of the physical characteristics associated with the skeletal density variations may be seen in Figure 3. For all of the corals examined the trabecular spacing (centre-to-centre separation) was relatively constant at 40 ± 5 trabeculae/cm (along the direction of polyp growth). This appears to be a genetically determined geometry, and is seemingly independent of geographical origin, growth rate, age of colony and density.

Density variations are associated with systematic variations in the thickness, and possibly the alignment, of the trabeculae. In Figure 3 it can be seen that 'dense' bands consist of at least 2 rows of more robust trabeculae, with correspondingly constricted intertrabecular compartments. One also has the visual impression of more uniform lateral alignment of trabeculae in the low-density regions; whether there is really more synchronism in trabecular deposition in these regions or whether the effect is an optical one resulting from differences in the 'hole' dimensions is difficult to judge. This dimensional difference, however, is associated with both the seasonal and the fine structure density variations.

In spite of the obvious advantage of being able to put known time-markers into the coral skeletal structure, there are difficulties involved in the unequivocal interpretation of the alizarin stain results. The stain is deposited over the entire surface of active calcification, and may therefore be more than 1 mm in vertical extent, limiting the resolution obtainable. In addition, both the existence and the intensity of the stains are dependent on conditions being sufficiently benign so that the bags remain intact and the corals calcify appreciably. Although all of our samples had been stained 4 times prior to collection, none showed more than 3 stain lines (and some only 1 or 2) when sectioned. Although field notes on the condition of the dye bags when revisited permit us to make tentative assignments of times to each observed stain line, this is a source of uncertainty in our monthly growth rate estimates. In spite of these problems, we are able to make some conclusions about both the periodicity of the density fine structure and variations in the rate of linear growth.

Most of the stained coral specimens have less consistent and pronounced density fine structure than the corals from the Line Islands (Figure 1). The larger of the corals shown in Figure 2 has the most pronounced fine structure record of the Kahe specimens. This colony was located under a slight overhang at the bottom of a large head; its living polypary surface was almost flat and oriented at a uniform upward angle. We surmise that this accident of 'unidirectional' growth geometry enhanced synchronism in the calcification of the polyps and made the systematic structural variations more noticeable. Unfortunately, the growth rate of this colony was one of the lowest and it only possessed one distinct stain.

However, the density band structures of 7 different coral slices representing 5 different specimens and containing from 1–3 alizarin stain lines each were compared to the corresponding stain patterns, and we consider the cumulative result conclusive in spite of individual uncertainties. In each case where density fine structure pattern occurred in the stained part of the coral, the alizarin stain most closely matched the low-density portion of the fine structure. Further, in corals with more than one stain the density bands aligned with the stains were

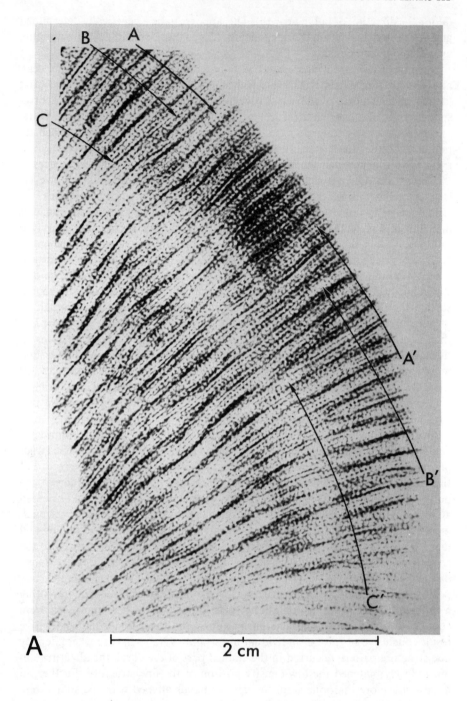

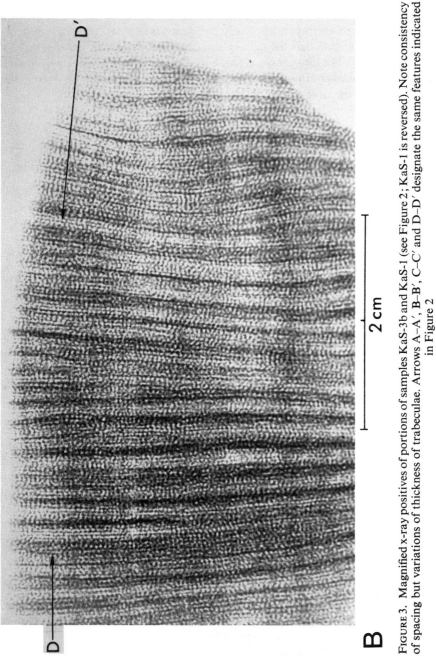

FIGURE 3. Magnified x-ray positives of portions of samples KaS-3b and KaS-1 (see Figure 2; KaS-1 is reversed). Note consistency of spacing but variations of thickness of trabeculae. Arrows A–A', B–B', C–C' and D–D' designate the same features indicated in Figure 2

either members of adjacent band pairs, or were separated by no more than one additional pair of fine structure bands. Since we believe that the experimental stains observed were separated in time of deposition by either 1 or 2 months, these results indicate that the density fine structure is lunar rather than tidal in origin, and that the low-density portion of the fine structure band pair is associated approximately (resolution in terms of time is probably no better than ±1 week) with the time of the full Moon.

TABLE 1. Growth data for *Porites lobata*

Sample	Average annual growth rate (maximum) mm/year	Alizarin stain positions, mm from surface	Density fine structure location/spacing
KaS-1	6·0	2·0 ± 1·0	Spacing ca. 0·5 mm
KaS-3a	5·0	2·0	0·2, 1·0, 2·2 mm from surface
KaS-3b	9·0	1·5, 3·5, 9·0	1·5, 3·5, 5·5, 7·0, 9·0 mm from surface
KaS-4	7·5	2·0, 8·0	1·0, 2·0 mm from surface
KaS-5	9·0	1·5, 5·5, 9·0	Spacing ca. 1·0–1·3 mm
KaS-6a	10·0	3·5	2·5, 3·5 mm from surface
KaS-6b	6·0	2·5	2·5 mm from surface
Ka-1	6·0	—	Spacing ca. 0·5 mm
KaF-1	8·0	—	Spacing ca. 0·5 mm
F-16/2	10·0	—	Spacing 0·5–1·0 mm
C-5	13·0	—	Spacing 0·8–1·5 mm

Table 1 summarizes the growth data from this study. Samples coded KaS are stained corals; letter subscripts denote different lobes of the same parent head. Ka-1 is the specimen collected just before the beginning of the staining experiments; KaF-1 is the fossil specimen taken from the poorly lithified face of the raised Pleistocene reef (age ca. 100,000 years). F-16/2 and C-5 are the comparison samples from Fanning and Christmas Islands respectively, whose x-radiographs are shown in Figure 1. The average annual growth rate along the axis of maximum growth was measured from the outermost several annual density band pairs on the x-radiograph. It should be noted that individual 'annual' growth increments measured in this fashion may differ from the average by a factor of as much as 2 (Buddemeier, Maragos and Knutson, 1974). Alizarin stain positions are measured from the surface of the coral to the centre of the stain line along the same growth axis used to determine the average annual growth rate. For corals which had discernible fine structure near the surface, the last column gives the distances from the surface to the centres of low-density fine structure bands along the same axis used for the other measurements. For specimens with no stains or where the fine structure growth pattern was not visible along the measured growth axis, the approximate spacing (centre-to-centre of adjacent low-density bands) of the observed fine structure is given.

Sample Ka-1, collected in early June, had 1–2 months of low-density growth (based on fine structure count) on top of a high-density band; the stained corals all showed full low-density bands and in some cases an incipient high-density increment just below their surfaces. We therefore conclude that low-density growth at the sample site occurs approximately from May through to October. Consideration of the stain and fine structure growth data in Table 1 shows that most of the corals grew at maximum linear rates of 1–2 (and in some cases possibly as high as 3) mm/month during the period of the staining experiments. The average annual growth rates, however, correspond to values of 0·5–0·8 mm/month. This indicates that most of the linear growth must occur during the summer months in the form of the low-density skeletal deposition, while the high-density growth during the winter occurs with very low rates of linear growth.

None of the Kahe corals studied showed more than 6–8 fine structure pairs per annual band, and fine structure is generally not visible within the narrow seasonal bands of highest density. This indicates that monthly variations either do not occur or are not visibly recorded when the rate of skeletal accretion falls below (or the skeletal density rises above) a certain critical value. This hypothesis is supported by the Fanning and Christmas Island corals (Figure 1) for which growth rates are higher than at Kahe, seasonal density variations (especially at Christmas I.) are less pronounced, and the fine structure patterns are correspondingly more visible and apparently more complete (e.g. F-16/2 shows an average of 10–11 fine pairs/year).

4. Discussion

The observations and results presented above have numerous implications for students of both contemporary and fossil growth patterns. One which has not been specifically alluded to is the problem of pseudo-periodicities. In the present case the combination of a constant trabecular spacing (ca. 40 cm) and a large number of samples with growth rates of 0·5 to 0·8 cm/yr presented a constant temptation to treat trabecular deposition as a fortnightly phenomenon. Self-deception is facilitated by the slight variations in trabeculae spacing and by the lack of clear-cut, objective demarcation lines in the approximately annual density variations. The staining experiments, however, conclusively show that although the trabecular structures are intimately involved in the nature of the density variations, the trabeculae themselves are not formed in synchrony with any external schedule.

It appears that a minimum of 2 adjacent rows of thick and thin trabeculae are needed to define 1 fine structure pair. This means that an actual (as opposed to an average) growth rate of 0·5 mm/month would be the slowest growth in which we could expect to observe monthly fine structure by macroscopic radiographic techniques, and in fact fine structure spacings of about 0·5 mm are the smallest we have observed.

One of the most interesting aspects of the *Porites lobata* fine-density variations is that they are apparently lunar in periodicity so long as growth conditions are near-optimal. In the case of the Kahe corals it was readily apparent that we were dealing with an incomplete monthly record; this was less obvious—because the record was more nearly complete—with the corals from more tropical locations. One could envision a confusion between a presumed decrease in an environmental periodicity (lunar in this case) and a gradual deterioration of the environmental conditions necessary for complete recording of that periodicity by the organism.

The mechanisms by which coral skeletal density responds to seasonal and monthly cycles remain undetermined. Seasonal variations are probably closer to being qualitatively described. It is clear that when water temperature drops below about 25°C for significant periods, as it typically does from November to April in Hawaiian waters (Maragos, 1972), coral growth is inhibited. However, corals retain seasonal density variations even in locations where temperature does not appear to be controlling (Buddemeier, Maragos and Knutson, 1974), and here the change in available light from the dry to the rainy season appears to trigger the change in calcification. This would suggest that the zooxanthellae play a sensitive and important role in controlling the rate and features of hermatypic coral calcification.

The monthly density variations may possibly arise in connection with reproductive cycles (Harrigan, 1972, has shown that another coral species releases its larvae with distinct lunar periodicity), availability of organic food (phytoplankton concentrations in the surface water vary systematically with the phase of the Moon), or the availability of low-intensity nighttime light and/or a lengthened photo-period during the time of the full Moon. Corals growing in relatively shallow water are clearly sufficiently photosensitive to detect moonlight, and even if it is not a significant source of photosynthetic energy it could still serve as an entraining signal for some particular biorhythm. Research into the mechanisms of response to environmental variations represents one of the most promising areas of coral research now being developed, and will have important implications for understanding the significance and reliability of the fossil record as well as the biology and ecology of living corals.

5. Conclusions

(1) Small cyclic variations in the skeletal density of *Porites lobata* have a lunar periodicity, with the low-density region of the skeleton probably deposited around the time of the full Moon.

(2) The density variations are structurally related to variations in the thickness, and possibly alignment of the trabeculae. A trabeculae count of about 40/cm appears to be a species characteristic, and the deposition of the trabeculae is not a periodic phenomenon.

(3) Skeletal deposition rates within a given colony are quite variable over the course of a year, and deposition of the lunar density records is apparently only possible when growth rate exceeds about 0·5 mm/month.

(4) Monthly growth records in *Porites* are therefore usually incomplete, and the apparent number of months per year as determined from the coral skeleton is a better indicator of the overall suitability of the coral's growth environment than of the periodicity of the astronomical stimuli.

Acknowledgments

The authors thank Carol-Ann Uetake for help in the field and in the laboratory, and R. H. Snider and S. V. Smith for their comments and suggestions. Part of this research was supported by the National Science Foundation under Grant GA 35836. This paper is Hawaii Institute of Geophysics Contribution No. 593.

References

Barnes, D. J. (1970). Coral skeletons: an explanation of their growth and structure. *Science*, 170, 1305–1308

Barnes, D. J. (1972). The structure and formation of growth ridges in Scleractinian coral skeletons. *Proc. Roy. Soc. Lond. B.*, 182, 331–350

Buddemeier, R. W. (1974). Environmental controls over annual and lunar monthly cycles in hermatypic coral calcification. In *Proceedings, Second International Coral Reef Symposium*, (in press)

Buddemeier, R. W., Maragos, J. E. and Knutson, D. W. (1974). Radiographic studies of reef coral exoskeletons: rates and patterns of coral growth. *J. Exp. Marine Biology and Ecology*, 14, 179–200

Dodge, R. E., Aller, R. C. and Thomson, J. (1974). Coral growth related to resuspension of bottom sediments. *Nature*, (in press)

Glynn, P. W. (1974). Rolling stones among the Scleractinia: mobile coralith communities in the Gulf of Panama. In *Proceedings, Second International Coral Reef Symposium*, (in press)

Harrigan, J. F. (1972). The Planula Larva of *Pocillopora damicornis*: Lunar Periodicity of Swarming and Substratum Selection Behavior, Ph.D. Dissertation, University of Hawaii, Honolulu. 303 pp.

Knutson, D. W. and Buddemeier, R. W. (1973). Distribution of radionuclides in reef corals: opportunity for data retrieval and study of effects, pp. 735–746 in *Radioactive Contamination of the Marine Environment*, International Atomic Energy Agency, Vienna. 786 pp.

Knutson, D. W., Buddemeier, R. W. and Smith, S. V. (1972). Coral chronometers: seasonal growth bands in reef corals. *Science*, 177, 270–272

MacIntyre, I. G. and Smith, S. V. (1974). X-Radiographic studies of skeletal development in coral colonies. In *Proceedings, Second International Coral Reef Symposium*, (in press)

Maragos, J. E. (1972). A Study of the Ecology of Hawaiian Reef Corals, Ph.D. Dissertation, University of Hawaii, Honolulu. 290 pp.

Moore, W. S. and Krishnaswami, S. (1974). Correlation of X-radiography revealed banding in corals with radiometric growth rates. In *Proceedings, Second International Coral Reef Symposium*, (in press)

Pannella, G. (1972). Paleontological evidence on the Earth's rotational history since early Precambrian. *Astrophysics and Space Science*, **16**, 212–237

Runcorn, S. K. (1966). Coral as paleontological clocks. *Scientific American*, **215**, 26–33

Scrutton, C. T. (1965). Periodicity in Devonian coral growth. *Palaeontology*, **7**, 552–558

Scrutton, C. T. (1970). Monthly periodicity in the growth of some corals, pp. 11–16 in *Palaeogeophysics* (Ed. S. K. Runcorn), Academic Press, London. 518 pp.

Wells, J. W. (1963). Coral growth and geochronometry. *Nature*, **197**, 948–950

Wells, J. W. (1966). Paleontological evidence of the rate of the Earth's rotation, pp. 70–81 in *The Earth–Moon System* (Ed. B. G. Marsden and A. G. W. Cameron), Plenum Press, New York

Wells, J. W. (1970). Annual and daily growth rings in corals, pp. 3–9 in *Palaeogeophysics* (Ed. S. K. Runcorn), Academic Press, London. 518 pp.

DISCUSSION

SCRUTTON: It would be nice if you could do a similar study on corals with epithecae. I have suggested a link between calcification and the lunar cycle for corals with epithecae. What do you think the link might be between calcification and the lunar cycles?

BUDDEMEIER: Some corals spawn on a lunar cycle although I am not certain about *Porites*. Some corals may be more autotrophic during the full Moon when there is more light for photosynthesis. However, *Porites* does not seem to care about the heterotrophy–autotrophy business, because it eats when food is available. I can't rule out growth entrainment of genetic rhythms which have no overt meaning in terms of sex, food, etc. While Hawaii tides are mixed semi-diurnal, I believe the reason for periodic growth variations in corals is the change in fullness of the Moon, not that the tides are different during the month.

RUNCORN: Could you specify a biological factor on which fullness of the Moon rather than tides affects growth?

BUDDEMEIER: The coral's potential food supply migrates up and down with fullness of the Moon, and this might have an effect on coral growth. Or, in terms of breeding, there might be a synchronization process. Corals might breed according to either a lunar illumination or tidal cycle, and then might not be able to generate another set of planulae in less than a month's time.

MADAME TERMIER: Have you compared your corals with *Porites* from other areas for latitudinal differences?

BUDDEMEIER: I studied specimens from Hawaii, Fiji and the Marshall Islands and over that latitudinal range, all had the same fine structure.

BIOLOGICAL CLOCKS AND SHELL GROWTH IN BIVALVES

IDA THOMPSON

*Department of Geological and Geophysical Sciences,
Princeton University, Princeton, New Jersey*

Abstract

Palaeontologists have estimated month and year length from incremental shell growth patterns in fossil bivalves. Such estimation has involved two assumptions, implicit or explicit, which were tested in the present study: (i) each growth increment represents a solar day; (ii) increment deposition is under the control of a biological clock. A process is by definition under biological clock control if it continues rhythmically in constant conditions.

Valve movement in the quahog *Mercenaria mercenaria* was monitored for periods ranging from 2–16 weeks. Experimental regimes varied from simulated intertidal conditions (semi-diurnal tides and day–night cycles) to constant conditions (no temperature, light, or water-level cycles). Incremental growth, inferred from alternation of light and dark bands, occurred in all regimes. Valve opening was found to correlate with, and thus measure, shell growth. From the correlation of valve opening with the experimental intertidal cycle it was concluded that growth increments in *M. mercenaria* probably record solar days and lunar days, at least under intertidal conditions. Valve opening showed circadian rhythmicity in all regimes, even constant conditions. It was therefore concluded that in *M. mercenaria* both valve movement and increment deposition are processes under the control of a biological clock. An incidental finding with implications for biological-clock experimentation was that valve opening in 'constant conditions' appeared to be responsive to both actual solar-day and lunar-day cycles.

> 'After I had been there about ten or twelve days, it came into my thoughts that I should lose my reckoning of time . . . but to prevent this . . . Upon the sides of this square post I cut every day a notch with my knife, and every seventh notch was as long again as the rest, and every first day of the month as long again as that long one; and thus I kept my calendar, or weekly, monthly, and yearly reckoning of time.'
> (DANIEL DEFOE, *The Life and Strange Surprising Adventures of Robinson Crusoe*)

1. Introduction

In the natural world outside the laboratory, biological clocks reflect geophysical cycles. Apparently all living things possess biological clocks which respond to such cycles as light and tide in their environments (Palmer, 1970). This chapter describes research on the bioclock which controls the rhythms of shell growth and valve movement in bivalves. Although the bivalve shell ultimately reflects geophysical cycles, that reflection is neither passive nor immediate. Between environment and shell there intervenes an organism. To adapt Arthur Koestler's trenchant image, there is a ghost in the machine, a rhythmic tissue which builds and dwells within a pair of curved levers. Like Robinson Crusoe, the hermit ghost records its own rhythmic history upon that shell.

At the outset three terms require definition. 'Cycle' will refer to a regularly repeating geophysical event such as the tide; 'rhythm' to a regularly repeating physiological event within a living organism, such as the opening and closing of a bivalve. The 'period' of a repeating event is the time from onset to onset. A bioclock is consensually defined as a timing device within an organism which maintains circadian rhythmicity of the organism's life processes in the presence or absence of such obvious geophysical cycles as light and tides. The test for a bioclock is to isolate the organism from obvious cycles and measure some life process; in an animal usually motor activity. If rhythmicity appears the organism is said to have a bioclock. In these so-called 'constant conditions', the rhythm may not reflect the period of any geophysical cycle. The period of the underlying clock is not known (Palmer, 1973). But in natural conditions rhythms phase with geophysical cycles. For example, an organism is usually active for only that part of the solar day during which it meets its needs such as feeding, and quiescent for that part of the day in which environmental stress, such as predation, is high. The phase relationship of the rhythms timed by the bioclock with geophysical cycles varies with environment and species.

Below, it will be shown that bivalves have bioclocks. These bioclocks use geophysical information to establish a rhythm of shell deposition. Bivalve shell consequently consists of a series of discrete increments, each increment representing approximately the same time interval. However, determining precisely which geophysical cycles the bivalves use to time increment deposition is not an open-and-shut matter. Increments have been interpreted to represent solar days (Barker, 1964; Pannella and MacClintock, 1968; Pannella *et al.*, 1968; House and Farrow, 1968; Berry and Barker, 1968) and lunar days (Evans, 1972). It is conceivable that shell growth responds either to tidal cycles or to a complex of solar and lunar-related events. Just which cycles shell growth records becomes critical when one attempts to determine from fossil shells the lengths of geophysical cycles in the past. So gradual have been the changes in cycle lengths that small errors in time estimation arising from misinterpretation of shell growth patterns may render conclusions from such studies meaningless. First we must know how shell growth records geophysical cycles. This in turn requires understanding of how bivalve bioclocks respond to these cycles.

The present chapter describes a series of eight experiments to determine how bivalve rhythms reflect geophysical cycles and, conversely, how independent these rhythms are from light, temperature and tidal cycles. Conclusions are based entirely on one species, *Mercenaria mercenaria*, but are probably transferable to other species with similar life habits. It will be concluded that (i) bivalves possess bioclocks, which has not been previously established; (ii) these clocks control the rhythms of valve movement; (iii) valve movement maps shell growth isomorphically, so that (iv) shell growth is also bioclock-controlled; (v) in simulated intertidal conditions shell growth occurs mainly at night and (vi) shows rhythms related to both the solar day and the lunar day; and (vii) subtidal and abyssal bivalves conceivably may also use both solar day and lunar day cycles to time shell growth.

2. Experimental design and rationale

To discover how a bivalve modifies geophysical cycles in the process of shell growth, one must monitor shell growth continuously. Some bivalve species deposit external growth increments; some deposit internal growth increments. External increment deposition can be directly observed. But, for reasons explained below, the experimental animal chosen was *Mercenaria mercenaria*, a species that deposits internal growth increments. The choice posed a problem: how to monitor shell growth?

One method of monitoring shell growth is to measure a rhythmic process in the organism that correlates with the shell growth rhythm. The process we measure must not interfere with either rhythms or shell growth, nor may it impart any cycle of its own. A suitable process is valve movement since it is readily measured automatically, and the monitoring process appears not to affect the activity pattern (Barnes, 1962). In addition, bivalve functional anatomy and chemistry both indicate that valve movement and shell growth are related processes. The outer lobe of the mantle deposits shell. When the bivalve closes, it draws this lobe into the mantle cavity, away from the growing edge of the shell. Withdrawing the lobe must inhibit shell growth, although some extrapallial fluid may still reach the growing edge. The position of the valves also affects tissue chemistry: pH falls during closed-valve intervals and rises during open-valve intervals (Dugal, 1939).

Experiments were designed to measure valve movement under conditions with simulated geophysical cycles and under constant conditions. The cyclic regimes simulated intertidal and non-tidal habitats. The constant-conditions regime was the specific test for bioclock control. Here the variables of light, temperature and water level were held constant. It was reasoned that if rhythmic valve movement and shell growth continued under constant conditions, then these processes would be bioclock-controlled. The principles of bioclocks learned from other organisms could then be applied to the study of shell growth.

3. Materials and methods

All experiments used the quahog *Mercenaria mercenaria*. This species inhabits the intertidal and subtidal zones of the eastern coast of North America. It was chosen for its clear growth increments (Pannella and MacClintock, 1968), ease of acquisition and hardiness. Specimens from 40–55 mm in height were obtained from Marine Biological Laboratories, Woods Hole, Massachusetts and from a local Chicago fish market. Woods Hole specimens were collected in a single locality in 2·5 m MLW and shipped by air to Chicago. Specimens from both sources yielded similar results.

Before a specimen was placed in the experimental tanks, its growing edge was rubbed to break the periostracal seal and inhibit growth for a few days. This left a notch in the shell to mark the start of the experiment (Pannella and MacClintock, 1968). Five to ten animals were kept without food in each tank. Each animal was secured to a drilled horizontal Plexiglass rack by a cork which was glued to the left valve and fitted into a hole in the rack. Thus the plane of commissure was horizontal. A light, narrow Plexiglass strip glued to the right valve extended dorsally. A nylon filament connected the distal end of the strip to a microswitch (Acro snap-action) above the tank. The strip acted as a lever to magnify bivalve gape 2·5 times. A gape of 0·6 mm or more depressed the lever and activated an automatic event recorder (Esterline Angus). Data from the event recorder were transferred to activity charts for each bivalve.

There were three experimental regimes: simulated intertidal, non-tidal and constant conditions. Two simulated intertidal experiments involved a total of 18 bivalves and lasted 36 and 45 days respectively. A 160-litre tank was used with an imposed semi-diurnal tidal cycle (De Santo, 1967). Each tidal cycle lasted 12·4 hr with the bivalves exposed for approximately half of this period. The light cycle was 12 hr light (269 lux) followed by 12 hr darkness. Temperature varied with the light cycle from 20–26° C.

The three non-tidal experiments were conducted in a 150-gallon Instant Ocean aquarium. They involved a total of 25 bivalves and lasted 81, 81 and 130 days respectively. Water level was constant. The light cycle was 12 hr light (258 lux) and 12 hr darkness. Temperature was maintained at $19 \pm 2°C$. The constant-conditions experiments were run in a similar Instant Ocean aquarium with constant water level, temperature ($19 \pm 2°C$) and light (86 lux). Three experiments involving 25 animals were run for 81, 81 and 130 days respectively. The first 81-day experiment was run with a light–dark cycle for the initial 32 days before switching to constant conditions. With the animals shielded from obvious environmental cycles, the conventional test for bioclock control of biological rhythms could be made (Aschoff, 1960). If the organisms maintain rhythmicity in these conditions, then by definition they possess a bioclock.

At the end of each experiment, the specimens were killed and the valve movement data analysed. The periods of valve movement rhythms were estimated

Biological Clocks and Shell Growth in Bivalves

from times of opening and closing relative to the solar day. Special attention was paid to phase relationships between valve movement rhythms and the cycles of solar day, lunar day and tides, whether these cycles had been intentionally imposed on the organisms or whether the organisms were supposedly shielded from them.

Acetate peels were made from the shells (Rhoads and Pannella, 1970), and then shell deposited distally from the notch was inspected for the number of increments. Counts of increments were made directly from the peels at magnifications of 100× and from photographs of the peels with magnification of about 500×. Each set of one light band and one dark band was counted as an increment. Complex increments as defined by Pannella and MacClintock (1968) were counted as two increments.

4. Results

Valve movement was clearly rhythmic in all experimental regimes. However, only in the simulated-intertidal regime was the pattern of rhythms similar among the individual bivalves. Under the non-tidal regime, variation in pattern of rhythms was greater, even among individuals known to have identical histories. Variation was even greater in the constant-conditions regime, although rhythmicity clearly continued, demonstrating the presence of a bioclock. Study of the shells revealed that the number of open intervals and the number of growth increments closely corresponded. Thus shell growth maps valve movement. By inference, shell growth is also bioclock-controlled.

The charts in Figure 1 were selected as representative of rhythmic patterns in the three regimes. Figure 1A is the record of a bivalve from a simulated-intertidal experiment. The pattern of opening and the phasing of opening with the light cycle are similar to the activity of 13 of the 18 bivalves in this regime. Valve-open intervals occur almost exclusively in high water and darkness. Time of opening is dependent upon both the light cycle and the tidal cycle. The result is a pattern of alternation in number of openings per day. Several days of one opening per 24-hr period are followed by several days of two openings. The pattern repeats every semi-month. Six of the bivalves in this regime were either light-active or showed no clear preference for light or dark activity.

When the tidal cycle was removed, simulating some aspects of subtidal conditions, valve movement was still rhythmic but varied greatly among individuals. Therefore it is not possible to generalize about the periods of these rhythms or their phase with the light cycle. However, all of the 45 bivalves had a rhythm with a 24-hr period. In 29 bivalves, the active, valve-open interval was at first in phase with dark hours, similar to the pattern in the intertidal experiments. But 26 of the 29 bivalves then rephased open intervals from dark to light hours. Opening appeared to be in response to lights-on at 6.00, as in Figure 1B. The rephasing occurred after a few days or weeks and may have been an artifact of

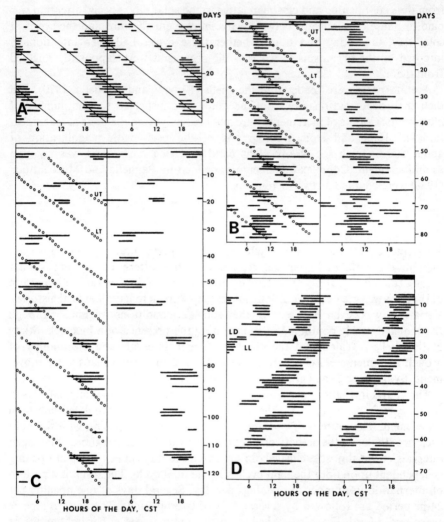

FIGURE 1. Activity records for four bivalves. Each record is shown twice, but moved up one day on the right to facilitate recognition of rhythms with non-24-hr periods. Bars at the top of each record show hours of light (white bars) and dark (black bars). Intervals of open valves are shown by heavy horizontal lines. A. Simulated-intertidal regime. Diagonal lines connect times of high tide; bivalve exposed about half of each 12·4-hr tidal cycle. B. Non-tidal regimes. Open circles indicate times of upper (UT) and lower (LT) lunar transits during the experiment. C. Constant-conditions regimes. Lunar transits indicated as in B. D. Non-tidal (LD) changing to constant conditions (LL) at small triangle. Bivalves producing records A, B and D from Long Island Sound via a Chicago fish market. Record C by a bivalve from 2·5 m MLW, Falmouth Harbour, Mass.

attachment to the Plexiglass rack: many animals were observed to extend their feet as in burrowing, presumably to escape the light. The remaining 16 bivalves in this regime were active during both light and dark hours.

Twenty-nine of the 45 bivalves in the non-tidal regime had a rhythm in addition to the 24-hr rhythm. For 22 of these bivalves, the period of the second rhythm was shorter than 24 hr, varying between 22 and 23·2 hr. For 7 other bivalves, the period of this second rhythm was longer than 24 hr, approximately equal to the lunar-day cycle, giving these bivalves a rhythmic pattern similar to the intertidal pattern. This lunar-day related rhythm was even found in bivalves known to come from a subtidal habitat (2·5 m MLW). Valve movement rhythms in subtidal *M. mercenaria*, then, may be a response to lunar-day as well as solar-day cycles.

Figure 1B is the activity chart of a bivalve that combined a 24-hr rhythm with a rhythm of approximately lunar-day period. Open intervals associated with the lunar-related rhythm tended to begin 2–4 hr before actual lunar transits at Chicago (indicated by the open circles). In Long Island Sound, origin of this specimen, maximum high tide regularly precedes lunar transit by this same interval. As the experiment progressed the lunar rhythm came to centre on lunar transits.

The same pattern of rephasing with the lunar day appeared in 5 of the 7 bivalves with lunar rhythms in the non-tidal regimes. Brown (1954) found a similar rephasing in the valve movement rhythms of oysters after they were transported from Woods Hole, Massachusetts to Evanston, Illinois. Thus bivalves may respond to actual lunar events even in the absence of tides or moonlight.

The constant-conditions regime tested specifically for bioclock control. All 35 bivalves in this regime were obviously rhythmic, indicating bioclock control of valve movement. The rhythms were in one respect similar, and in another dissimilar, to the non-tidal patterns. Like the non-tidal patterns, the constant-conditions patterns generally contained two rhythms in all but two of the 35 records. The two rhythms were a 24-hr one and another with a period approximately 1 hr shorter or longer than 24 hr.

Unlike the non-tidal patterns, the constant-conditions patterns were produced in the absence of a light cycle. Without the light cycle the open-interval phase of the 24-hr rhythm tended to shift more frequently relative to the actual solar day. For example, in Figure 1C open intervals initially centred on midnight, then shifted to noon, then to morning, before stabilizing in the evening hours. This 24-hr rhythm, open intervals phasing with various solar-day times, was apparently joined by a 24·8-hr rhythm. The interaction of the two rhythms during the last two months produced bursts of activity every 15 days. These bursts occurred when actual lunar transits coincided with evening hours from about 16.00 to 20.00.

One experiment designed to test response in constant conditions began with the animals exposed to a light cycle for the first three weeks. Figure 1D is an example of activity in this experiment. With the light cycle present, activity originally phased with night, but a lights-on response developed on day 10. In constant conditions (LL), a 24-hr rhythm remained, inhibiting activity around midnight during the last 45 days of the experiment. This was combined with a 23·2-hr rhythm, scanning to the right in the chart but disappearing as it approached midnight. This rhythm was common in all but the intertidal experiments. The 23·2-hr rhythm is of special interest because it is the mirror-image of the 24·8-hr lunar-day rhythm and will produce a semi-monthly (if bimodal) or monthly (if unimodal) rhythm in combination with a 24-hr rhythm. Possibly,

TABLE 1. Analysis of shell growth and valve openings during experimental intervals

Experiment	Bivalve	No. of increments	Total no. of openings	No. of days in experiment	No. of days bivalve opened	Accretion in height of shell during experiment
Constant conditions I	#2[a]	85	86	81	64	1·9 mm
	#3[a]	74	69	81	62	0·9 mm
	#4[a]	80	88	81	64	1·2 mm
	#12[b]	43	42	71	38	0·8 mm
	#13[b]	36	42	71	38	0·6 mm
	#14[b]	50	51	71	46	1·4 mm
Non-tidal I	#6	78	78	81	59	1·3 mm
	#8	70	71	81	67	0·6 mm
	#10	70	81	81	65	0·7 mm
	#8	83	72	81	65	1·9 mm

[a] Bivalve in light cycle for first 32 days, in constant conditions for last 49 days.
[b] Bivalve in light cycle for first 22 days (added 10 days after start of experiment), in constant conditions for last 49 days.

adaptation by a bivalve to a cycle 0·8 hr longer than the solar day somehow facilitates the expression of a rhythm with a period 0·8 hr shorter than the solar day when tides are removed.

Shell growth was found to continue incrementally in all regimes. However, in most specimens the increments were narrow, difficult to differentiate, and recessed as in winter shell deposition (Pannella and MacClintock, 1968). Only ten shells from two experiments (one non-tidal and one constant-conditions experiment) had countable increments. The number of increments deposited during the experimental periods was compared to number of days in the experiments, number of days on which the bivalve opened, and total number of openings (Table 1). Number of increments closely corresponded only to total openings (Figure 2). Thus valve movement maps increment deposition. The implication is that shell growth has the same rhythms and phase relations with geophysical cycles as valve movement. Since valve movement is bioclock-controlled, shell growth is also bioclock-controlled.

Two questions remain. First, what is the influence of length of open-intervals and closed-intervals on the quality of associated increments? In these experiments it was not possible to establish a one-to-one correspondence between valve-

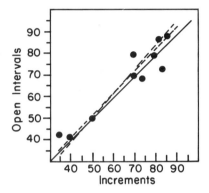

FIGURE 2. Relation between increments deposited and the number of valve openings during experimental interval. Each dot represents one bivalve. The solid diagonal line indicates a hypothetical one-to-one correspondence between increments and openings. Regression lines are dashed. Testing the null hypothesis that there is no difference between the solid line ($a = 0$, $b = 1$) and the regression lines gives F values (2° and 5° of freedom) of 0·1708. The null hypothesis is accepted ($p \gg 0 \cdot 1$)

movement interval and increment because of crowding of increments at the beginning and end of experimental runs. Second, does a shell record days when no opening occurs? Perhaps very narrow increments are deposited even with the valves closed, as Pannella and MacClintock have observed (1968, p. 75). Answers to these questions may come from future experiment in which the bivalves will be fed, which should improve the quality of shell growth.

5. Conclusion

Valve-movement rhythms in *M. mercenaria* can be used to hypothesize what increment patterns bivalves from various marine habitats will display. These hypothetical patterns can be compared with actual patterns in recent shells from known habitats and perhaps elucidate how geophysical cycles are recorded.

This in turn may aid the interpretation of fossil shells. Three habitats related to the experimental regimes will be considered: the intertidal, subtidal and abyssal.

The intertidal

If combined solar-day and tidal cycles entrain bivalve rhythms in natural habitats as powerfully as in the experimental regime, then it is clear that shells grown intertidally will record both solar and tidal cycles. Growth, instead of being continuous except for low-tide interruptions, will occur mainly at night. Valves will be closed during daylight hours, pH will fall in the tissues, and conchiolin will concentrate on the growing edge, forming the dark band of the increment. This concentration of conchiolin may result from dissolution of calcium carbonate by falling pH at the growing edge. Alternatively, conchiolin concentration may result from the juxtaposition of the inner side of the accreting periostracum, where conchiolin has not yet entirely polymerized, with the growing shell edge as the mantle is withdrawn into the shell.

Knowledge of the ecology of intertidal bivalves helps to explain the night-active pattern. Light-active birds are major predators of intertidal bivalves (Thorson, 1971). Concealment of siphons as well as shell below the substrate during light hours should increase chances of survival when the water is shallow. Therefore night activity would appear to be adaptive in shallow water when a major element in predation is light-active animals which locate prey visually.

The addition of a tidal cycle to a solar-day rhythm of shell growth suggests a model for the formation of semi-monthly tidal clusters. Tidal clusters appear in several bivalve species (Barker, 1964; Pannella and MacClintock, 1968; Pannella et al., 1968). Just as open intervals in the experimental regime were interrupted periodically by low tides (Figure 1A), so growth of nightly increments should be interrupted during certain times of the semi-month. The resulting increment may be narrower than usual, or the interruption may leave a dark band in the increment similar to the light-deposited dark band. But because the low-tide hiatus will be shorter than the light-time hiatus, the band should be either lighter or narrower, or both. This dark, low-tide band should scan across the nightly increment as the tide advances at the rate of 50 min per solar day. Thus each semi-month of shell growth should consist of a group of narrow or complex increments formed when low tides interrupt nightly growth and a group of simple increments formed when low tides occur outside the nightly growth period.

The above model can be used to explain the pattern of growth increments seen in the Plexiglass peel of a *Tridacna squamosa* shell in Plate 7 of Pannella and MacClintock (1968). In this case, however, active time of growth may have been during light hours because the presence of zooxanthellae in this species may make light-time calcification more efficient. If tides are regular semi-diurnal, day and night will share almost identical tidal patterns. There must, of course, be caution when generalizing about increments from one species to another.

The present finding of the way bivalves map solar and lunar cycles allows prediction of just when during the synodic month complex increments will be deposited. If tides are equilibrium (that is, if high tides occur when the Moon is in upper and lower transit), then low tide will occur at midnight during first and third lunar quarters. Deposition of complex increments will occur on days around the lunar quarters. Equilibrium low tides occur in the morning and evening at full and new Moons, so the deposition of simple increments will centre on syzygy. The same relation of kind of increment deposited and lunar phase will occur if shell growth is restricted to daylight hours.

The subtidal

In the non-tidal experiments, which simulate some aspects of subtidal habitats, a general pattern of response is more difficult to find, for two reaons. First, the experimental regime itself apparently caused a shift to light activity for reasons not presently understood. Second, there was more variability in the periods of the non-24-hr rhythms and in the phasing of the rhythms with solar day hours. This variability, both interindividual and intraindividual, precludes any firm conclusions about natural period and phase of the valve-movement rhythm and thus of the shell-growth rhythm in subtidal habitats. However, the majority of bivalves did exhibit two independent but interacting rhythms with the number of open-intervals per 24-hr period alternating from one to two, as in Figure 1B. This pattern is similar to the intertidal pattern and suggests that rhythms may involve the interaction of solar day and tidal cycles in subtidal as well as intertidal habitats. The failure of the bivalves to establish some permanent and consistent rhythmic pattern indicates that light cycles alone are poor entrainers of bivalve rhythms. The tidal cycle, even in subtidal habitats, is probably an important input in establishing organismic rhythms in harmony with the environment. Extensions of this research will include experiments with water levels that fluctuate without exposing the animals, more closely simulating subtidal conditions. Valve-movement measurements will be made with the animals covered by sediment and the planes of commissure vertical, in an attempt to eliminate the lights-on response.

The abyss

In the constant-conditions experiments, the absence of light cycles, tidal cycles, or temperature cycles invites comparison with the abyss. The abyss receives cyclic input, as do the experiments in so-called 'constant conditions'. Cyclic variations in gravitational and electromagnetic fields are present and could entrain abyssal biological rhythms just as these same variations may have entrained the rhythms in my constant-conditions experiments and in other experiments (Brown, 1960, 1969, 1970). Bivalve shells grown at abyssal depths do have growth increments but apparently no tidal clustering (Rhoads and Pannella, 1970). Do these increments measure geophysical cycles and, if so, which ones? Knowledge of biological rhythms in organisms living where the 'obvious' geophysical cycles do not penetrate will help to elucidate the true nature of bioclocks.

Bioclocks, then, do not appear to be completely independent variables even in constant-conditions regimes. These experiments and others (see Brown, 1970, for references) show the apparent influence of actual lunar and solar events on the clock-controlled valve-movement rhythm. What happens to the rhythmic organization in 'true' constant conditions is still not known.

However, this much is clear: in natural conditions bioclocks and shell growth are dependent variables, dependent for their timing on the phased interactions of several variables in the geophysical environment. These variables are in turn timed by the great clock we call the solar system. Empirical investigation of the bioclocks of bivalves, then, may lead us a little closer to an understanding of the past, the present and the future of that great clock.

Acknowledgments

My sincere thanks to Professors Ralph G. Johnson and Franklin H. Barnwell for help with many aspects of this study, and to Professor Alfred G. Fischer for critical reading of the manuscript.

References

Aschoff, J. (1960). Exogenous and endogenous components in circadian rhythms. *Cold Spring Harbor Symp. Quant. Biol.*, **25**, 11–28

Barker, R. M. (1964). Microtextural variation in pelecypod shells. *Malacologia*, **2**, 69–86

Barnes, G. E. (1962). The behavior of unrestrained *Anodonta*. *Anim. Behav.*, **10**, 174–176

Berry, W. B. and Barker, R. M. (1968). Fossil bivalve shells indicate longer month and year in Cretaceous than present. *Nature*, **217**, 938–939

Brown, F. A., Jr. (1954). Persistent activity rhythms in the oyster. *Am. J. Physiol.*, **178**, 510–514

Brown, F. A., Jr. (1960). Response to pervasive geophysical factors in the biological clock problem. *Cold Spring Harbor Symp. Quant. Biol.*, **25**, 50–72

Brown, F. A., Jr. (1969). A hypothesis for extrinsic timing of circadian rhythms. *Can. J. Botany*, **47**, 287–298

Brown, F. A., Jr. (1970). Hypothesis of environmental timing of the clock, pp. 15–59, in *The Biological Clock, Two Views* (Ed. J. D. Palmer), Academic Press, New York, 94 pp.

De Santo, R. S. (1967). An aquarium tide time—theory and operation. *Ecology*, **48**, 668–670

Dugal, L. P. (1939). The use of calcareous shell to buffer the product of anaerobic glycolysis in *Venus mercenaria*. *J. Cell. Comp. Physiol.*, **13**, 235–251

Evans, J. W. (1972). Tidal growth increments in the cockle *Clinocardium nuttalli*. *Science*, **176**, 416–417

House, M. R. and Farrow, G. E. (1968). Daily growth banding in the shell of the cockle, *Cardium edule*. *Nature*, **219**, 1384–1386

Palmer, J. D. (1970). Introduction to biological rhythms and clocks, pp. 1–14, in *The Biological Clock, Two Views* (Ed. J. D. Palmer), Academic Press, New York. 94 pp.

Palmer, J. D. (1973). Tidal rhythms: the clock control of the rhythmic physiology of marine organisms. *Biol. Rev.*, **48**, 377–418

Pannella, G. and MacClintock, C. (1968). Biological and environmental rhythms reflected in molluscan shell growth. *Mem. J. Palaeo.*, **42**, 64–80

Pannella, G., MacClintock, C. and Thompson, M. N. (1968). Paleontological evidence of variations in length of synodic month since Late Cambrian. *Science*, **162**, 792–796

Rhoads, D. C. and Pannella, G. (1970). The use of molluscan shell growth patterns in tion of the Pismo Clam (*Tivela stultorum*). *J. Mar. Res.*, **9** (3), 188–210

Thorson, G. (1971). *Life in the Sea*, McGraw-Hill, New York. 256 pp.

DISCUSSION

CLARK: How might you try to eliminate the possibility of variation in gravity influencing growth? By how much would the tanks have to be jiggled while the animal was alive to throw random noise into the system so the animals couldn't pick up anything happening to the Earth's gravitational field?

THOMPSON: I don't know, but the clams I studied are sensitive to certain intensities and frequencies, and as long as the cycle continued, no matter how much noise, given enough time the organism may still be able to pick up the cycle. (There was also discussion on the means of eliminating electromagnetic influence from the organism. Metal shielding or a Helmholtz coil were suggested.)

BROSCHE: It seems an open question whether the animals learned the cycles when they were living in the natural habitat, or whether they were able to pick up cycles from other information in the laboratory, even though you tried to control the laboratory habitat.

THOMPSON: The rhythms my clams showed couldn't be a memory of their previous tidal regime, because the tidal rhythm did not appear immediately in the experiment; they got stronger with the duration of the experiment. The question remains whether the gaping rhythms are controlled by an internal clock, or are cued to change in environmental parameters not immediately detectable.

EVANS: What is the periodicity of the increments in *Mercenaria*?

THOMPSON: The darkest lines are solar daily, the lightest are lunar daily.

LATITUDINAL VARIATION IN SHELL GROWTH PATTERNS OF BIVALVE MOLLUSCS: IMPLICATIONS AND PROBLEMS

CLARENCE A. HALL, JR.
Department of Geology and Institute of Evolutionary and Environmental Biology, University of California, Los Angeles, California, U.S.A.

Abstract

During the growth of the shell of the California surf clam *Tivela stultorum* (Mawe, 1823) and the subtidal bivalve *Callista chione* (Linnaeus, 1758) from the northern Adriatic Sea, a period of slower growth occurs annually. This period of growth is manifested by a macroscopic growth band or biocheck, consisting of relatively thin and closely spaced shell growth increments and a relatively thick band of fast growth: together they form an annual growth band. The number of growth increments in biochecks and bands of fast growth varies with age, latitude and water depths at which populations of the two taxa are living. Examples are: (i) there is a maximum of 307 daily growth increments in the annual growth bands of 3-year-old individuals of *C. chione* and only a maxmum of 210 daily shell growth increments present in 12-year-old individuals; (ii) there is an average of 82 daily growth increments in the fourth annual biocheck of shells of *T stultorum* from near 35° N latitude (eastern Pacific Ocean), while near 26°30′ N latitude there is an average of 111 daily growth increments in biochecks of the same age- and year-class. Latitudinal gradients in shell growth increments and patterns could be an aid in suggesting positions of geographic poles during the past; however, comparisons should be made between the same age-classes.

Comparing the same age-classes of Pliocene and Holocene individuals of *C. chione*, the average number of daily growth increments is greater in biochecks of Pliocene individuals (114 and 86, 1- and 4-year age-classes) than in biochecks of extant individuals (86 and 64, 2- and 4-year age-classes) from near 45° N latitude, northern Italy. Latitudinal gradients in shell growth of the extant *C. chione* must be documented, as they were for *T. stultorum* Preliminary studies suggest that ecological factors such as depth also influence shell growth. Thus, caution should be exercised when formulating palaeolatitudinal or palaeoclimatological interpretations.

Introduction

Shell growth responds to a number of environmental factors (Weymouth, 1923; Smith, 1928; Davenport, 1935; Pratt, 1953; Green, 1957; Barker, 1964; Lammens, 1967; Montfort, 1967; Clark, 1968; House and Farrow, 1968; Pannella and MacClintock, 1968; Kennedy *et al.*, 1969; Rhoads and Pannella, 1970; Farrow, 1971, 1972; Pannella, 1972), and the interpretation of shell growth increments and cyclic shell growth bands must be approached with caution. However, once shell growth diversity gradients can be determined in modern organisms, it may be possible to: (i) follow the development of these gradients through time and to suggest palaeolatitudes; (ii) note significant differences in shell growth patterns, which reflect broad-scale environmental changes, such as in marine palaeoclimate, through time and across time boundaries between geological epochs; (iii) delimit present-day and fossil molluscan provinces and (iv) apply diversity gradients in shell growth to structural geology problems such as displacements of shell growth diversity gradients measured in terms of hundreds of miles along major faults (Hall, 1960).

The purpose of this paper is to present incremental shell growth data relative to marine bivalve molluscs: *Callista chione* (Linnaeus, 1758) and *Tivela stultorum* (Mawe, 1823). Specimens of *Callista chione* studied were collected from the northern Adriatic Sea and from Pliocene and Pleistocene rocks of Italy. Specimens of *Tivela stultorum* were collected alive along the west coast of North America over a latitudinal range of approximately 9°. Analysis and interpretation

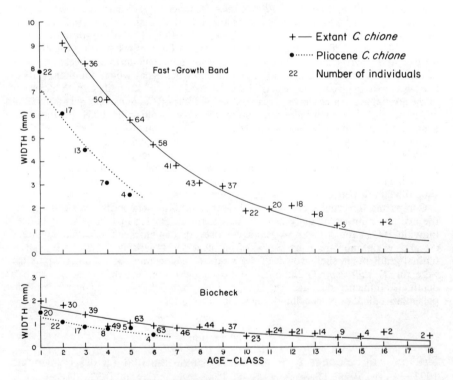

FIGURE 1. Growth (in mm) within bands of fast growth and biochecks of individuals of *Callista chione* from 1–18 years of age. The mean thickness of the growth band of a particular age-class within all year-classes studied is shown by a plus (+) for extant individuals and by a dot (·) for Pliocene individuals. The numbers refer to the number of individual bands measured. For example, the mean width of 30 second-year biochecks is 1·88. The data for the extant individuals are from Hall (1974)

of the incremental shell growth data from these taxa suggest that: (a) age must be considered when evaluating periodicity of growth increments; (b) there are latitudinal differences in growth patterns in shells of individuals from the same bathymetric zone, year- and age-class and (c) the direct or indirect affects of substrate and depth on shell growth are not sufficiently understood.

Growth increments

Extant Callista chione

An analysis of 474 bands of slow growth or biochecks and 414 bands of fast growth in 137 individuals of *Callista chione* (Linnaeus, 1758) from 5–10 m of

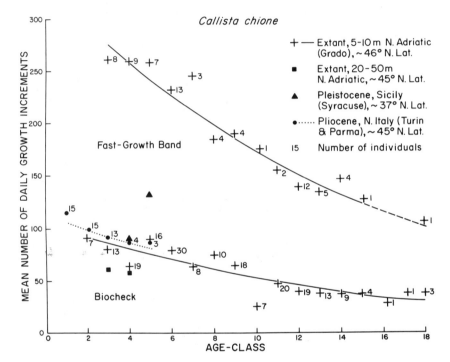

FIGURE 2. Number of shell growth increments in the band of fast growth and the biocheck plotted against the age-class of individuals of *Callista chione*. Example: the mean number of daily shell growth increments within seven 2-year-old biochecks is 88·8. The numbers beside these averages refer to the number of growth bands counted. The data for the curves for extant individuals collected from 5–10 m of water in the northern Adriatic Sea, near Grado, are from Hall (1974). Pliocene specimens were collected by P. G. Caretto near Turin and by C. A. Hall near Parma. Both Pliocene collections are from upper Pliocene rocks. The extant specimen from 20–50 m of water was collected by L. M. J. U. van Stratten in the northern Adriatic Sea, and the single specimen from the Pleistocene rocks of Sicily was collected by Italo Di Geronimo

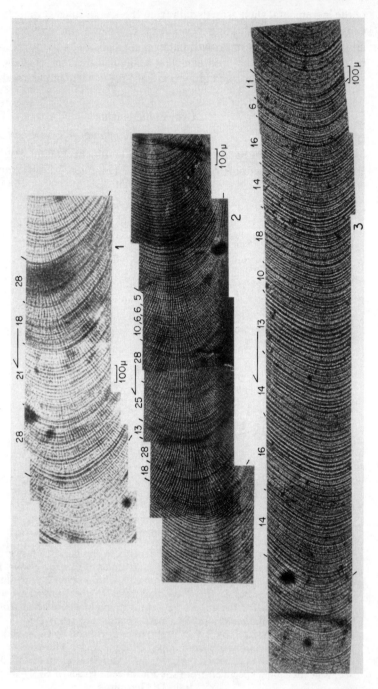

Latitudinal Variation in Shell Growth Patterns of Bivalve Molluscs

FIGURE 3. Numbers refer to the number of daily growth increments in bundles of growth increments. There is no significance to bundles or groupings of growth increments, which are only provided for purposes of understanding what is being counted as a presumed daily shell growth increment. Arrows indicate direction of growth

1. *Callista chione* (Linnaeus, 1758). Collected live 23 February 1971, from 5–10 m of water off Grado, northern Adriatic Sea. UCLA hypotype No. 49035. Fifth annual biocheck of a 9-year-old individual. Approximately 90 daily shell growth increments are present in the biocheck.

2. *Callista chione* (Linnaeus, 1758). Collected live from silty sand, 0·8 km south of Badagnano (approximately 7 km southwest of Castell' Arquato, near 45° 51′ 00″ N latitude, northern Italy. Locality is between Piacenza and Parma in the Picacentine Hills of the Apennines of northern Italy. Piacenzian Stage, upper Pliocene rocks. UCLA hypotype No. 49036. Fifth annual biocheck in an 8-year-old individual. Approximately 139 daily shell growth increments are present in the biocheck.

3. *Callista chione* (Linnaeus, 1758). Collected from silty sand near Floridia (Syracuse), Sicily. Sicilian Stage, Pleistocene. UCLA hypotype No. 49037. Fifth annual biocheck of an 8- or 9-year-old individual. Approximately 132 distinct shell growth increments are present in the biocheck; not all are presumed to be daily

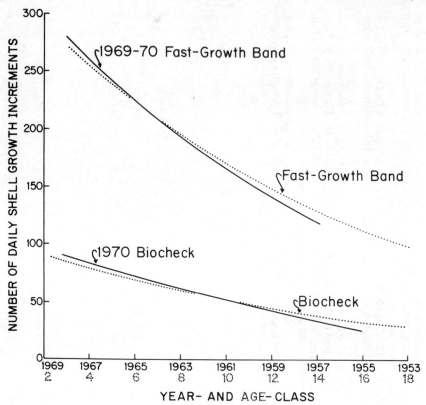

FIGURE 4. Average number of growth increments in the band of fast growth and the biocheck, *Callista chione*, deposited during 1969–1970 and 1970, respectively, plotted against the year-class (solid lines). The solid-line curves (1969–1970 and 1970) represent the average number of daily growth increments of a particular age-class within the same year-class. The other curves (dotted lines 'fast-growth band' and 'biocheck') show the average number of growth increments in the band of fast growth and the biocheck plotted against the age-class. The data from both curves are from Hall (1974) and illustrate that a study of shell growth increments in age-classes would not significantly differ from interpretations based on year-classes of fossil populations if year-classes could be determined in fossil populations

water in the northern Adriatic Sea (~45° N latitude) has been made (Hall, 1974). From this study of *C. chione*, it is known that a period of slower growth occurs annually and is manifested by a macroscopic, seasonal growth band or biocheck, 0·9–3·60 mm wide. The biocheck is not formed during the months when sea surface temperatures are at their lowest, nor is it formed during known periods of spawning. Based on limited evidence, the biocheck begins abruptly, follows the period of spawning, and precedes the onset of the coldest winter sea surface temperatures. The width of the band of fast growth is from 0·60–11·80 mm. The biocheck and fast-growth band form an annual growth band composed of 13–121 daily growth increments in the biocheck and 73–307 daily growth increments in the band of fast growth. Older individuals, i.e. from 4–18 years of age, do not deposit as great a thickness of shell nor as great a number of daily shell growth increments as do younger individuals (Figures 1 and 2). Note also in Figures 1, 2 and 3 that there are relative differences in shell growth and numbers of shell growth increments between Pliocene, Pleistocene and extant *C. chione*. There is close agreement between the number of daily growth increments in the biochecks of individuals of the same age-class from the same locality, collected at the same time of the year. There is also close agreement between the number of daily growth increments in biochecks of the same age-class but different year-classes of extant individuals (Figure 4).

The size of a specimen and the width of biochecks and fast-growth bands are smaller in a single sample of *C. chione* (supplied by L. M. J. U. van Stratten) from 20–50 m of water, than are the same features of an average specimen of the same age from 5–10 m of water. All specimens being compared in this case are from the northern Adriatic Sea. There are also fewer daily shell growth increments in the biochecks of the specimen from 20–50 m of water than there are in the average biochecks of the same age-classes collected in 5–10 m of water near Grado, at the head of the Adriatic Sea (Figure 2). From the paucity of material collected from relatively deep water, it is not possible definitively to suggest the direct or indirect influence of depth on shell growth. However, without knowledge of shell growth gradients related to bathymetry and substrate (Figure 5), analyses of the shell growth increments of fossil populations of *C. chione* are incomplete.

Extant Tivela stultorum

Some 115 biochecks and 74 bands of fast growth were counted in 100 individuals of *Tivela stultorum* (Mawe, 1823) collected along the coast of California and Baja California (~35° N latitude to ~26° 30′ N latitude) (Hall, 1974). Weymouth (1923) studied more than 2000 individuals of *T. stultorum* over a period of more than two years and conclusively documented the fact that a relatively narrow growth band appears annually. This narrow growth band is referred to here as the biocheck. The annual growth band, i.e. fast-growth band and a biocheck, commonly contains more than 300 growth increments, 5–200 μm thick. The more-than 300 growth increments in the biocheck and band of fast growth are assumed to be daily growth increments.

Growth in *T. stultorum* may exceed 20 mm per year in young clams and is less than 1 mm per year in older individuals. In this taxon 80% of the total length is reached by the end of the seventh or eighth year of growth, and some individuals reach an age of 50 or more years (Weymouth, 1932; Coe and Fitch, 1950; Fitch, 1965). Within shells of 12–23-year-old individuals of *T. stultorum* collected from Baja California, Mexico, and near Santa Monica, California, there is an average of 60 daily growth increments in the biocheck and less than

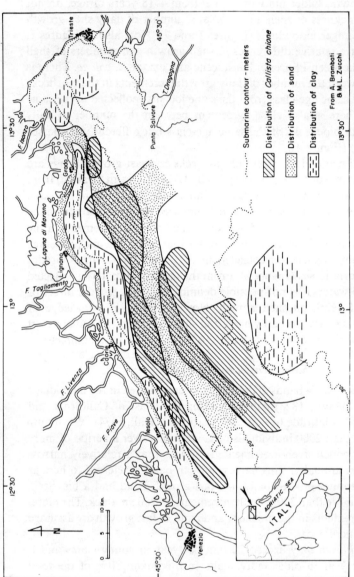

FIGURE 5. Distribution of *Callista chione* in the northern Adriatic Sea based on the work of Brambati and Zucchi (1969) and Brambati (written communication, 1970). The individuals of *C. chione* used by me for the study of this taxon from the northern Adriatic Sea (Hall, 1974) were living in sandy substrates. Brambati and van Stratten (written communication, 1970) show that *C. chione* also lives in silty sand, but rarely in clay. The relation of substrate to shell growth patterns in this taxon is not known, nor are the relationships between shell growth patterns and bathymetric gradients clear. Note that rocks termed 'clay' are called 'pelite' by Brambati. According to Brambati (written communication, 1974), these sediments have a grain size of less than 50 μm and include silty to sandy pelite facies. Unpatterned areas on the map are an intermediate facies between the argillaceous or mud facies and the sand. Reproduced from Brambati, A. and Zucchi, M. L. (1969), *Studi Trentini di Scienze Naturali*, Sez, A, **46**, 30–40 by permission of the Museo Tridentino di Scienze Naturali

200 daily growth increments in the fast-growth band; in some cases there are less than 100 (Hall, 1974). A maximum of 239 and a minimum of 164 daily growth increments were present in an annual band of growth in individuals older than 12 years. There is an absence of 126–201 or more daily growth increments in an annual band of growth in individuals of *T. stultorum* older than 4–6 years. In order to study the number of days in a year during the geological past, such as the study reported by Pannella *et al.* (1968), the application of suitable techniques to account for the trend of slowing growth is required.

Near the southern limit of the range of *T. stultorum* within individuals, for example, of the year-class of 1965, there is an average of 111 daily growth increments in the fourth-year biocheck, whereas near the northern limit of the range of this species, there is an average of 82 daily growth increments. Until more individuals of the same age- and year-class from different microhabitats of the same latitude and different latitudes are studied, there are reservations about the

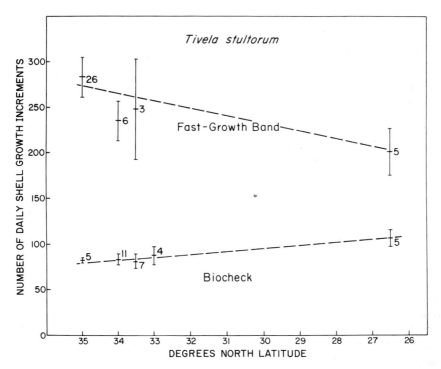

FIGURE 6. Number of daily shell growth increments per biocheck and bands of fast growth in the year-classes 1961 through 1968 and age-classes 1–8 of *Tivela stultorum* from the west coast of North America plotted against latitude. The numbers refer to the number of individual bands from which growth increment counts were made. Diagrammatic curves show increasing numbers of daily shell growth increments in biochecks with decreasing latitude and decreasing numbers of daily shell growth increments in fast-growth bands with decreasing latitude

acceptance of the observed relationship between latitude and the number of daily growth increments in the biocheck (Figure 6).

Fossil *Callista chione*

Pliocene specimens of *C. chione* have been collected from near the same latitude as that from which the living specimens of *C. chione* were taken. The Pliocene shells are from near Turin (Castelnuovo, Valle Andona, late Pliocene, collected by P. G. Caretto) and Parma (Castell'Arquato and Badagnano, late Pliocene, collected by C. A. Hall) in northern Italy. Based on 76 observations of 1–6-year-old individuals of *C. chione*, the width of the biocheck and band of fast growth is less than that of extant specimens from near the same latitude collected from 5–10 m of water (Figure 1). The average number of daily growth increments in 50 Pliocene individuals of *C. chione* is greater than in biochecks of extant individuals of the same age-classes. For example, there is an average of 114–86 daily shell growth increments in biochecks within 1–4-year-old Pliocene individuals versus averages of 88–64 growth increments in extant 2–4-year-old individuals collected during 1970 and 1971 (Hall, 1974; Figure 2). If there is a latitudinal gradient in the number of shell growth increments of *C. chione*, as there seems to be in *T. stultorum*, then it is possible that the greater number of shell growth increments in biochecks of Pliocene individuals from near 45° N latitude reflects warmer marine climate during the Pliocene than at that latitude today.

A single specimen of *C. chione* from Sicilian Pleistocene rocks (collected by Italo Di Geronimo near Syracuse, Sicily) has a significantly greater number of shell growth increments in its biochecks than biochecks of the same age-classes of extant specimens from the northern Adriatic and Pliocene specimens from northern Italy (Figure 2). Again, the sample size is inadequate, but the relatively large number of growth increments in the biocheck is provocative and should stimulate further investigation.

Summary

Measurement and growth-increment counts of *C. chione* (Hall, 1974) show unmistakably that, with increasing age, there are fewer growth increments in both the biocheck and the band of fast growth in the annual growth band. Clearly, age must be considered when trying to determine the number of days in a year based on shell counts of growth increments in molluscs.

Although not entirely conclusive, there seem to be more daily growth increments in biochecks of *Tivela stultorum* near the southern limit of its range than in biochecks of the same age-classes near the northern limit of its range. If true, *T. stultorum* and other taxa may provide a means for suggesting the positions of palaeolatitudes and palaeoclimate. For example, the relatively greater number of daily shell growth increments in biochecks of *C. chione* from Pliocene rocks near 45° N latitude in northern Italy, as compared with daily growth increments in biochecks of the same age-classes of extant individuals from the Adriatic Sea, may reflect warmer sea temperatures during the Pliocene than today. However, other factors, such as the effect of substrate, depth, etc., on shell growth in *C. chione* are not sufficiently well enough understood to accept such an interpretation based on shell growth alone.

Acknowledgments

Thanks are extended to Professor E. Montanaro Gallitelli, Instituto di Paleontologia, Università di Modena, for assistance and encouragement during the course of this study. The research was supported in part by a Fulbright Research Grant: Italy, 1970–71.

Those who helped me obtain or collect fossil material from Italy are: P. G. Caretto and Professor C. Sturani, Università di Torino; Professor G. Pelosio, Università di Parma; P. Rompianesi, Università di Modena; and Italo Di Geronimo, Università di Catania: their assistance is very much appreciated.

References

Barker, R. M. (1964). Microtextural variation in pelecypod shells. *Malacoloqia*, **2** (1), 69–86

Brambati, A. and Zucchi, M. L. (1969). Relazioni tra granulometrie e distribuzione dei Molluschi nei sedimenti recenti dell'Adriatico settentrionale tra Venezia e Trieste. *Studi Trentini di Scienze Naturali*, Sez. A, **46**, 30–40

Clark, II, G. R. (1968). Mollusk shell: daily growth lines. *Science*, **161** (3843), 800–802

Coe, W. R. and Fitch, J. E. (1950). Population studies, local growth rates and reproduction of the Pismo Clam (*Tivela stultorum*). *J. Mar. Res.*, **9** (3), 188–210

Davenport, C. B. (1935). Growth lines in fossil pectens as indicators of past climates. *J. Palaeontol.*, **12**, 514–515

Farrow, G. E. (1971). Periodicity structures in the bivalve shell: experiments to establish growth controls in *Cerastoderma edule* from the Thames Estuary. *Palaeontology*, **14** (4), 571–588

Farrow, G. E. (1972). Periodicity structures in the bivalve shell: analysis of stunting in *Cerastoderma edule* from the Burry Inlet (South Wales). *Palaeontology*, **15** (1), 61–72

Fitch, J. E. (1965). A relatively unexploited population of Pismo clams, *Tivela stultorum* (Mawe, 1823), (Veneridae). *Proc. Malacol. Soc. Lond.*, **31**, 144–159.

Green, J. (1957). The growth of *Scrobicularia plana* (da Costa) in the Gwendraeth Estuary. *J. Mar. Biol. Assoc. U.K.*, **36**, 41–47

Hall, Jr., C. A. (1960). Displaced Miocene molluscan provinces along the San Andreas Fault, California. *Univ. Calif. Pubs. in Geol. Sci.*, **34** (6), 281–308

Hall, Jr., C. A. (1974). Shell growth in *Tivela stultorum* (Mawe, 1823) and *Callista chione* (Linnaeus) (Bivalvia): Annual periodicity, latitudinal differences, and diminution with age. *Palaeogeogr., Palaeoclimatol., Palaeoecol.*, **15** (1), 33–61

House, M. R. and Farrow, G. E. (1968). Daily growth banding in shell of the cockle *Cardium edule*. *Nature*, **219**, 1384–1386

Kennedy, W. J., Taylor, J. D. and Hall, A. (1969). Environmental and biological controls on bivalve shell mineralogy. *Biol. Rev.* **44**, 499–530

Lammens, J. J. (1967). Growth and reproduction in a tidal flat population of *Macoma baltica* (L.). *Netherlands J. Sea Research*, **3**, 315–382

Montfort, A. F. (1967). Edad y crecimiento de *Cardium edule* de la ría de Vigo. *Invest. Pesq.*, **31** (2), 361–382 (summary in English)

Pannella, G. (1972). Paleontological evidence of the earth's rotational history since early Precambrian. *Astrophys. Space Sci.*, **16**, 212–237

Pannella, G. and MacClintock, C. (1968). Biological and environmental rhythms reflected in molluscan shell growth. *J. Palaeontol.*, **42** (2), 64–79 (supplement to No. 5, Mem. 2)

Pannella, G., MacClintock, C. and Thompson, M. N. (1968). Paleontological evidence of variation in length of synodic month since Late Cambrian. *Science*, **162**, 792–796

Pratt, D. M. (1953). Abundance and growth of *Venus mercenaria* and *Callocardia morrhuana* in relation to the character of bottom sediments. *J. Mar. Res.*, **12** (1), 60–74

Rhoads, D. C. and Pannella, G. (1970). The use of molluscan shell growth patterns in ecology and paleoecology. *Lethaia*, **3**, 143–161

Smith, G. M. (1928). Food material as a factor in growth rate of some Pacific clams. *Trans. Roy. Soc. Canada*, Section V, Biol. Sci., Vol. XXII, Pt. II, sect. II, p. 287–291

Weymouth, F. W. (1923). The life-history and growth of the Pismo Clam (*Tivela stultorum* Mawe). *Fish. Bull. Calif. Fish Game Comm.*, **7**, 1–120

DISCUSSION

SCRUTTON: How do you define the boundaries of the biocheck? Biochecks look rather diffuse.

HALL: Biochecks can be distinctly recognized on the surface of the shell, and the growth patterns in thin section can be matched with the external pattern. Biochecks are distinguished by an abrupt change from widely spaced to narrow increments.

EVANS: If you have demonstrated fewer daily increments to be deposited by older animals, how could the increments indeed by daily?

HALL: In 1–4-year-old specimens. I could count nearly 365 daily lines by calling them daily. In older individuals, the animal may not have added increments every day.

CLARK: If there are fewer lines in older individuals, when does growth stop being invariably daily? Do you mean the clam follows a daily periodicity for only the first part of its life, then drifts?

HALL: No. There is probably an equilibrium between deposition and non-deposition which changes.

DOLMAN: Why should there be such an abrupt difference between the biocheck and the fast growth series? Could they be caused by storms?

HALL: I don't know.

CLARK: As far as the influence of spawning is concerned, it is important to note that not all species have a spawning check.

EVANS: And not all species have a winter slowdown.

BUDDEMEIER: Perhaps different environmental parameters in different habitats are important. In the corals I studied, temperature seems to control growth increment production. Once outside the normal temperature range, I can't be so certain. The limiting parameter shifts.

PANNELLA: We should also consider sexual differences: male and female bivalves have different spawning series.

HALL: We must also consider age differences. *Tivela* don't spawn in their first year.

TIME, TIDE AND THE COCKLE

MARTIN A. WHYTE
Geology Department, University of Hull, England

Abstract

Studies of cockle populations in Poole Harbour, Dorset, show that *Cerastoderma edule* (L.) and *C. glaucum* (Poiret) calcify at night. Consideration of the probable interaction between this calcification cycle and seasonal, tidal and diurnal environmental cycles permits the development of a nomothetic approach to the study of periodicity structures in cockles. The comparison of predicted growth patterns with growth patterns recorded from natural situations allows refinement of the model which can also be used to highlight suitable areas for future research.

Seasonal growth variation is interpreted as the result of interaction between temperature and light cycles. Tidal growth patterns whose minima may coincide with either spring or neap tides depend on the position of the cockle within the intertidal zone and on the tidal characteristics of the locality. The effects of random events may be strongly influenced by the tidal phenomena.

Introduction

The existence of periodicity structures in the shell of the cockle, *Cerastoderma edule* (L.), was first demonstrated by House and Farrow (1968) and the potential of these growth bands in the study of the relationships between shell growth and environmental parameters has been demonstrated by Farrow (1971, 1972). These works have been of an essentially idiographic nature: that is to say that they are case studies of particular situations in which the detailed growth of the shell is related to one or more environmental parameters whose variation is well-documented. A more general nomothetic approach containing generic concepts and predictive qualities can be evolved from consideration of the probable interaction of the calcification cycle with environmental factors. Such an approach must be tested against the results of idiographic studies but provides these results with a context and can also be used to highlight useful areas for further study.

Calcification in *Cerastoderma*

Each growth band in the shell of the cockle consists of a relatively thick carbonate-rich zone accompanied by a thin organic-rich zone but the time of deposition of these players has never been determined. House and Farrow (1968) and Farrow (1971) have suggested that in *Cerastoderma edule* the carbonate layer forms during the day while Peterson (1958) has conjectured that in the related species *Cardium lamarcki* Reeve (=*Cerastoderma glaucum* (Poiret)) it is laid down at night. In order to elucidate the calcification sequence cockles were collected from the Redhorn Lake area of Poole Harbour, Dorset (Gt. Britain National Grid SZ021853) where a mixed population of *C. edule* and *C. glaucum* occurs over a wide area of uniform depth (−0·3 to −1·0 m O.D.) and substrate (soft mud). Two series of collections were made at 3-hr intervals over 24 hr, on 27–28 October 1971 and 19–20 July 1972, at neap tides when the cockle beds are never exposed. Both series started and ended with a collection at 17.00 hr G.M.T. Each collection was made from a previously undisturbed station and the cockles were immediately killed and fixed in 10% phosphate-buffered formalin. Growth surfaces of up to five specimens per collection were examined, using a Cambridge Stereoscan 600 electron microscope in the Department of Geology, University of Hull.

The results of the two species are identical and since the growth surface is covered at all times by a membrane, secretion of the organic matrix or conchiolin appears to be more or less continuous. When this membrane is thinnest the underlying cross-lamellar shell structure, which often has an open porous character, is expressed at the surface as a lineated texture (Figure 1A). This is interpreted as the result of a phase of active calcification within and behind an accreting conchiolin matrix and is found in specimens collected during the night (October, 23.00–05.00 hr; July, 20.00–02.00 hr). Secretion of carbonate behind organic membranes has also been recorded as associated with nacreous and foliated shell structures (Bevelander and Nakahara, 1961). During the day (October, 08.00–17.00 hr; July, 11.00–17.00 hr), when the membrane is relatively thick, it has a rough or pitted surface texture and overlies a compact carbonate structure (Figure 1B). At these times calcification is believed either to have ceased or to be minimal. During transitional stages the two textures may occur together (October 11.00 hr; July, 05.00 and 08.00 hr) giving variable results.

Nocturnal calcification has also been recorded in *Mercenaria mercenaria* (MacClintock and Pannella, 1969) which is the only other bivalve whose calcification sequence has been studied though growth patterns in *Clinocardium nuttalli* and *Protothaca staminea* (Evans, 1972a) could possibly be reinterpreted as indicating nocturnal calcification.

Incremental shell growth is generally regarded as the result of fluctuations in the composition (Pannella and MacClintock, 1968; Wilbur, 1972) or the nature (Wilbur, 1972; Digby, 1968) of the extrapallial fluid. During the second series of collections an attempt, based on the methods of Digby (1968), was made to measure the variation in the pH of the extrapallial fluid of both *C. edule* and *C. glaucum*. Drops of indicator fluid (B.D.H. Universal Indicator) were placed on shell commissures which were then nicked with a scalpel, allowing the indicator to mix with the extrapallial fluid. The shells were then opened and the reaction colours compared with those of a set of buffered solutions of known pH. In both species the pH of the extrapallial fluid showed a small diurnal range with

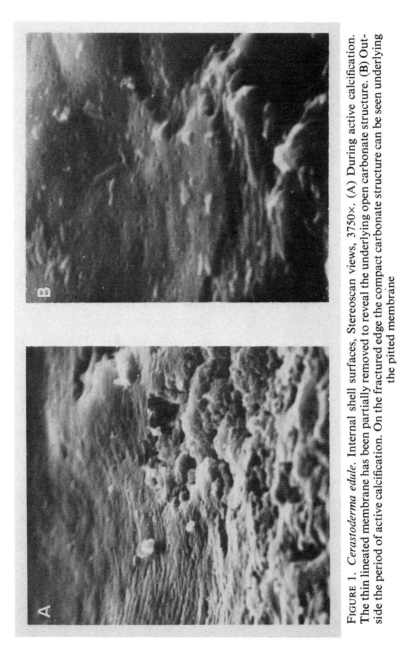

FIGURE 1. *Cerastoderma edule*. Internal shell surfaces, Stereoscan views, 3750×. (A) During active calcification. The thin lineated membrane has been partially removed to reveal the underlying open carbonate structure. (B) Outside the period of active calcification. On the fractured edge the compact carbonate structure can be seen underlying the pitted membrane

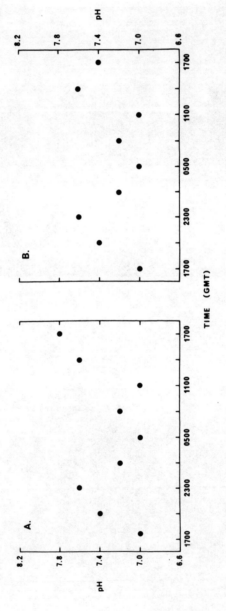

FIGURE 2. Diurnal variation in pH of the extra-pallial fluid. (A) *C. edule*. (B) *C. galucum*

maximum values being recorded during the night (Figure 2). The pH variation in *Cerastoderma* spp. is smaller than the range recorded by Digby (1968) for *Mytilus edulis* (pH 7·3–9·4) though similar to values obtained by other workers (Wilbur, 1964; pH 7·1–8·3) and the results have probably been affected by the rapid fall in pH which is known to occur on shell closure (Wilbur, 1964; Crenshaw, 1972). Crenshaw (1972) has recently shown that this is accompanied by an increase in the calcium content of the extrapallial fluid.

In *Mercenaria mercenaria* the phases of maximum adductor activity and valve movement occur at high tide, in the afternoon and at neap tides (Bennett, 1954) and calcification as suggested by Pannella and MacClintock (1968) thus appears to coincide with a nocturnal quiescence. The mutual relationship of the valves during this phase is not described by Bennett (1954) but in several other bivalve species, which also exhibit a diurnal rhythmicity with a nocturnal quiescent phase, the shell normally gapes slightly (Morton, 1971; Salanki, 1966). The tidal rhythm of adductor muscle activity which has been recorded in *C. edule* (Morton, 1970), does not preclude the existence of a diurnal rhythm, since both tidal and diurnal rhythms occur in *Ostrea edulis* (Morton, 1969) and *M. mercenaria* (Bennett, 1954) and nocturnal quiescent phases can be detected in part of a kymograph trace illustrated by Marceau (1909). Incremental growth of the shell may thus be a response to, and to some extent indicative of, this rhythm of adductor muscle activity with calcification being inhibited during the day due to frequent valve movements.

The quiescent phase is normally interrupted by a number of strong muscle contractions which in *O. edulis* have a frequency of 0·1–2·3/hr (Morton, 1969). Such events will cause fluctuations in the pH of the extrapallial fluid and break the calcification sequence, giving rise to the subdaily (fifth order (Barker, 1964) or complacent (Pannella, 1972)) banding which is well displayed by specimens of both *C. edule* and *C. glaucum* from Poole and which has also been recorded in specimens of *C. edule* from Burry Inlet (House and Farrow, 1968). Tidal exposure during the calcification period will also interrupt the calcification sequence and produce a second type of subdaily banding, the effects of which are discussed below.

Calcification and environmental factors

Introduction

Understanding of the calcification sequence allows consideration of the probable effects of environmental parameters on the calcification cycle. These effects will be discussed under the four headings, annual bands, seasonal cycles, tidal cycles and random fluctuations, which correspond broadly to the various components of a time-series analysis. The discussion will deal exclusively with *C. edule* though, in many cases, similar effects might be expected in *C. glaucum*, especially where this species occurs in the intertidal zone.

Seasonal cycles

It has been suggested that the timing of the nocturnal quiescence in *Pecten jacobaeus* is related to the diurnal variation in light levels (Salanki, 1966) and it is possible that in *Cerastoderma*, which, though not possessing well-developed mantle eyes, does have light-sensitive tentacles (Johnstone, 1899), the same feature may occur. The seasonal variation in the relative lengths of the day and night may thus control the potential time available for calcification which will reach a maximum in the winter and a minimum during the summer. The results of the Poole experiments show that the length of the calcification period in July was indeed slightly shorter than the length of the period in October.

The seasonal light cycle will, however, interact with other seasonal cycles, notably that of temperature, which may control the amount of growth actually achieved. Since chemical reaction rates increase with temperature, and since the solubility of carbon dioxide decreases with increase in temperature, it might be expected that secretion of carbonate would be favoured by warmer waters. Temperature-controlled growth rates would thus have maxima in the summer and minima in the winter.

The results of the interaction of these two antagonistic variables may possibly be seen in cockles examined by Farrow (1971) who found that, while shell growth paralleled temperature changes between January and April, it became relatively retarded in the summer. This may also explain the seasonal bimodality of some growth plots (Farrow, 1971) which show maximum growth in April or May and a secondary maximum in August or September.

These two factors both show latitudinal variation and thus latitudinal variation in the seasonal pattern may also occur. However, since temperature appears to exert a more dominant influence on the pattern and is less rigorously entrained to latitude this may not be a well-developed feature.

Tidal cycles

A highly pervasive feature of the habitat of *Cerastoderma edule* is the rise and fall of the tides. When a low tide causes exposure or near exposure of a cockle during the calcification period, then the calcification sequence will be disrupted. Depending on the time of this low water the growth may be truncated at either one or both ends of the calcification period, giving rise to a partial band; may be interpreted giving rise to a double growth band; or may even be completely suppressed.

In areas with semi-diurnal tides such as the British Isles (Russell and McMillan, 1952), the amplitude or range of tides and the times of high and low water vary according to two fortnightly cycles. The amplitude cycle gives maximum tidal range at spring tides, which occur a few days after the new and full Moons, and minimum tidal range at neap tides, which occur a few days after the quarter Moons. Concomitantly with this, the second cycle varies in such a way that at any locality, at any given age of the Moon, high and low waters tend to recur at approximately the same time of day. In particular, high water on the day of new or full Moon at any locality always occurs at about the same time of day. The actual time, however, varies from locality to locality, depending on their tidal establishments. As a consequence, the extent to which calcification is disrupted and the sequence of growth band types which thus develop in any shell will vary with the position of the animal within the intertidal zone and with the local tidal establishment. Conversely, the correct interpretation of growth banding patterns should allow determination not only of the position at which the cockle lived within the intertidal zones but also of the local tidal establishment. Sequences will, however, also be complicated by the effects of the parallactic inequality and other longer term lunar cycles.

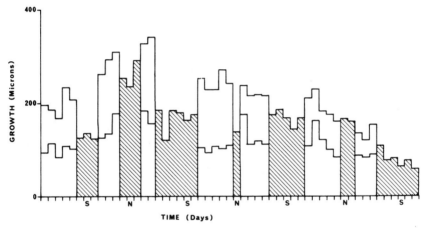

FIGURE 3. *Cerastoderma edule*, Poole Harbour. Shell growth, from late July to early September 1971, of an individual collected from just below low water mark of neap tides. Interrupted increments are unornamented and the position of the interruption is indicated. The days of spring (S) and neap (N) tides are indicated

Predictions of how the calcification cycle might be affected could be made for all levels on the shore and for all tidal establishments. In general, though, the degree of disruption will vary with tidal establishment; the upper shore will sustain maximum disturbance at neap tides while levels on the lower shore will be most disturbed at spring tides.

The fortnightly changes in tidal effects are well displayed in the growth of a cockle collected from below the low-water mark of neap tides in Poole Harbour (Figure 3). Here, at spring tides, low waters occur in the morning and evening between six and seven o'clock and give rise to partial growth bands. At neap tides, however, low waters occur between midnight and one o'clock, the animal is not exposed and a complete growth band forms. At intermediate tides of moderate range exposure occurs during the calcification period but is of limited duration and a double band forms. Occasionally this double banding may be noticeable even at neap tides, showing that complete exposure is not necessary for its formation. The tides at Poole are abnormal in character and the high waters have a double character. At neap tides these are separated by a tidal stand but at spring tides there is a slight drop in water level which occurs during the calcification period. This appears to have caused a further complication of the growth record since, even though exposure has not occurred, some of the spring tide bands and even, on occasion, some of the double bands, exhibit a very faint subdivision of the growth band. For the sake of clarity this feature has been omitted from Figure 3.

The phenomenon of suppressed bands is exhibited by cockles from about high-water mark of neap tides in Burry Inlet (Farrow, 1972). Though up to eight consecutive bands may be suppressed (Farrow, 1972), the tidal establishment is such that for this level on the shore calcification is relatively less seriously disturbed than it would be under other tidal establishments. The apparent monthly pattern of lower shore cockles at Burry Inlet (House and Farrow, 1968; Farrow, 1972) is probably an artifact resulting from failure to recognize the existence of double bands. A probably fortnightly variation in growth band type is present in shells from the lowest shore levels (Farrow, 1972, plate 10). A similar confusion may also have occurred in the analysis of cockles from Southend (Farrow, 1971) and may, at least in part, explain the variation in the number of days growth recorded per year. The tidal establishment of Southend is such that one would predict that the double banding phenomenon would be less strongly developed than at Burry.

Random fluctuations

Random fluctuations in the environment may cause either enhancement or reduction of growth and may affect a single growth band (e.g. frosts) or a number of growth bands (e.g. spawning). The time when fluctuations occur relative to the fortnightly tidal cycles may affect the extent to which they influence shell growth and a correlation of this type can be seen in the early winter growth of a Southend cockle originally discussed by Farrow (1971). Reinterpretation of this specimen (Figure 4) suggests that low temperatures are most effective in causing growth reduction and stoppage when they occur on dates where a low water coincides with the calcification period, though since these are neap tides the animal is in fact not being exposed. Such phenomena might thus impose a secondary pseudocyclicity on the growth banding.

Annual bands

Absolute growth curves for *C. edule* can be determined using the winter rings (Walton, 1919), though disturbance rings (Orton, 1926; Kreger, 1940) and indistinctness of the first winter ring (Stephen, 1932) are complicating factors. Such curves generally indicate that, though initially high, the growth rate decreases

Time, Tide and the Cockle

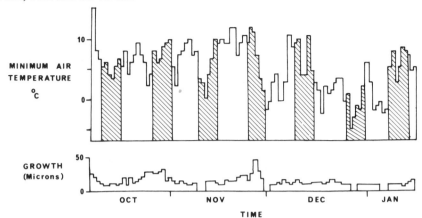

FIGURE 4. *Cerastoderma edule*, Southend. Correlation between minimum temperatures, times when low tides occur during the calcification period (ornamented) and growth disturbances. Reproduced from Farrow, G. E. (1971) *Palaeontology*, **14**, by permission of the Palaeontological Association

with age in the pattern typical of many bivalves (Hallam, 1967; Levington and Bambach, 1969). While the actual growth rates attained will be dependent on environmental factors, the underlying pattern is clearly biological and it is interesting to note that other physiological processes show similar effects since oxygen uptake in *C. edule* decreases geometrically with size (Newell and Northcroft, 1967; Boyden, 1972).

It is at this level of study that the effects of essentially uniform environment factors will be most evident and interesting results showing the effect of substrate (Farrow, 1971) and salinity (Eisma, 1965) on cockles from different localities have been obtained. The effects of more variable environmental factors such as the seasonal and tidal phenomena previously discussed will, however, also be evident and must be allowed for in any comparisons. The relative effects of tidal immersion within the intertidal zone of single localities have been studied by Farrow (1971, 1972) and by Stephen (1930a, 1930b, 1932) and of particular significance may be the fact that, at Burry Inlet, two distinct regimes of growth, above and below midtide level, were recognized (Farrow, 1972). These may be related to the upper shore (neap tide) and lower shore (spring tide) disturbance patterns previously recognized. In addition to growth variation up and down the shore there should also exist, as was first suggested by Orton (1934), a variation in growth along the shore which is related to differing tidal establishments.

Weymouth and Rich (1931) have shown that in *Siliqua patula* there is an inverse relationship between growth rate and longevity and that longevity is greater and growth rate less in cold waters than in warm waters. This relationship has not been detected in *C. edule* though the occurrences of giant cockles (Winckworth, 1921) may indicate its existence in a form which has been highly modified by tidal establishments.

Conclusion

The consequences of nocturnal calcification and its interaction with environmental cycles are such that the growth banding patterns developed are dominated by tidal effects with contributions from seasonal cycles. The patterns deduced refer essentially to a particular species and its habitat range and though similar effects appear to be present in other bivalves from similar environments, e.g. *M. mercenaria* (MacClintock and Pannella, 1969), differing patterns are clearly produced under differing environmental regimes, e.g. the mixed tidal cycles in *Clinocardium nuttalli* (Evans, 1972a) and the sublittoral growth of *Nucula proxima* (Rhoads and Pannella, 1970). Different modes of life will also affect patterns, a feature well shown by the growth of the rock-boring clam *Penitella penita* (Evans, 1972b) and also suggested by the overall shell growth of the deposit feeder *Tellina tenuis* which increases in size up the shore (Stephen, 1928) rather than down the shore as in *C. edule*.

Whatever the ecology of an animal has been, it is clear that the retrieval of any part of the information encoded within its shell depends on a complete interpretation of the growth-banding sequences. Thus neither the geophysicist nor the palaeontologist can afford to carry out his studies without consideration of the other's discipline and of the information relating most directly to it.

Acknowledgments

This work forms part of a project supported by the Natural Environment Research Council which is being carried out at the University of Hull under the general guidance of Professor M. R. House. I am greatly indebted to Dr. G. E. Farrow for the free use that he has allowed me to make of his data.

References

Barker, R. M. (1964). Microtextural variation in pelecypod shells. *Malacologia*, **2** (1), 69–86

Bennett, M. F. (1954). The rhythmic activity of the quahog, *Venus mercenaria*, and its modification by light. *Biol. Bull. mar. biol. Lab., Woods Hole*, **107**, 174–191

Bevelander, G. and Nakahara, H. (1961). An electron microscope study of the formation of the nacreous layer in the shell of certain bivalve molluscs. *Calc. Tiss. Res.*, **3**, 84–92

Boyden, C. R. (1972). Aerial respiration of the cockle *Cerastoderma edule* in relation to temperature. *Comp. Biochem. Physiol.*, **43**A, 697–712

Crenshaw, M. A. (1972). The inorganic composition of molluscan extra-pallial fluid. *Biol. Bull. mar. biol. Lab., Woods Hole*, **143**, 506–512

Digby, P. S. B. (1968). The mechanism of calcification in the molluscan shell. *Symp. zool. Soc. Lond.*, **22**, 83–107

Eisma, D. (1965). Shell-characteristics of *Cardium edule* L. as indicators of salinity. *Netherlands Journal of Sen Research* **2** (4), 493–540

Evans, J. W. (1972a). Tidal growth increments in the cockle *Clinocardium nuttalli*. *Science, N.Y.*, **176**, 416–417

Evans, J. W. (1972b). Functional micromorphology and circadian growth of the rock-boring clam *Penitella penita*. *Can. J. Zool.*, **50**, 1251–1258

Farrow, G. E. (1971). Periodicity structures in the bivalve shell: experiments to establish growth controls in *Cerastoderma edule* from the Thames Estuary. *Palaeontology*, **14**, 571–588

Farrow, G. E. (1972). Periodicity structures in the bivalve shell: analysis of stunting in *Cerastoderma edule* from Burry Inlet (South Wales). *Palaeontology*, **15**, 61–72

Hallam, A. (1967). The interpretation of size-frequency distributions in molluscan death assemblages. *Palaeontology*, **10**, 25–42

House, M. R. and Farrow, G. E. (1968). Daily growth banding in the shell of the cockle, *Cardium edule*. *Nature*, **219**, 1384–1386

Johnstone, J. (1899). Cardium. *L.M.B.C. Mem. typ. Br. mar. Pl. Anim.* **2**, 92 pp.

Kreger, D. (1940). On the ecology of *Cardium edule* L. *Archs. neerl. Zool.*, **12**, 351–453

Levington, J. S. and Bambach, R. K. (1969). Some ecological aspects of bivalve mortality patterns. *Am. J. Sci.*, **268**, 97–112

MacClintock, C. and Pannella, G. (1969). Time of calcification in the bivalve mollusk *Mercenaria mercenaria* (Linneaus) during the 24-hour period. *Abstr. Ann. Meeting Geol. Soc. Am. for 1969*, **140**

Marceau, F. (1909). Recherches sur la morphologie, l'histologie, et la physiologie comparées des muscles adducteurs des molluscs acéphales. *Archs Zool. exp. gén.* ser 5, **2**, 295–469

Morton, B. S. (1969). Feeding and digestive rhythms in the mollusca. *Science Chelsea*, **3**, 23–29

Morton, B. S. (1970). The tidal rhythm and rhythm of feeding and digestion in *Cardium edule*. *J. mar. biol. Ass. U.K.*, **50**, 499–512

Morton, B. S. (1971). The diurnal rhythm and tidal rhythm of feeding and digestion in *Ostrea edulis*. *Biol. J. Linn. Soc.*, **3**, 329–342

Newell, E. R. and Northcroft, H. R. (1967). A reinterpretation of the effect of temperature on the metabolism of certain marine invertebrates. *J. Zool., Lond.*, **151**, 277–298

Orton, J. H. (1962). On the rate of growth of *Cardium edule*. Part 1. Experimental observations. *J. mar. biol. Ass. U.K.*, **14**, 239–279

Orton, J. H. (1934). Bionomical studies on *Cardium edule* with special reference to mortality in 1933. *James Johnstone Memorial Volume*, 97–120

Pannella, G. (1972). Paleontological evidence on the Earths rotational history since Early Precambrian. *Astrophysics and Space Science*, **16**, 212–237

Pannella, G. and MacClintock, C. (1968). Biological and environmental rhythms reflected in molluscan shell growth. *J. Palaeontol.*, **42** (5), (suppl.), 64–79

Petersen, G. H. (1958). Notes on the growth and biology of the different *Cardium* species in Danish brackish water areas. *Meddr. Danm. Fisk. -og Havunders*, **2** (2), 31 pp.

Rhoads, D. C. and Pannella, G. (1970). The use of molluscan shell growth patterns in ecology and palaeoecology. *Lethaia*, **3**, 143–161

Russell, R. C. H. and Macmillan, D. H. (1952). *Waves and Tides*, Hutchinson, London 348 pp.

Salanki, J. (1966). Daily activity rhythm of two Mediterranean lamellibranchia (*Pecten jacobaeus* and *Lithopaga lithophaga*) regulated by light-dark period. *Annls. Inst. Biol. Tihany*, **33**, 135–142

Stephen, A. C. (1928). Notes on the biology of *Tellina tenuis* Da Costa. *J. mar. biol. Ass. U.K.*, **15**, 683–702

Stephen, A. C. (1930a). Studies on the Scottish marine fauna. Additional observations on the fauna of the sandy and muddy areas of the tidal zone. *Trans. R. Soc. Edinb.*, **56**, 521–535

Stephen, A. C. (1930b). Notes on the biology of certain lamellibranchs on the Scottish coast. *J. mar. biol. Ass. U.K.*, **17**, 277–300

Stephen, A. C. (1932). Notes on the biology of some lamellibranchs in the Clyde area. *J. mar. biol. Ass. U.K.*, **18**, 51–68

Walton, C. L. (1919). On the shell of *Cardium edule*. *Lanc. Seafish. Rep.* no. 28, 47 pp.

Weymouth, F. W. and Rich, W. H. (1931). Latitude and relative growth in the razor clam, *Siliqua patula*. *J. exp. Biol.*, **8**, 228–249

Wilbur, K. M. (1964). Shell formation and regeneration, pp. 243–282 in *Physiology of Mollusca* (Ed. K. M. Wilbur and C. M. Yonge), Academic Press, New York

Wilbur, K. M. (1972). Shell formation in mollusks, pp. 103–146 in *Chemical Zoology* (Ed. M. Fiorkin and B. T. Scheer), vol. VII *Mollusca*, Academic Press, New York

Winckworth, R. (1921). Gaint race of *Cardium edule* L. *J. Conch., Lond.*, **16**, 157

The great morphological variation of *Cerastoderma* revealed by European workers presents a considerable taxonomic problem. It has been suggested that the variants, of which more than 40 types have been recognized, fall into two species—*C. edule* (L.) and *C. glaucum* (Poiret) (Mars, 1951; Petersen, 1958)—and recent investigations appear to substantiate this. The lines of evidence used include geographical distribution (Russell, 1971), ecology (Boyden and Russell, 1972), shell and soft tissue characters (Boyden, 1971a; Russell, 1972), behavioural and physiological differences (Boyden, 1972; Rygg, 1970), isoenzyme taxonomy (Jelnes *et al.*, 1971) and differences in the reproductive cycle (Boyden, 1971b; Rygg, 1970) and spermatozoan morphology (Rygg, 1970). These papers also include references to earlier works and ideas on this problem.

Boyden, C. R. (1971a). A note on the nomenclature of two European cockles. *J. Linn. Soc. (Zool)*, **50** (3), 307–310

Boyden, C. R. (1971b). A comparative study of the reproductive cycle of *Cerastoderma edule* and *C. glaucum*. *J. mar. biol. Ass. U.K.*, **52**, 661–680

Boyden, C. R. and Russell, P. J. C. (1972). The distribution and habitat range of the brackish water cockle (*Cardium (Cerastoderma) glaucum*) in the British Isles. *J. Anim. Ecol.*, **41**, 719–734

Jelnes, J. E., Petersen, G. H. and Russell, P. J. C. (1971). Isoenzyme taxonomy applied on four species of *Cardium* from Danish and British waters with a short description of the distribution of the species. (Bivalvia.) *Ophelia*, **9** (1), 15–20

Mars, P. (1951). Essai d'interprétation des former généralement groupées sons le nom de *Cardium edule* Linné. *Bull. Mus. Hist. nat. Marseille*, **11**, 1–31

Petersen, G. H. (1958). Notes on the growth, and biology of the different *Cardium* species in the Danish brackish water areas. *Meddr. Danm. Fis, -og Havunders* (Nye serie), **2** (22); 1–31

Russell, P. J. C. (1971). A reappraisal of the geographical distributions of the cockles, *Cardium edule* L. and *C. glaucum* Bouguière. *J. Conch.*, **27**, 225–234

Russell, P. J. C. (1972). A significance in the number of ribs on the shells of two closely related *Cardium* species. *J. Conch.*, **27**, 401–409

Rygg, B. (1970). Studies on *Cerastoderma edule* (L.) and *Cerastoderma glaucum* (Poiret). *Sarsia*, **43**, 65–80

DISCUSSION

WILSON: What is the bearing of your work on the *C. edule–C. glaucum* species problem?

WHYTE: Despite the similar growth patterns which I have reported for the two species, I believe Boyden had good reason for describing them as different species, and further work might show differences in growth patterns.

(Whyte also stated that it might eventually be possible to correlate salinity and temperature variations with specific effects on growth increments.)

A TECHNIQUE FOR THE EXTRACTION OF ENVIRONMENTAL AND GEOPHYSICAL INFORMATION FROM GROWTH RECORDS IN INVERTEBRATES AND STROMATOLITES

J. DOLMAN

School of Physics, University of Newcastle upon Tyne, U.K.

Abstract

Geophysicists have postulated that the Earth–Moon distance and the rotational velocity of the Earth have not remained constant through geological time. During the last decade several palaeontologists have produced direct evidence for this by counting the number of daily or tidal growth increments between fortnightly, monthly or annual series of growth increments in bivalves, branchiopods, corals and stromatolites. The analysis of such data by the traditional biostatistical methods is, however, rarely valid for the purpose of obtaining accurate geophysical periodicities preserved in an animal's growth record. It is therefore the purpose of this paper to outline a more suitable technique of analysis. *Cardium[Cerastoderma]edule* (L.) is used to illustrate this method. An optical microdensitometer and an automatic micrometer-eyepiece attached to a microscope were used for obtaining the raw data.

The growth of many invertebrates and stromatolites is preserved as a series of growth increments. In order to obtain environmental or geophysical information from this it is necessary to construct a time series of measurements of such features as growth increment width or distinctness of the growth lines bounding the increments. To achieve this one must identify features in the growth record that have been deposited at regular intervals in time, e.g. diurnal or tidal markings. This may be performed either by direct observations or from microdensitometer traces. In either case it is necessary to define precisely the criteria for the recognition of such markings. Using these markings the new series may be constructed of the variations in the different growth parameters, e.g. distance between successive markings and the distinctness of the markings.

After subtracting the influence of the animal's ageing on its growth patterns (i.e. growth increments narrow with age in all organisms studied except stromatolites) the remainder of the series is Fourier-analysed to give the 'apparent periodicities' in the growth of the animal. By recombining the low harmonics of the transform it is possible to observe the seasonal and tidal influences affecting the animal, and to see if there have been dramatic changes in the pattern of growth. In order to obtain the 'apparent periodicities' within a habitat it is necessary to measure several specimens and obtain a mean power spectrum. By removing the peaks formed by these periodicities from this spectrum and recombining the remainder of the spectrum, the effect of random weather events may be determined.

Specimens living in intertidal environments above the level of the low high water neaps (L.H.W.N.) may often have discontinuities in their records, which result in the periodicities observed in the specimens having higher frequencies than the actual environmental or geophysical periodicities. To guard against a faulty interpretation of these periodicities it is advisable to obtain the coherences between specimens living at a range of intertidal positions.

In conclusion, this technique offers a more objective determination of the number of days per fortnight and days per month, given suitable fossil material. In addition the method may indicate the following information, (a) the tidal position of a specimen (given specimens from only one site), (b) the range of heights of the tides (given specimens from a selection of known positions), and (c) the phase relationship between the tide and the time of the day.

The growth record of *Cardium edule* (L.)

Acetate peels of specimens of *Cardium[Cerastoderma]edule* (L.) were prepared from specimens, cut and sectioned perpendicularly along the middle rib of the valve (in a similar manner to that outlined by Farrow, 1971). Maximum growth rate in this species occurs along the reflex zone (Figure 1), referred to by Pannella and MacClintock (1968) as the *surface of maximum growth*. In this zone external influences on the growth of the shell are found to be at their greatest. From a series of three experiments conducted near neap tide level over 24-hr periods, at spring tide, midtide and neap tide at Holy Island, Northumberland, England, it was found that growth lines were formed every time an animal was exposed by the tide. These growth lines delimit the boundaries of the semi-diurnal tidal growth increments.

In the peels the growth lines appear as thin dark layers that may vary in width (Figure 1). They are bordered by dark Becke lines that are often of different distinctness. The boundaries of these growth lines are sharp, in contrast to the solar diurnal growth lines found in other species, which are diffuse (see Pannella, this volume).

In stereoscan photographs of etched polished sections, the growth lines formed in this zone consist of a coarse cross-lamellar structure of laths (Taylor *et al.*, 1969) of crystals (about 1 μm in length). In this reflex zone the distinctness of successive growth lines often does not appear to vary greatly. However, when the growth lines are followed from this area towards the pallial line alternate growth lines are usually found to be more distinct than intervening members.

As the animals were subjected to an oscillation of tides showing mainly semi-diurnal components, specimens were uncovered approximately twice daily. From the experiments it was concluded that the growth lines formed at night were those that appeared the more distinct. These growth lines were often twice the thickness and brightness of the intervening members. Many of these growth lines consisted of 2-5 finer lines. The alternate growth lines set between these members were formed during the animals' exposure in the daytime, these growth lines containing no more than three finer lines.

As the successive 'daytime' growth lines are a result of the animal being uncovered at successively later times on successive days, the animal is uncovered first in the morning, then on successive days in the afternoon and in the evening.

Growth Records in Invertebrates and Stromatolites 193

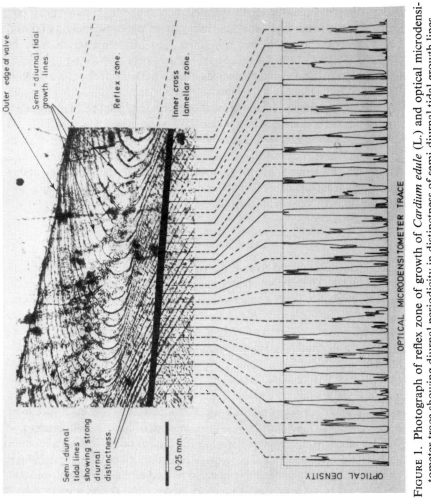

FIGURE 1. Photograph of reflex zone of growth of *Cardium edule* (L.) and optical microdensitometer trace showing diurnal periodicity in distinctness of semi-diurnal tidal growth lines

In this manner there is a change of 'daytime' growth lines to 'nighttime' growth lines every fortnight. The time that it takes for the switch in the distinctness of the growth lines to occur appears to be extremely variable. In some cases, two less distinct growth lines ('daytime' growth lines) may be observed between the more distinct 'nighttime' growth lines (Figure 1). In other specimens the 'daytime' and 'nighttime' growth lines may not be easily discernible from each other during most of the fortnightly tidal cycle. In specimens from the high water neap tide level this switch occurred when the low tide was at approximately 11.00 and 23.30 hr G.M.T. (i.e. the specimens were uncovered between 8.00 hr–14.00 hr and 20.30 hr–02.30 hr G.M.T.). In some specimens there was found to be considerable switching backwards and forwards over 3–5 days before the phase shift was stabilized.

Examination of specimens taken from the Burry Inlet, South Wales (taken from various tidal positions between mean low water neaps (M.L.W.N.) and mean high water neaps (M.H.W.N.)) showed a similar switching over from 'daytime' to 'nighttime' growth lines. In these specimens it occurred approximately 3 days after the neap tide. This switching occurred at the same time of day as that found in specimens from Holy Island. The relevance of these results will be dealt with later in this paper.

Extraction of data

Growth data may be extracted from the growth records in two basic ways.

(1) *Phenomenologically.* This is the counting of the occurrence of one order of feature with respect to another and will be referred to as 'feature-ratio' counting. This type of data collection is the easiest to obtain, as it does not involve data measuring, and the counted number is usually an integer. It has hence been the most popular method of data collection (Barker, 1964; Berry and Barker, 1968; Mazzullo, 1971; Pannella, 1972; Pannella *et al.*, 1968; Pannella and MacClintock, 1968). It suffers, however, because the mathematical analysis that can be applied to this sort of data is usually very limited, as normally few data counts (i.e. less than 100 separate counts) can be taken from each specimen. Often it can also easily be statistically misinterpreted due to modulation and aliasing (these terms being explained later in the paper) occurring in the counts.

(2) *Construction of time series.* The growth record of an observed specimen is spatial in form (i.e. the information is linearly related to space and not to time), but it can be said to be a temporal memory of growth, as the growth rate is proportional to the events to which the animal is subjected. Variations in growth-line thickness and growth-increment thickness (distance between growth lines deposited at equal intervals in time) are therefore frequency-modulated, while variations in growth-line distinctness are both amplitude- and frequency-modulated. Unfortunately, it is difficult to study the periodicities in the growth record of the spatial series. It is therefore necessary to construct a time series.

In order for this to be achieved, equal-time markers in the growth record must be subjectively identified by the observer. If the growth lines can be assumed to be deposited at approximately equal intervals in time, measurements of the increment thickness or the growth-line distinctness may be plotted as successive ordinals against the increment thickness or the growth-line distinctness (Clark, 1968; Farrow, 1971, 1972; House and Farrow, 1968).

Although considerable labour is usually required in measuring the lines and increments, the main advantage of this method is that it leaves itself open to a more stringent analysis than 'feature-ratio' counting.

Because of the great statistical advantages in the use of increment measurement and growth-line distinctnesses, these methods were chosen for extracting the main growth record data.

Growth-line distinctness was measured from thin-sections and micrographs by the use of an optical transmission microdensitometer (automatic recording microdensitometer model IIIb made by Joyce Loebl and Co.). Measurements of the thickness of the increments were made by a digitized micrometer eyepiece attached to an ordinary microscope (Figure 2). The instrument was constructed as an inexpensive method of taking accurate rapid increment measurements of

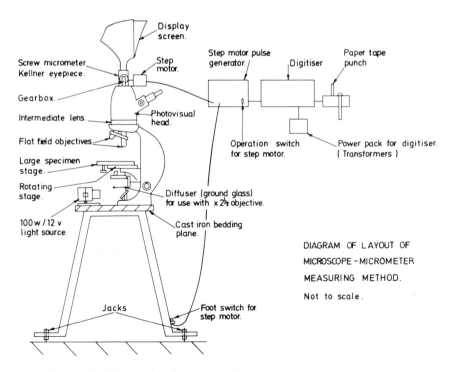

FIGURE 2. Diagram of microscope-micrometer measuring instruments

growth-line spacings. It consists of a pulse generator connected to an 8-pole step-motor. The step-motor drives (via a 10:1 reduction gear-box) the cross-wires across the field of view in the micrometer eyepiece. All three of these components are mounted on the same eyepiece stage. Approximately 6000 pulses are required to move the cross-wires across the complete field of view. The pulses generated were counted in a standard 4-figure digital display unit, and could be recorded on a 8-track paper-punch tape when the cross-wire position in the eyepiece was at some ordinate to be measured.

In the *Cardium edule* (L.) specimens studied from Holy Island and the Blyth Estuary, Northumberland and the Burry Inlet, South Wales, growth lines were deposited at semi-diurnal tidal intervals, during most of the summer months. In the winter, growth lines appeared to be deposited at diurnal tidal intervals, but this might be due to lack of optical resolution when studying the specimens. Diurnal tidal growth lines were also found in specimens that were not uncovered or subjected to disturbance at low tides (Evans, 1972). When there is a mixture of semi-diurnal and diurnal tidal growth lines or semi-diurnal tidal 'daytime' growth lines are only faint it is preferable to construct a time series against the animal's diurnal rate of growth. This can be achieved with the *Cardium edule* increment thickness measurements by summing successive pairs of semi-diurnal tidal increment measurements, and adding to the odd semi-diurnal tidal interval the sum of half the increments to either side. The resulting time series may then be treated as if the new increments are diurnal.

Present methods of analysis

Estimations of the length of a fortnightly tidal cycle, the month and the year have all been determined by demodulating the growth record by eye (e.g. the occurrence of one order of growth feature between a second order of growth features), using feature-ratio counts of either the original growth record (Pannella, 1972) or growth increment plottings (Farrow, 1972).

The estimates of periodicities have been based on the use of basic statistics using:

(i) the arithmetic mean in corals (Wells, 1963), bivalves and a cephalopod (Pannella, 1972 and Pannella and MacClintock, 1968);
(ii) the mode in corals (Scrutton, 1970);
(iii) the maximum of the groupings in corals and brachiopods (Mazzullo, 1971) and bivalves (Clark, 1968).

Errors have been estimated by the use of the standard error (standard deviation of the mean) and the standard deviation (Pannella, 1972; Pannella and MacClintock, 1968; Scrutton, 1970).

At first sight it might appear that these techniques were adequate. For this reason it is best to understand exactly what each of the above estimates means.

Arithmetic mean $\bar{X} = \left(\dfrac{\Sigma x}{n}\right)$ (where x is a value in the data sequence of n values)

This as an estimate of the values expected (i.e. fortnightly, monthly or annual) assumes that each feature-ratio count has an equal chance of being higher or lower than the actual value. Therefore the integrated areas lying to both sides of the arithmetic mean value on a histogram should be equal. The data must satisfy the following requirements.

(1) There is no trend in the values produced from successive feature-ratio counts, e.g. decreasing with age.

(2) In each case the same feature types are being equated in the counts, e.g. only semi-diurnal tidal increments or diurnal increments are measured, but not both at the same time.

(3) Only one periodicity exists in the grouping counts of feature ratios, e.g. fortnightly or monthly, but not both.

(4) Frequency modulation does not occur, i.e. the periodicity being counted does not change frequency with time in a rhythmic manner.

(5) Aliasing is not present. (The phenomenon of periodicities produced by aliasing is discussed later in this paper.)

(6) There is no systematic error in the counts, i.e. missing increments do not occur in the growth record.

In estimates of arithmetic mean counts of growth lines in modern bivalves the calculated period for the fortnightly, monthly or annual pattern has always been different from the geophysical rhythms causing the rhythmicity in the growth of the animal. Four factors appear to be mainly responsible for this.

(1) During ageing of the animal, it does not always grow at all stages of the rhythm, which causes an increase in the apparent frequency of the rhythm.

(2) More than one frequency is counted. A count of 21 days might be used to lower the mean value for the number of days/month or might be used to increase the number of days/fortnight.

(3) Systematic errors occur due to occasional missing lines. This is especially found in intertidal bivalves that live at high latitudes.

(4) When growth lines are formed in animals (e.g. bivalves) for different reasons within the same specimen, counts may contain more than one type of growth increment (e.g. growth lines may sometimes be formed at intervals of a lunar day, while a short time later they may form at semi-lunar daily intervals (Evans, 1972).

Estimations of distribution of counts

The standard deviation (σ) as an estimate

$$\left(\sqrt{\dfrac{(x-\bar{x})^2}{n}}\right)$$

assumes that there is an approximately Gaussian (normal) distribution of values with no skew, if tests are to be applied to the estimate. Since the shape of the distribution of feature-ratio counts should not vary with different numbers of counts from the same specimen, the standard deviation can be useful for comparing counts made from different parts of the same specimen. This has not been done in published papers and therefore the author would suggest that this type of data may be tested by the use of the F test (Moore and Edwards, 1965), i.e.

$$\text{where } F = \frac{(\sigma \text{ of 1st half of the count values for a specimen})^2}{(\sigma \text{ of 2nd half of the count values for a specimen})^2}$$

In a similar manner the arithmetic mean could be tested to see if there are significant differences in the counts at differing parts of the records, by the use of the t-test (Moore and Edwards, 1965). That is,

$$\text{where } t = \frac{\bar{x}_1 - \bar{x}_2}{\sqrt{\left(\frac{(n_1 - 1)\sigma_1^2 + (n_2 - 1)\sigma_2^2}{n_1 + n_2 + 2}\right)}\sqrt{\left(\frac{1}{n_1} + \frac{1}{n_2}\right)}}$$

and

$\bar{x}_1, \bar{x}_2$ are the arithmetic mean of two sets of counts at different parts of the same specimen;

n_1, n_2 are the number of counts in each set;

σ_1, σ_2 is the standard deviation of each set of counts.

Reference to tables (e.g. Pearson and Hartley, 1954) would then given an indication of whether there were significant differences.

The standard error (S.E.)

$$\left(\sqrt{\frac{\Sigma(x - \bar{x})^2}{n(n - 1)}}\right)$$

is a function distribution (given by the standard deviation) and the number of counts that were made. Therefore as long as each individual count is governed by the same rules, the S.E. decreases with increase of the number of counts made.

When looking at the published results of Pannella (1972), this does not appear to occur. The S.E. remains nearly constant with increasing counts, the former being compensated for by the increasing of the standard deviation (Figure 3). It is clear in this case that counting must be affected in varying ways in different parts of the growth record. The reason for this is, however, lost in all published data, because the feature-ratio counts have not been individually shown as an ordinal series (i.e. the original successive individual counts have not been listed).

In addition, histograms of feature-ratio counts usually show a marked right-hand skew in the distribution (Pannella, 1972), with often systematically lower count values than those expected (Farrow, 1972).

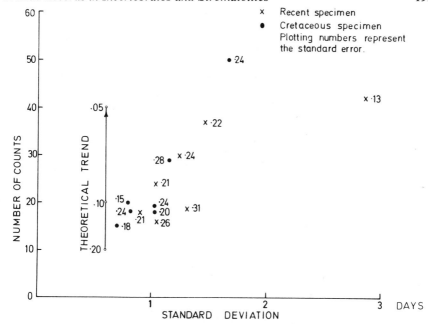

FIGURE 3. Data extracted from Pannella (1972) showing increase in standard deviation with increasing counts

Mode

Because of the skewing of the data, Scrutton (1970) suggested that the mode (the most frequent value counted) lies nearer the actual value than the arithmetic mean. For a moderately skewed distribution the mode may be related as below:

mode ≃ arithmetic mean −3 (arithmetic mean − median)

where the median is the value midway between the highest and lowest value obtained.

In the early stages of growth of an animal the trend in the feature-ratio counts is usually small, but as the animal becomes older there is often an approximately exponential decrease in the values of the counts. For this reason the mode of the counts made in a specimen is likely to be nearer to the arithmetic mean of the counts taken only from the early stages of growth than to the arithmetic mean of all the counts. The counting of occasionally different frequencies to that being studied (e.g. counting a monthly period of 25 days and grouping it with fortnightly counts) is unlikely to effect the mode since the relative number of occurrences of such counts will be low. However, using either the arithmetic mean or the mode there is the possibility that the pattern being counted is a beat frequency.

In most instances, though, Scrutton's observation would appear true, except in stromatolite counts (Pannella, 1972). Unfortunately the mode lends itself to little if any statistical scrutiny, i.e. there is no error estimate.

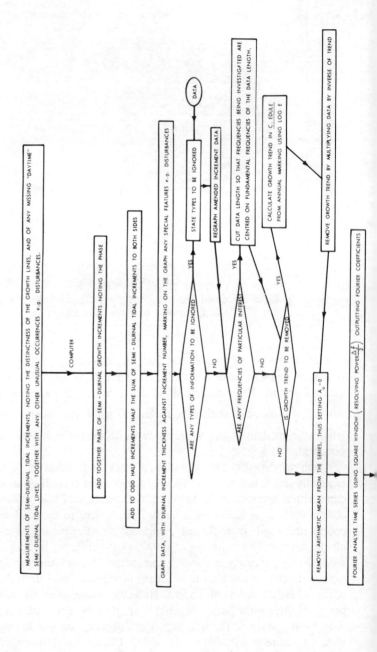

Growth Records in Invertebrates and Stromatolites

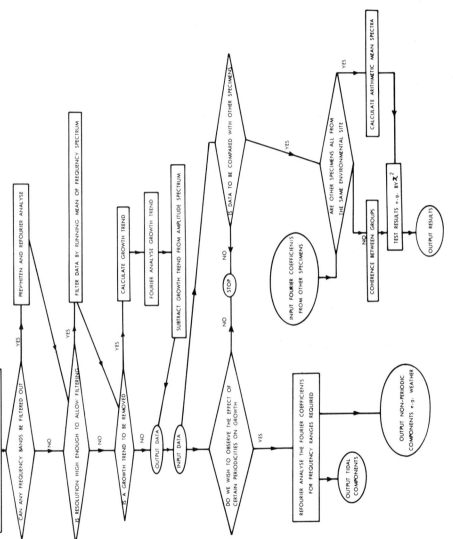

TABLE 1. Flow diagram outlining the method of analysis of increment thickness measurements.

Maximum number counted

Mazzullo (1971) believed that maximum counts in groupings produced truer values than the averages used above by other authors. Unfortunately this is not true, as random fluctuations occur in the exact timing of the deposition of the features (e.g. due to weather). More importantly, the 'maximum count' is more sensitive to the occasional incorrect identification of a higher order feature as being of the same order of feature as the one being studied (e.g. a semi-diurnal tidal feature being identified as a diurnal tidal feature). For these reasons the maximum count is of little use and should therefore be avoided.

Other statistical estimates (e.g. median, geometric mean or harmonic mean) require each data count to be independent and unrelated to the counts on each side. This is clearly not the case with growth counts.

The type of analysis used until now is clearly suitable only for the analysis of a simple regular cyclic pattern of growth lines. It is therefore necessary to have a more flexible method of analysis which analyses the growth pattern in greater depth. These requirements have been studied and developed in the fields of communications theory and geophysics (Barber, 1966). Fourier analysis has for many years formed the centre of such analysis and the following sections are based upon its application in growth record analysis.

Fourier analysis

A summary of the method of analysis is seen in the flow-diagram (Table 1).

Analysis of periodicities in records was first undertaken by Lagrange (1772) and the original mathematical concepts used today follow from Fourier (1825), Scheister (1898) and Stokes (1879). During the first half of this century more rapid methods of analysis were developed, such as autocorrelation (Blackman and Tukey, 1958). However, until the advent of digital computers even the most rapid methods of calculation often took many hours of tedious computation. This situation changed with the development of algorithm computer programs to do first autocorrelation and, more recently, Fourier analysis.

A Fourier transform of a discrete time series or data sequence analyses the input data into a set of frequency components so that values of the sequence can be described as

$$\aleph_t = \sum_{k=0}^{n/2} a_k \cos(2\pi kt/((n-1)\Delta t)) + \sum_{k=1}^{n/2-1} b_k \sin(2\pi kt/((n-1)\Delta t))$$

where a_k and b_k are the cosine and sine Fourier coefficients (Zygmund, 1935);

n is the number of data values;

$\Delta t = 1$ (interval in time between successive data values);

and $t = 0, 1, 2, 3, \ldots, n-1$ (ordinal number of successive data values).

The a_k and b_k coefficients were obtained from results of using a Fast Fourier Transform subroutine (commonly known as FFT) (Cooley and Tukey, 1965) written by Singleton (1968).

Harmonic (k)	1	2	3	4	5	6	7	8	9	10	11	12	13	14	15
Frequency (f)	.005	.010	.015	.020	.025	.030	.035	.040	.045	.050	.055	.060	.065	.070	.075
Period (1/f)	200	100	66	50	40	33.3	28.6	25	22.2	20	18.2	16.6	15.3	14.3	13.3 days
							Monthly				Fortnightly				

TABLE 2. Diagram showing the decreasing resolution of a period with increase in its length as estimated in a single Fourier transform

From the coefficients can be obtained
(1) The amplitude (A) for a harmonic k of the number of data values analysed in the transform

$$A_k = \sqrt{a_k^2 + b_k^2}$$

Sometimes Power (amplitude^2) is used in preference to amplitude
(2) Phase of harmonic k of the number of data values analysed

$$\theta_k = -\tan^{-1}\left(\frac{b_k}{a_k}\right)$$

The estimations of periods are assessed in a single transform at harmonics (k) of the original length of the data sequence. Normally rhythms in data sequences are studied in relationship of frequency (k/n) as the estimations are evenly spaced ($\Delta f = 1/n\Delta t$). In growth line studies, though, we wish to know the period ($n\Delta t/k$) (i.e. the number of days it takes for the pattern to repeat itself). It will be seen that if the period is long, the spacing between successive estimations can be undesirably coarse, e.g. the uninterrupted record length for *Cardium* data was usually 6–8 months (approximately 200 'daily' increments). This resulted in the assessment of periods of the following lengths, when only using a single Fourier transform as seen in Table 2.

It is clear from the above that for short records the resolution of the determination of the monthly and fortnightly periods is nowhere near that required for geophysical purposes.

The subroutine FFT obtains estimations at frequency intervals of Δf using a square data window, which produces a frequency window with steep sides and sizable side lobes (Figure 4). However, estimates of power taken at intervals of $\Delta f/2$ would have approximately half the total power of successive frequency estimates independent of adjacent estimates (Figure 4), but if line spectra occur at the frequency estimations, leakage into adjacent frequencies will not occur. Estimations of power at frequency intervals of less than $\Delta f/2$ would clearly be of no use, because of the leakage of power from adjacent frequencies. Estimation of power at intervals $\Delta f/2$ may be obtained by finding the period in a mean data sequence that is cut so that the periods of this new data length approximately coincide with the periods obtained from $\Delta f/2$ for which one is interested, in the original length data sequence. For example, if the original data sequence is 192 days long, the 6th harmonic corresponds to 32 days, the 7th harmonic to 27·4 days, the 13th harmonic to 14·8 days and the 14th harmonic to 13·7 days. It may thus be desirable to estimate the power at harmonics of 6·5 (corresponding to a period of 29·5 days) and of 13·5 (corresponding to 14·2 days). If, therefore, the data sequence is shortened to 178 days the 13th harmonic of this new series corresponds to a period of 14·2 days and the new 6th harmonic corresponds to a

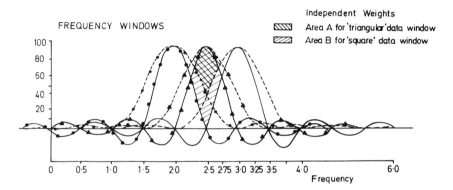

FIGURE 4. Diagram illustrating the degree of independence of estimations of power at frequency intervals of $\Delta f/2$ for 'square' and 'triangular' data windows

period of 29·7 days. It should be noted that, if a frequency window is plotted for a shorter record length on the frequency window of the original length of data, the window for the shortened record length has a broader and lower profile than that of the original length of data. It is therefore important that the shortened data length should be as close as possible to that of the original length of data, and that the estimations of the power of harmonics from the shortened record length should be increased in proportion to the shortening of the data length.

It will be noted (Figure 4) that considerable leakage may occur, if the power of one frequency far exceeds those of adjacent frequencies. In growth increment studies, though, it has been found that the amplitude of the prominent frequency, such as the fortnightly or monthly, rarely exceeds by ten times that of the amplitude of the background and therefore such leakage is minimal. The zero harmonic (k_0), whose power is an estimation of the mean, has no real value and therefore its power may be set to zero by subtracting the arithmetic mean value from all data values in the data sequence.

As seen in Figure 4 the value at point Δf away from the centre of the peak frequency is for both square and triangular data windows equal to zero. For the square window the value is also zero at $\Delta f/2$. Therefore if a data series contains only one periodicity and this is exactly equal to a frequency of the Fourier transform, the values at all other estimated frequencies will be near zero. (If the wave form is sinusoidal in shape all frequency estimates will be zero, except at the fundamental frequency of the periodicity.) For this reason it is preferable to have the length of the record cut so that a periodicity being studied is a multiple of this length. As the subroutine FFT required that the length of data has all its prime numbers below 20, the length of data (n) for studying tidal affects in modern bivalves should be 297, 250, 221, 192 or 105 days. Since continuous data lengths were normally approximately 200 days, these records were cut to 192 days.

Analysis of *Cardium edule* (L.)

The increment thickness between growth lines has been found to result as a sum of the following parameters.

(1) *Trends* in growth due to the animal ageing
   (a) in the running mean;
   (b) in the amplitude of periodicities;
   (c) in the interrelationship of the amplitudes of the various periodicities (e.g. the amplitude of the fortnightly periodicity varying in relationship to the amplitude of annual periodicity with the ageing of the organism);
   (d) in the frequency, due to the number of missing increments increasing with age.

(2) Periodicities (including aliasing) may have both amplitude and frequency modulation. In different stages of life the periodicities may change.

(3) Non-periodic events, e.g. changes in weather such as storms.

(4) Noise, e.g. error in measurements.

Trends in growth

This appears to be of no geophysical use. The growth trends in the running mean and in amplitude are subtracted from the time series by multiplying the latter by the reciprocal of a fitted growth trend obtained from macromeasurements of the shell growth between annual winter checks. For geophysical purposes nothing can be done to remove the trend of increasing frequency with ageing due to the increasing number of days the animal does not grow. It is therefore important that the section of record to be analysed does not show this trend. It was found in *Cardium* that specimens did not grow for progressively longer periods of time each winter. From high tidal localities in the Burry Inlet, South Wales, specimens stopped growing for approximately 4–6 weeks, at intermittent times (see Figure 10) during their winter (specimens from low tidal positions showed approximately a 1-week loss of increments during the winter). This gradually increased to approximately a 4-month stop in growth in the 5th winter. For this reason, and because the determination of 'what is an increment' in the winter is very subjective, only summer growth increments, approximately between April and November, could be used for geophysical purposes. The size of the growth increments in the 2nd and 3rd summers of growth (i.e. when the animal is 1- and 2-years-old respectively), usually ranged from 10–150 μm, making their measurement preferable to those of successive summers of growth when the growth increments rarely exceeded 20 μm, and were often only 2–3 μm thick.

Periodicities in growth increment thickness

If the time series of diurnal growth rate is Fourier-analysed and a periodogram plotted (Figure 5) a series of 'apparent' peaks are observed. These peaks may occur

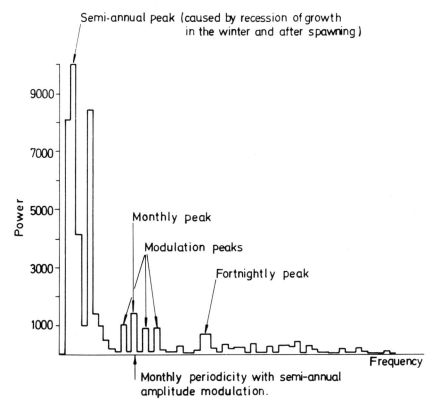

FIGURE 5. Power spectrum of specimens from site SW4, Llanrhidian Sands, Burry Inlet, South Wales

for a variety of reasons, and therefore require careful scrutiny before one deliberates on the periodicities present in the time series. 'Apparent' peaks may be due to the following reasons.

(1) A *periodicity* present in the time series that has the particular frequency indicated by the periodogram. This might or might not have occurred due to a random set of fluctuations in the growth.

(2) *Harmonics* of the periodicity. If the fundamental real periodicity does not have a sinusoidal waveform (as in *Cardium*), peaks will occur at the harmonics

of the fundamental frequency at which the periodicity occurs, e.g. if there is a periodicity of a month, harmonics will occur at a fortnight, approximately 10 days, a week, etc. The amplitude of the harmonics (if the periodicity is centred on a frequency estimation), will depend on the shape of the waveform. If the waveform is nearly sinusoidal in shape in the time series, the harmonics in the periodogram will have low amplitudes, but if in the time series the wavelengths are other shapes, the harmonics in the periodogram may be of a similar amplitude to the fundamental frequency. In *Cardium edule* the harmonics were found to be less than half the amplitude of the fundamental.

(3) *Leakage* from nearby frequencies. As has already been discussed, leakage can occur due to one periodicity swamping the frequency estimations to each side of it. Although this caused no problems in *Cardium* data, strong periodicities in growth may significantly affect the frequency bands to either side of this, when analysing some growth data.

(4) *Amplitude modulation.* The animal's responses to the tidal fluctuations change during the year. Similarly the tides vary in their amplitude. These two factors result in the animal showing an oscillation in the amplitude of the monthly periodicity.

That is: if $\alpha \sin(2\pi\theta t + \phi)$ is the monthly periodicity
and $\beta \cos(2\pi\theta t)$ is the equinoctial or annual periodicity
Then $A = \alpha(K + \beta \cos 2\pi\mu t) \sin(2\pi\theta t + \phi)$
where K is a constant
θ and μ are integer frequencies
α and β are the amplitude for the separate periodicities
t = time
ϕ = phase

Hence $A = \alpha \sin(2\pi\theta t + \phi) + \dfrac{\alpha\beta}{2}\sin(2\pi(\theta + \mu)t + \phi) + \dfrac{\alpha\beta}{2}\sin(2\pi(\theta - \mu)t + \phi)$

Therefore the affect of the modulation on the original $\alpha \sin(2\pi\theta t + \phi)$ periodicity is to produce a pair of components (called sidebands) with amplitudes of $\alpha\beta/2$ and differing in frequency from the original periodicity by plus and minus the modulating frequency. The total power due to these frequencies is therefore

$$\alpha^2 + 2\left(\frac{\alpha\beta}{2}\right)^2 = \alpha^2\left(1 + \frac{\beta^2}{2}\right)$$

therefore the power in the sidebands due to the modulation is

$$\frac{\alpha^2 \beta^2}{2}$$

(5) *Aliasing.* If the sampling interval is greater than half the period of the signal, periodicities will appear in the periodogram that are a function of the difference between the periodicity and the sampling interval (as illustrated in Figure 6).

Growth Records in Invertebrates and Stromatolites

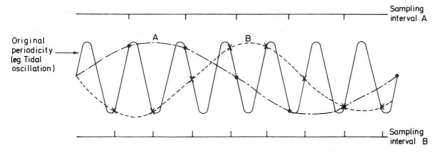

If sampling or measurements occur at periods at greater intervals than twice the period being studied, aliasing will be found to occur.

Pattern A is an aliasing wave form produced if sampling interval is longer than the wavelength being studied.

Pattern B is that produced when the sampling interval is slightly shorter than the wavelength being stretched.

FIGURE 6. Illustration of aliasing

Examples of this have not been found in *Cardium edule* when analysing a time series of estimated diurnal growth increments. It may be found when looking at a time series of semi-diurnal tidal increments or at a time series of semi-diurnal tidal growth-line distinctness. In both cases there appears to be a strong fortnightly periodicity, but this is caused mainly in the latter, and partly in the former, by aliasing of a diurnal rhythm by the semi-diurnal tidal estimations. In the former it is because the animal finds that it is more suitable to grow at a particular time of the day, and in the latter the growth-line distinctness is possibly determined either by the light, which has entrained the animal to a diurnal rhythm, or by the fact that the animal possesses a circadium rhythm. Therefore another reason for analysing a growth record of estimated diurnal increments rather than semi-diurnal tidal increments is to separate from the time series aliasing frequencies that will affect the estimations of amplitude of true fortnightly and monthly periodicities.

(6) *Missing increments.* If missing increments occur, the interactions of frequencies due to phase changes can result in spurious peaks. The missing increments may be classified into three groups.

(a) *Missing increments that cause frequency modulation.* This is when the number of missing increments fluctuates with a periodicity (i.e. the instantaneous frequency varies with the modulating frequency). In *Cardium edule* it was found that the frequency modulation had two frequencies, annual (due to the changes in weather, e.g. temperature and the occurrence of storms) and equinoctial (due to the modulation in the number of days that specimens near high water neaps were not covered by the tide each fortnight). Unlike a continuous frequency modulation which would add increments in parts of the data series and subtract them in other parts of the series, the growth record can only subtract growth increments. This therefore results in the overall length of the data being shortened by the frequency modulation, and therefore the periodicity in the periodogram tends to the arithmetic mean periodicity resulting from the modulation. This frequency represented by the mean periodicity may not always be of as great an amplitude as the modulation peaks in the periodogram resulting from the modulations. Since I have not been able to estimate the frequency modulation from this periodogram, I have had to take the Fourier transform of the modulation. This may be done simply by counting the number of increments between the neap tides and

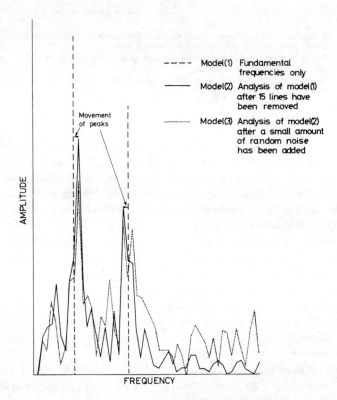

FIGURE 7. Effect of removing increments at random from a data sequence. Frequency analysis of models of 100 increments in length and having fundamental periods of 12·5 and 5·0 increments

then taking a Fourier transform in the normal manner. The frequency modulation of the original time series is then given by the periodicities in this transform.

(b) *Systematically occurring missing increments.* When an animal lives above high high water neaps (H.H.W.N.) missing increments will occur every fortnight during the times of low tidal fluctuation. The periodogram for such a specimen will therefore show the fortnightly and monthly periodicities to be less than

Growth Records in Invertebrates and Stromatolites 211

14·7 days, but will not tend to produce additional peaks beyond those normally formed in an animal not losing any increments. As these missing lines are systematic, their number may not be estimated unless they are compared with specimens from a tidal locality below H.H.W.N.

Normally specimens above H.H.W.N. show a combination of systematically occurring and frequency modulated missing lines, since the shape of the high tides changes as already described. It has been found in specimens from Holy Island, Northumberland, where specimens approximately 10 cm above H.H.W.N. systematically lost 2 semi-diurnal tidal increments per fortnight and lost up to 15 further semi-diurnal tidal increments per fortnight due to the frequency modulation.

(c) *Random missing increments.* This effect only slowly changes the frequency, either to display single peaks in the periodogram or to split them (Figure 7). In a series, originally consisting of a line spectra, and from which random increments have been removed, the position of the missing increments may easily be estimated. If, however, a 10% random noise is added to the series (Figure 7, model (3)), the harmonics resulting from the missing lines are swamped, and it becomes impossible to estimate with any degree of certainty the position or number of these missing lines. Therefore for geophysical purposes there is little use in inserting 'estimated' positions of random missing increments, such as might occur in specimens in the winter due to storms.

Results of geophysical significance

Arithmetic mean spectra and coherence

From the above we can see that it is both difficult and dangerous to analyse the growth record of one specimen out of context and to make conclusions about the tides. For this reason a series of specimens from the same site should be analysed, and the arithmetic mean spectrum for the site obtained, by which random peaks in the periodogram are eliminated, and the degrees of freedom in χ^2 increased (Blackman and Tukey, 1958). The stability of the periodicities may be found, by comparing sets of specimens from different tidal positions through finding the degree of coherence.

Sets of 12 1-year-old specimens taken from a series of six sites along transect F, on the Llanrhidian Sands, the Burry Inlet, South Wales, described by Hancock and Urquahart (1966) show that the monthly and fortnightly periods in specimens from below low high water neaps (L.H.W.N.) are coherent, with periods of 26·5–28·4 days and 14·2–14·8 days respectively (Figure 8). At site SW5 (near mean high water neaps (M.H.W.N.)) the apparent fortnightly and monthly periodicities are found to shift upwards in frequency by $\Delta f/2$, while at site SW6 (near H.H.W.N.) both the tidal periodicities shift by $3\Delta f/2$. The reason for these shifts in frequency at high tidal positions is due to the time series being discontinuous,

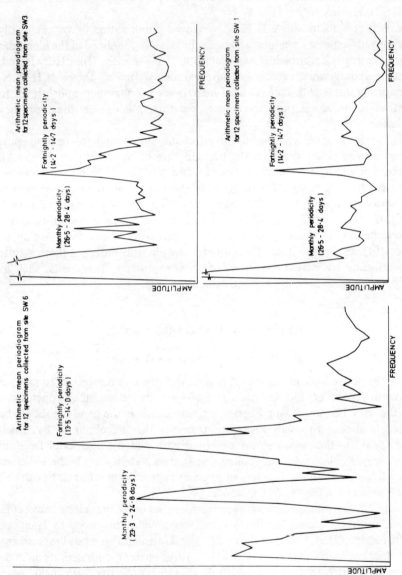

FIGURE 8. Mean Arithmetic Fourier Transforms of specimens collected from tidal positions approximately along transect F (Hancock and Urquhart, 1966) Llanrhidian Sands, Burry Inlet, South Wales

i.e. the animal not always depositing a growth line with each tide, due to the animal not being covered by all tides. The proportional shifts in frequency of the apparent monthly periodicity at sites SW5 and SW6 are greater than that of the apparent fortnightly periodicity, so that the shifting in the latter moves its 1st harmonic progressively away from the frequency of the apparent fortnightly

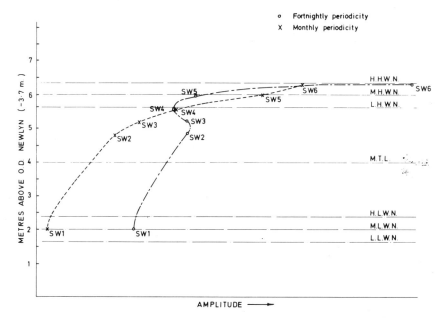

FIGURE 9. Diagram showing Arithmetic Mean amplitudes of fortnightly and monthly periodicities found in *Cardium edule* (L.) from different tidal positions. Section at approximately transect F (Hancock and Urquhart, 1966), Llanrhidian Sands, Burry Inlet, South Wales

periodicity. Therefore it would appear that the missing increments are distributed in a rhythmical manner that effects the monthly periodicity to a greater extent than that of the fortnightly. The cause of the fortnightly periodicity is normally found to have a greater amplitude than that of the monthly. Although part of the amplitude of the fortnightly periodicity is a result of the 1st harmonic of the monthly frequency, the true fortnightly periodicity still has a greater amplitude than the monthly, between M.L.H.N. and approximately 1 m below L.H.W.N. and also near H.H.W.N. The reflex in the amplitude of the fortnightly periodicity, that results in the monthly periodicity being stronger than the fortnightly between approximately L.H.W.N. and M.H.W.N., appears to be due to the affect of the alternate neap tides being higher than the intervening ones.

Specimens sampled below H.H.W.N. at Holy Island and from the Blyth Estuary show a similar pattern of periodicities to those specimens from the

LOCATION	SITE No.	TIDAL POSITION (At Local Chart Datum) (L.C.D.)	SUBSTRATE	MEAN SPECTRA Of	GEOPHYSICAL PERIODS Wavelength in Terrestrial Days Amplitude As Peak To Local Mean Background Ratio (In Brackets)			Ratio of Peak Amp. of Fortnightly : Monthly Period	Variations in Peaks of Geophysical Periods (See Key below)
Llanrhidian Sands, Burry Inlet, South Wales	SW 1	2.0 ± 0.1m L.C.D. (Approx M.L.W.N.)	Fine Muddy Sand	12 Specimens	Fortnightly	14.2 - 14.8	Days (12 : 4)	116 : 50	
					Monthly	26.5 - 28.4	Days (5 : 4)		
	SW 2	4.8 ± 0.1m L.C.D.	Medium Grain Sand	12 Specimens	Fortnightly	14.2 - 14.8	Days (15 : 5)	152 : 91	
					Monthly	26.5 - 28.4	Days (9 : 6)		
	SW 3	5.2 ± 0.1m L.C.D.	Medium Grain Sand	12 Specimens	Fortnightly	14.2 - 14.8	Days (16 : 6)	156 : 120	
					Monthly	26.5 - 28.4	Days (12 : 7)		
	SW 4	5.6 ± 0.1m L.C.D. (Just Below L.H.W.N.)	Fine Grain Sand	12 Specimens	Fortnightly	14.2 - 14.8	Days (15 : 6)	146 : 50	
					Monthly	26.5 - 28.4	Days (15 : 10)		
	SW 5	6.0 ± 0.1m L.C.D. (Approx M.H.W.N.)	Fine Grain Sand	12 Specimens	Fortnightly	13.9 - 14.2	Days (16 : 7)	163 : 14	
					Monthly	25.4 - 26.5	Days (21 : 12)		
	SW 6	6.3 ± 0.1m L.C.D. (Just Below H.H.W.N.)	Fine Grain Sand	12 Specimens	Fortnightly	13.5 - 14.0	Days (33 : 7)	332 : 249	
					Monthly	23.3 - 24.8	Days (25 : 6)		
Blyth Estuary, Northumberland	BE 1-2	1.6 ± 0.1m L.C.D.	Fine Estuarine Mud	6 Specimens	Fortnightly	14.2 - 14.8	Days (7 : 4)	67 : -	
					Monthly	Not identified			
	BE 5	1.9 ± 0.05m L.C.D.	Fine Estuarine Mud	6 Specimens	Fortnightly	14.2 - 14.8	Days (8 : 4)	75 : -	
					Monthly	Not identified			
	BE 14-16	2.2 ± 0.05m L.C.D.	Fine Estuarine Mud	6 Specimens	Fortnightly	14.2 - 14.8	Days (9 : 3)	86 : 35	
					Monthly	26.5 - 28.4	Days (3.5 : 2)		

Location	Tidal Level	Sample	Period	Range	Days (ratio)	Ratio	Pattern
HL 1	4.24 ± 0.01m L.C.D. (Above H.H.W.N.)	12 Specimens	Fortnightly	7.9 – 8.7	Days (12 : 8)	125 : 168	→ ←
			Monthly	15.3 – 16.6	Days (16 : 10)		
HL 3	4.00 ± 0.01m L.C.D. (Above H.H.W.N.)	12 Specimens	Fortnightly	11.3 – 12.0	Days (17 : 10)	173 : 237	–·– → ← –·–
			Monthly	21.3 – 22.6	Days (23 : 11)		
HL 5	3.88 ± 0.01m L.C.D. (Approx. H.H.W.N.)	12 Specimens	Fortnightly	13.2 – 13.7	Days (38 : 10)	381 : 185	– – – –
			Monthly	23.3 – 24.8	Days (18 : 12)		
HL 8	3.80 ± 0.01m L.C.D (Approx. Midway between L.H.W.N. and M.H.W.N.)	12 Specimens	Fortnightly	13.7 – 14.2	Days (14 : 6)	142 : 133	·········→
			Monthly	25.6 – 27.4	Days (15 : 7)		
HL 11	3.44 ± 0.01m L.C.D. (Approx. 0.2m below L.H.W.N.)	12 Specimens	Fortnightly	14.2 – 14.8	Days (14 : 5)	147 : 108	
			Monthly	26.5 – 28.4	Days (10 : 6)		
HL 11A	3.10 ± 0.01m L.C.D.	12 Specimens	Fortnightly	14.2 – 14.8	Days (12 : 6)	125 : 82	
			Monthly	26.5 – 28.4	Days (8 : 6)		

Snook House, Holy Island, Northumberland.

Key to Variations in Peaks of Geophysical Periods

Monthly Period of Greater Amp. than Fortnightly Period	Increase in Amp. of Fortnightly Period	Increase in Amp. of Monthly Period
– – – –	–·– →	→

Increase in Frequency Modulation ·········→

TABLE 3. Summary of apparent periodicities found in *Cardium edule* (L.)

Llanrhidian Sands, as seen in Table 3. On the Llanrhidian Sands specimens are not found above H.H.W.N., but at Holy Island specimens may be found to live at approximately 0·35 m above this level. In these specimens it is found that the amplitude of the fortnightly periodicity immediately decreases with increase in tidal position. There is a slower response in the monthly periodicity, which increases in amplitude up to 0·2 m above H.H.W.N., after which it decreases again. The result is that the monthly periodicity gains a greater amplitude than the fortnightly periodicity. The cause of this would appear to be the severe frequency modulation that occurs.

Through most of the middle intertidal range (i.e. between M.L.W.N. and L.H.W.N.) in specimens from all three sites the apparent fortnightly periodicity (14·2–14·8 days) and the apparent monthly periodicity (26·5–28·4 days) are coherent with the predicted tide. The apparent fortnightly periodicity therefore represents the true synodic fortnightly tidal pattern produced mainly by the beat frequency of the semi-diurnal tidal components, and to a smaller degree by the shallow water tidal component *MSf* (resulting from $S_2 - M_2$). The tidal component *Mf* and the beat frequency produced from the M_2 and K_2 tidal components both have a period of a tropical fortnight and therefore this would probably have led to the spread in the peak at the synodic fortnight.

The apparent monthly periodicity would appear to represent the anomalistic month resulting from the tidal component *Mm* and the semi-diurnal beat frequency produced between M_2 and N_2 tidal components.

It is important to note that it may be expected that specimens subjected to a diurnal tidal pattern will show a prominent tropical fortnightly periodicity (resulting from a beat frequency between K_1 and O_1 diurnal tidal components) rather than a synodic fortnightly periodicity.

Recombining time series

Time series are often smoothed by the use of a running mean in order that noise will be levelled out. When looking at a time series of growth increments, we may wish to see visually the amount of daily growth that results from certain frequency bands, or variations in growth that result from other than tidal or annual periodicities. The ease of recombining certain frequency ranges, or post-whitening or filtering the Fourier transform, and then reforming the time series makes this method a valuable tool.

(1) *Recombining Fourier coefficients to look at the effect of tidal periodicities*

If a certain frequency range over a series of tidal frequencies is recombined, a series of approximately sinusoidal curves might well result, when the true periods may be of a square wave type. This is because the harmonics of the fundamental frequencies of the periodicities have been omitted from the recombination. In the *Cardium edule* growth increments studied this could be ignored.

Growth Records in Invertebrates and Stromatolites

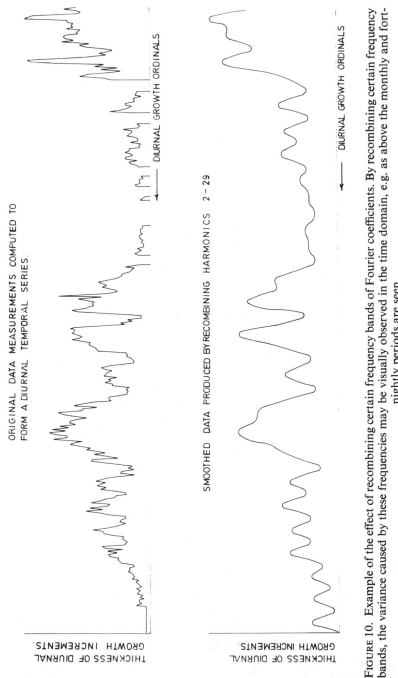

FIGURE 10. Example of the effect of recombining certain frequency bands of Fourier coefficients. By recombining certain frequency bands, the variance caused by these frequencies may be visually observed in the time domain, e.g. as above the monthly and fortnightly periods are seen

Figure 10 shows the original time series with diurnal intervals in time, and the estimated position of randomly occurring missing winter increments. Below this is the time series produced from recombining Fourier coefficients of all periods from 6 months to approximately 13 days. Approximately half the total power is recombined here. From such a reconstructed time series may be seen the times at which the amplitude of periodicities 'appear' strongest. The ends of the time series should, however, be ignored, as they often show spurious effects, that are a result of the assumption that the time series will continue in the same manner past the end of the real time series.

(2) *Recombining Fourier coefficients to look at random effects on the animal's growth*

One can study the effect on the animal of weather, which over short periods may be considered to be mainly random, around the British coast.

First the tidal periodicities, their modulation frequencies and their harmonics, are reset in the Fourier transform, so that their amplitudes are similar to those of the surrounding frequencies. If the Fourier coefficients are then recombined, most of the effect due to the tidal oscillation is removed from the data, and the resulting time series may be correlated with meteorological records.

Phase of tides with respect to the day

Where specimens are mainly subjected to semi-diurnally controlled tides, spring tides will occur at approximately the same time in the day each fortnight. Therefore by comparing the time of the 'switch over' in the semi-diurnal line distinctnesses each fortnight, with the phase of the tide, the time of spring and neap tides may be estimated. We may obtain a greater accuracy than that described at the beginning of the paper by first obtaining the phase of the fortnightly tides from the Fourier transform of the growth increment thicknesses, and then comparing this with the phase taken from the aliasing fortnightly frequency resulting from the transform of the growth-line distinctnesses.

Summary of results obtainable from this method

By using the Fourier analysis methods outlined, the following information may be extracted and should therefore be of equal value to both geophysically and ecologically orientated growth increment 'enthusiasts'.

(1) The length of the anomalistic, tropical and synodic month.
(2) The shape and range of the tides.
(3) The rate of movement of the tidal masses (by plotting the times of the spring tides at respective sites along a coast).
(4) The tidal position of a specimen.
(5) The affect of random weather changes on the animal (e.g. such as storms).

Given only the length of the tropical month and that of the synodic month from specimens, in theory it is possible to estimate approximately the length of the sidereal year. The tropical month approximately equals the nodical month (27·2122 days at present) and more importantly the sidereal month (27·3216 days at present), therefore the length of the year may be obtained from the difference in frequency of the synodic and sidereal months. Unfortunately, however, several continuous years of growth records are required for Fourier analysis to be able to define the length of the two types of month with sufficient precision to be able to estimate usefully the length of the year.

Fourier analysis also seems unsuitable for estimating the length of the tropical year from the seasonal changes in the rates of growth of organisms (with the possible exception of stromatolites). For example, with a complete growth record of a length of 10 years, with no frequency trend such as caused by ageing, the resolution for the number of days for the present year would be between 356–374. The annual periodicity in specimens is often controlled by seasonal variations in the environment, the timing of which in temperate latitudes may vary by more than 6 weeks. Consequently, the length of the year as recorded by an organism may appear to vary more than 6 weeks. The maximum resolution with which the length of the year may be theoretically calculated is equal to:

the maximum variation in the timing of the seasons ÷ the number of years

Therefore for 10 years of *continuous records*, where the length of the seasonal year may vary by up to 6 weeks, the theoretical resolution of determining the length of the tropical year is 42 days/10, i.e. ±2·1 days. The theoretical estimation of the length of the tropical year would then lie between approximately 363–367 days. Although this resolution in low frequencies such as length of the year cannot be obtained by the use of Fourier analysis, estimations of even greater precision have been claimed in the use of Maximum Entropy Power Spectrum analysis (Lacross, 1971; Ulrych, 1972a,b) in the analysis of geophysical data. It has yet to be seen if this method of analysis of periodic data is applicable to the analysis of annual growth cycles: even if it is, however, it must be remembered that estimations of greater precision than the theoretical maximum have no real geophysical meaning.

Unfortunately, bivalves with more than 5 years of *uninterrupted continuous* growth records do not appear to occur, as most specimens reaching the adult stage grow only slowly and discontinuously. It may therefore be difficult to find suitable specimens that show long enough growth records for a reasonable estimation of the length of the tropical year. Stromatolites might therefore be more suitable, as they show no ageing trend with growth, although they often suffer numerous disturbances.

If, however, we are left only with being able to determine the shorter periods such as fortnightly and monthly tidal with precision, the only geophysical estimation of any accuracy that may be made is the number of terrestrial days at that geological period occurring in each anomalistic, tropical and synodic month. We may therefore not be able to estimate the detailed history of the Earth–Moon system over short geological time spans. In the more distant geological past, less precise estimations of the length of the year and the inclination of the lunar orbit are still of great value since they should be able to indicate the long-term relationship between the Earth and the Moon.

Acknowledgments

The author is indebted to N.E.R.C. and to the N.A.T.O. Advanced Study Institute for the studentships under which this work was conducted. My thanks also go to Mrs. D. Hewett and Mrs. E. Thompson for producing the diagrams.

References

Barber, N. F. (1966). Fourier methods in geophysics, pp. 123–204 in *Methods and Techniques in Geophysics*, Vol. 2 (Ed. S. K. Runcorn), Interscience London

Barker, R. M. (1964). Microtextural variation in pelecypod shells. *Malacologia*, 2, 69–86

Berry, W. B. N. and Barker, R. M. (1968). Fossil bivalve shells indicate larger month and year in Cretaceous than present. *Nature*, 217, 938–939

Blackman, R. B. and Tukey, T. W. (1958). *The Measurement of Power Spectra*, Dover, New York

Clark II, G. R. (1968). Mollusk shell: daily growth lines. *Science*, 161, 800–802

Cooley, J. W. and Tukey, J. W. (1965). An algorithm for the machine computation of complex Fourier series. *Math. Computation*, 19, 297–301

Evans, J. W. (1972). Tidal growth increments in the cockle *Clinocardium nuttalli*. *Science*, 176, 416–417

Farrow, G. E. (1971). Periodicity structures in the bivalve shell: experiments to establish growth controls in *Cerastoderma edule* from the Thames Estuary. *Palaeontology*, 14, 571–588

Farrow, G. E. (1972). Periodicity structures in the bivalve shell: analysis of stunting in *Cerastoderma edule* from the Burry Inlet (South Wales). *Palaeontology*, 15, 61–72

Fourier, (1825). Mémoire sur la théorie analytique de la chaleur. *Mémoires de l'Académie Royale des Sciences de l'Institut de France*, 8, 581–622

Hancock, D. A. and Urquhart, A. E. (1966). The fishery for cockles (*Cardium edule* L.) in the Burry Inlet, South Wales. *Fishery Investigation* Ser II, 25 (3), pp. 32

House, M. R. and Farrow, F. E. (1968). Daily growth banding in the shell of the cockle, *Cardium edule*. *Nature*, 219, 1384–6

Lacross, R. T. (1971). Data adaptive spectral analysis methods. *Geophysics*, 36, 661–675

Lagrange, (1772). *Oeures* 6:605 *fide* Granger, C. W. J. (1964). Spectral analysis of economic time series. (Princeton studies in mathematical economics, no 1), Princeton University Press

Mazzullo, S. J. (1971). Length of the year during the Silurian and Devonian periods. *Geol. Soc. Amer. Bull.*, 82, 1085–1086

Moore, P. G. and Edwards, D. E. (1965). *Standard Statistical Calculations*. Pitman, London

Pannella, G. (1972). Palaeontological evidence of the Earth's rotational history since Early Precambrian. *Astrophysics and Space Science*, 16, 212–237

Pannella, G. and MacClintock, C. (1968). Biological and environmental rhythms reflected in molluscan shell growth. *J. Palaeontology*, **42**, 64–80

Pannella, G., MacClintock, C. and Thompson, N. (1968). Palaeontological evidence of variations in length of synodic month since Late Cambrian. *Science*, **162**, 792–796

Pearson, E. S. and Hartley, H. O. (1954). *Biometrika Tables for Statisticians*, Vol. 1, Cambridge University Press

Scheister, A. (1898). On the investigation of hidden periodicities. *Terr. Mag.*, **3**, 13

Scrutton, C. T. (1970). Evidence for a monthly periodicity in the growth of some corals, pp. 11–16 in *Palaeogeophysics* (Ed. S. K. Runcorn), Academic Press. 518 pp.

Singleton, R. C. (1968). Multivariate complex Fourier transform, computed in place using mixed-radix fast Fourier transform algorithm, *Stanford Research Institute*

Stokes, G. G. (1897). Notes on searching for periodicities. *Proc. Roy. Soc.*, **29**, 122

Taylor, J. D., Kennedy, W. J. and Hall, A. (1969). Shell structure and mineralogy of the bivalvia (Nuclucea–Trigonacea), *Bull. Brit. Mus. Nat. Hist. (Zool.)* Suppl. 3

Ulrych, T. J. (1972a). Maximum entropy power spectrum of truncated sinusoids. *J. Geophysical Research*, **77**, 1396–1400

Ulrych, T. J. (1972b). Maximum entropy power spectrum of long period geomagnetic reversals. *Nature*, **235**, 218–219

Zygmund, A. (1935) (1955 reprint). *Trigonometrical Series*, Dover, New York

DISCUSSION

HIPKIN: I do not understand how you compensated for missing increments in your growth analyses.

DOLMAN: I counted the number of increments between neaps, plotted the counts, and then Fourier-analysed them. Peaks in the periodogram then give the frequency modulation, that is the variation in number of missing increments with time. I also recognized two other types of missing increments.

(1) Systematic: animals above neap tide level have growth increments missing on a regular basis, as they are regularly left exposed, except at the highest tides. One cannot estimate the number of missing increments simply by looking at one animal, one needs a suite of animals obtained from different tide levels. For geophysical purposes, one needs to obtain clams with the most complete tidal record possible; the procedure of restoring increments is of value only for, say, growth analyses related to ageing of the animal, and studies of environmental influences on growth.

(2) Random: During winter, or during storms, the animal may occasionally miss an increment. For geophysical purposes, it is impossible to predict the number of random missing increments.

EVANS: What was the frequency of the increment which you measured?

DOLMAN: They are semi-diurnal, tidal. House and Farrow thought they were daily, consequently their measurements showed only monthly (and with further analysis by me) 2-monthly periodicities. I took some of House and Farrow's material and measured between the end of a growth record of a specimen collected in September and the beginning of fast growth in spring, finding 330 increments. Thus, I concluded the increments were definitely not diurnal—they were semi-diurnal tidal. When Fourier-analysed the data then showed fortnightly and monthly periodicity.

FARROW: Then the bivalves must stop growing for half the year.

DOLMAN: No, although there are a lot of missing days/year, much of the discontinuous growth occurred during winter.

APPROACHES TO CHEMICAL PERIODICITIES IN MOLLUSCS AND STROMATOLITES

G. D. ROSENBERG AND C. B. JONES

School of Physics, University of Newcastle upon Tyne, U.K.

Abstract

The electron microprobe is used to study chemical distribution in living and fossil molluscs and stromatolites. Regular fluctuations in calcium and sulphur in living *Cardium edule* can be correlated with tidal structural series. Repeating rhythms of calcium, magnesium, iron and silicon are found in stromatolites. The results may enable determination of the rate of the Earth's rotation in the distant past, using chemical rhythms as well as structural patterns.

Our studies also have implications for fossil metabolism and diagenesis investigations.

Introduction

Accretionary skeletons of marine organisms record periodic changes in the organisms' environment and in their metabolism. Corals, bivalves and stromatolites, among others, grow in response to diurnal, tidal and seasonal changes in such factors as temperature, light intensity, nutrution, pH, oxygen and carbon dioxide concentration. Changes in growth increment patterns, crystallography and mineralogy of the skeletons correlate with environmental changes (Barker, 1964, and Dodd, 1967, are offered as only two recent examples of many related studies).

With such structural periodicities in mind, it is reasonable to assume the existence of periodicities in chemical concentration. Rosenberg (1973) suggested that calcium concentration varies rhythmically in the shell of the bivalve *Chione undatella* Sowerby, perhaps according to the season of deposition. Bryan (1973) investigated seasonal variations in trace element concentration in the soft tissues of scallops. Both workers suggested that elemental concentration generally increases toward the winter months.

We report here additional evidence of chemical cycles in the shells of molluscs and in stromatolites. Results are reported for electron microprobe analyses of the shells of the living bivalves *Cardium* (*Cerastoderma*) *edule* (9 specimens), *Spisula* sp. (1 specimen), Jurassic belemnites (cf. *Megateuthis*) from Skye (5 specimens), and one Precambrian stromatolite from the Pethei Gp. of the east arm of the Great Slave Lake, Canada. Calcium and sulphur variations are reported for the molluscs, calcium, magnesium, iron and silicon for the stromatolite.

Tidal rhythms in calcium and sulphur concentration were found in some of the bivalves. Diagenetic trends were also found in the belemnites. The recognition of tidal chemical series may provide an additional means of measuring the rate of the Earth's rotation in the past.

Methods

Molluscs

The organisms studied were sectioned, ground and polished, and then analysed with the electron microprobe (Cambridge Geoscan) according to standard techniques (Moberly, 1968; Schopf and Allan, 1970; Rosenberg, 1972). The ventral-most 2 cm of the bivalves and sections of the belemnites were mounted in Specifix cold-mounting resin (Struers, Copenhagen). If the resin is prepared to a thin consistency and is cooled slowly, it has the advantage of being nearly bubble-free; bubbles in the plastic mount act as traps for grit during grinding and polishing. With continued polishing, the grit plucks and scratches the specimen. The mounts are ground on carborundum grit (successively finer grades down to 600). After each grinding and polishing the mounts are cleaned with a sonic vibrator. The mounts are then polished with 5/20, 3/50 and finally γ-alumina on a cloth lapping wheel. Use of a wax wheel instead of a cloth wheel may help to prevent scratching. Washing the specimens with soap solution after each grade of grit and polish also helps to minimize scratching. Coating of the specimens with bakelite and use of carborundum paper instead of grit further ensure a finely polished surface.

Following the mounting and polishing of the specimens, the samples were coated with aluminium in vacuum. Aluminium has two advantages over carbon (previously used by Rosenberg for coating): (i) carbon coatings are readily burned through by the microscope beam, whereas aluminium coatings are more resistant to the intense electron beam; (ii) charges are more liable to build up on a carbon-coated specimen than on an aluminium-coated sample. These charges repel electrons and can result in anomalous readings (Satter, 1970). It was also hoped that use of an aluminium coating would enable more accurate analysis of carbon concentration in the specimens; counts from carbon coating could interfere with actual carbon readings from the specimen. Microprobe accuracy was subsequently found to be insufficient to obtain reliable estimations of carbon concentration and variability, and analyses of carbon simultaneous with calcium were abandoned in favour of sulphur and calcium. A quartz crystal was used for calcium ($K\alpha_1$) analyses, a mica crystal for sulphur ($K\alpha_1$). The probe was operated at 15 kv with a specimen current of 50 mμ A on aragonite. A spot size of approximately 15 μm was used. This was found to be the minimum size which could be used, to prevent burning of the specimen. Each reading was taken for 10 sec and each specimen was analysed along the same traverse three times. Standard readings on aragonite and pyrite were taken before and after each traverse. Data were then corrected on the School of Physics Electron Microprobe Program.

Drift, dead time, background, fluorescence, absorption, atomic number are all corrected for in this program. Finally, the data were refined and graphed using Dolman's Fourier Analysis Program (this volume). The theory is discussed by Barber (1966).

The data in this report have been frequency smoothed by the use of a Fourier analysis in which more than 50% of the total power spectrum is recombined. However, it is noted that, in some cases (especially with sulphur), repeatable patterns are apparent without frequency smoothing.

Stromatolites

Stromatolite specimens were prepared for the microprobe by cutting them into rectangular rods perpendicular to the lamination. One long side was flattened and polished, using diamond pastes. The polished surface was made conducting by coating it with a film of carbon.

The analyses illustrated here are taken from pen recorder traces of the count rates of Ca and Mg, Mg and Fe, and Mg and Si along traverses made by moving the specimen underneath the electron beam at a speed of 30·5 μm/min, in a direction perpendicular to the lamination. The gun potential was 15 kv and the specimen current about 50 mμ A on aragonite. No corrections have been applied. Drift over the length of the traverses did not exceed 5%. The form of the resulting graphs was found to depend considerably upon the size and shape of the electron beam, or spot, at the specimen surface. If the spot is circular and of a diameter which is small relative to the increment thickness then the graph will be saw-toothed with narrow peaks, and parallel traverses will be markedly dissimilar.

Larger spot sizes result in broader, more rounded peaks, and parallel traverses are comparable. The optimum spot size is the smallest for which parallel traverses are easily correlatable, since at this level the maximum number of significant linear features on the surface will be detected. Traverses using a rectangular, 'line spot', oriented parallel to the lamination may record greater significant detail.

Error

The problem of determining error of procedure can be approached both statistically and experimentally. In statistical terms each measurement of the microprobe has an error of $\pm\sqrt{n}$ where n is the count rate for that measurement. In a series of measurements in a traverse across a sample most of the measurements could be expected to vary within $\pm 3\sqrt{\bar{N}}$ where $\bar{N}$ is the traverse mean; this variation is expected even if the element under consideration were distributed uniformly throughout the specimen.

In the data reported here (for the molluscs), the variations in calcium are small relative to the expected, statistical variations; actual variation in element distribution lies close to the statistical limits, for the mollusc shells are composed of nearly uniformly distributed calcium carbonate. Consequently, traverses are repeated three times, to distinguish actual variations in chemical concentration from random effects of 'noise'.

In the molluscs the analyses of sulphur have been found to be more repeatable than the analyses of calcium. This is despite the very low concentration of sulphur (usually less than 0·5%) compared with the high calcium concentration (usually greater than 50% in terms of CaO). Simply stated, the repeatability is a function of the actual range in count rate for a traverse and the expected variation of the measurements about the mean:

$$R \propto \frac{\Delta n}{6\sqrt{\bar{N}}}$$

where R is repeatability, Δn the range in measurements for a traverse, $\bar{N}$ the traverse mean. If $R > 1$ the traverse can be expected to be repeatable; repeatability is greater with increasing R. For a hypothetical (but typical) traverse calcium counts averaged 27,000 cts/10 sec and the range was 1500 cts/10 sec. For the same traverse sulphur counts averaged 200 cts/10 sec and the range was 200 cts/10 sec. $R_{Ca} = 1·5$; $R_S = 2·4$. In other words, although the concentration of sulphur is low in the specimens the actual variability is high compared with the statistically expected variation about the mean, hence the repeatability is high.

A more accurate estimation of repeatability may be given by:

$$R = \frac{\sigma}{6\sqrt{\bar{N}}}$$

where σ is the standard deviation of the population. For the purpose of cursory estimation the first formula suffices in this case.

Results

Molluscs

There is a high concentration of CaO in the mollusc shells studied, ranging from 51% up to about 55%. Calcium distribution varies regularly within these limits, although the variability is clearly small relative to the actual concentration. Both the uniformity and the small but regular variability are important to discussing the physiology of the molluscs.

There is a low concentration of S in the molluscs studied, ranging from 0·00% to about 1·5%. Variation in sulphur distribution is regular within these limits. Patterns of calcium distribution are comparable in at least two of three traverses of a specimen, while sulphur patterns are comparable in all three traverses of the same specimen. At present it is difficult to correlate patterns of elemental distribution from specimen to specimen.

Figure 1 graphs CaO distribution along the ventral-most section of *Cardium* specimen 1/1/2–. Each point represents one measurement; measurements were taken at 50 μm intervals, approximately one reading per growth increment along a series of growth increments apparently representing tidal series (distinguished by repetitive narrowing and widening of increments). The upper graphs A, B, C, represent data which have been corrected for drift, background, etc. but which have not been Fourier-analysed. The heavy dashed line is an estimated trend in calcium distribution determined by averaging three consecutive points. The lower graphs A', B', C', represent the Fourier analysis of the microprobe-corrected data; in each case the Fourier analyses graph compares favourably with the estimated trend, and suggests that Dolman's Fourier analysis program is applicable here. The letters on the lower graphs designate oscillations in calcium distribution which repeat from traverse to traverse. Repeatability is best in graphs A' and B'. Graph C' is less similar to the other two. In part, it is believed this is due to the difficulty of locating the spot on the sample with the optics of the microprobe. The result is that similar points or patterns do not repeat at similar positions on the microprobe traverse. In addition, there is the problem of repeatability discussed in the Methods section.

(The specimen preparation, especially polishing, is designed to remove topographical differences in the specimen. At present the recognition and analysis of tidal series in molluscs with the microprobe relies on first locating regular variations in growth increment width with microprobe optics; care is taken to place the spot only on smooth areas. Chemical patterns are compared with the known tidal structural increments to determine whether chemical periodicities do indeed exist. At present, continuous traverses of molluscs are not reported, as one must continually monitor the position of the spot on specimens, some of which are soft and do not always polish uniformly.) There are approximately four tidal series represented in the graphs of *Cardium* 1/1/2–: they are (approximately); a–d, d–h, h–l, l–p. It is not possible at this time to state whether high concentrations of calcium occur at any particular part of the tidal cycle—indeed, this may never be possible because there are almost certainly other factors besides tides affecting calcium concentration. Figure 2 shows calcium distribution along a section of *Cardium* 1/1/1– at 50 μm intervals. Again a tidal series is depicted. Graphs B and C repeat well, graph A is less similar to the other two. There are two tidal series represented, approximately from a–b and from b–f. Carbon was run simultaneously with calcium but is not reported.

Figure 3 graphs A, B, C and graphs A', B', C' show sulphur and calcium distribution respectively in three traverses of *Spisula* specimen 4/5/14. In this case, measurements do not represent tidal series, as they have been taken every 200 μm. The ventral decrease in average sulphur concentration is noteworthy. This contrasts with possible ventral increase in calcium concentration in graphs A' and C'; this calcium trend also correlated with that reported for calcium distribution in 44 analyses of *Chione undatella* Sowerby previously reported (Rosenberg, 1973).

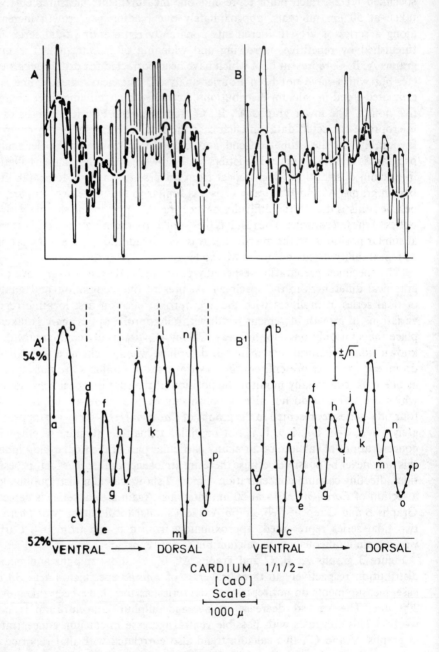

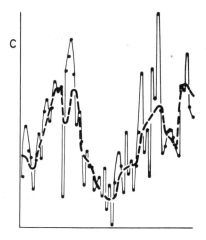

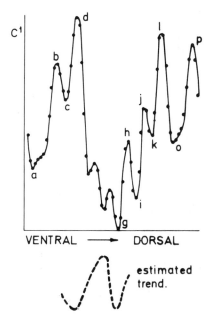

FIGURE 1. CaO distribution in *Cardium* 1/1/2–. Vertical dashed lines indicate the boundaries of tidal series seen with microprobe optics. Letters indicate patterns of calcium distribution which repeat from traverse to traverse of the specimen

Again, letters designate patterns which repeat from traverse to traverse. Letters a, c, e, g, i, k denote maxima of sulphur concentration and letters b, d, f, h, j, minima of sulphur concentration; the repeatability of this pattern from traverse to traverse is high. Comparing patterns for calcium distribution, it is clear that repeatability in B' is low, indeed the average calcium concentration decreases ventrally whereas in A' and C' the trend is quite different. Such a result is clear warning that a single traverse is inadequate for determining variations in element (at least calcium) distribution. On the other hand, non-repeatable variations in element distribution may be more apparent than real, especially when the actual variability is small relative to statistically expected variation. Additional measurements in the ventral-most area might detect high concentrations of calcium; a ventral increase in calcium might not be apparent here as the interval between measurements is large. In any event, additional measurements of calcium traverses are in order.

It is possible to compare calcium and sulphur concentrations in traverses A and A' and C and C' (see also the three traverses for specimen *Cardium* 4/5/11). Sulphur distribution is sometimes, but not always, inversely proportional to calcium concentration. The relatively high concentration of sulphur in region c contrasts with the relatively low calcium concentration in region o, approxim-

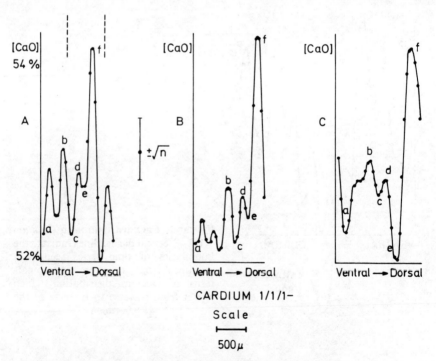

FIGURE 2. CaO distribution in *Cardium* 1/1/1–. Dashed lines as before

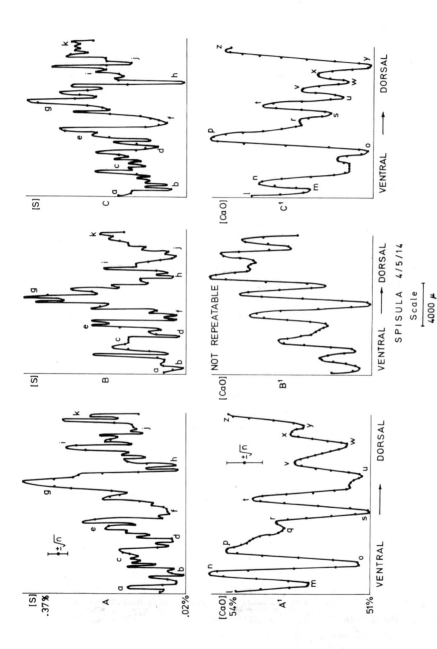

FIGURE 3. CaO and S distribution in *Spisula* 4/5/14. Sulphur at top, calcium below

ately at the same location on the traverses. Similarly the high concentration at region i contrasts with the low CaO at region w. It may be that there are different reasons for various fluctuations in element concentration. One may suppose that in regions where both elements increase simultaneously and decrease simultaneously that what is being measured is change in porosity. But even this is not certain without knowing the behaviour of the other elements making up the rest of the skeleton (especially carbon, which forms a sizeable percentage of the skeleton, and whose abundance and variations are certain to influence the abundance of the other elements).

What is looked for in correlating charts is the appearance of a high or low concentration of an element on a certain position on the traverse, and the height of the high or depth of the low relative to surrounding oscillations. The relative heights of the peaks of sulphur increase from a to g, then drop from g to i and k. The occurrence of a peak in element concentration in the case of an element such as calcium, which has low variability relative to the absolute concentration, is often more important than the relative height of the peak. This is because slight changes in operating characteristics of the probe (filament, spectrometer, etc.) are more liable to affect the patterns of elements which have low variability relative to mean concentration.

Figure 4 compares *Spisula* shell structure with variations in sulphur concentration (Figure 3, graph A). The peak of sulphur concentration corresponds to the lightly coloured series of summer growth increments; the areas of low sulphur concentration correspond to the dark coloured series of winter growth increments.

With the successful analysis of sulphur in *Spisula*, subsequent analyses all combined sulphur and calcium measurements. Figure 5 shows analysis of calcium and sulphur distribution along a series of tidal increments in *Cardium* specimen 4/5/11. While approximately 48 increments could be recognized with the optics (3·5 tidal series), a maximum of 42 measurements could be taken with the probe, due to narrow width of some of the increments, or to topographic differences. Thus, about 6 increments distributed throughout the traverse are not measured. The position of the tidal series' boundaries is also indicated in Figure 5, graph A.

The repeatability of both the calcium and sulphur graphs is good. The increase in calcium concentration and the decrease in sulphur concentration toward the ventral margin is apparent in all traverses. There are 6 repeatable peaks in sulphur concentration (a, c, e, g, i, k) and 5 peaks in calcium concentration (o, s, u, w, y). It should be noted in graph B′, the variability in calcium measurements is less than that for traverses A′ and C′. Variability of measurements is believed to increase with age of the filament, and it is believed filament age could thus especially affect apparent distribution of elements with low variability, such as calcium. This may in part account for the anomalous traverse of calcium in Figure 3, graph B′.

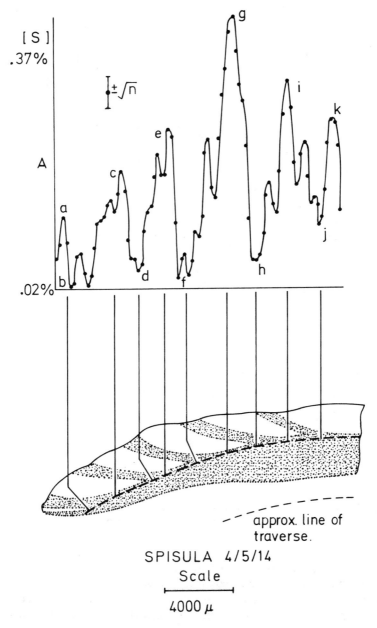

FIGURE 4. Sulphur concentration in *Spisula* 4/5/14 correlated with variations in shell structure. Dashed line in cross-section traces line of traverse, at base of outer shell layer, where structural differences are minimal

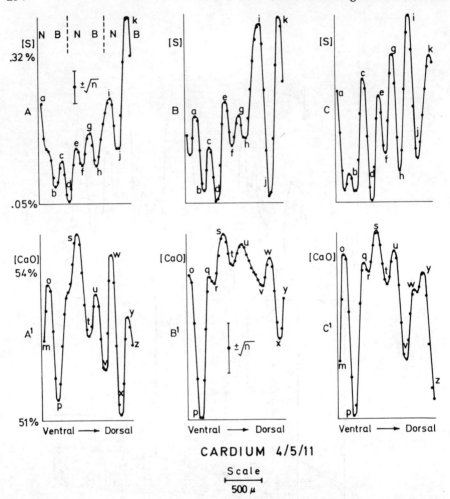

FIGURE 5. CaO and S distribution in Cardium 4/5/11. Dashed lines indicate tidal series boundaries; N = narrow increments, B = broad increments

There is no simple correlation with increment width and concentration of either calcium or sulphur. This complexity is significant in that it indicates the probe is not simply measuring porosity differences or anomalous differences due to topography, assuming that narrow increments and wide increments have different structures.

The point of the graphs is that the traverses are repeatable. As they represent the chemical correlates of structural differences in growth increment width, and as the increment pattern is due to tidal influence, evidence for chemical

Approaches to Chemical Periodicities in Molluscs and Stromatolites

variation with the tides is presented. Hence, there is reason to be optimistic that an alternative method of measuring growth periodicities for geophysical purposes is available.

In Figure 6, sulphur and calcium results are presented for belemnite Skye 6/2. An inverse relationship between sulphur and calcium is apparent, at least for the average trend across the specimen. The repeatability is good in all cases (the high variability in C' is perhaps due to instability of an old filament). Of note is the pronounced decrease in sulphur concentration, and inversely the increase in calcium concentration, toward the perimeter. Within the minor oscillations, an inverse correlation between calcium and sulphur is clearly seen at points b, c, d for sulphur and r, s, t for calcium.

In two of the five belemnite studies, the decrease in sulphur concentration from the centre toward the perimeter is accompanied by a slight increase in sulphur concentration at the perimeter of the specimen. This is believed to be a diagenetic trend; the specimens are found in a dark grey, organic-rich silty sediment. The diagenesis of the sediment is inferred to have affected the composition and diagenesis of the belemnites. In the center of all the belemnites, pyritic sediment reflects the high sulphur concentration, and the formation of pyrite probably influenced the chemical composition of the belemnite. Additional belemnites must be studied before any firm conclusions on diagenesis can be reached.

Stromatolites

Electron microprobe analysis of the variation in concentration of the elements Ca, Mg, Si and Fe across the lamination of stromatolites from the Lower Proterozoic of the Great Slave Lake area of Canada was carried out in co-ordination with the analysis of periodicities in the thickness of successive growth increments. It was envisaged that the chemical data might provide independent support for the results obtained from the increment measuring technique. (For a description of the latter method see Dolman, this volume)

The material examined is from an approximately 3-m thick horizon within the Hearne Fm. of the Pethei Gp. (Hoffman, 1968, 1974). The stromatolites are in the form of laterally linked domes (often triangular in cross-section), the orientation of which is related to local current directions. The lamination consists of an alternation of coarse, microspar to spar, and fine, micritic, quartz-rich dolomitic limestone. Etching indicates that the coarse increments are calcite-rich and the fine increments dolomite-rich. Haematite is disseminated throughout the rock, and it is particularly concentrated within many of the fine increments. The thickness of the coarse–fine couplets is generally between 45–200 μm.

Traverses I–VII (Figures 8 and 9) are parallel and were taken on a specimen from the area marked A in Figure 7 while VIII–XI (Figure 10), also parallel, were taken on a specimen from the equivalent, but more closely spaced, increments in another part of the stromatolite, marked B. All parallel traverses are within 2·5 mm of each other. I, II and III employ a large spot, of diameter 125,

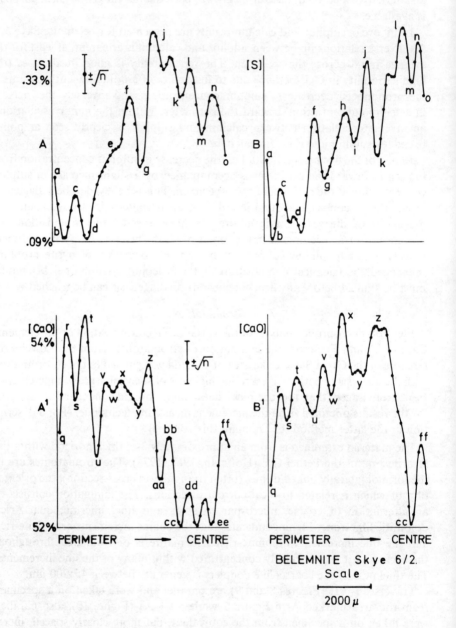

BELEMNITE Skye 6/2.
Scale
2000 μ

Approaches to Chemical Periodicities in Molluscs and Stromatolites

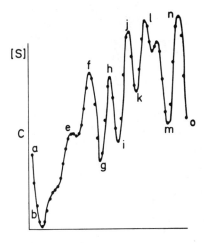

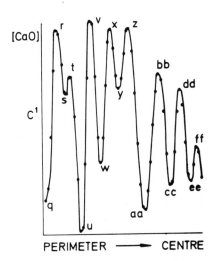

FIGURE 6. CaO and S distribution in belemnite 6/62 from Skye. Traverse starts from perimeter of specimen (at left) and proceeds to edge of central canal

FIGURE 7. Precambrian stromatolite from the Pethei Group, East Arm of Great Slave Lake, Northwest Territories. The positions of electron microprobe traverses A and B are indicated

120 and 105 μm respectively, and indicate a regular variation in Ca and Mg concentration with a usually negative correlation between the two, corresponding to variation in the ratio of calcite to dolomite. Traverses IV, V and VI employ a line spot of 260 × 40, 260 × 40 and 180 × 30 μm respectively, oriented parallel to the lamination, and demonstrate the greater sensitivity of this shape in comparison with the circular spots of similar area used in I, II and III. Traverse VII (Mg, Fe) was made with a round spot of diameter 125 μm. The Fe is seen to be preferentially associated with the Mg. Traverses VIII and IX (Ca, Mg) use a round spot of diameter about 60 μm. These two differ noticeably from the Ca and Mg traverses I to VI in that Ca and Mg appear to be positively correlated, though the ratio of Ca to Mg varies with a tendency to be higher in the troughs. This form of graph results from the locally high concentration of quartz which, as a comparison of traverses X (Ca, Mg) and XI (Si, Mg) demonstrates, occurs in high concentration where Ca and Mg are low. Traverses X and XI were made along the same line and employ a round spot of diameter 120 μm.

It was found that traverses repeated along the same line under the same operating conditions were practically identical, but those made along adjacent parallel lines were not, though they were correlative to varying degrees, as the traverses figured here illustrate.

Approaches to Chemical Periodicities in Molluscs and Stromatolites 239

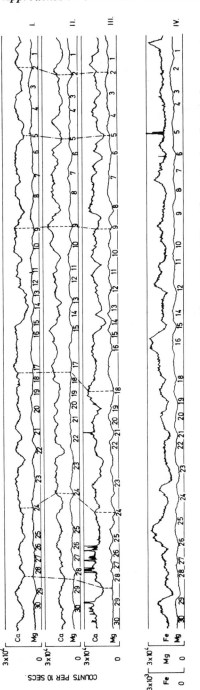

FIGURE 8. Electron microprobe analyses of Ca and Mg (I, II, III) and of Fe and Mg (IV), using a circular electron beam spot, on continuous traverses, 7·5-mm long, perpendicular to the lamination of a Precambrian stromatolite (position A in Figure 7). The spikes in III, about 26 and 21, and in IV at 5 and 6 are due to interference from spurious external signals

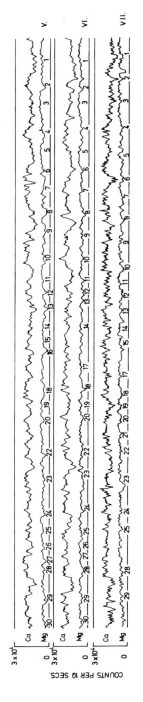

FIGURE 9. Electron microprobe analyses of Ca and Mg using a rectangular electron beam spot on continuous traverses, 7·5-mm long, perpendicular to the lamination of a Precambrian stromatolite (position A in Figure 7)

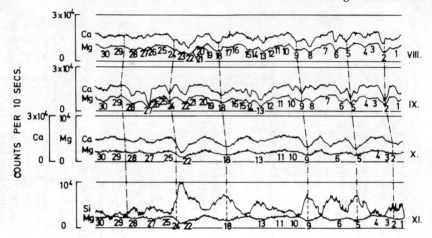

FIGURE 10. Electron microprobe analyses of Ca and Mg (VIII, IX, X) and Si and Mg (XI), using a circular electron beam spot, on continuous traverses, 3·1-mm long, perpendicular to the lamination of a Precambrian stromatolite (position B in Figure 7)

Some of the main correlatable features of the traverses have been labelled (1 to 31). Comparing parallel traverses, relative heights of adjacent peaks are in many cases not preserved, and the separation of identifiable peaks is also not constant. This is understandable in view of the small-scale irregularities in increment thickness and continuity apparent under microscope examination of thin sections.

Conclusions

Rhythmic variations in chemical distribution in living and fossil invertebrates may eventually be used in the same way as structural growth increment studies are used to determine length of the day in the distant past. Electron microprobe techniques reported here show that such chemical rhythms are both recognizable and repeatable. The suggestion of fortnightly calcium and sulphur rhythms in bivalves permits optimistic pursuit of an independent chemical alternative to structural growth increment studies. The repeatability of chemical cycles in stromatolites indicates that these rhythms may be preserved back into the Precambrian.

These chemical studies are potentially valulable in other areas of research: fossil metabolism, especially calcium metabolism, may be studied with the electron microprobe. Variations in calcium concentration may be correlated with Schmalz and Swanson's (1969) demonstrated diurnal variations in pH, dissolved CO_2 and carbonate saturation of seawater. Palaeoenvironmental information may emerge from chemical analysis. Additionally diagenetic processes may be revealed by these analyses.

Acknowledgments

We are grateful for John Dolman's assistance with the Fourier Analysis, Bill Davison's help on the electron microprobe, and Susan Hofmann's assistance with the computer. Deborah Skeen typed the manuscript and Dorothy Hewett prepared the figures.

References

Barber, N. F. (1966). Fourier methods in geophysics, in S. K. Runcorn (ed.), *Methods and Techniques in Geophysics*, vol 2., Interscience, London, 123–204

Barker, R. M. (1964). Microtextural variation in pelecypod shells. *Malacologia*, 2, 69–86

Bryan, G. W. (1973). The occurrence and seasonal variation of trace metals in the scallops *Pecten maximus* (L.) and *Chlamys opercularis* (L.). *J. Mar. biol. Ass. U.K.*, 53, 145–166

Dodd, J. R. (1967). Magnesium and strontium in calcareous skeletons: a review. *J. Palaeontology*, 41, 1313–1329

Hoffman, P. F. (1968). Stratigraphy of the Lower Proterozoic (Aphebian), Great Slave Supergroup, East Arm of Great Slave Lake, District of Mackenzie. *Geol. Surv. Can.*, Paper 68–42

Hoffman, P. F. (1974). Shallow and deep water stromatolites in lower Proterozoic platform-to-basin facies change, Great Slave Lake, Canada. *Am. Assoc. Petrol Geologists Bull.*, 58 (5), 856–867

Moberly, R. Jr. (1968). Composition of magnesian calcites of algae and pelecypods by electron microprobe analysis. *Sedimentology*, 11, 61–82

Rosenberg, G. D. (1972). Patterned Growth of the Bivalve *Chione undatella* Sowerby Relative to the Environment PhD Dissert., Geology Dept. University of California, Los Angeles

Rosenberg, G. D. (1973). Calcium concentration in the bivalve *Chione undatella* Sowerby. *Nature*, 244, 155–156

Satter, W. J. M. (1970). *A Manual of Quantitative Electron Probe Microanalysis*, London Structural Publications, 151 pp.

Schmalz, R. F. and Swanson, F. J. (1969). Diurnal variations in the carbonate saturation of seawater. *J. Sed. Petrol.*, 39 (1), 255–267

Schopf, T. J. M. and Allan, J. R. (1970). Phylum Ectoprocta: Order Cheilostomata: Microprobe analysis of calcium magnesium, strontium, and phosphorus in skeletons. *Science*, 169, 280–282

DISCUSSION

PANNELLA: Gary, why did you use spot readings instead of continuous traverses of your specimens?

ROSENBERG: I wanted to make certain the chemical variations were not due to spurious counts from topographic variations. While it would be a good idea to use continuous readings, I chose instead to monitor the position of the electron spot at all times (as the shells studied were very soft, and consequently sometimes polished unevenly). Checking the position of the spot before each reading prevented continuous recordings because of the construction of the optics of the machine.

EVANS: What spot size did you use?

ROSENBERG: The spot size was 15 μm, the width of the growth increments 50 μm or less. I tried to place the spot between rather than on the increment boundaries in each case.

HEWITT: I agree with your belemnite diagenetic trends. When I stained belemnites with alizarin red and K Ferro Cyanide, I found the outer part was usually iron-rich carbonate. Also I found pyrite near the edge of my specimens, presumably from sulphur added during decay of organic matter and subsequent diagenesis. In the centre I sometimes found glauconite and pyrite, present because of the organic matter in the centre of the specimen which calcifies later on. Do you agree with my interpretations?

ROSENBERG: In a few cases sulphur concentration also increased just at the perimeter of a specimen, supporting your observation of pyrite near the edge of the specimen. But I don't want to generalize on the basis of a few specimens.

MOHR: Mr Jones, what was the iron concentration in your stromatolites?

JONES: Up to 5–6% (presumably haematite).

JOHNSON: Do the stromatolites show banding according to a definite pattern?

JONES: Yes, but my probe data merely show a regular alternation of calcite- and dolomite-rich areas. While laminae were obvious under the microscope, with the spot size used I could not determine regular variations in width. Eventually it may be possible.

THE SEASONAL PERSPECTIVE OF MARINE-ORIENTATED PREHISTORIC HUNTER-GATHERERS

P. J. F. COUTTS

State Archaeologist, Archaeological and Aboriginal Relics Office, Melbourne 3000, Australia

Abstract

The problems inherent in trying to reconstruct settlement patterns of prehistoric hunter-gatherers are well known. The principal difficulty hinges on the fact that we have not yet developed techniques to date or conjunct occupation sites with sufficient accuracy to enable the system structures to be clearly defined. Central to this problem is the discovery of the time of the year and the duration of occupancy associated with archaeological layers.

Current techniques of determining these last two settlement pattern parameters have been briefly reviewed. Capitalizing on the results of recent research on the growth dynamics of the bivalve *Chione stutchburyi*, which are summarized here, new approaches are outlined and illustrated with data from several archaeological sites in southern New Zealand. While these recent developments promise to aid substantially in elucidating important aspects of marine-orientated patterns of settlement, these same results indicate that more detailed research is required on biological facets of molluscan growth dynamics before these techniques can be refined further.

Introduction

In general, ethnographic studies of hunter-gatherer societies suggest that their mobility, seasonality and territoriality are three important characterizing features. The degree to which these features can be defined depends on the data available. Since historical and ethnographic records are often incomplete, archaeological evidence can sometimes be used to complete and verify these data.

In my own recent studies of European–Maori contact situations in southern New Zealand, ethno-historic data were used to define the general patterns of exploitation and other salient features of the indigenous societies (Coutts, 1972). Subsequently a project was set up to look for archaeological evidence of these features and it was found that few of them could be identified archaeologically, although additional features were discovered which were not known from the ethno-historical facts. It was concluded that archaeological studies of hunter-gatherer societies, in the absence of complementary data from the other sources, are likely to be fraught with difficulties. This is because the archaeologist's interpretive licence is greatly increased when he cannot confidently impose constraints on the cultural system he is studying. In these circumstances the archaeologist is forced to start with the assumption that his site is a part of a much wider system; he will examine the problem of settlement at the site in the hope of discovering its role and how it is related to the wider settlement pattern. In this respect he seeks to establish the function of the site, the *length of occupancy* and the *seasonality of the occupation*. His ultimate aim will be to build up an integrated view of the system structure (Binford, 1969; Struever, 1971). Indeed, it is incumbent upon the archaeologist to devise techniques to enable him to assess these last two parameters as accurately as possible if we are to make real progress in elucidating system structures. In this chapter I intend to probe the ramifications of seasonality since it is one of the most important settlement pattern parameters, to a large extent related to and determining the duration of site occupancy.

Until recently the methods of discovering the season of occupation of sites have relied, in the main, on identifying the presence of seasonal fauna at a site. In general, this method can lead only to broad estimates of seasonality and in some cases may give unreliable or, at best, misleading results. Another promising, but as yet technically difficult, method makes use of palaeotemperature spectrums from marine shells deposited in ancient rubbish dumps (McBurney, 1967). A further approach, the main subject of a discussion in this chapter, which is applicable only to archaeological sites yielding marine mollusca, has been developed as a corollary to recent studies on the growth dynamics of intertidal bivalves. (Weide, 1969; Coutts, 1970; Coutts and Higham, 1971; Koike, 1973.)

Previous experimental work on *Chione stutchburyi*

During 1969–70 living populations of the cockle *Chione stutchburyi* were studied in their natural habitats. Samples were collected periodically and various measurements and observations were made on them as described elsewhere (Coutts, 1970; Coutts and Higham, 1971). The main finding of this analysis was that the growth rates varied regularly with a maximum during the summer period and no perceptible growth during the colder parts of the year. The implication of this finding was that properly quantified results could be applied to appropriate mollusca from archaeological sites to help determine at what time of the year they were collected. Consequently, three methods of seasonal dating were proposed and have been applied to archaeological sites from several areas in southern New Zealand. These methods depend on two major hypotheses; that the major rings observed on the exterior shell surface, here called macrobands, form annually during the coldest months of the year, and that black bands observed in sections of the shells cut through the umbos and traversing the axes of maximum shell diameter, here called growth lines, are daily phenomena. The accuracy of the method depends to a large extent on the reliability of estimates of when the macrobands were formed.

To test these hypotheses, a further series of experiments were conducted on approximately 300 specimens of *Chione* (Coutts, 1974). Briefly, each shell was marked with a saw-cut, placed in a wire cage and partly buried in a sand bank. The growth dynamics of these mollusca were observed over a period of 12 months. Samples were collected at varying intervals and appropriate measurements taken. At the end of 12 months the specimens were removed from the cage and killed.

The results of these experiments confirmed that macrobands form annually during winter. However, due to inherent difficulties in supervising the experiment, the exact time in winter when the macrobands are formed was not determined. Consequently, it still has to be assumed that they form when air and sea temperatures drop to a minimum, though there is some evidence that they may form following periods of spawning and preceding the onset of the coldest winter surface temperatures (Hall, this volume).

When the average numbers of growth lines formed were counted, it was found that from the beginning of the experiment to the point where the macrobands started to form, the number was 160 ± 25, and from the point where the macrobands ended to the ventral margins the average number was 120 ± 25. The cockles continued to grow whilst the macrobands formed and an average number of 80 ± 25 growth lines were counted for that period. Doubt is therefore thrown on the validity of the hypothesis that the growth lines are *all* daily phenomena.

If one growth line is formed each day, then approximately 240 (cf. 160 in fact) should have been formed between the beginning of the experiment and the time when the macroband began to appear, and thereafter up until the end of the experimental period about 120 (cf. 200 in fact). The difference between the expected and actual numbers of the growth lines cannot be accounted for solely by counting errors and other reasons must be sought.

The formation of the growth lines bounding increments may be controlled by tidal cycles as proposed by Evans (1972) for the cockle *Clinocardium nuttalli*. However, even if this were the case for *Chione stutchburyi*, then the difference between the numbers of growth lines formed during the summer–autumn and winter–spring periods would still need to be explained. Possibly the animals feed more frequently during the winter–spring periods, thereby increasing the number of grown lines. Certainly it has not been possible to separate out semi-daily winter growth lines to date and, if they exist, then the excessive numbers of winter–spring growth lines could be explained. Again, during summer–autumn, the animals may feed less frequently, thereby reducing the numbers of growth lines formed. Feeding habits could also be linked with the amount of local pollution, which is often high, the cockle beds lying beside a sea lane and adjacent to an industrial port.

The results of these investigations indicate that much more research is needed on the growth of dynamics of *Chione stutchburyi* before the ramifications of the proposed seasonal dating methods can be assessed. The assumption that the growth lines are all daily phenomena is not borne out in this study though this does not mean that it is impractical to use growth-line counts to determine the seasonality of archaeological sites. Indeed, once the factors which control the formation of the increments are better understood it will probably be possible to make seasonal determinations with considerable accuracy.

Recent research by Koike (1973) is of considerable interest in this respect. She has examined the growth lines formed by the clam *Meretrix lusoria* and shown that they can be classified into five groups; the two most prominent types (Koike Type A and B) can be easily seen under low-power magnification and it happened that the sum of types A and B equalled the number of days duration of the experiment. The other types are generally much fainter and can easily be distinguished from types A and B. These results strongly suggest that not all lines are daily, but if the most prominent are counted, then the total should closely approximate the number of days growth. Again, she has shown that for *Meretrix* at least, counting errors would normally be over- rather than underestimates.

Unfortunately Koike's experiment ran only for 16 days and it remains to be seen whether or not she can get the same results for a similar experiment conducted over a 12-month period. Nevertheless, her approach shows much promise and could well be applied to *Chione* data in the near future, with fruitful results.

Meanwhile, observations of the variations in the distances between the growth lines will give the archaeologist a good idea of seasonality. Regardless of whether growth lines are all daily phenomena or not, counts can be used for estimating how long sites have been occupied. If all the shells from a site exhibit approximately the same numbers of growth lines between their last macrobands and their ventral margins, then the site has probably been occupied for a very short period of time. If the line counts vary, then a more extensive period of occupation is indicated. Again, the archaeologist may use growth-line counts to establish the 'relative seasonality' of sites, which must be proportional to the 'real seasonality' of the site.

The method in use

For purposes of illustration of both the utility of this method and its inherent problems, we will consider the analysis of relevant data from a European–Maori contact site (SP/1) in Fiordland, New Zealand (Coutts, 1972). This site is a small cave, with well-preserved varieties of faunal and organic remains, occupied some time between 1820–1830. The seasonal data for select archaeological layers are summarized in Figure 1. Determinations labelled 'shell' are estimates derived by the methods described in this paper. It is appropriate to interpret these data in conjunction with man-days estimates for each archaeological layer (Table 1) since this will facilitate discussions of occupancy times and population sizes. The estimates were derived from a careful analysis of the economic data, tem-

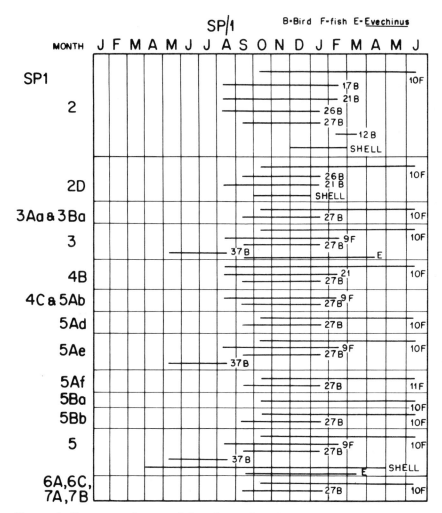

FIGURE 1. Summary of seasonal data for each archaeological layer of site SP/1, Southport, Fiordland, New Zealand. Codes are as follows:

FISH
10 Barracouta — *Thyrsites atun*
9 Southern Kingfish — *Jordanidia solandri*

BIRD
17 Southern Blue Penguin — *Eudyptula m. minor*
21 Southern Diving Petrel — *Pelecanoides urinatrix chathamensis*
26 Fluttering Shearwater — *Puffinus g. gavia*
27 Fairy Prion — *Pachyptila turtur*
12 Paradise Duck — *Tadaorna variegata*
37 Antarctic Prion — *Pachyptila d. depolata*

OTHER
E. Sea Urchin (see Coutts and Jones, 1973) — *Evechinus chloroticus*

TABLE 1. Estimated percentage numbers of man-days of food based on the probable daily calorific food intake, provided by the bird, fish and shellfish components for each archaeological layer of site SP/1, Southport, Fiordland, N.Z.

Layer	Min	Mean	Max
2	21	67	135
2A	1	3	7
2D*	6	19	38
3Aa	3	9	18
3Ba	4	14	28
3Ca	2	5	11
3Cb	1	4	7
"3"	77	277	577
4Aa	1	2	5
4B*	3	11	23
4C	12	46	97
4D*	2	6	13
5Ad	3	7	13
5Ac	14	44	90
5Af	2	5	11
5Ag		1	1
"5"	57	205	427
Totals, 5Ad–"5"	76	262	542
5Ba*	5	16	32
5Bb	6	21	44
5Ca	1	2	4
6A & 6C*	20	68	140
6D		1	1
7A*	5	17	34
7B*	1	4	8

" " = pooled layers, i.e. "3" = 3 + 3Bb; "5" = 5 + 5Aa and 5Ac + 5Ah. * = Midden deposit.

pered by a number of qualified assumptions which have been discussed elsewhere (Coutts, 1972). The calculations take into account a number of possible cumulative errors and the results have been expressed in ranges.

The data in Figure 1 give an approximate indication of the periods of the year when the site was occupied. Summer occupation seems to be important, but the shell data indicate that the site was occupied during winter as well.

Now consider the number of man-days of food represented by each midden layer, and in particular, by way of example, by layer 4B. The midden refuse from

The Seasonal Perspective of Marine-Orientated Prehistoric Hunter-Gatherers 249

this layer represents a maximum of 23 man-days food and the seasonal chart suggests that the site was occupied during the period August–June. Even if we were to double the maximum estimate of man-days to allow for uncontrollable factors, one man could have occupied the cave for about $1\frac{1}{2}$ months and not 11 months as indicated on the seasonal chart. Hence the first step in interpreting the seasonal charts is to reduce the seasonal dates to the range of overlap of all the individual estimates. This would lessen the possible period of occupation associated with layer 4B to September–January. The range is still too large, but it is the best estimate that can be derived from the available data.

When faced with the possibility of all-year occupancy (such as with layers 3 and 5), the problems are slightly different. The data are best interpreted within the framework of the hypothesis that the cave was tenanted by more than two but less than X persons, where X has yet to be determined. X would have to be limited by a number of factors, including the size of the cave and its 'living area', by local food availability and by the numbers and sizes of the boats available. The upper limit for SP/1 is probably *circa* ten, allowing a minimum area of 3 m^2 per person and taking account of waste and storage areas. (This area estimate may be conservative. Naroll (1962) has suggested that normally one person requires an average living space of 10 m^2.)

The lower limit (set at two persons) was probably dictated by the minimum amount of labour required to handle an ocean-going boat. Since the inhabitants of SP/1 almost certainly came to the area in boats, group sizes of more than two are probable. Hence, the population limits can be set at a minimum of two and a maximum of ten.

Using these limits the seasonal and man-days data for layer 5 can be reviewed in juxtaposition. At first sight there appears to be contradiction between the two data sets. The seasonal charts indicate all-year-round occupation, while the number of man-days are much less than would be expected if this were the case. The man-days estimate for layer 5 can be increased by adding the contributions from sublayers. This increases the mean to 262 man-days, which still falls short of the minimum expected figure of *circa* 700 man-days. There are a number of possible explanations for this anomaly. Firstly, since the man-days calculations did not take into account potential contributions from vegetable foods, crayfish, sea urchin, seal and dog remains, it may be argued that such omissions have drastically reduced the estimated numbers of man-days. However, the small volumes of waste material from crayfish, sea urchin and dog rule out the possibility of large contributions from these sources and only a modest contribution from vegetable sources seems likely.

The question of the role of seal in the diet is a different matter. It is relevant to consider the magnitude of the errors involved by omitting seal from these calculations. An adult female southern fur seal (*Arctocephalus fosteri*) would provide about 50 kgm of flesh (Bonner *et al.*, 1964, p 171) and by assuming that 1 gm of flesh provides 3 calories (after Osmonde and Wilson, 1961, p 73, for cooked meat with medium fat) and that a man-day is equivalent to a unit of 3000 calories, such a seal is equivalent to 50 man-days of food.

Assuming that SP/1 was occupied throughout the year by two persons, and that the error in the number of man-days for layer 5 is due to the omission of seal from these calculations, then between 4–13 seals (each yielding 50 kgm of flesh) would be enough to account for the discrepancy. These are not unrealistic estimates providing seals were plentiful in the area at the time. However, when the group size is increased to 10, the number of seals has to be increased to between 20–65. The fact remains that during the late 1820s seals were not plentiful in this area and were certainly never numerous in the immediate vicinity of the site. Again, even if numbers of seals were being caught and butchered locally, and taking into account the possibility that portions of carcasses may have been left at the kill sites, far more skeletal evidence could have been expected from the midden layers. Hence, while some significant error may have been introduced by the omission of seal from the calculations, this cannot be the full explanation.

A further potential source of error arises from the fact that some of the midden components in layer 2, and possibly layer 3, may be contemporary with material in layer 5. However, even when the man-days contribution for layer 2 is added to layer 5, the difference between expected and actual numbers of man-days remains large.

The third, and perhaps the most plausible, alternative is that the cave was occupied intermittently throughout the year; certainly the structural evidence from within layer 5 supports this hypothesis.

The seasonal and man-days data have been interpreted within the framework of the foregoing discussion. Consequently, it is held that the site has been occupied intermittently, generally during warmer months of the year, and sometimes through the year.

These data are consistent with a hypothesis of small socio-economic units engaged in very low levels of industrial activity (the evidence for which is discussed in Coutts, 1972).

Settlement pattern parameters

Earlier in this chapter, difficulties inherent in utilizing counts of growth lines to indicate, in absolute terms, the occupancy times of sites were outlined. However, it was suggested that these data may still be used to characterize an occupation horizon; that is, they may be used as measures of an artificial settlement pattern parameter.

Their utilization in conjunction with all other sources of seasonal information may be illustrated by referring to data from approximately 30 archaeological sites in southern New Zealand. Occupation horizons were divided into two chronological divisions, late prehistoric and post-European–Maori contact, and seasonal determinations were made for each archaeological layer. Histograms were constructed for these data as shown in Figure 2, for each of the two chronological horizons. In addition, following the procedure described in the last section, the estimated man-days equivalent for every layer at each archaeological site was calculated and is also presented in histogram form.

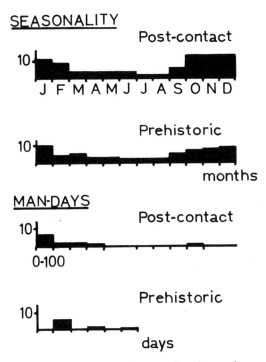

FIGURE 2. Histograms of seasonal and man-days data from coastal sites in Fiordland, New Zealand. Number of archaeological layers in each category

Clearly, there do not seem to be any outstanding differences between late-prehistoric and post-European–Maori contact settlement patterns judged on the basis of seasonal data alone. However, differences can be seen in the man-days histogram and it would appear that either larger parties went to Fiordland in the late-prehistoric era or else sites tended to be occupied for longer periods. Unfortunately, the nature of the surviving archaeological evidence is insufficient to allow independent checks of these alternative hypothesis. However, one plausible explanation is that greater mobility was induced by the presence of Europeans in some of these areas during the post-contact period, thereby reducing the time spent at any particular camp site.

Summary

I began this paper by outlining some of the difficulties which can be encountered in studying prehistoric hunter-gather societies, and in particular I stressed the importance of being able to find reliable methods of documenting settlement patterns. Germane to this problem is the determination of site seasonality. Some new seasonal dating methods have been reviewed and their limitations discussed. The utility, potential and shortcomings of these methods have been illustrated with real archaeological data from several sites in Fiordland, New Zealand. Results are sufficiently encouraging to suggest that these techniques should be developed further, following more intensive biological research on molluscan species such as *Chione*.

References

Binford, L. R. (1969). Archaeological perspectives, in *New Perspectives in Archaeology* (Ed. S. R. Binford and L. R. Binford), pp. 5–32, Aldine Publishing Company, Chicago

Bonner, W. N. and Laws, R. M. (1964). Seals and sealing, in *Antarctic Research* (Ed. R. Priestly, R. Adie and G. de J. Robin), Butterworths, London

Coutts, P. J. F. (1970). Bivalve growth patterning as a method of seasonal dating in archaeology. *Nature*, **226**, 874

Coutts, P. J. F. (1972). The Emergence of the Foveaux Strait Maori from Prehistory: A Study of Culture Contact. PhD. Thesis (unpublished) University of Otago, Dunedin, New Zealand

Coutts, P. J. F. (1974). Growth characteristics of the bivalve *Chione stutchburyi*. *N.Z. Journal of Marine and Freshwater Research* (in press)

Coutts, P. J. F. and Higham, C. F. W. (1971). The seasonal factor in Prehistoric New Zealand. *World Archaeology*, **2** (3), 266–277

Coutts, P. J. F. and Jones, K. L. (1973). A method of deriving seasonal dates from sea urchin (*Evechinus chloroticus*) fragments in archaeological deposits. *American Antiquity* (in press)

Evans, J. W. (1972). Tidal growth increments in the cockle *Clinocardium nuttalli*. *Science*, **176**, 416–417

Koike, H. (1973). Daily growth lines of the clam *Meretrix lusoria*: a basic study for the estimation of Prehistoric seasonal gathering. *J. Anthropological Society of Nippon*, **81** (2), 122–138

McBurney, C. B. M. (1967). *The Hauah Fteah (Cyrenaica) and the Stone Age of the South-East Mediterranean*, Cambridge University Press, Cambridge

Naroll, R. S. (1962). Floor area and settlement population. *American Antiquity*, **27** (4), 587–89

Osmonde, A. and Wilson, W. (1961). *Tables of Composition of Australian Foods*, Commonwealth Department of Health, Canberra

Struever, S. (1971). Comments on archaeological data requirements and research strategy. *American Antiquity*, **3** (1), 9–19

Weide, M. L. (1969). Seasonality of pismo clam collecting at ORA-82. *Archaeological Survey Annual Report*, 127–141. Dept. of Anthropology, University of California, Los Angeles

PALAEONTOLOGICAL CLOCKS AND THE HISTORY OF THE EARTH'S ROTATION

GIORGIO PANNELLA

Department of Geology, University of Puerto Rico, Mayaguez, Puerto Rico 00708

Abstract

The work of the last 10 years on palaeontological clocks should be considered as the preliminary phase of a field of research that will provide direct evidence of the evolutionary trends in the Earth's rotation. It was based on oversimplifications which are gradually corrected as our knowledge of biological growth increases.

To the complexity of the growth processes, resulting from the interaction of internal and external rhythms, there corresponds the complexity of growth patterns visible in the skeletal structures of many invertebrates. Precise information on environmental astronomical cycles is directly tied to the correct interpretation of growth patterns. Future data will have to come from fossils which lived in both intertidal and relatively deep subtidal environments. Intertidal animals reflect in their growth the ebb and flow of the tides (generally two cycles with a period of 12 hr 25 min which makes a lunar day), the lunar fortnightly tides (lowest period of 13·66 days), the lunar month (27·32 day period), but less precisely the solar semi-annual effects. Temperate subtidal animals provide a better record of the solar semi-annual effects on growth. Essential to the accuracy of the information is the feasibility of separating the semi-diurnal lunar from the diurnal solar effects on the characteristics of growth increments. There are some criteria that can be used for it, at least for mollusc growth patterns. The major difficulty in the application of palaeontological data to geophysical problems lies in finding exceptionally well-preserved fossils with continuous growth sequences.

An analysis of published data in the light of the most recent advances suggested that there are many potential sources of error in the counts and that a complete restudy of all the material is needed. A preliminary revision of part of the data indicates that some published figures are spurious because intertidal growth patterns were not separated from the subtidal ones. Figures from molluscs appear to be more reliable than those from corals.

Stromatolites are, with only one exception so far, very poor 'clocks', but they provide the only source of palaeontological evidence of the Precambrian history of the Earth's rotation. It is likely that they will continue to provide qualitative rather than quantitative information. They will be invaluable in reconstructions of the characteristics of palaeotides.

A great deal of work remains to be done before we can confidently accept the use of palaeontological clocks for geophysical implications: the only hope of success rests on a concerted effort by many researchers.

Introduction

One name is inseparably associated with the subject of this paper: J. W. Wells. He was the first to propose the use of fossils as 'geochronometers', and 10 years after his suggestion was made seems an appropriate time to assess the results of research on palaeontological evidence of the evolution of the Earth's rotation, its validity, implications and unsolved problems. The need to examine the work of the last 10 years is indicated by the fact that three review papers (Scrutton and Hipkin, 1973; Clark, 1974; Pannella, 1974) have been, or are in the process of being, published. This conference, involving most of the researchers in the field, should, hopefully, be the turning point from the pioneering stage characterized by necessary but naive individual efforts to a new stage in which a concerted effort to gather together and study large quantities of well-preserved fossil material will provide incontrovertible variations in the rotation of the Earth through geological time. This first stage has produced some positive achievements which must be mentioned here because they permit optimisim as to the future of palaeontological clocks.

The early suggestion as to how these clocks were going to work overlooked the fact that nature rarely confirms our wishful thinking and also left the impression that the method was going to be simple and unambiguous: it was just a question of counting growth layers, each representing a day, from one seasonal band to the next. But this simplistic approach proved inadequate, even in living organisms, to interpret the complexity of the code of growth. It is this complexity which makes us suspicious of the early interpretations and data but, at the same time, instils confidence in the accuracy of the biological time-keeping mechanisms. Notwithstanding the many problems still be be solved, growth patterns will produce the direct evidence for tracing a reasonably reliable history of the rotational changes of the Earth.

One cannot help but suspect that the interpretation of fossil growth patterns was strongly influenced by the current theory of the deceleration of the Earth's rotation due to tidal friction. Wells (1963) did not intend to prove the theory; he only suggested that, if the theory is correct and if animals with continuously growing exoskeletons deposit daily increments, it would be possible to determine the exact geological age of the fossilized skeletal parts simply by counting the number of daily growth bands per annual band. Obviously, if the method works, the theory is correct. However, because of the complexity of growth patterns and the possibility of formation of double increments in the 24-hr period, it is not unlikely that an observer can count more growth layers than days and reach the conclusion that in the past the length of the day was shorter. To avoid this danger, certain precautions should be adopted and precise criteria should be used for recognizing a daily band from others. But this goal can only be achieved after the mechanism of band formation has been fully understood.

Although other groups will possibly be used as palaeontological clocks, mollusc bivalves and corals have been actively studied and are relatively well-known and therefore are the most likely groups to provide information in the Phanerozoic. Of the two, mollusc bivalves are in the writer's major field of studies and will therefore receive a more detailed treatment than corals. The writer's conclusion is that mollusc bivalves have some definite advantages over the corals and should be preferred to them when feasible.

The use of stromatolites for determination of long-term changes in the rotation rate of the Earth will be treated briefly since it has been the subject of other papers (Pannella, 1972a,b, 1974). The importance of the stromatolites must not

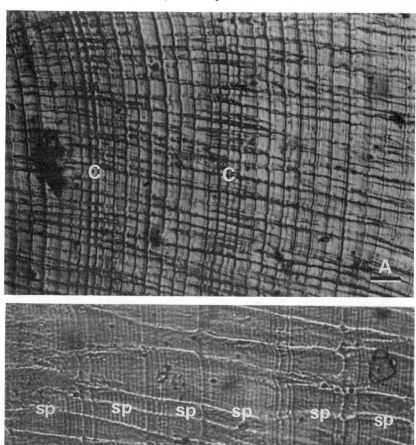

FIGURE 1. Growth patterns, lunar increments and subdaily patterns in subtidal *Mercenaria campichiensis* from 20 feet depth in Chesapeake Bay. Scale: 100 μm.
A. Spring and neap tidal modulations affect the thickness of lunar increments, so that the spacing between organic lines changes with a 6–8-day periodicity (C).
B. Lunar asymmetrical increments showing very regularly spaced subdaily patterns (sp) and organic lines (○). Detailed view of A

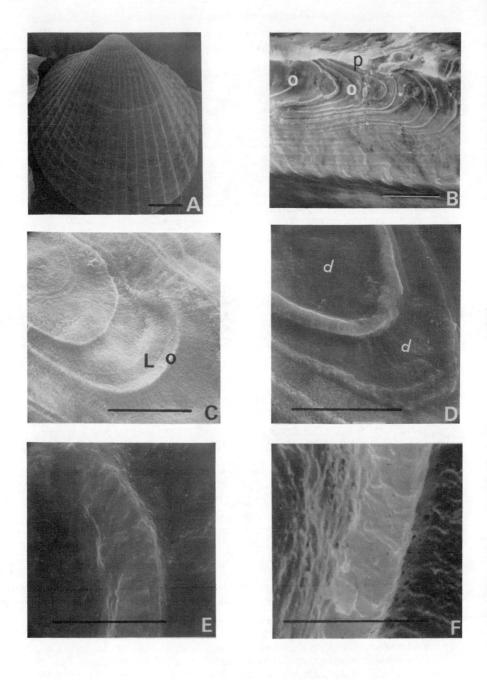

be overlooked. Although most of the known stromatolites have been examined and found unsuitable as palaeontological clocks, there are still large areas of the Earth where the Precambrian has not been studied and these may well yield promising material. More work on modern analogues of stromatolites is necessary, for the accepted mechanism for forming laminations cannot clearly account for all the stromatolite types found in the Precambrian (Walter, 1972; Walter et al., 1972).

At the time of writing a complete restudy of the material and measurements from which the published data were derived (Pannella et al., 1968; Pannella, 1972) is under way. In selecting terms for the description of growth patterns two options are open: one is to set up a hierarchy of orders quite independent of their chronological significance, the other is to choose terms separating different patterns on the basis of either their periodicity or their genesis. The first option was partially adopted by Barker (1964) who listed five orders of patterns in bivalve shells but also added a chronological interpretation: the smallest unit (V-order) represented subdaily growth layers; the basic unit, representing a 24-hr period, was called IV-order, the monthly was the III-order, higher orders were used to represent 6-month (II-order) and 12-month (I-order) growth zones. This terminology did not fit the different patterns found in other bivalves by Pannella and MacClintock (1968) who selected the second option and proposed the terms: daily, bidaily, fortnightly, monthly, spawning, seasonal and annual. Later, when dealing with growth patterns in stromatolites, in which the interpretation is often, if not always, based on numerological relationships, I suggested the adoption of the first option and proposed the following orders: I-order for the smallest unit represented by one organic-rich and one organic-poor layer, II-order and III-order for periodical patterns within seasonal (IV-order) and annual (V-order) bands (Pannella, 1972a). This solution is not completely satisfactory and the hierarchy will be upset as soon as other patterns are discovered. The danger of the second option is that an observer will be forced to attribute a chronological significance to a pattern even when it cannot be proved. In this way speculation assumes the false aspects of fact. Table I presents a list of recommended terms and definitions adopted in this paper.

FIGURE 2. Growth patterns of intertidal *Cerastoderma edule* (etched specimen) from Holy Island (England) seen at SEM. Direction of growth from left to right.
A. General view of left valve of a juvenile specimen. Scale: 1 mm; × 7·2.
B. Strongly reflected *organic lines* (○). Each line represents the beginning of one episode of tidal inundation. P, periostracum. Scale: 100 μm; × 528.
C. Detailed view of a lunar increment (○) showing the sign of the diurnal slowing down of calcification (L). Scale: 50 μm; × 1260.
D. *Organic lines*. The etching has created a differential relief between calcified layers (d) and organic lines (○). Scale: 50 μm; × 1500.
E, F. High magnification pictures of organic lines. Note the vacuolar and flaky structures of the organic material. Scale: 5 μm; × 9·6 (E); × 10·5 (F)

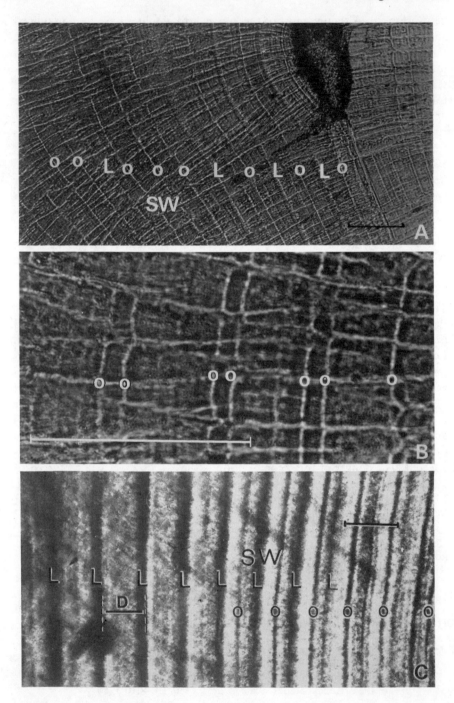

TABLE I. Growth pattern terminology

Internal Patterns

Growth Layer: The result of a discrete episode of shell growth represented by a very thin organic-rich (*organic layer*) or relatively thick inorganic-rich (*inorganic layer*) deposit. Minor fluctuations in growth rate may be expressed in fine sublayering. (Figure 1, A 2).

Growth Line: in Clark's definition 'Abrupt or repetitive changes in the character of accreting tissue' (1974, p. 1).

Organic Line: 0·5–3 μm-thick organic layer, structurally similar to the periostracum and chemically different from organic matrix. It differs from the *organic layer* for its purely organic composition, and from the *growth line* because of its thickness (Figure 2, E, F).

Doublet: a pair of layers representing one episode of prevalent calcification and one of prevalent organic deposition. A doublet may consist of one calcified layer and one organic line.

Daily Increment: a set of growth layers and growth lines that have been proved to represent one day's shell deposition. *Diurnal increment* represents a period of 24 hr. Lunar increment represents a period between two Moon culminations, the lunar day. *Simple increment* consists of two doublets and can be subdivided into a *symmetrical increment* when the two doublets have the same thickness, *asymmetrical* when the thickness is unequal (Pannella MacClintock, 1968; Pannella, 1972) (Figure 3, A, B).

Growth Patterns: sequences of growth layers and growth lines recurring periodically. The periodicity can be expressed by adding adjectives such as fortnightly, monthly, seasonal. Particular patterns within daily increments are called *subdaily* patterns (Figure 1, A, B).

Winter Check: pattern characterized by a gradual decrease in the thickness of growth layers up to a point (winter break) followed by a gradual increase.

FIGURE 3. Types of growth increments in living mollusc bivalves, seen in acetate replicas (A, B) and thin section (C). Direction of growth from right to left. All scales: 100 μm.

A. Symmetrical increments (S) and switch zone (SW) in intertidal transplanted *M. mercenaria*. Organic-rich layers (L) sandwiched between two organic lines (○) are the characteristic growth structures of the increments. In the switch zone the patterns shift from symmetrical to asymmetrical by losing one or two half-lunar days so that a fortnight is represented by 13·5 lunar increments.

B. Asymmetrical increments in same individual: three organic lines (○) define a lunar increment.

C. 'Solar' increments (d) and switch zone in *Tridacna squamosa*. Thick dark lines represent periods of slow calcification (organic layers: L). Organic lines (○) are due to interruptions of calcification due to spring low tides; they are not as sharp as in *M. mercenaria* because *Tridacna* does not close the valves tightly

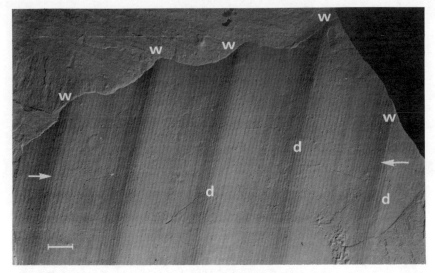

FIGURE 4. Outer topography of a Middle Devonian unidentified Cephalopod. Major modulations (annulations) correspond to periods of slow growth (W) in which growth bands are thin and disturbances (d) more frequent. The darker line in the middle of W could correspond to winter breaks. There are 106 bands between the arrows which give an average of 26·50 bands for the four annulations. Longer sequences would provide more reliable averages. If the annulations represent annual periodical events, the number of bands/annulation strongly suggests a fortnightly periodicity and 26·5 tidal neap–spring cycles in Middle Devonian. Scale: 1 cm. Direction of growth from left to right

Biochecks: macroscopic growth band, consisting of relatively thin and closely spaced increments, representing a period of relatively slow growth occurring after spawning and preceding the onset of the coldest winter sea surface temperatures (Hall, this volume).

External Patterns

Fine Growth Ridges: topographical expression of growth layers on the coral epitheca, or on any external exoskeletal surface.

Growth Bands: sequences of growth ridges between periodically recurring constrictions (Figure 4).

Annulations: prominent annular ridges related to seasonal variations in growth rate.

Stromatolites (from Pannella, 1972a)

First-Order: basic bipartite time-unit, made of a clastic (corpuscular) and organic-rich layer. Thickness 1–500 μm (Figure 5, B, C).

Second-Order: bipartite unit made of two groups of laminae with different thickness. Thickness 0·1–3 mm (Figure 5).

Third-Order: Succession of two second-order bands, Thickness 0·2–5 mm (Figure 5).

Fourth-Order: Consisting of several second-order bands grouped in zones of similar textural and compositional differences. Can

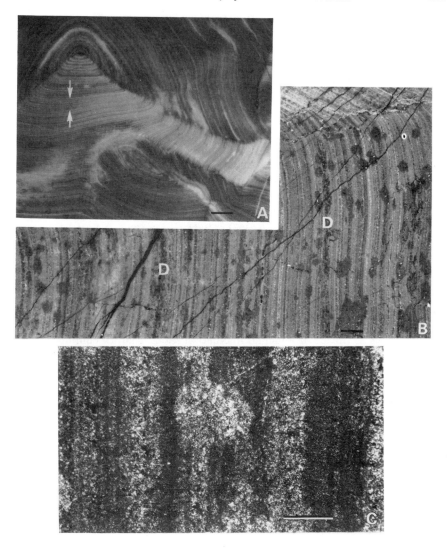

FIGURE 5. Growth pattern in 2000-My-old stromatolites from Great Slave Lake (Canada).
A. General view of a polished surface, showing the general structure and the tidal banding. Ten tidal bands are visible between the two arrows. Scale: 1 cm.
B. Print from a thin section (used as negative) of the same specimen. Corpuscular and organic layers are recognizable in the tidal bands marked by checks. D, disturbances in the growth sequence. Scale: 400 μm.
C. Detailed view of tidal bands in thin section, showing corpuscular and organic layers. Scale: 200 μm

be detected visually because of colour differences. Thickness 0·5–25 mm (Figure 5).

Fifth-Order: unit made of two fourth-order bands, one organic-rich and dark, the other organic-poor and light. Thickness 0·5–45 mm (Figure 5).

Daily growth increments: the elusive unit

Crucial to the accuracy of information of biological growth patterns is the possibility of identifying discrete, recognizable, unambiguous and consistent growth layers representing regular time units. The fact that most biological systems oscillate with daily frequency provides a solid ground for the idea that growth, and consequently growth increments, follows the daily rhythms and represents a 'clock' (Aschoff, 1966; Neville, 1967). There is, however, a basic difference in the use of the term *daily* when applied to biological systems and when referred to an astronomical cycle. A biological system can perceive the cycle only through its environmental effects, that are basically daily but no not always occur at exactly the same time. For instance, the temperature fluctuations have daily frequency with diurnal lows and highs but with variable time spans between peaks. A marine invertebrate tends to synchronize its physiological and metabolic activities to this environmental fluctuation rather than to culminations of celestial bodies. Biological clocks are obviously as accurate in recording the cycle as the environmental events are in reflecting it. It is not only convenient but also necessary for organisms to synchronize their behaviour and internal functions to the strongest environmental oscillations. The universal short-term oscillations are those caused by the rotation of the Earth on its axis. Light, temperature, barometric pressure are just a few of the many oscillations to which an organism can gear its behaviour. Another rhythm with 24 hr 50 min or 12 hr 25 min periods, the ebb and flow of the tide, imposes ecological fluctuations to which most marine organisms must adapt.

Organic growth is two-phased: one phase of energy accumulation and storage and the other of energy consumption. Food supply is an obvious factor that controls the amount of energy available for growth. Rhythmic feeding will produce rhythmic growth; rest must follow activity and continuous growth is a physiological impossibility. The alternate aspects of growth are recorded in the deposition of skeletal parts as 'growth layers' in which the period of active growth has created a structure different from that of rest. The fact that growth is not continuous but is affected by slowing-downs or stoppages, compensates for the variations in the time of occurrence of environmental fluctuations. Fast growth will be triggered by certain environmental stimuli occurring once or more times per day at approximately the same time. A record of the episode or episodes will be left in the growth patterns. Since the episodes are not continuous it is the frequency and not the time of their occurrence which is recorded. The activity and rest cycle follows the oscillations of the environment and also its disturbances. A certain amount of noise should be expected in the record. Traumatic and catastrophic events will tend to disrupt growth and thus its record. The importance of the noises must be evaluated in order to assess the degree of accuracy of growth patterns as clocks. But, before this, it is advisable to analyse what is known of the factors and mechanisms involved in the formation of growth bands. Growth-layer formation in plant and animal groups has been studied and the best understood mechanisms are those of insects (Neville, 1963) and molluscs (Newcombe, 1935; Orton, 1923; Petersen, 1958; Davenport, 1938; Pannella and MacClintock, 1968; Clark, 1968; House and Farrow, 1968; Koike, 1973). Neville recognizes two types of daily layers, one created as direct response to environmental fluctuations, the other as direct expression of internal circadian oscillations. Growth patterns would, therefore, be the result of the effects on growth of exogenous rhythms interfering with exogenous rhythms.

Of the animals listed by Neville (1967) as known to contain growth layers, I should like to discuss the ones that, abundant in the fossil record, are most likely to be most used as palaeontological clocks: namely *molluscs*, with particular emphasis on bivalves, and *corals*. The special case of stromatolites will also be discussed.

Molluscs

With few exceptions, molluscan shells are external and function as supporting and protective shields for the soft parts. They are deposited by a modified part of the body wall called 'mantle' and are calcium carbonate structures made of sequentially deposited layers. The basic mechanics of external shell deposition is similar in most molluscs. Inorganic material deposits when the mantle is extended over the surface of accretion. Any external or internal stimulus that forces the withdrawing of the mantle also interrupts the calcium carbonate deposition. Prolonged interruptions are recorded in growth sequences as crystalline surfaces of structural discontinuity. In mollusc bivalves the withdrawal of the mantle is generally connected to the tight closing of the valves and causes the formation of surfaces of discontinuity. Bivalves are the best known molluscs as regards shell deposition and calcification because they are easy to experiment with and have a relatively simple geometrical form.

The formation of growth layers in mollusc bivalves

A bivalve shell consists of an outer organic horny layer (periostracum), an inner layer made of calcium carbonate and an organic matrix. It is secreted at the contact between the margin and the epithelial tissues of the outermost of three mantle folds. The inner face of this fold secretes the periostracum, whereas the outer face is thought to be responsible for the deposition of the calcium carbonate. The precise modalities and causes of the inorganic deposition are still a controversial subject that will not be discussed in this paper. One fact appears undeniable: calcification is always preceded by the secretion of the

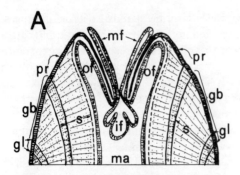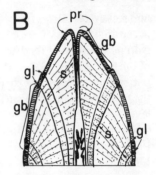

FIGURE 6. Diagrammatic representation of mantle–shell relationships of mollusc bivalves with non-reflected edge at the ventral margin during gaping of the valves and shell deposition (A) and during tightly closed valve and interruption of growth (B). The *organic lines* are possibly deposited in the early stage of valve gaping after prolonged closings. Key: if, internal mantle fold; mf, middle mantle fold; of, outer mantle fold; ma, mantle cavity; pr, periostracum; gb, surface expression of internal growth increments; s, internal growth increments; gl, growth lines due to interruptions of calcification

organic matrix which consists of a wide array of structually different proteins, the peptide fraction of which could be responsible for the enucleation of inorganic ions. (Degens et al., 1967). Some differences between the organic material of the periostracum and that of calcified shell have been noticed. In *Mytilus californianus* the calcitic layers have a higher ratio of acidic to basic amino acids than the periostracum and outer ligament (Hare, 1963); periostracal proteins are hardened by quinone-tanning agents apparently secreted only by the cells underlying the inner face of the fold (Hillman, 1961). Large amounts of phenolic amino acids are present in the periostracum.

The relationship between marginal mantle folds and the shell surface of accretion is shown in the idealized marginal cross-section perpendicular to the growth bands and to the shell represented in Figure 6. At the contact between the shell and the outer face of the outermost fold, calcium deposition occurs, the periostracum being secreted from the inner face of the same fold. Some calcium carbonate deposition also occurs in the inner area delimited by the myostracum. The maximum deposition takes place along a surface (surface of maximum growth, SMG) that is either immediately below the periostracum or in the middle of the shell outer layer, according to the attitude of the mantle margin. Some bivalve groups have a non-reflected margin and some a reflected margin (Pannella and MacClintock, 1968).

The mantle fold is withdrawn from the shell margin when the shell is tightly closed, resulting in the interruption of calcium carbonate deposition and apposition of periostracal flap against the shell. To each tight, prolonged closing there will correspond an interruption in the deposition of the inorganic material which should be recorded in the growth patterns. In a tightly closed shell, also, the acidity of the internal environment increases and, in extreme cases, not only is deposition stopped, but also part of the shell is resorbed to buffer the decreasing pH (Dugal, 1939). It must also be mentioned that feeding can only take place when the shell is gaping and the food-gathering structures are extended.

Any environmental or internal stimulus that forces the animal to close for a certain length of time during calcification interrupts calcium deposition and creates a 'growth line'. The widely adopted techniques of analysing growth patterns by means of acetate replicas and scanning electron microscope pictures show that etching with 1% HCl aqueous solution produces a differential relief between the growth lines and the growth layers, and that the 'line' is actually a thin layer of organic material with a flaky structure similar to that of the periostracum (Figure 2, E, F). The material of the line is water-insoluble and different from the organic matrix in the calcified layers which appears to etch even more than the associated calcium carbonate. Stain reactions of the two materials in *Cerastoderma edule* are different (Dolman, personal communication). There are no published data on the amino-acid composition of the two materials, because the entire shell is considered homogeneous as far as the type of the organic material is concerned. Indeed, analyses of the different matrices within these layers are needed. Internal growth layers of bivalve shells can be separated into two broad categories: one characterized by layers distinguishable from each other because of a gradual change in the colour which corresponds to a change in organic content and/or crystalline structure (Figure 3, C), the other by layers separated by organic lines (Figure 3, A, B). After the study of many species and populations from different ecological situations it became apparent that the first type tend to be found in bivalves living in subtidal areas, whereas intertidal animals generally show the second type. Very significant for determining the origin and mechanism of organic line formation were the experiments carried out in natural conditions using juvenile *Mercenaria mercenaria* (MacClintock and Pannella, 1969). The experiments consisted of collecting one or more specimens every hr for 52 consecutive hr which had been previously notched and transplanted, and studying the amount of shell material deposited at the margin since the previous collection. The experimental animals were kept 1 ft below the mean water level in an area of semi-daily tides and where tidal fluctuations ranged, during the periods of the experiment, ±3·5 ft with respect to the mean water level. During spring tides, the animals were uncovered for up to 3 hr. The results of the experiments justified the following conclusions on layer-forming mechanisms in intertidal *M. mercenaria*.

(1) The organic lines are formed during periods of shell closing at low tides. Thus to each low tide there corresponds a line, and the periodicity of line formation is a lunar one: 12 hr 25 min in semi-daily tides or 24 hr 50 min in daily tides. Because of the bimodal lunar-day rhythm a *lunar increment* may consist of two growth layers and two organic lines (when two low tides interfere with internal calcification rhythms).

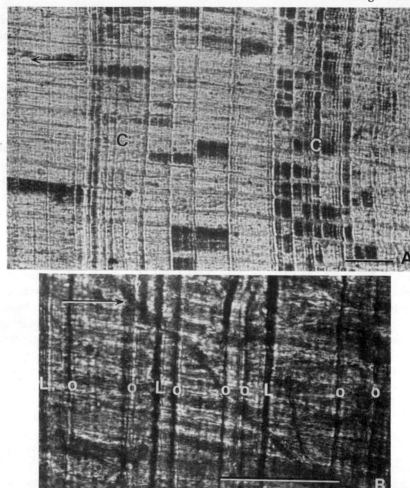

FIGURE 7. Types of growth increments in living *M. mercenaria*. Direction of growth is shown by arrow. Scale: 100 μm.

A. The effect of tidal modulation on increment thickness in tide-pool *M. mercenaria*. The clustering (*C*) is related to neap-tides periods when water exchange and food supply is reduced.

B. The shift of phase between diurnal slowing down or stoppage of calcification (*L*) and tidal uncovering (*O*) produces, in some instances, lunar 'triplets'. Note the shift of line *L* through the organic lines

(2) The growth layer is deposited rapidly in 2–3 hr during rising tides in what can be called rebound acceleration of calcification after a period of enforced retardation. This compensatory rebound is common in biological systems (Figure 7). The fastest rate of deposition was noted during evening rising tides.

(3) There is a propitious time for calcification during the 24-hr period dictated

by internal circadian rhythms and by external parameters such as light and temperature. Growth patterns in intertidal *M. mercenaria* are not the results of purely tidal rhythms and can only be explained by the interplay of solar, circadian and lunar rhythms.

At low tide, intertidal bivalves are forced to close their valves if uncovered and to interrupt calcium carbonate deposition. The characteristics of the growth patterns are controlled by: (a) the position of the animal with respect to mean low water, (b) tidal amplitude and type, (c) the time of day during which spring and neap tides occur.

At the incoming tide, filter-feeder bivalves can be seen actively feeding with their syphons emergent from the substratum. Rao (1954) reported tidal rhythms in the pumping rates of intertidal *Mytilus edulis* and *M. californianus*. *Cerastoderma edule* show tidal rhythms in shell gaping and feeding activity (Morton, 1970). *Mercenaria mercenaria* and *Ostrea virginica* show increased shell gaping at times of high tides (Palincsar, 1958; Brown, 1954). Similar lunar rhythms are commonly found in animals living along coastlines (see Palmer, 1973 for a review on the subject). Some of these rhythms persist in constant conditions in the laboratory and in the absence of tides. By analogy to circadian rhythms, Palmer (1973) has suggested the term circalunadian. Independently of the problem of the endogeneity of these rhythms, it would appear only natural that intertidal bivalves synchronize their physiological activity to lunar rhythms and that growth patterns record this rhythmicity. 'The periodic wetting', writes Palmer, 'is not an important entraining factor for most intertidal organisms'. For bivalves, however, exposure to air definitely interrupts growth and mechanical agitation due to the incoming tide plays an important role in the resumption of calcification.

Growth patterns in subtidal bivalves still show the influence of lunar periodicities but less markedly: organic lines are replaced by organic-rich layers. The deepest specimen the writer ever found with fortnightly recurrent patterns was dredged alive from a depth of 100 fathoms.

That rhythmic environmental stimuli which cause the valves to close play an essential role in the formation of organic lines is demonstrated by the growth patterns of bivalves which are unable to close completely.

Pectinidae belong to one group of such bivalves: they are subtidal, capable of movement either by flapping their inequal valves or by using the foot, and cannot close their shell tightly. Clark's work on this group (1968, 1969, 1974) demonstrated that they do not have sharply defined internal patterns. The reason for this is not only the foliated shell structure which masks growth structures, but also the inability to close their shell tight.

The discovery of the role of tidal fluctuations in the formation of growth patterns has very important implications. First it explains why growth patterns in animals living in areas of different tide types, or in the same area but at different water levels, are different. Growth patterns of *Clinocardium nuttalli* living along

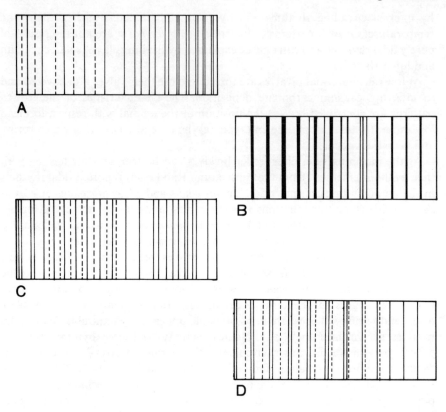

FIGURE 8. Common types of 'switch zones' in bivalves. Each type is the result of the tidal effects with semi-daily, diurnal, or mixed inundations on the 24-hr rhythms of calcification.

A. Type found in semi-daily tides, with one uncovering at neap tides and two at spring tides (e.g. *M. mercenaria* in Milford Harbor, Connecticut). Symmetrical increments, with the central dashed line representing the 'solar' or diurnal slowing down of calcium carbonate deposition, are deposited at neap tides, asymmetrical at spring tides. In the switch zone the 'solar' diurnal stoppage line disappears because it falls at the same time as low tides.

B. The solar periodicity is mirrored by calcification rate: during the slowing down of calcification an organic-rich layer is deposited (heavy lines); with fortnightly frequency, low tides interrupt calcification (organic lines are represented by light lines) (e.g. *Tridacna squamosa* from North Australia).

C. The effects of mixed diurnal and semi-daily tides produce a succession of symmetrical increments with internal layers due to diurnal slowing down of calcification (represented by dashed lines) and asymmetrical ones (e.g. *Clinocardium nuttalli* living along Oregon shores).

D. Dashed lines represent the diurnal slowing down of calcification; the thin lines shifting with respect to the dashed lines represent organic lines (e.g. *Cerastoderma edule* from intertidal of Holy Island, England the Oregan coast (Evans, 1972) are different from those of *Cerastroderma edule* living along the coasts of Great Britain (Farrow, 1971, 1972); those of *M. mercenaria* in New England from those of the same species living in Florida. All these patterns, however, have something in common which allows them to be treated in the same way when interpreted for geophysical information. Without the effect of tidal inundations the bivalves would deposit calcium carbonate only once a day, following environmental temperature, light and food supply fluctuations which quite generally have 24-hr frequency. Calcification would reach a peak and then gradually decrease. The two resulting growth layers would be an inorganic-rich thick increment and a relatively thin organic-rich increment, but because of the tidal fluctuations some of the environmental parameters such as temperature and food supply acquire 'lunar' periodicities. The interplay of the two groups of lunar and solar periodicities which are 50 min/24 hr out of phase, mixed to internal rhythms, results in a complexity of patterns. Animals living in intertidal areas show a 'switch zone' (Figure 8) which is the effect of the tidal phase 50-min shift with respect to the sidereal day. The shift compresses the time of calcium carbonate deposition in one direction until it snaps back, missing a day every fortnight, consisting of $13\frac{1}{2}$ lunar days and corresponding to 14·8 sidereal days. The different types of 'switch zones' represented in Figure 8 have in common the fact that one growth line used as a reference to measure the 'daily' spacing suddenly repeated at half the previous or subsequent distance. When this happens the sequence of symmetrical increments changes to sequences of asymmetrical increments (e.g. in *M. mercenaria*, living 1 ft below mean water in Milford Harbor, Connecticut). In patterns of *Cerastroderma* described by Farrow (1971) the change is from thick symmetrical increments with thick organic lines to thin symmetrical or asymmetrical ones with thin organic lines, whereas in *Clinocardium nuttalli* (Evans, 1972) the change is from simple to asymmetrical. In *Tridacna squamosa*, from the Southern Pacific, the switch from the solar day doublets to symmetrical increments with sharp tidal lines is another example of the interplay of the solar and tidal effects on patterns (Figure 3, C).

The effects of tidal rhythms and solar rhythms on growth patterns were not fully understood in the early work on bivalves. The underlying conviction was that light or temperature were basic factors involved in growth patterns. It is interesting to look at the results discussed in these papers, while keeping in mind the tidal effects. Pannella and MacClintock (1968) transplanted subtidal *Mercenaria mercenaria* into an intertidal environment and obtained growth bands that matched in number the days of transplant. They concluded the bands were daily. The position of the animals with respect to the mean water level was such that only the highest of the two tides triggered calcification, except every fortnight when two inundations were recorded and interpreted as two daily increments. The coincidence between the number of counted bands and the number of days was fortuitous and masked the true chronological meaning of the bands. On the contrary, the interpretation of the patterns in *Tridacna*, which consist of solar-day bands periodically affected, every fortnight, by a tidal interruption which split the band into a 'complex increment' was correct (Pannella and MacClintock, 1968).

House and Farrow (1968) studied the growth patterns of *Cerastoderma edule* and wrote: 'The result of the counts of the banding seen in shells killed at successive dates is to show conclusively that most fine banding is daily' (1968, p. 1384), and a little further: 'A daily increment seen on the acetate peels consists of a narrow and a wide band'. This definition, however, does not apply to all increments, because in the intertidal *C. edule* the 'narrow band' is the tidal organic line and two narrow 'bands' are deposited every 24 hr 50 min. Doubling the number of days by counting tidal layers as daily, House and Farrow interpret the fortnightly patterns as monthly. The same error affected the interpretation of patterns in Farrow's subsequent papers (1971, 1972). The error led to the conclusion that *Cerastoderma* has the tendency to miss several days of growth every 'month'. In reality, it is only in the winter band that some days are probably missing, but it is difficult to establish how many because of the closely packed layers that can neither be counted nor measured with precision.

In conclusion, while for palaeoecological work the loose definition that one finds in the literature of daily increments is acceptable (Rhoads and Pannella, 1970), a strict separation of 'solar' and 'lunar' increments is necessary for geophysical information. Once this separation is recognized, then the accuracy of the biological record in reflecting environmental events becomes apparent. Mollusc bivalves seem the best candidates to provide precise information, but other taxa should not be neglected. Gastropods and cephalopods should be further explored. The above description of the basic mechanism of growth pattern formation could apply to other taxa.

Corals

While the suggestion that the conspicuous external growth bands in fossil corals were the reflection of growth-rate variations related to environmental fluctuation was proposed a long time ago by Ma (1934), Krempf (1934) and Wells (1937), it is only recently that experimental work attempting to clarify the relationship between environmental factors and growth patterns in modern corals has been carried out (Barnes, 1970, 1971, 1972; Hipkin, 1972). The time lag between the early suggestion and Wells' proposal of using corals as geochronometers (1963) and between this and the recent experimental work is an indication of the difficulties of gathering information on the mechanism of coral growth. The Goreau's work (1959, 1959a, 1959b) is of fundamental importance in the attempt to understand it, but was not specifically aimed at the discovery of the chronological meaning of growth increments. It clearly shows that one of the most important factors of growth is light and justifies the conclusion that daylight changes could be the rhythmic basis of coral growth ridges.

It is Barnes (1970, 1971, 1972), however, who attempts to decipher, using modern hermatypic corals as experimental animals, the mechanism of ring formation. Laboratory and *in situ* observations show that growth ridges are formed as a result of the changes in the shape and extension of the soft parts depositing the inorganic material forming the epitheca. In particular, the ridge is deposited when the outer wall is enveloped by the distended and folded tissue called 'lappet', in a fashion somewhat similar to the bivalve mantle, since calcification occurs when soft parts envelop the growing edge. The enclosing of the epitheca growing margin is rhythmic and apparently controlled by daylight. The fold is withdrawn during the day and extended at dusk, and it is at this time the ridge is formed. The calicoblastic epidermis deposits an inner layer that progressively thickens the epitheca from the inside, whereas the external epidermis deposits the epitheca. The position of the lappet varies according to environmental changes. Barnes concludes 'any stimulus-feeding, long-term rhythmical events, anything which shifts the position of the free body wall can affect the position, it is impossible to assign a temporal significance to growth ridges' (1971). Information not provided by Barnes' experiments (in part due to the fact that they were carried out in a laboratory and in an area of minimal tides), which could be important in deciding whether or not to accept his conclusion, is with regard to the effect of tidal fluctuations on the rhythmic withdrawal of the lappet. The multiplicity of events that can cause the withdrawing of the depositing tissues is the same in the mollusc bivalves: yet their patterns are quite clear and the noise easily distinguished. The basic difference lies in the mechanics of growth line formation, which, in bivalves, is related to closing of the valves. Because growth is closely related to light, the solar component should always be recognizable in coral patterns, and if the tidal fluctuations are recorded in shallow water corals, they should be very good clocks. The major drawback of coral ridges lies in the lack of internal record that is very useful in resolving ambiguities and in studying important details and clues; were the mollusc bivalves growth patterns expressed only as external ridges, their usefulness and accuracy would be greatly impaired. The other drawback is that the epitheca is not protected from damage and often shows signs of breaking.

Hipkin (1972) stressed the unlikelihood of post-mortem preservation of the delicate leading edge of the epitheca and also the fact that corals respond by distending and retracting soft parts not only to daily cycles but also to other unpredictable events which do not necessarily follow the same cycles. More interested in the methods of objective analysis and counting of coral growth pattern than in gathering experimental data on ring formation, Hipkin described in detail the annulations of some Devonian *Endophyllum archiaii* (Billings). The annulations consist of two zones: one about 2 mm thick, containing ridges up to 10/mm, which represents rapid growth; followed by a second zone containing about 150 ridges averaging 50 μm in thickness. The shape of the ridges changes through the zones from double ridges to regularly spaced ridges, then simple ridges with double crest and back to single-crested ridges. It is tempting to compare this description with that of mollusc bivalves (Pannella and MacClintock, 1968; Pannella, 1972) and speculate that the patterns are the results of solar and tidal effects.

Hipkin found it very difficult to identify the different types of ridges on an irregular curved surface. He counted 35 ridges from paired type to single ridges with double crest and suggested the possibility of a tidal cycle interfering with the diurnal cycle. He also recognized two more zones, one characterized by very fine ridges, the other by alternating fine- and normal-type ridges. The variety of ridge types, not to be unexpected from the descriptions of the early work, suggests that similar environmental cycles act on both coral and mollusc shell deposition. It is unfortunate that Barnes did not pay attention to the ridge morphology in his experiments. From the available pictures (Barnes, 1971) it seems that at least a single and a double ridge type are present on the epitheca of *Montastrea annularis*. This fact immediately arouses some suspicion that the early interpretation of 1 ridge = 1 day could have been naive and that the published figures from corals may be erroneous. Hipkin could not eliminate the subjectivity in identifying growth ridges; as an example of how drastic the bias could be he mentions that he counted 253 and 359 ridges in the same specimen at different times. The difficulty of separating disturbances from daily ridges, the obvious discontinuity of the record, and the lack of internal growth patterns, together, at this point, make the use of corals as palaeontological clocks certainly problematical, if not impossible. Undoubtedly, before discarding them, more work should be done on both living and fossil corals. While many questions could be answered with further research, the use of corals for tracing the Earth's rotational history will possibly be limited to a few geological periods and then only as supporting evidence for data from other taxa.

Stromatolites

Work on the chronological significance of stromatolite laminations is rather scanty, notwithstanding the important role stromatolites can play in the reconstruction of the Earth–Moon system evolution in the Precambrian. Almost all the described Precambrian types of stromatolites were examined for growth patterns and most of them were found unsuitable as palaeontological clocks. The major drawback of stromatolites is the incompleteness of their sequences. Because of their mode of growth (Gebelein, 1969) stromatolites generally show very discontinuous records and can provide, in the most optimistic of the hypotheses, *minimum* values (Pannella, 1972a, 1972b). For this reason, only the highest figures from different contemporaneous stromatolites can be considered as more representative and closer to the actual values. When seasonal (IV-order) zones are present, stromatolites may provide the most useful information on the presence of seasonal bands. Figure 5 illustrates an example of this qualitative clue. The very regular II-order bands are interpreted as tidal because they are subseasonal and quite numerous (10–12 bands/IV-order). The number of laminations within these bands, however, is small (maximum number encountered) and it is unlikely to represent the true number of days/tidal band. The fact that tidal bands are present in 2000-My-old fossils eliminates all theories that imply a late capture of the Moon. Obviously this information is acceptable only if it does not remain isolated and only if it is supported by other data. A string of similar data, distributed through the Precambrian, will give more ground to this type of interpretation. In the future, subtidal and special types of stromatolites may permit more precise information (Walter, 1972; Walter et al., 1972). The Mink stromatolites (Pannella, 1972b) may be one of these special types, that could provide unique information on the length of the day and of the synodic month 2000 My ago. I'd like to point out here that the value of the 448-day year, being a minimum value, led to the compelling conclusion that the slowing down of the Earth's rotation has been going on since Gunflint time and that its rate is much lower than the one accepted for the secular acceleration.

Perhaps there are other stromatolites like the 'Mink' type waiting for discovery by the palaeontologists, and with the deepening of our knowledge on modern and fossil stromatolites we will be able to determine which types are the best recorders of the environmental rhythmic changes; perhaps we will be able to fill some of the gaps between Gunflint time and the Phanerozoic. The search must continue. So far, however, only the 'Mink' stromatolites have fulfilled the hope for Precambrian palaeontological clocks.

Growth patterns: the cadences of the environment

From the characteristics of the growth layers and growth lines it is possible to decide whether the rhythmic mechanism of valve closing is lunar or diurnal. It is clear that intertidal and subtidal animals living in areas affected by strong tides will time their activities and growth to tidal rhythms. Where tidal range and effects are small, animals are generally synchronized to solar rhythms (Pannella, unpublished observations).

As pointed out in the preceding pages, solar or diurnal increments are separable from the lunar ones because they are delimited not by organic lines but by colour changes to which chemical and calcium-organic matrix ratio changes correspond. Solar increments make up relatively simple growth patterns. Purely solar growth patterns probably do not exist in marine animals, with the exception of deep-water ones. The lunar effects are generally manifested either in the thickness of the doublets or in colour changes.

Also inescapable are the solar effects on tidal growth patterns. These patterns are made of tidal and lunar increments bound by organic lines. In *M. mercenaria* and, judging from pictures or direct observations, in many intertidal bivalves, the shortest periodical grouping consists of 6–8 simple bands alternating with a similar number of double bands, the latter deposited during spring tides, the former during neap tides. The two groupings make up the fortnightly pattern. In bivalves living in zones of very strong fluctuations the fortnight consists of $13\frac{1}{2}$ increments corresponding to 14·8 solar days. The Moon completes one revolution of the Earth in space—that is, relative to the stars—in approximately 27 days, 7 hr, 43 min 11·5 sec. Since the Earth is also going around the Sun, it takes the Moon one synodic month or almost 29 days, 12 hr, 44 min and 28·8 sec to go around the Earth relative to the Sun. Every 29·5 days an intertidal organism will be subjected to a similar tidal situation. This periodicity is reflected in patterns of 27 lunar days.

Actual tide records depart in many details from the predicted curves: many irregular oscillations related to wind and pressure changes introduce noises and distortions. An intertidal bivalve records, obviously, not only the tide *line spectrum* but the total *tide noise spectrum*. It is possible, as Munk and Cartwright (1966) have done, using time series analysis of 50-year-long tide records, to extract the true tidal periodicities from the growth patterns, notwithstanding the noise (Dolman, 1974).

The tidal potential is higher when the Sun is over the Equator during the equinox than during the solstices. The tidal range will, consequently, be higher and the patterns will generally be recording the differences. Barker (1964) has mentioned the possibility of some effect of equinoctial tides on growth patterns, on theoretical grounds. There is, however, a strong effect on shell growth due, at least in part, to the fact that these are also periods of intense storms. The effect of equinoctial tides on growth patterns is expressed by a 6-month periodical change in the colour and organic content of the shell and, in *M. mercenaria*, at least, in the formation of cross-lamellar sublayers.

Finally, the other rhythm recorded in both 'solar' and 'lunar' patterns is seasonal. The orbit of the Earth–Moon system around the Sun is completed in 365 days, 5 hr, 48 min, 45·7 sec. During this time the Earth is subjected to seasonal changes expressed by temperature, daylight, rainfall and many other environmental parameter variations. In midlatitude regions these seasonal changes are pronounced: all biological systems are affected by them. Marine animals and molluscs show clear changes in the depositional patterns. For example, *Mercenaria mercenaria* living in the Cape Cod area show winter bands of 80–140 thin, organic-rich growth increments different from the thick summer increments. Winter growth sequences are made of units that progressively thin to a winter break (or winter check) coinciding, in intertidal animals, with the first freezing spell (Pannella and MacClintock, 1968). By counting lines from winter check to winter check one obtains a variable number of growth increments, depending on the year. Only averages of long, continuous sequences of annual patterns provide the exact number of days per tropical year. In fossils, long, well-preserved continuous annual sequences are extremely rare and it is unlikely that one could determine with the necessary accuracy the length of the year in the past. One way to overcome this difficulty is to determine the length of the synodic month and

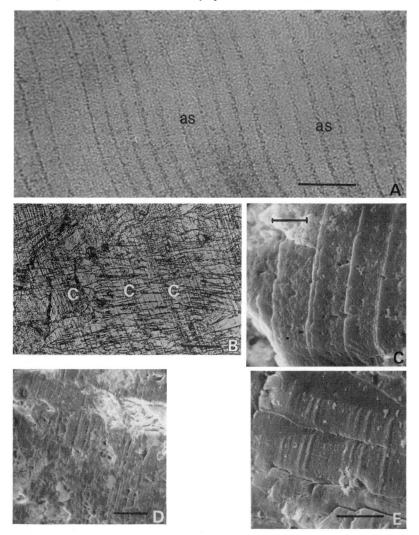

FIGURE 9. Growth patterns and increments in fossil bivalves. Direction of growth shown by arrow.
A. Acetate replica of an Oligocene *Crassatella* sp. cross-section. Asymmetrical increments (*as*), and organic lines are well preserved. Scale: 100 μm.
B. Acetate replica of Mississipian *Conocardium herculeum* Konick, showing clusters of thin (C) and thick increments. The increments are lunar. Scale: 100 μm.
C. SEM picture of growth increment in etched *Conocardium herculeum*. Scale: 7 μm; × 2100.
D. SEM picture of patterns shown in B. The organic lines have disappeared during fossilization and left a porous layer. Scale: 50 μm; × 440.
E. Detailed view of patterns in same specimen of C, showing 'triplet' increments. Scale: 10 μm; × 2025

the number of tidal fortnights per year. The latter information is often preserved even when the single increments have been destroyed. Occasionally it is possible to use the outer topographical expression of these fortnights for the same goal. In Figure 4 the outer topography of a Devonian cephalopod is the only record left of the periodical growth of the shell. Two patterns are well defined: major modulations, which represent periodical changes in the shell diameter, and minor bands, the thickness of which decreases when the diameter is at a maximum. The regularity of the patterns and the comparison to other cephalopods' outer topography support the interpretation that the modulations have annual, and the bands fortnightly, periodicities. The average number of bands/year observed from this specific example is 26·50. Multiplying this figure by the number of days/fortnight (obtained from another shell of the same age or, even better, from the same assemblage) the approximate length of the Devonian year could be arrived at. Berry and Barker (1968) used a similar method to evaluate the length of the year in the Late Cretaceous. This method, if properly applied, offers the most promising development in the field of palaeontological clocks.

Scanning electron microscope is often helpful in deciphering the intricacies of growth. In Figure 9, B depicts a sequence of fortnightly growth patterns of a Mississippian *Conocardium herculeum* Konick, as seen in a replica under optical microscope; D is the SEM view of similar patterns in the same specimen. The depression in D corresponds to the dark lines in B: they represent the vestigia of organic lines. The increments are obviously lunar and are complicated: asymmetrical (C), complex asymmetrical (E) and symmetrical (D). Because of the presence of many strong subdaily patterns the measurement of spacing is often difficult in this sequence.

The analysis of growth patterns

Growth records expressed only on the outer surface, as in many corals, are difficult to analyse because of scarcity of details to support the interpretation and because they are difficult to study microscopically and they are readily weathered.

Hipkin (1972) has tried to improve the visual identification of patterns by using a mechanical profiling apparatus which magnifies the original epithecal topography but produces curves which preserve noises without preserving the clues to recognize them as such. The intervention of mechanical devices may be useful in many instances, generally when the record is very clear and clean (in which case the interpretation is also clear) but they are not likely to provide the ultimate solution of growth pattern reading.

Internal records of growth such as those in most molluscs have the advantage over the external ones of being more detailed and more readily measured. Thin sections and acetate replicas have been successfully used in the study of bivalves (Pannella and MacClintock, 1968). In replicas, more than in thin sections, the contrast between the organic-rich layer and the calcified layers is reinforced and facilitates the reading of the patterns.

The readings, aimed at extracting the periodicity of regularly recurrent patterns, were obtained either by counting the number of a selected basic unit thought to represent daily growth per repeated pattern (Barker, 1964; Berry and Barker, 1968; Pannella and MacClintock, 1968) or by measuring the thickness of the basic unit and determining the frequency of the periodical changes in thickness (Clark, 1968; House and Farrow, 1968; Farrow, 1971, 1972). Although neither method eliminates the subjectivity of defining and selecting the basic unit, the latter is to be preferred because it preserves detailed information which can be subjected to harmonic analysis.

Of the many problems related to palaeontological clocks, the one most stressed by critics is the subjectivity of the readings. The term *subjectivity* is used derogatorily on the basis of what Michael Polanyi has called *Laplacean delusion* of identifying accuracy and science with observational objectivity. So far, however, both experimental and theoretical science have relied on unspecifiable arts and intuitive judgements. Without underplaying the possible danger of subjectivity, I would like to stress here that *subjectivity* implying 'depending on a subject' is not negative *per se*, what is deleterious to science is the *biased subject*. Undoubtedly the readings of growth patterns often depend on critical interpretation of ambiguous areas and could be seriously affected by a biased observer. The erroneous readings that one finds in the literature, however, are due more to misunderstanding of the basic mechanism of shell growth than to forcing observations into preconceived models. This is not to say that methods in which the subject intervention and his potential bias are either limited or eliminated, should not be devised. All attempts to find a mechanical reading technique have failed, because often the output was even more difficult to interpret than the input. Many insuperable obstacles were found in the use of the negative pictures of coral growth patterns as diffraction gratings because of the irregularities, incompleteness and curvature of the sequences. Hipkin (1972) used the negative pictures of coral epithecal features as Fraunhofer gratings through which a laser beam was diffracted into a photographic plate which, in theory, should have recorded 'demodulated' sequences. Hipkin's conclusion was that the noise present in the record made this method unsuitable.

The most promising suggestion, based on measuring the spacing between growth lines, is offered by Dolman (1974). He added to a standard microscope with a standard rotating and x-y movement stage, a projection screen provided with a micrometer screw eyepiece. The micrometer, driven by a motor, is attached to a digitizer which records the motor movements, The digitizer values are transferred to paper tape that can later be analysed by a computer. This method does not eliminate the potential bias of subjective selections of what to measure, but definitely expedites the gathering of data and their harmonic analysis and preserves much of the information stored in growth patterns. It also reduces the danger of unconscious bias of a subject trying to match periodicities.

Another problem concerns the accuracy and continuity of the biological growth record. I have tried to show that the idea of 'missing lines' often recurring in the literature originated from unwarranted generalizations or misinterpretation of experimental results. Clark (1968) carried out laboratory experiments on *Pecten diagensis* to check the chronological significance of growth ridges and concluded that the ridges were daily but that most of the experimental animals were missing some ridges during the 51 days of the experiment, and that the highest number of ridges and not the average was representative of the true number of days. This conclusion is now quoted as if it were generally applicable (e.g. Mazzullo, 1971). However, warnings have been issued about extrapolating from laboratory results to natural animal behaviour (Pannella and MacClintock, 1968; Rhoads and Pannella, 1970). In Clark's case the extrapolation is tentative since Pectinidia do not adapt to laboratory conditions and die rapidly. That the health of Clark's animals was rapidly deteriorating is shown by the gradual decrease in their shell growth rate. Furthermore, this is an unusual group of bivalves and its behaviour cannot be extended to an entire class, phylum or kingdom. The other evidence for 'missing line' derives from work on *Cerastoderma edule*, the intertidal cockle abundant along the English shores (House and Farrow, 1968; Farrow, 1971, 1972). The true role of tidal fluctuations in the mechanism of the shell deposition escaped both workers; and their definition of the daily layers led to the misinterpretation of growth patterns. The definition applies to the layers deposited during a tidal inundation and thus represents not a 'daily' unit but a 12 hr 25 min period. This error explains why the fortnightly patterns, which are strongly developed, were interpreted as monthly patterns and why missing lines were added in order to match the supposed daily doublets with the calendar days. The need for adding 'missing lines' disappears once the true chronological meaning of the doublets is recognized (Dolman, 1974a). The patterns then appear to record the tidal inundations with extreme accuracy, also the spring and neap tidal modulations. Lunar periodicities are easily and accurately determined. What is more difficult to determine in *Cerastoderma* and, possibly, in most intertidal bivalves living at high latitudes, is the number of lunar or solar days in a year. During winter months growth rate is either drastically reduced or shell calcification stops altogether. In arctic bivalves growth records extend only from 3–5 months (Pannella, unpublished data) and the length of growth stoppages is obviously a function of latitude, at least at latitudes higher than 40–45 N. Maine's intertidal bivalves (latitudes 43°50′) miss about a month of growth in the winter. There was no stoppage of growth in bivalves south of Boston's latitude along the Atlantic coast (Pannella, unpublished observations). These are instances of growth stoppage due to cold climates which should be kept in mind but should not be considered representative of a frequent climatic situation in geological time. Of the 3000 fossil bivalve specimens of different ages and geographical areas from the United States and Europe I have examined, none appeared to have lived in high latitudes or in areas where the winters were so harsh as to stop growth. In fact many indicate a very even climate and very little seasonal changes in growth rate.

Palaeontological Clocks and the History of the Earth's Rotation

This does not imply that missing record does not occur in biological growth but shows that this is more the exception than the rule and that the evidence supposed to prove it is immaterial, at least for mollusc bivalves living in natural conditions.

The basic weakness of the palaeontological data is due more to the paucity of growth sequences from which they were derived than to intrinsic inaccuracy of the biological record. In theory one *perfect* sequence for, say, every 10 My could provide the accurate figures needed for geophysical implications, but in practice there is no way to prove the 'perfection' of the sequence and it is very dangerous to use only scant sequences instead of a good statistical sample. Unfortunately, if this has not been done yet it is because of the difficulty in finding enough fossils with well-preserved growth patterns. I am now attempting to apply for the Jurassic period what I would call 'the total approach' which consists of examining *all* the taxa which present growth layers in a given assemblage. I am examining not only bivalves, but all the molluscs, the brachiopods, corals, coralline algae and so forth with the hope that each one of them will provide some clues on the periodicities environmental cycles. Because of the dependence of growth on the environmental fluctuations, an assemblage should provide a representative sample of these periodicities, relatively independently from taxonomic differences. The total approach could be the answer to one of the basic limitations of palaeontological data—the paucity of good material. The study of fossil periodicities must become more and more a matter of statistics.

Palaeontological data and the history of the lunar orbit

As will appear evident to the reader of this book and this chapter, the field of palaeontological clocks is still suffering from 'growth pains'. At this stage, although the pressure from geophysicists demanding reliable figures is strong, it is premature to consider the published figures as definitive or even representative. There are still too many uncertainties to be sure that the early interpretations of growth patterns were correct.

My less-than-enthusiastic attitude should not be interpreted as rejection of the method or of all the work done, but is the natural reaction to the uncritical use of data and to the tendency of nonchalantly publishing isolated figures based on weak evidence and accidental countings which just happened to agree with the previously published figures obtained in a similar fashion. This tendency obviously establishes precedents which do not contradict each other and that give the false appearance of 'compelling evidence'. One wonders what would be the status of the field had Wells published a different number, say 365 days for the Devonian year! But is it possible that Wells and everybody else following him were so wrong and that the trend in the slowing down of the Earth's rotation is not real?

While a critical discussion of all the palaeontological data is being published elsewhere (Pannella, 1974), I would like to restrict this last part of the paper to a few considerations on the validity and meaning of the data for the publication of which I am directly responsible (Pannella *et al.*, 1968; Pannella, 1972a). How accurate these data are will become apparent in the very near future when the ex-examination, now in progress, of all the material on which they were based will be completed. In restudying the material, I have adopted a very strict discriminatory technique first by separating fossils with *solar patterns* from those with *lunar patterns*, then measuring the spacings between growth lines of organic lines and compensating for the effect of the longer lunar day with respect to the sidereal day in the switch zones. This is a licit compensation, especially in the Phanerozoic, during which the Earth–Moon system is not supposed to have changed its relationships with respect to the Sun. All the ambiguous material and sequences are being eliminated so that only clear and noise-free sequences are studied. The measurements are then analysed through a computer using the same programs adopted by Dolman (1974a) for harmonic analysis and coherence. The strict selection has further reduced the amount of fossils available and requires the addition of new material to obtain large enough statistical samples for each geological period. The quantity of material that can be amassed by an individual is limited. Hence it would be appropriate to create a centre of growth patterns to which many workers could send material for study. The advantages of such a centre would be manifold. To mention just a few: a comprehensive worldwide vertical and horizontal representation of fossils, the standardization of analysis, the possibility of selecting only the best material, the storage in one place of all material, the possibility of creating a training centre.

Of all the data on the length of the synodic month, those from the Cainozoic appear to be the most reliable because they are derived from good material that often presents no problem of interpretation (e.g. Figure 9, A). These data indicate a rapid deceleration since the Palaeocene. Figures for the Late Cretaceous will have to be supported by additional data since long sequences from which they were derived were rare. However, if one accepts the validity of the Cainozoic figures, then one should not expect that the future data will change the figures much.

For the pre-Cretaceous figures one has to wait for results of the restudy of the old material. Additional material will also be necessary, because only the Mississipian *Conocardium herculeum* Konick has well-preserved, though occasionally ambiguous, patterns (Figure 9, B–E).

On the basis of palaeontological data two conclusions seem inevitable. Even if inaccurate, the data do show a consistent trend which indicates that the deceleration of the Earth's rotation has been an active phenomenon for at least 2000 My. As for the rate of deceleration, one cannot calculate it with precision but, if the Mink stromatolite values are correct (even if taken as minimum values) the slowing-down rate accepted for the secular variation cannot be extrapolated back into the geological time. The fact that tidal bands exist in 2000-My-old stromatolites (Figure 5, A–C) confirms this conclusion.

These two conclusions, of course, have important implications on the evolution of the Earth–Moon system. So far they can only be accepted on a qualitative basis, but the time when palaeontological clocks will also provide quantitative information is not far away.

Acknowledgments

This work forms part of a project, supported jointly by the Minna–James–Heineman Foundation and by NATO, which was carried out at the School of Physics (University of Newcastle upon Tyne).

Thanks are due to Professor S. K. Runcorn, Dr. G. Rosenberg, Mr. C. Jones and all those who helped during my stay in England. Special thanks are due to Dr. J. Dolman for suggestions and assistance. Dr. C. A. Hall, P. F. Hoffman, Dr. A. McGugan, Dr. J. Pojeta, Dr. C. Scrutton and Dr. M. N. Walter kindly provided some of the material used in this study. Finally C. MacClintock must receive full credit for his collaboration, assistance and suggestions which were instrumental in the development of the ideas developed here. Dr. J. D. Weaver kindly read the manuscript.

References

Aschoff, J. (1966). Circadian activity patterns with two peaks. *Ecology*, **47**, 657–62

Barker, R. M. (1964). Microtextural variation in pelecypod shells. *Malacologia*, 2 (1), 69–86

Barnes, D. J. (1970). Coral skeletons: an explanation of their growth and structure. *Science*, **170**, 1305–1308

Barnes, D. J. (1971). *A Study of Growth, Structure and Form in Modern Coral Skeletons*. Ph.D. Thesis, University of Newcastle upon Tyne.

Barnes, D. J. (1972). The structure and formation of growth-ridges in scleractinian coral skeletons. *Proc. Roy. Soc. Lond.*, B. **182**, 331–350

Bennett, M. F. (1954). The rhythmic activity of the quahog, *Venus mercenaria*, and its modification by light. *Biol. Bull.*, **107**, 174–191

Berry, W. B. N. and Barker, R. M. (1968). Fossil bivalve shells indicate longer month and year in Cretaceous than present. *Nature*, **217**, 938–939

Brown, F. A., Jr. (1954). Persistent activity rhythms in the oyster. *Am. J. Physiol.*, **178**, 510–514

Clark II, G. R. (1968). Mollusk shell: daily growth lines. *Science*, **161**, 800–802

Clark II, G. R. (1969). Shell characteristics of the Family Pectinidae as environmental indicators, Ph.D. Thesis, Caltech, Pasedena, 181 pp

Clark II, G. R. (1974). Growth lines in invertebrate skeletons (in press)

Davenport, C. D. (1938). Growth lines in fossil Pectens as indicators of past climates. *J. Paleon.*, **12**, 514–515

Degens, E. T., Spencer, D. W. and Parker, R. H. (1967). Palaeobiochemistry of molluscan shell proteins. *Comp. Biochem. Physiol.*, **20**, 533–579

Dolman, J. (1974a). *An Investigation of Growth in Bivalves*. Thesis, University of Newcastle upon Tyne

Dolman, J. (1974b). A technique for the extraction of environmental and geophysical information from growth records in invertebrates and stromatolites. (This volume)

Dugal, L. P. (1939). The use of calcareous shell to buffer the product of anerobic glycolysis in *Venus mercenaria. Cellul. Comp. Physiol.*, **13**, 235–251

Evans, J. W. (1972). Tidal growth increments in the cockle *Clinocardium nuttalli*. *Science*, **176**, 416–417

Farrow, G. E. (1971). Periodicity structures in the bivalve shell: experiments to establish growth controls in *Cerastoderma edule* from the Thames estuary. *Palaeontology*, **14**, 571–588

Farrow, G. E. (1972). Periodicity structures in the bivalve shell: analysis of stunting in *Cerastoderma edule* from the Burry Inlet (South Wales). *Palaeontology*, **15**, 61–72

Gebelein, C. D. (1969). Distribution, morphology, and accretion rate of recent subtidal stromatolites (Bermuda). *J. Sed. Pet.*, **39**, 49–69

Goreau, T. F. (1959). The physiology of skeleton formation in corals. I. A method for measuring the rate of calcium deposition by corals under different conditions. *Biol. Bull.*, **116**, 59–75

Goreau, T. F. and Goreau, N. I. (1959a). The physiology of skeleton formation in corals. II. Calcium deposition by hermatypic corals under various conditions in the reefs. *Biol. Bull.*, **117**, 239–250

Goreau, T. F. and Goreau, N. I. (1959b). The physiology of skeleton formation in corals. III. Calcification rate as a function of colony weight and total nitrogen content in the reef coral *Manicinarare lata* (L.). *Biol. Bull.*, **117**, 419–429

Hall, C. A. (1974). Latitudinal variations in the shell growth patterns of bivalve molluscs: implications and problems. (This volume)

Hall, C. A., Dollase, W. A. and Corbato, C. E. (1974). Shell growth in *Tivela stultorum* (Mawe, 1823) and *Callista chione* (L., 1785) (Bivalvia): annual periodicity, latitudinal differences and diminution with age. *Palaeogeogr., Palaeoclimatol., Palaeocol.*, **15** (1), 33–61

Hare, P. E. (1963). Amino acids in the proteins from aragonite and calcite in the shell of *Mytilus californianus*. *Science*, **139**, 216–217

Hillman, R. E. (1961). Formation of the periostracum in *Mercenaria mercenaria*. *Science*, **134**, 1754–55

Hipkin, R. G. (1972). *Some Aspects of the Dynamical History of the Earth–Moon System.* Thesis, University of Newcastle upon Tyne, 220 pp.

House, M. R. and Farrow, G. E. (1968). Daily growth banding in the shell of the cockle, *Cardium edule*. *Nature*, **219**, 1384–1386

Koike, H. (1973). Daily growth lines in the clam, *Meretrix lusoria*. A basic study for the estimation of prehistoric seasonal gathering. *J. Antrop. Soc. Nippon*, **81**, 122–138

Krempf, A. (1934). Inscription marégraphique des cycles des retrogradation des noaeds de la lune par certains coraux constructeurs de récifs. *Comp. R. Acad. Sci. Paris*, **198**, 1708–1710

Ma, T. Y. H. (1934). On the seasonal change of growth in a reef coral, *Favia speciosa* (Dana) and water temperature of the Japanese Seas during the latest geological times. *Proc. Imp. Acad. Tokyo*, **10**, 353–356

MacClintock, C. and Pannella, G. (1969). Time of calcification in the bivalve mollusk *M. mercenaria* (L.) during the 24 hour period. Abstract Annual meeting *Geol. Soc. Am.*, p. 140

Mazzullo, S. J. (1971). Length of the year during the Silurian and Devonian periods: new values. *Geol. Soc. Am. Bull.* **82**, 1085–1086

Millar, R. H. (1968). Growth lines in the larvae and adults of bivalve molluscs. *Nature*, **217**, 683

Morton, B. (1970). The tidal rhythm and rhythm of feeding and digestion in *Cardium edule*. *J. Mar. Biol. Ass. U.K.*, **50**, 488–512

Munk, W. and Cartwright, D. E. (1966). Tidal spectroscopy and prediction. *Phil. Trans. R. Soc. London*, A **259**, 533–581

Neville, A. C. (1963). Growth and deposition of resilin and chitin in locust rubber-like cuticle. *J. Insect. Physiol.*, **9**, 265–278
Neville, A. C. (1967). Daily growth layers in animals and plants, *Biol. Rev.*, **42**, 421–441
Newcombe, C. L. (1935). Growth of *Mya arenaria* L. in the Bay of Fundy region. *Can. J. Res.*, **13**, sect. *D*, 97–137
Orton, J. H. (1923). On the significance of 'rings' on the shells of *Cardium* and other molluscs. *Nature*, **112**, 10
Palincsar, J. S. (1958). *Periodism in Amounts of Spontaneous Activity in the Quahog Venus mercenaria*. Doctoral dissertation, Northwestern University
Palmer, J. D. (1973). Tidal rhythms: the clock control of the rhythmic physiology of marine organisms. *Biol. Rev.*, **48**, 377–418
Pannella, G. (1972a). Precambrian stromatolites as palaeontological clocks. 24th IGC, sect. 1; 50–57
Pannella, G. (1972b). Palaeontological evidence on the Earth's rotational history since Early Precambrian. *Astrophys. Space Sci.*, **16**, 212–237
Pannella, G. (1974). Biological time, palaeontological clocks and the history of the Earth–Moon system. *Geophysical Reviews* (in press)
Pannella, G. and MacClintock, C. (1968). Biological and environmental rhythms reflected in molluscan shell growth. *J. Palaeont., Mem.*, **42**, 64–80
Pannella, G., MacClintock, C. and Thompson, M. M. (1968). Palaeontologic evidence of variations in length of synodic month since Late Cambrian. *Science*, **162**, 792–796
Petersen, G. H. (1958). Notes on the growth and biology of the different *Cardium* species in Danish brackish water areas. *Meddr. Danm. Fisk. og Havunders, N.S.*, **2**, 31
Petersen, G. H. (1966). *Balanus balanoides* (L.) (Cirripedia): Life cycle and growth in Greenland, *Meddr. Grønland*, **159**, 1–114
Rao. K. P. (1954). Tidal rhythmicity of rate of water propulsion in *Mytilus* and its modificability by transplantation. *Biol. Bull.*, **106**, 353–359
Rhoads, D. C. and Pannella, G. (1970). The use of molluscan shell growth patterns in ecology and paleoecology. *Lethaia*, **3**, 143–161
Runcorn, S. K. (1964). Changes in the Earth's moment of inertia. *Nature*, **204**, 823–825
Runcorn, S. K. (1968). Fossil bivalve shells and the length of month and year in the Cretaceous. *Nature*, **218**, 459
Scrutton, C. T. (1965). Periodicity in Devonian coral growth. *Palaeontology*, **7**, 552–558
Scrutton, C. T. (1970). Evidence for a monthly periodicity in the growth of some corals. In *Palaeogeophysics* (Ed. S. K. Runcorn), pp. 11–16, Academic Press, London
Scrutton, C. T. and Hipkin, R. G. (1973). Long-term changes in the rotation rate of the Earth. *Earth Science Reviews*, **9**, 259–274
Thompson, I. L. and Barnwell, F. H. (1970). Biological clock control and shell growth in the bivalve *Mercenaria mercenaria*. *Geol. Soc. Am., Abstr. with Programs*, **2**, 704
Walter, M. R. (1972). A hot spring analog for the depositional environment of Precambrian Iron Formations of the Lake Superior Regions. *Econ. Geol.*, **67**, 965–980
Walter, M. R., Bauld, J. and Brock, T. D. (1972). Siliceous algal and bacterial stromatolites in hot spring and geysers effluents of Yellowstone National Park. *Science*, **178**, 402–405
Wells, J. W. (1937). Individual variations in the rugose coral species *Heliophyllum halli* (Edwards & Haine) *Paleont. Amer.*, **2**, 1–22
Wells, J. W. (1963). Coral growth and geochronometry, *Nature*, **197**, 948–950
Wells, J. W. (1970). Problems of annual and daily growth rings in corals. In *Palaeogeophysics* (Ed. S. K. Runcorn) Academic Press, London, 3–9

DISCUSSION

MULLER: I like the conservatism of your data. It is interesting that all investigators seem to be getting the same answer for the accleration of the Earth; either everyone is wrong (which I doubt) or your data is settling around the true value. Even if, as you propose, the biological worker uses growth data to try to disprove the geophysical hypothesis, I believe the data probably would give the same answer for the Earth's deceleration.

BUDDEMEIER: What are the causative mechanisms of periodic growth increment production?

PANNELLA: There is no one parameter because nutrition, temperature changes, current flow, and probably other environmental parameters are operative. When the animal's valves close, it deposits a line. Experiments can neither prove nor disprove any hypothesis related to periodic growth as the animals were inevitably disturbed.

PALAEONTOLOGICAL AND ASTRONOMICAL OBSERVATIONS ON THE ROTATIONAL HISTORY OF THE EARTH AND MOON

S. K. RUNCORN
School of Physics, University of Newcastle, U.K.

Abstract

An angular acceleration in the longitudes of the Moon and Sun of non-gravitational origin was first discovered from early observations of eclipses. This was interpreted as a secular lengthening of the terrestrial day of about 2 msec/cy and was explained in terms of tidal friction. The telescopic observations of the longitudes of the Moon in addition have shown irregular fluctuations and these were proved to arise from irregular fluctuations in the length of the day by the demonstration that they also appeared, reduced in the ratio of the mean motions, in the longitudes of the Sun, Mercury and Venus. These variations in the speed of the Earth's rotation of the order of 1 part in 10^7 on a time scale of a few years have been explained by the coupling of the magnetohydrodynamically turbulent core to the weakly conducting mantle. It is of major geophysical importance to revise the observations of historical times, which is being done both by the present use of atomic clocks to monitor the rotation of the Earth, and by the more thorough study of Chinese, Babylonian and Greek data on the eclipses seen from various known sites: some discrepancies remain unresolved.

The changes in the Earth's rotation rate resulting from the interchange of angular momentum between the core and the mantle, or from a secular change in the moment of inertia of the Earth (possible consequences of theories of the expansion or contraction of the Earth or of the growth of the iron core), or from solar tidal friction, have no effect on the orbit of the Moon. The lengthening of the day by lunar tidal friction does result in the transfer of angular momentum to the Moon's orbit resulting in a secular increase in the Earth/Moon distance and a real deceleration in the Moon's longitude (as measured in Ephemeris time). Extrapolation backwards in time therefore predicts that the Moon was much closer to the Earth in its earlier history, a matter of considerable importance for theories of lunar origin. The present value of lunar tidal friction, if assumed to be typical of geological time, in fact poses a very considerable problem for the early history of the Earth/Moon system, an attempted solution of which by Gerstenkorn leads to the assumption of the Moon being captured in a retrograde orbit and a 'catastrophic' approach of the Moon to the Earth, which must have resulted in a 'thermal event' in the Precambrian.

The further development of the palaeontological method of determining the number of days and months in the year at different geological times to a very much greater accuracy is therefore ciritical for the discussion of the origin of the Earth/Moon system and for the study of various hypotheses of the Earth's evolution.

The fact that the Earth's day is slowly lengthening has been known for quite a long time. Edmund Halley, over 2½ centuries ago, inferred from studies of ancient eclipses that there was a secular acceleration in the longitude of the Moon, later to be interpreted as evidence that the day was gradually lengthening.

Spencer Jones (1939) demonstrated that there were, in addition, changes in the length of the day of a more complicated kind. It had been known that there were differences in the observed longitude of the Moon from those calculated on the basis of Newtonian mechanics over the last $2\frac{1}{2}$ centuries. Spencer Jones showed that the similar discrepancies in the longitudes of Mercury, the Sun and Venus, when divided by the ratio of their mean motions to that of the Moon, agreed with the anomaly curve for the Moon. As the Moon moves 13·4 times as fast as the Sun across the celestial sphere the detail of the latter curve was known much better than the others. But the reasonably good agreement between these weighted discrepancies in the longitude of these bodies in the solar system showed that the true cause of the discrepancy was that the unit of time then used by astronomers was not the invariable time assumed in the dynamical calculations. These are called the irregular fluctuations in the length of the day. A change in the slope of the discrepancies in longitude shows that a change has occurred in the rotation rate of the Earth. This interpretation of the data in terms of changes in the Earth's rotation can be thought of as follows. Imagine a clock with a second hand, a minute hand and an hour hand. If the angular movements of the second and minute hands are divided by the ratios 720 and 12 they should all be exactly in agreement with the hour hand if they are attached to the same clock. Thus the agreement between these various bodies, or the equal movements of these various hands of the celestial clock, show that Mercury, the Sun, the Moon and Venus all tell the same time. But these movements are being compared with the movement of a clock based on the Earth's rotation. The fact that there are discrepancies between this and the former is therefore due to the fact that the time which astronomers used in their analysis of the positions of these bodies was until recently based on submultiples of the rotation period of the Earth. The discrepancy has reached about 15 sec of arc in the longitude of the Moon in the observations since about 1650. The variation that was observed just before 1900 is a lengthening of the day of about 4 msec, which became established within a few years. These variations were a great puzzle to people trying to explain them by meteorological effects, or changes in the moment of inertia of the Earth due to tectonic movements, e.g. associated with earthquakes.

Since the introduction of crystal and atomic clocks, it has been discovered that there is an annual variation in the length of the day with an amplitude, varying somewhat from year to year, of 1·8 msec, and the inference that it is of meteorological origin has been quantitatively proved (Munk and McDonald, 1960). The irregular fluctuations on a time-scale of 10–100 years, however, have proved more difficult to explain as they are large and persist for 10–20 years. While this suggests that they might be explained by movement of mass within the crust or mantle, it turns out that this explanation is not quantitatively plausible. A meteorological cause might be possible, for the changes are about twice the amplitude of the annual term, but it is hard to suppose such a large change in the circulation pattern could persist over a decade or so. I conclude that causes at the surface of the Earth are unlikely to explain the irregular changes.

A proposal was made some years ago that the irregular fluctuations arose from the exchanges of angular momentum between the solid mantle and the fluid core. We have evidence from the study of geomagnetism that the core is rotating more slowly than the mantle. Many comparisons of isomagnetic maps and isoporic charts illustrate this. Bauer (1895) drew maps showing lines of equal difference between the observed angle of magnetic dip and that calculated on the basis of the geocentric dipole formula. Examination of those diagrams for 1780 and 1885 show these lines of equal geomagnetic dip anomaly shift to the west at about $\frac{1}{5}°$ per year. This old discovery in geomagnetism was rediscovered about 1950 by Vestine et al (1948) and, by the principle of magnetohydrodynamics, which has increasingly been applied in this subject, it is inferred that the fluid iron core where the geomagnetic field is generated is rotating westwards relative to the mantle. That the core is more slowly rotating than the mantle cannot be the result of tidal friction. The westward drift must therefore be explained in relation to the irregular rotation. Therefore an interchange of angular momentum between the core and mantle due to some kind of electromagnetic coupling between the two is an adequate cause of both phenomena. There are many unresolved difficulties about the quantitative discussion of the electromagnetic torques between the core and mantle. Rochester (1960) has concluded that the calculated torques based on knowledge of the geomagnetic field and its secular variation fail by about a factor of 5 to be sufficient to cause the accelerations deduced from the data. However, the fact that there is an interchange of angular momentum between the core and the mantle now seems established by the work of Vestine and his colleagues. In Figure 1 the full line shows the variation in the length of the day essentially derived from data similar to that of Spencer Jones, but actually due to a rediscussion of the fluctuations

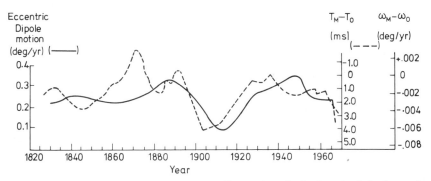

FIGURE 1. Correlation between the irregular fluctuations in the length of the day and the westward drift of the geomagnetic field

in the longitude of the Moon by Brouwer. The relative rotation of the core westwards with respect to the mantle is shown by the dotted line. This latter is assumed to be given by the rate of change in longitude of the position of the eccentric dipole which best represented the field at different epochs. This position is determined from the spherical harmonic analyses for those epochs. There is a reasonably good correlation, bearing in mind that in the last century neither set of data was very accurate. The clear inference can be drawn that equal and opposite torques are acting on the mantle and on the core and no other agency apart from induction by the varying geomagnetic field generated by the core has been established. While the irregular fluctuations in the length of the day are due to an internal redistribution inside the Earth of its angular momentum, the slow secular deceleration of the Earth arises from an external cause. Halley first drew attention to an acceleration in the Moon's longitude of about 10 arcsec/cy required by the records of ancient eclipses, but about half was later explained by gravitational effects. This use of ancient eclipse records has established that the day has gradually lengthened by something like 2 msec/cy over the last 2000 years, and Muller and Stephenson have critically reassessed the evidence. From Spencer Jones' analysis one can also determine this secular acceleration: it is separable from the irregular fluctuations because it affects the Moon's orbit and thus there is a real (angular) acceleration of the Moon as well as the apparent one due to the lengthening of the day. The value which is obtained for this from Spencer Jones' data is $21''/cy$ but is now suspected of being inaccurate perhaps because the data in the earlier (17th and 18th) centuries is less accurate than the later data and yet is critical for the determination of the quadratic term.

Some years ago I argued that the core of the Earth was growing due to settling of iron from the mantle. This arose from a very speculative attempt to explain the fact that there were three main periods in the Precambrian based on radiometric dating. Igneous and metamorphic rocks were formed, presumably at the time of worldwide tectonic movement. Work on the palaeomagnetic polar wandering path of the Precambrian shows that the interpretation I proposed, that there were three earlier periods of large continental displacements prior to that discovered by Wegener, is reasonable. I interpreted these as due to three changes in the convection pattern in the mantle, successively from a 1st, to a 2nd, to a 3rd and 4th harmonic pattern. To explain this I assumed that the core of the Earth has grown gradually as a result of iron collecting in the centre. A change in the moment of inertia through time, an idea first put forward by Urey (1952), would therefore be brought about. This would result in a secular change in the length of the day distinguished from that due to the gradual change in the angular momentum of the Earth and from the irregular changes due to the angular momentum of the Earth being redistributed between core and mantle. Other theories have been put forward, which predict changes of either sign in the moment of inertia of the Earth, for instance if the Earth has expanded through

Observations on the Rotational History of the Earth and Moon 289

time, because of a secular decrease in the gravitational constant (G) or if the radius of the Earth changes due to internal heating or cooling, the length of day would gradually change, but the change would be only a fraction of that produced by tidal friction.

Unless G changes, the year remains constant as the orbital angular momentum of the Earth cannot change. Thus Wells' (1963) observation from corals that in the Devonian (450 My ago) there were 400 days in the year, means that the Devonian day had 22 hr (defined in terms of the year). If the changes in the moment of inertia of the Earth due to these various possible theories are calculated, they affect the length of the day at most by a fraction of that now occurring by tidal friction. Thus from Wells' observations alone it is not possible to separate the two types of change. Later, however, Scrutton (1964), studying the Devonian corals, saw bands which he postulated were monthly and found 30·6 days in the month. With the month as well as the day it is possible to distinguish between the changes.

Runcorn (1964) determined a relation between the palaeontological data and the astronomical if the number of days in the year is ω and the number of days in the sidereal month is S, when ω/S is related to the change in orbital angular momentum of the Moon (L) (Figure 2). This is obtained by writing down Kepler's 2nd and 3rd laws of motion, neglecting the eccentricity, of the lunar orbit. With the two observations, the number of days in the month and the number of days in the year, another quantity of geophysical interest can be found. The difference between the present angular momentum of the Moon L_0 and the past L equals

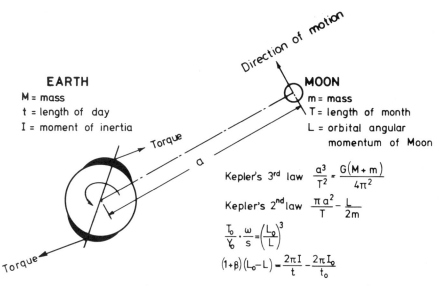

FIGURE 2. Tidal recession of the Moon from the Earth. Y_0 = number of days per year; T_0 = number of days per month

the angular momentum that the Earth has lost to the Moon. If β is the ratio between the solar tide and the lunar tide, then

$$(1 + \beta)(L_0 - L) = 2\pi I/t - 2\pi I_0/t$$

when t is the length of the day in the past. This gives the moment of inertia change with time as well as the loss of angular momentum of the Earth and it is these two quantities which, of course, are so interesting to the geophysicist. The loss of angular momentum from the Earth as a result of tidal friction, whether at a constant or variable rate, is relevant to the problem of whether tidal friction occurs only in the oceans or whether some part of tidal losses is occurring in non-elastic behaviour in Earth's mantle; a matter which has not been resolved and which cannot be resolved from present-day observations. Also, of course, information about changes in the moment of inertia of the Earth over long periods of time provides possible tests of theories of the Earth's evolution. In discussing the various palaeontological data, I have ignored the interchange of angular momentum between the core and mantle because I assumed that over very long periods the interchange of angular momentum between core and mantle average out. They may also average out when in the eclipse data which takes us back to a period of about 2000 years.

In the short-term changes in the length of the day and in the Chandler wobble, I think one is dealing with the turbulence of the core coupling to the mantle. Over long periods of time the question is whether the changes in the rate of rotation of the Earth and Moon are due to tidal friction, which may be slowly varying over time, or whether changes are occurring in the moment of inertia of the Earth.

References

Bauer, L. A. (1895). *Amer. J. Sci.*, **50**, 109, 189, 314
Brouwer, D. (1952). *Astron. J.*, **57**, 125
Munk, W. H. and MacDonald, G. J. F. (1960). *The Rotation of the Earth: A Geophysical Discussion*, Cambridge University Press, London
Rochester, M. G. (1960). *Phil. Trans. Roy. Soc. Lond.*, **A252**, 531–555
Runcorn, S. K. (1964). *Nature*, **204**, 823–825
Scrutton, C. T. (1964). *Palaeontology*, **7**, 552
Spencer Jones, H. (1939). *Mem. R. astr. Soc.*, **99**, 541
Urey, H. C. (1952). *The Planets: Their Origin and Development*, Yale Univ. Press, New Haven, 224 pp.
Vestine, E. H., Laporte, L., Cooper, C, Lange, I. and Hendrix, W. C. (1948). Carnegie Institute of Washington Publication 578, Washington, D.C., p. 532
Wells, J. W. (1963). *Nature*, **197**, 948–950

DISCUSSION

ROSENBERG: Whereas biologists are used to working with material they can observe, measure and describe, geophysicists seem to be working with an unknown quantity in tidal friction. What is tidal friction and how is it measured?

RUNCORN: I am glad biologists are so pragmatic; I didn't know biologists could see evolution but they still study it. Admittedly, however, the nature of tidal friction is a mystery. We can observe the effects of the tides—tidal bulges in the sea, the atmosphere and the land itself, and tides in the Moon's mass induced by the Earth; torques result which slow the Earth down. We can show there are energy losses—waves beating on a beach, for example. But energy must be conserved. When we look at the Moon, and satellites of Jupiter and Saturn which always keep one face to their planets, we cannot help but postulate that they were once rotating, but have been slowed down by tidal friction. The Earth does raise tides on the solid Moon. If the Moon were at one time rotating, there would probably have been a lag in the lunar tidal bulge (because all solids have some non-elastic behaviour). A torque would have resulted which would have slowed down the satellite. But we cannot determine at this time what part of the slowing down of the Earth results from oceanic tides and what part from Earth tides. Atmospheric tides probably influence the Earth's rotation, but the effects are probably small.

PROFESSOR TERMIER: How do we know the length of the year is constant through time? If the length of the year were to change, the number of growing days/year would also change.

RUNCORN: If the gravitational constant remains the same, then the orbital angular momentum of the Earth must remain constant (except for very small perturbations produced by the planets and the Sun's tide) and one would expect the year to remain constant. The effects of gravity change referred to have been calculated and are sufficiently small to be unimportant, even when considering 3000 My.

DOLMAN: This is valid for the sidereal year, not the tropical year. Hipkin believes there would be a significant difference between the tropical and sidereal year when the Moon was much closer to the Earth, due to changes in precession of the equinox.

RUNCORN: I recognize that, due to changes in the Moon's orbit, there is constant change in inclination of the axis of rotation of the Earth to the point of the ecliptic, but the rate of precession is still long. When you talk of approach of the Moon to the Earth of 2 or 3 radii, you are talking about a period of Earth–Moon history prior to that investigated in this research.

ON A TENTATIVE CORRELATION BETWEEN CHANGES IN THE GEOMAGNETIC POLARITY BIAS AND REVERSAL FREQUENCY AND THE EARTH'S ROTATION THROUGH PHANEROZOIC TIME

KENNETH M. CREER

University of Edinburgh, Department of Geophysics, Edinburgh, Scotland

Abstract

The polarity bias of the geomagnetic field, estimated for each geological subperiod through the Phanerozoic from the integrated percentage of reversed magnetization reported in all available published research, exhibits four marked discontinuities. The strongest of these occurs at about −220 My, at the end of the Kiaman Reversed Interval. Others occur in the Early Cenozoic at about −55 My and in the Devonian and Silurian at about −375 and −425 My. Counts of growth rings on fossil shells allow estimates of the number of days in the month and in the year to be made. The rate of change in the length of the day deduced from counts of the former exhibits pronounced discontinuities at −50 My, −215 My, −360 My and −420 My. Similar discontinuities at −65 My, −245 My, −360 My and −415 My are indicated from counts of the latter. The apparently very good correlation between the two phenomena is tentatively accepted as real and is suggestive of a common source: mass transport in the lower mantle on the one hand creating 'bumps' on the surface of the 'hole' in the mantle wherein sits the convecting liquid iron core, thus unsettling the action of the geomagnetic dynamo and, on the other hand, affecting the moment of inertia of the mantle and hence the spin velocity.

Moreover, the Earth's angular velocity appears to have slowly accelerated over pro-prolonged lengths of time, e.g. during most of the Mesozoic, a condition which cannot be accounted for by a purely dissipative mechanism such as tidal friction. Possible causes are examined.

1. Introduction

Although many theories of the origin and maintenance of the geomagnetic field have been considered for perhaps a century, none has yet reached a satisfactory stage of development. The most promising theory attributes the field to magneto-hydrodynamic processes in the fluid outer core and in recent years much effort has been applied in this direction. None of the observed features of the reversing geomagnetic field are understood: for example neither the morphology nor the timing of individual reversals can be explained nor can reasons be given why the frequency of reversals has varied through geological time.

Some intriguing suggestions have nevertheless been made, relating the frequency of reversals to other geological or geophysical phenomena. For instance, Irving (1966) suggested that reversals occurred frequently during periods of continental drift or polar wander, but it now turns out that no strong correlation exists between the two sets of phenomena. Hide (1967) introduced the idea of 'bumps' on the core–mantle interface, partly to explain the coupling between core and mantle. Changes in the topography of the core–mantle interface, or alternatively of the temperature distribution over its surface, would probably affect the frequency of reversals and could well be brought about by mass transport, possibly convective overturn of the lower mantle. The latter would change the moment of inertia of the Earth and hence the length of the day. By such a chain of events we could link two apparently unconnected phenomena, viz changes in reversal frequency with changes in the rate of change of the length of the day (l.o.d.).

This paper discusses whether any correlation can be established between these two phenomena and explores some of the consequences.

2. Geomagnetic reversal frequency and polarity bias

During the past 80 My the geomagnetic field has undergone frequent reversals and there has been no bias in polarity. However, the frequency of reversals has increased from one every 0·94 My on average from −75 to −45 My to one every 0·33 My from −45 to −10·5 My to one every 0·23 My from −10·6 My to the present (Cox, 1969). These data have been obtained from studies of sea-floor magnetic anomaly 'stripes'. Recently, the record has been extended back to −160 My (Larson and Pitman, 1972), who detect no reversals between −82 and −112 My, in agreement with measurements of continental rocks by Irving and Couillard (1973) although Pechersky and Khramov (1973) show two short events of reversed polarity in this 30 My interval of steady *normal* polarity. It was preceded by a slightly longer interval of about 36 My when at least 40 reversals occurred without any noticeable polarity bias. Before this, terminating at −148 My, there was another long interval of less steady *normal* polarity which can be traced back to −160 My by the sea-floor magnetic anomaly method which is as far back as it has been taken.

The reversal chronology can only be extended further back in time by palaeomagnetic studies of continental rocks. Burek (1970) and Johnson, Nairn and Peterson (1972) have demonstrated that this interval of normal polarity began at about −200 My, but that, although it was 52 My long, it was broken by about 9 very short reversed events of less than 2 My duration. Pechersky and Khramov (1973) recognize in data from the U.S.S.R. a short mixed interval between about −165 and −180 My. While the reversal chronology of the Late Triassic and Jurassic is not yet well established it seems clear that polarity of the field was then strongly biased in the normal sense.

The interval between about −200 and −230 My was one of frequent reversals and followed the longest interval of stable polarity recorded in the whole of the Phanerozoic discovered by Irving and Parry (1963), who named it the Kiaman

Interval of *reversed* polarity. It runs from about −240 My to −300 My and, as yet, only two or three short normal events have been recognized within it.

There are fewer reliable data from which a reversal chronology for the rest of the Palaeozoic can be drawn up. An attempt to do this using measurements on U.S.S.R. rocks has been described by Khramov, Rodianov and Komissarova (1966) and their reversal chronology section has been incorporated into the

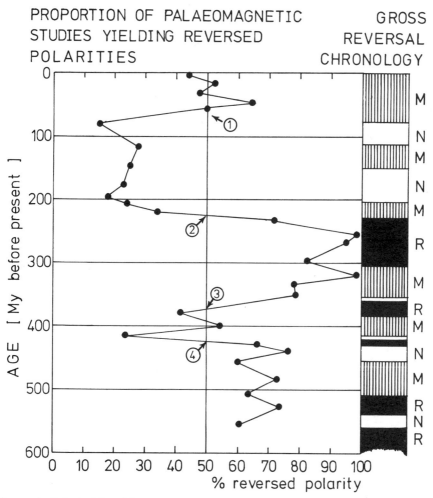

FIGURE 1. Polarity bias of the Geomagnetic Field during the Phanerozoic. A 'section' illustrating the grosser polarity characteristics is shown on the right. The graph is compiled from global palaeomagnetic data contained in Hicken et al.'s (1972) catalogue: the section is compiled from sea-floor spreading data back to −160 My and thereafter from a number of land-based palaeomagnetic sections. The breaks referred to in the text are labelled 1, 2, 3 and 4

section for the whole of the Phanerozoic given in Figure 1. This section illustrates the distribution in time of (i) predominantly normal geomagnetic field polarity, (ii) predominantly reversed polarity and (iii) mixed polarity with frequent reversals and no obvious bias to either polarity sense. Only the grosser features of the polarity sequence are shown and no attempt is made to illustrate the finer structure.

Another way of presenting information about polarity bias of the geomagnetic field, again on the grosser time scale, is to estimate the proportion of time the field possessed each polarity (Figure 1). Geological subperiods have been chosen as the time units over which these time averages should be made. This allows a

TABLE 1. Properties of palaeomagnetic studies in which reversed polarities are measured for each subperiod of Phanerozoic Time

Geological subperiod	Age [My]		Number of studies	Reversed polarities (%)
	range	mean		
Pliocene	2–7	5	48	44·3
Miocene	7–26	17	40	52·3
Oligocene	26–38	32	7	47·1
Eocene	38–54	46	28	64·3
Palaeocene	54–65	60	5	50·6
Ur. Cretaceous	65–100	82	28	15·2
Lr. Cretaceous	100–136	118	29	28·1
Ur. + M. Jurassic	136–162	149	9	25·0
Lr. Jurassic	162–190	176	10	23·1
Ur. Triassic	190–205	198	23	19·1
M. Triassic	205–215	210	11	24·5
Lr. Triassic	215–225	220	30	34·5
Ur. Permian	225–240	232	27	71·9
M. Permian	240–265	253	10	98·6
Lr. Permian	265–280	272	43	95·0
Ur. Carboniferous	280–315	297	52	82·6
M. Carboniferous	315–325	320	22	98·5
Lr. Carboniferous	325–345	335	32	78·1
Ur. Devonian	345–359	352	18	79·3
M. Devonian	359–370	365	12	62·9
Lr. Devonian	370–395	382	14	41·9
Ur. Silurian	395–410	402	11	54·5
M. Silurian	410–420	415	5	24·0
Lr. Silurian	420–435	428	3	66·7
Ur. Ordovican	435–445	440	9	76·3
M. Ordovican	445–470	458	8	60·2
Lr. Ordovican	470–500	485	9	73·1
Ur. Cambrian	500–515	507	11	64·4
M. Cambrian	515–540	527	5	73·4
Lr. Cambrian	540–570	555	9	61·0

significant amount of data to be included in each population and moreover the relevant age data are not available to allow their classification into smaller time units. A recently published compilation of palaeomagnetic data (Hicken et al., 1972) was used. These data include the ratio of normal to reversed magnetization within each study. The basic assumption made in the preparation of this diagram is that formations studied are uniformly distributed in time within each averaging interval. This is not strictly true since palaeomagnetic sampling by the international scientific community at large has been carried out in an almost random manner both in space and time. Were the sampling sites really distributed randomly through each averaging period used, we could suppose, with some justification, that the distribution of sampling was also effectively uniform through time for sufficiently large populations.

Note that the graph (main part of Figure 1) is compiled from all available palaeomagnetic data while the 'section' on the right-hand side is compiled from sea-floor anomaly data back to −160 My and before that from local stratigraphic sections. So they illustrate different data.

The Middle Cretaceous Interval of normal polarity ended at about −85 My, but the field continued to show a strong bias to normal polarity until about −75 My therefore in the 'section' in Figure 1, the current interval of mixed polarity is shown as starting at the latter date. The switch (labelled 1) to about 50% reversed polarity shows up rather sharply on the graph.

The mixed polarity interval in the Cretaceous, between −112 and −148 My, sandwiched between intervals of predominantly normal polarity (shown in the 'section' at the right of Figure 1) does not register strongly on the graph showing the proportion of studies exhibiting reversed polarity. There are two reasons for this: (i) the coarseness of the 'filter' through which we are looking at the data and (ii) this interval of mixed polarity is contained partly in the Lower Cretaceous and partly in the Upper Jurassic subperiod.

The most striking feature of Figure 1 is the rapid switch (labelled 2) from a strong bias to reversed polarity in the Late Palaeozoic, culminating in the 60-My-long Kiaman Interval of almost exclusively reversed polarity to a strong bias to normal polarity in the Mesozoic. The cross-over through 50% polarity bias is at −220 My and coincides with the 30-My-long interval of mixed polarity in the Early and Middle Triassic (Figure 1).

According to the U.S.S.R. reversal section (Khramov et al., 1966) the field had a strong preference for reversed polarity back to about −310 My. This was preceded by a mixed interval in which there appears little polarity bias, although the reversals illustrated are not frequent, appearing on average every 2 My or so (see Figure 2 of Khramov et al., 1966). However, the global palaeomagnetic data (Table 1 and Figure 1) show a bias to reversed polarity back to about −350 My. The Middle Carboniferous (−315 to −325 My) point which incorporates data from 22 different reports indicates a 98.5% preference for reversed polarity. A possible explanation is that this relatively short (10-My-duration) subperiod coincided with an epoch of reversed polarity contained within this mixed interval.

On the whole, the U.S.S.R. section, i.e. the part before about −300 My, seems to run some 20 My ahead of the graph plotted from the global data: for instance, the normal intervals between −350 and −450 My do not line up with the minima of the graph. This could be (i) due to a systematic difference in the radiometric age scale adopted by the Russians from the Geological Society of London Scale (Harland et al., 1964) or (ii) simply be a reflection of the inadequacy of the data.

The graph shows two sharp swings, labelled (3) and (4) in Figure 1, at −375 and −425. However, the Silurian points (at −402, −415 and −428 My) are based on rather few data points. Combining all the Silurian results into a single control point, we find a 48·4% preference for reversed polarity, so the swing towards a bias towards normal polarities at these times would appear real.

The zig-zag appearance of the graph for Ordovican and Cambrian times can probably be attributed to insufficiency of data to resolve fluctuations in bias of the order of 10%. It is clear, however, that the field maintained a fairly strong preference for reversed polarity in the Early Palaeozoic, the Ordovician and Cambrian averages for bias in that sense being 70·2 and 65·0% respectively.

We would expect the stability of the geomagnetic field to be influenced by topographical (Hide, 1967) or temperature (Doell and Cox, 1971) anomalies at the core–mantle interface. Quite small topographical features, as yet undetectable seismically, might, in the present epoch, locate the foci of the non-dipole field: larger features (perhaps detectable by seismic techniques!) might trigger off reversals. On the other hand, a smooth, but not necessarily symmetrical core–mantle interface (Lilley, 1970 and section 8) might constitute necessary boundary conditions for well-settled dynamo action. While these ideas must remain speculative with our limited knowledge of these palaeo-geophysical phenomena, it is useful and illuminating to enquire whether the sharp swings in polarity bias of the field can be correlated with any other geophysical observations. Since changes in core–mantle interface topography or temperature distribution imply mass or heat transport in the lower mantle, we might expect to find accompanying changes of the Earth's moment of inertia and hence of its spin rate. With these thoughts in mind we proceed to discuss the palaeontological evidence about the length of the day during Phanerozoic time.

3. Palaeontological 'clock' data

Pannella, MacClintock and Thompson (1968) concluded that the deceleration of the Earth's spin angular velocity has varied throughout Phanerozoic time as a result of studies of tidally controlled growth patterns exhibited by fossil molluscs and stromatolites. They fitted a polynomial to their 9 data points and found empirically that the curve for degree 4 fitted best. Their Triassic control point did not lie on this curve and they regarded it as less secure than their others. Nevertheless, their curve showed that the deceleration of spin was negligible between the Pennsylvanian and Upper Cretaceous and they suggested that this might have been due to abnormally low tidal friction brought about as a result of sea-floor spreading which changed the area and distribution of shallow seas where most of the retarding couple is thought to act (Miller, 1966).

If, however, the reliability of a control point is related to the magnitude of its associated standard deviation and error, the Triassic point would appear no less reliable than the others. An important consequence of accepting this point is that we are faced with an apparent *acceleration* of the Earth's spin angular velocity, i.e. of a modest shortening of the length of the day (l.o.d.) during the

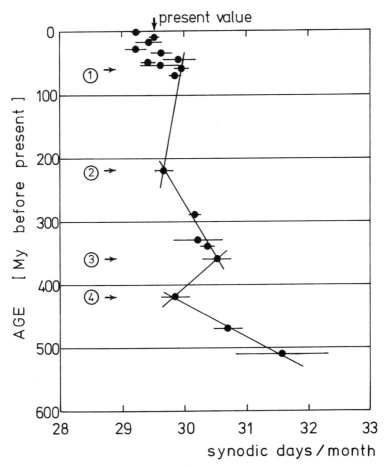

FIGURE 2. Number of synodic days/synodic month estimated from studies of growth rings exhibited by fossil shells. Data source: Pannella (1972). The breaks referred to in the text are labelled 1, 2, 3 and 4

Mesozoic (Creer, 1972). A purely dissipative mechanism such as tidal friction is of no direct help, although if it were minimal during the Mesozoic, the accelerating effect we must find need not be so large as if it had first to overcome the braking effect of, say, the present-day tides. Since the angular momentum of the Earth must remain constant unless acted on by some external agency, the problem reduces to an investigation of how the Earth's moment of inertia might be caused to fluctuate by the required proportions over characteristic times of the order of 100 My. This is discussed in sections 5–7.

A more recent account of palaeontological evidence on the Earth's rotational history (Pannella, 1972) describes many new measurements of growth rings and also a review of data obtained by other independent research groups. These data relate not only to the number of days/month, but also to the number of days/year. They are plotted respectively in Figures 2 and 3. The shortening of l.o.d. in the Mesozoic and the above average increase in l.o.d. in the Cenozoic and Late Palaeozoic suggested by the earlier data are confirmed. New data for the Early and Middle Palaeozoic suggest a further short interval of shortening of l.o.d., lasting about 60 My in the Silurian–Devonian, evident both from counts giving days/month and days/year.

The Phanerozoic can thus be divided into five regimes corresponding to periods of above average deceleration of spin velocity and periods of modest acceleration. The transitions between these regimes (labelled 1, 2, 3 and 4, Figures 2 and 3), appear quite sharp and a linear relationship of l.o.d. with time within a particular

TABLE 2. Data on number of days/year obtained from palaeontological data

Period	Days/Year	Source
Cretaceous	375	Pannella (1972)
Cretaceous	370·3	Berry and Barker (1968)
Triassic	371·6	Pannella (1972)
Pennsylvanian	380–390	Wells (1966, 1970)
Pennsylvanian	383	Pannella (1972)
Mississippian	398	Wells (1966, 1970)
Mississippian	398	Pannella (1972)
Devonian	398	Wells (1966, 1970)
M. Devonian	405·5	Pannella (1972)
Lr. Devonian	410	Mazzullo[a] (1971)
Silurian	400	Wells (1963)
M. Silurian	419	Mazzullo (1971)
Lr. Silurian	421	Mazzullo (1971)
Ur. Ordovician	412	Wells (1966, 1970)
M. Cambrian	424	McGugan[b] (1967)

[a] Mazzullo (1971) uses the maximum number of counts.
[b] Data from stromatolites.

The Earth's Rotation through Phanerozoic Time

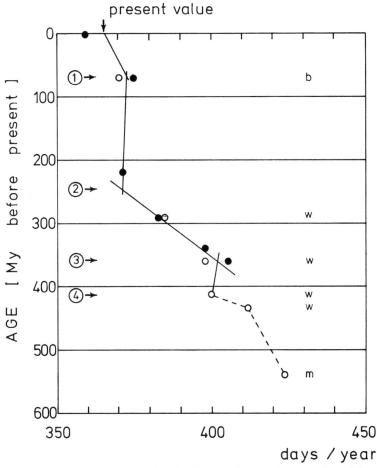

FIGURE 3. Number of synodic days/year estimated from studies of fossil shell growth rings. Data source: b = Berry and Barker (1968), w = Wells (1966, 1970), m = McGugan (1967) and unlabelled = Pannella (1972) where these data are all reviewed. The breaks referred to the text are labelled 1, 2, 3 and 4

regime seems to fit the presently available control points sufficiently well. Hence it is not worth while constructing a more sophisticated (but still empirical) curve through them: it would not affect the recognition of these sharp transitions of which the ages are listed in Table 2. The agreement with the ages of transition in geomagnetic polarity bias (Figure 1) is remarkable.

We may now proceed further and attempt, following Runcorn (1964), to use the data on monthly and annual banding to separate the effects of the lunar tidal torque on the Earth's rotation rate from those due to changes in its internal state. The ratio between past and present moments of inertia I/I_0 for those geo-

TABLE 3. Change in Earth's moment of inertia calculated from palaeontological 'clock' data

Age (My)	Period	Days/Month[4]	Days/year	$10^4 \Delta I/I_0$
70	Ur. Cretaceous	29·85	375[a]	+07
70	Ur. Cretaceous	29·85	370·3[e]	−11
220	Triassic	29·66	371·6[a]	+44
290	Pennsylvanian	30·16	383[a]	−40
340	Mississippian	30·37	398[a,b]	+99
360	Devonian	30·53	405·5[a,b,c]	+73
420	Silurian	30·64	413[b,c]	+577
470	Ordovician	30·64	414[b]	+156
510	Cambrian	31·56	424[d]	−142

Data sources: [a] = Pannella (1971), [b] = Wells (1966, 1970), [c] = Mazzullo (1971), [d] = McGugan (1967), [e] = Berry and Barker (1968).

logical periods for which counts of both days/month and days/year have been made are presented in Table 3. The ratio, β, between the solar and lunar retarding torques has been taken to be 1/3·4. No clear trend emerges and it would appear that the pairs of data are not precise enough. Scrutton and Hipkin (1973) also concluded that the computations of I/I_0 for the Devonian made by Runcorn (1964) were unreliable due to the statistical uncertainty of the basic data.

4. Present and past rates of tidal dissipation

Miller (1966) has estimated the energy dissipation at the present time by tidal currents in the sea, particularly the deep oceans at $1·5 \times 10^{19}$ erg/sec ($\pm 50\%$). This would cause the day to lengthen at the rate of just less than 1 msec/cy.

From a satellite determination of the phase lag of the Earth's tidal bulge, Newton (1968) has estimated the tidal retardation of the Earth's spin angular velocity at between 3·9 to 4·95 rad/sec^2. This gives an increase in l.o.d. of between 1·44 and 1·84 msec/cy. Hence, it seems that we cannot quite account for currently observed average rate of increase in l.o.d. of 2 msec/cy by tidal friction and fall considerably short of some of the larger values deduced for the geological past. The average rate of increase in l.o.d. from the Devonian to the present, using growth ring data, is 2·7 msec/cy, while for the Late Palaeozoic regime the rate appears to have been as high as 5·5 msec/cy (calculated from days/year shown in Figure 3). To what extent can we explain the variations on the rate of change of l.o.d. by possible variations in tidal friction in the geological past?

In principle we should be able to account for decelerations but not accelerations of spin angular velocity. Miller (1966) argues that the main sources of friction at present are the shallow seas. Because of sea-floor spreading and changes in sea level due to glaciation we should expect the energy dissipated in these places to have varied appreciably from one epoch to another. Can we detect any such variation?

Valentine and Moores (1970) studied the diversity of marine fauna through Phanerozoic time and these studies are relevant to our problem because they can be related to patterns of continental fragmentation and reassembly. The species-area effects of the expanding and shrinking of the epicontinental seas act in concert with the effects of widespread transgressions and regressions which respectively moderate the seasonal climates and enhance continentality.

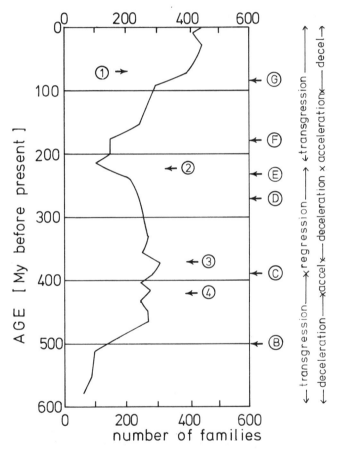

FIGURE 4. Number of families of skeletonized benthonic fauna through the Phanerozoic as counted by Valentine and Moores (1970). Intervals of marine transgression and regression and of acceleration and deceleration of Earth's spin deduced from Figures 2 and 3 shown. For the significance of ages marked B to G, see section 4 of the text. Redrawn from Valentine, J. W. and Moores, E. M., *Nature*, **228**, 657–659, 1970, by permission of Macmillan (Journals) Ltd.

In Figure 4 Valentine and Moores' plot showing the number of skeletonized benthonic shelf families is reproduced and they argued that the major trends correlate well with the major surface events which they describe as follows: 'The emergent Eo-Cambrian supercontinent (Pangaea I) fragmented (event A, not shown in Figure 4) and was then transgressed during Cambrian time (event B). Caledonian-Acadian and Hercynian-Appalachian suturings (events C and D) were accompanied by regressions and the Permo-Triassic continental assemblage (Pangaea II) was accompanied by a profound regression (event E). Subsequent transgressions may mark episodes of fragmentation (events F and G)' Event G correlates well with a worldwide pulse of rapid spreading at most of the ocean ridges at -110 to -85 My (Hays and Pitman, 1973) who also maintain that the subsequent regression was caused by a reduction in spreading which began at -85 My.

The intervals of transgression and regression are shown at the side of Figure 4 where they may be compared with intervals of acceleration and deceleration of the Earth's spin velocity (boundaries labelled 1, 2, 3 and 4 as in Figure 3). The Upper Palaeozoic regime of rapid deceleration was accompanied by regression while the Mesozoic regime of modest acceleration was accompanied by transgression. This correspondence does not hold in the Lower Palaeozoic, however, the deceleration in the Cambrian and Ordovician being accompanied by transgression. Also the transition from acceleration to deceleration at the end of the Mesozoic (labelled 1) falls within an interval of transgression.

Thus, it appears that while marine transgressions and regressions might have some influence on the Earth's rotation, they are almost certainly not the only influencing factor.

Moreover, we cannot account for even the most modest acceleration in spin velocity such as that of 0·5 msec/cy which apparently persisted for more than 100 My in the Mesozoic by a purely dissipative mechanism such as tidal friction. It is necessary, therefore, that we should now direct our attention to a consideration of the possibility of long-term fluctuations in the moment of inertia of the Earth.

5. Thermal expansion of the mantle

Temperature within the mantle, especially the upper mantle, is undoubtedly one of the most important factors contributing to the mechanism driving plate motions. The onset of rapid sea-floor spreading at the start of the Mesozoic may be taken as an indication of above average geothermal flux and hence that the upper mantle was cooling. We may speculate that this thermal energy had been stored up during the Late Palaeozoic, possibly as a consequence of the considerable blanketing effect of Pangaea: the crustal radiogenic heat sources must then have been concentrated within a single large surface area.

We are thus led to investigate the effects on the Earth's rotation of a gradual overall increase (or decrease) of the temperature of the mantle, noting that we require that it should have warmed up during the Late Palaeozoic to account for the rather large rate of increase of l.o.d. and then that it should have cooled down during the Mesozoic so as to account for the decrease in l.o.d.

To make the argument quantitative, albeit on an order of magnitude basis, let us consider the consequences of an overall decrease in mantle temperature of 200°C. On the time scales of Figures 2 and 3 this would take place during an interval of about 100 My. The most obvious consequences would be a decrease in moment of inertia due to contraction and hence by conservation of angular momentum, an increase in spin angular velocity. Taking a value of 6×10^{-6} as coefficient of expansion and 0.91×10^{27} cm^3 as the volume of the mantle, we find that the decrease in volume ΔV is 1.09×10^{24} cm^3. The proportional decrease in global volume is then 1.01×10^{-3} (taking the volume of the whole earth to be 1.08×10^{27} cm^3 and the consequent proportional decrease in moment of inertia $\Delta I/I$ would be 6.7×10^{-4}. By conservation of angular momentum the change in angular velocity is $\Delta\omega/\omega = 6.7 \times 10^{-4}$ and the corresponding change in length of the day ΔT is given by

$$\Delta T = -\frac{2\pi}{\omega^2} \cdot \Delta\omega = 58 \sec$$

since ω, the Earth's angular velocity, is 7.3×10^{-5} rad/sec.

Supposing that this occurs in 100 My we obtain a rate of decrease in l.o.d. of only 0.05 msec/cy which is too small by a factor of about 10. It would be necessary to decrease the overall temperature of the mantle by some 2000°C to decrease l.o.d. by 0.5 msec/cy, the value indicated for the Mesozoic regime! Also, it is implicit that if the mantle temperature fluctuated by 1000°C up and down, it would have been molten at the end of the Palaeozoic.

To attain some feel of the problem of by how much we might reasonably allow the mantle temperature to increase, let us suppose we lock in a thermal flux of the present value of 10^{28} erg. We take the specific heat of mantle material to be 0.17 cal/g/deg (the value of igneous rocks at NTP) and the mass of the mantle to be 4.1×10^{27} g. The amount of energy absorbed by the mantle in raising the temperature even by 200°C would be 5.85×10^{16} erg so that 585 My would be required, and this is longer than the typical length of a regime, about 100 My (Figure 3).

Thus any attempt to account for the decrease in l.o.d. deduced for the Mesozoic by thermal expansion of the mantle would appear to be sterile.

6. Temperature-induced phase changes in the mantle

In the Bullen (1963) model, the lower mantle (region D) is separated from the uppermost mantle (region B) by a transition zone, C, at depths between 900 and 350 km approximately. Zone C is characterized by discontinuities in seismic velocities occurring principally at about 400 and 600 km (Figure 5). Bullen and

Haddon (1967), for their Earth model HB1, computed sharp increases in compressed density (Figure 6) of 7% at 350 km and 1% at 650 km depth, and a total increase of about 25% across the whole of zone C. Phase changes in mantle materials constitute the main contribution to the 25% change, but we must allow for the increased compression of each phase with depth. We estimate this for the materials of layer B and estimate that phase transformations in zone C produce $\Delta \rho/\rho = 20\%$. To be on the safe side, we use 15% in the calculation which follows.

Akimoto and Fujisawa (1967) have investigated the olivine to spinel transition in the pressure range 43–96 kb at 800°C, 1000°C and 1200°C in the laboratory and have extrapolated their experimental results to higher pressures. Fujisawa (1967, 1968) argues that the upper mantle is composed of 80% olivine with about 20% pyroxene so that investigations of this transition are directly relevant to our problem. The experimental data suggest that the composition of mantle olivine is $(Mg_{0.9}Fe_{0.1})_2SiO_4$ and that the temperature at depth 370 km where one of the discontinuities in seismic velocity occurs (Figure 5) is between 1150°C and 1530°C. The thickness of the transition region is directly related to the temperature distribution: with uniform temperature it would be 50 km, with an increase of 100°C across the zone it would be 80 km. Thus, the location and sharpness of this discontinuity deduced from high-pressure mineralogy shows substantial agreement with the seismic evidence.

Several seismic discontinuities occur within the C region of the mantle and

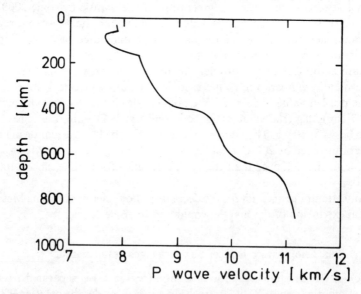

FIGURE 5. Variation of the velocity of seismic compressional waves with depth in the vicinity of transition zone C (after Kanamori, 1967 and Johnson, 1967)

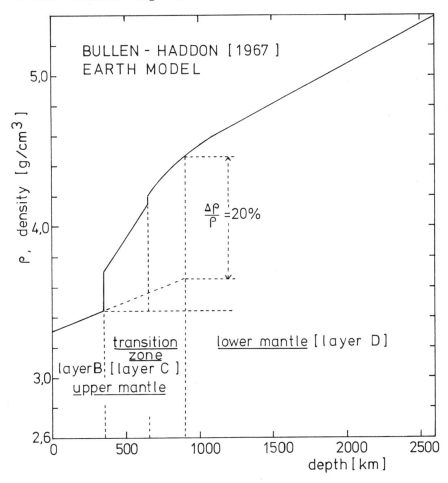

FIGURE 6. Variation of compressed density with depth through the mantle computed by Bullen and Haddon (1967) for their Earth model HB1

this is in agreement with results of experiments on phase changes of mantle materials which show that a particular transition region should not be spread over so wide a thickness as the whole C zone. Other phase transitions occurring with the zone are of the spinel structure to an ilmenite structure with the formation also of FeO, MgO and SiO_2, the latter with a rutile structure corresponding to stishovite.

In the following we consider the effect of an overall heating or cooling of the mantle on the depth of the olivine–spinel transition. From the solvus curves (Figure 7) we find that at 800°C the transition starts at 323 km depth (107 kb), at 1000°C at 350 km (115 kb) and at 1200°C at 390 km (130 kb). Thus we expect the depth in the mantle at which the transition occurs to sink by about 27 km

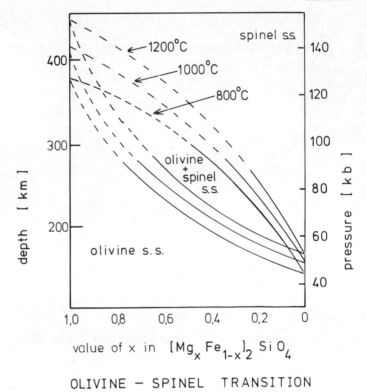

OLIVINE — SPINEL TRANSITION

FIGURE 7. Solvus curves for olivine-spinel transition (after Akimoto and Fujisawa, 1967 and Fujisawa, 1968)

(or rise by about 40 km) in response to a decrease (or increase) in temperature of 200°C. We use 35 km in our calculation below and hence estimate the increase in the length of the day consequent upon a supposed drop in mantle temperature of 200°C. We make the assumption that the influence of temperature and pressure on the other transitions occurring within layer C are similar to those found for the olivine–spinel transition.

We have already estimated that the density change across layer C due to phase transitions (of which the olvine–spinel transition is one) is about 15%. The effect of an overall drop in temperature of 200°C would therefore be roughly equivalent to that of decreasing the specific volume of a 35 km thick shell of mantle material at a mean depth of about 600 km (say at radius 5750 km) by 15%. This decrease in volume, Δv_s amounts to $2 \cdot 2 \times 10^{24}$ cm^3 and causes a proportional decrease in volume of the whole earth, $\Delta v_s/V_e$ of 2×10^{-3}. The corresponding decrease in moment of inertia $\Delta I/I$ is $1 \cdot 3 \times 10^{-3}$ and this is equal to the increase in angular velocity $\Delta\omega/\omega$. The resulting decrease in the length of the day would be $-2\pi\Delta\omega/\omega^2 = 112$ sec, total. This must persist for about 100 My

(Figure 3) in which case we obtain a decrease in the length of the day of 0·1 msec/cy which is about 20% of the required decrease. To increase l.o.d. by 0·5 msec/cy it would be necessary to force the transition zone down by about 170 km and this would require a very large temperature increase of nearly 1000°C, using the crude extrapolation mentioned above of the experimentally determined temperature and pressure coefficients. Possibly a somewhat smaller increase in temperature might be sufficient because the experimental data (Figure 7) suggest that the rate at which the solvus curve moves to greater depths (pressures) increases with increasing temperature: between 800°C and 1000°C it is 0·135 km/deg and between 1000°C and 1200°C it is 0·205 km/deg so that at 1600°C the rate might be faster than we supposed above (0·175 km/deg) by a factor of 3. This would allow us to lower transition zone C by the required 170 km with an increase in temperature of about 325°C which would appear reasonable on the following grounds:

(i) The temperature we have assigned to the top of the transition zone at 370 km, 1150°C to 1530°C is rather lower than that for some other estimates, e.g. 2000°C (Stacey, 1969, convecting model).

(ii) The mantle melting points at 370 km depth is about 2250°C (Stacey, 1969), so we could allow a temperature increase considerably in excess of 325°C without causing the mantle to melt at these depths.

(iii) The free energy associated with the olivine–spinel phase transition is about 3 kcal/mole, i.e. $7·6 \times 10^8$ erg/g and hence the total free energy used in lowering the top of the transition zone by 170 km would be $1·9 \times 10^{35}$ erg. If a supply of heat equal to the whole of the present geothermal flux were available (10^{28} erg/year), sufficient free energy could be supplied in about 19 My: note that we have allowed 100 My.

Hence this approach, in which we attempt to account for a slow acceleration in the Earth's spin angular velocity deduced for the Mesozoic, appears promising.

Birch (1968) calculated that if the upper 600 km of mantle were converted to the high-pressure forms of the lower mantle with a 20% increase of density, the shrinkage would be about 5% of the whole volume or a radial decrease of about 100 km. This would be accompanied by a total decrease in the length of the day of 2724 sec, giving a rate of 2·7 msec/cy, the change occurred in 100 My. This gives us an upper limit to the effects of the mechanism.

7. Changes in the coefficient of moment of inertia or of the ellipticity

The coefficient of moment of inertia, $C/Ma^2 = J_2/H = 0·33$. Here, C is the moment of inertia about the polar axis; M, the mass of the earth; a, the equatorial radius; J_2, the dimensionless coefficient of second degree which is a measure of the oblate ellipsoidal form of the geoid, and H, the dynamical ellipticity. The present value, 0·33, departs from the value of 0·4 for a homogeneous sphere by about 20%, and in Earth models is used as a boundary condition in the computation of density with depth in the mantle. None of these models, e.g. the Bullen–Haddon (1967) model assumes any lateral inhomogeneities. Since the possible existence of such inhomogeneities has not yet been adequately explored no estimate can be made of their contribution to the coefficient of inertia. We have seen (section 6) that a rate of change of l.o.d. of 1 msec/cy persistent for 100 My would require a 0·7% change in moment of inertia, i.e. about 3 to 4% of the 20% mentioned above, most of which is due to layering in the mantle (the core, because of its high density, only makes a 10% contribution to the total moment of inertia because of its smaller radius). Hence, we require that slowly varying lateral inhomogeneities amounting to about 3 to 4% of the permanent inhomogeneities due to the radial pressure and temperature gradients should have existed through Phanerozoic time. We can only speculate as to whether this might have been so.

Another possibility is that the oblateness of the Earth's figure might have slowly fluctuated. The reason for supposing this is that the present ellipticity, determined from gravity and satellite measurements of the geoid, exceeds the value suggested by hydrostatic theory. The two values are 1/298·3 and 1/299·7 respectively, a difference of 0·5%. It was formerly thought that the excess ellipticity was a measure of the viscosity of the mantle delaying the response of the equatorial bulge to the Earth's deceleration (Munk and MacDonald, 1960). However, more recently Goldreich and Toomre (1969) have pointed out that the excess ellipticity is to be expected in a randomly evolving spheroid. Alternatively we might suppose it is not a static phenomenon and accept that the present excess of 0·5% might be typical of long-term fluctuations brought about by, for example, convection in the mantle. Since the moment of inertia is about proportional to the ellipticity, we should expect variations of the same proportion. i.e. 0·5%, to result. This is of the order required to produce a rate of change in l.o.d. of about 1 msec/cy in 100 My.

8. Displacement of geomagnetic centre

Creer, Georgi and Lowrie (1973), by carrying out a spherical harmonic analysis of Upper Tertiary and Quaternary palaeomagnetic directions, showed that the geomagnetic centre was persistently displaced some 200 km from the geocentre during those epochs, towards the Pacific hemisphere, i.e. away from the mass of continental crust.

The validity of the application of the spherical harmonic method to palaeomagnetic directions from older epochs is somewhat doubtful because of the poor global coverage of control points which are restricted to the major land areas. The Permo-Carboniferous epoch is a very interesting one to apply the method to because we are reasonably sure of the arrangement of the continental land-

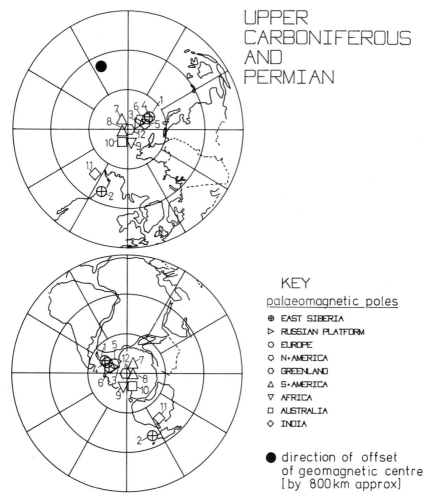

FIGURE 8. Palaeomagnetic poles averaged for Permo-Carboniferous data for each of the major landmasses, plotted on the map of Pangaea. The large dot represents the radial direction along which the geomagnetic centre is displaced from the geographic centre (Creer, 1972)

mases as parts of Pangaea. When palaeomagnetic poles are plotted on a map of Pangaea, they do not form a well-defined group (Figure 8) and this is not considered to be due to experimental error (Creer, 1973). While it could possibly be due to a different palaeogeography than that shown in Figure 8, it is worth while exploring the other possibility that the geomagnetic field was not symmetrical about the geographic axis.

Using palaeomagnetic directions which are regional averages for the control

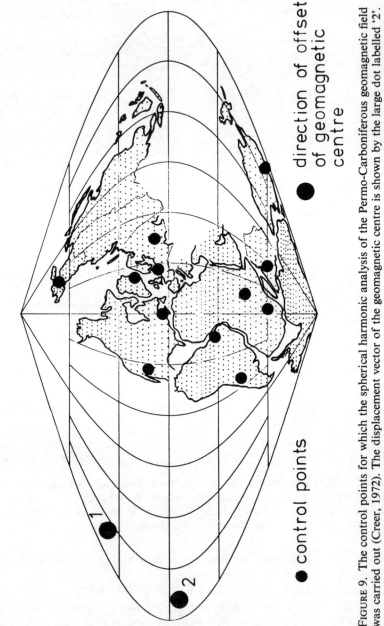

FIGURE 9. The control points for which the spherical harmonic analysis of the Permo-Carboniferous geomagnetic field was carried out (Creer, 1972). The displacement vector of the geomagnetic centre is shown by the large dot labelled '2'. The other large dot, labelled '1', illustrates another quite independent computation of the displacement of the geomagnetic centre for a different distribution of control points by Benkova et al. (1973)

The Earth's Rotation through Phanerozoic Time

points (regionally averaged coordinates of collecting sites) illustrated in Figure 9 as input data, the geomagnetic centre for the Permo-Carboniferous is found to be displaced some 710 km in the direction shown in Figure 8 and Figure 9 (point 1). As for the Upper Tertiary and Quaternary, the geomagnetic centre appears to have been pushed into the hemisphere opposite to the continental crust.

Using different groupings of Late Palaeozoic palaeomagnetic directions Benkova *et al.* (1973) independently found that the geomagnetic centre was displaced some 860 km from the geocentre in substantially the same direction (point 2, Figure 9).

There must be some doubt even about this crude picture of the shape of the palaeo-geomagnetic field by the above approach for the following reasons.

(i) At best the control points can only cover about one third of the globe because the landmasses formed a single supercontinent.

(ii) Our reconstruction of Pangaea may not be correct in detail.

(iii) The palaeomagnetic directions contain appreciable experimental error.

Nevertheless, an inspection of the grouping of palaeomagnetic poles on the map of Pangaea for epochs from the Early Carboniferous to the Early Jurassic suggests that the field was abnormally assymetric between the Late Carboniferous and Early Triassic (Creer, 1973).

The highly settled polarity of the geomagnetic field in the Late Palaeozoic, taken together with its apparently assymetric (non-axial) shape and the suggestion of Lilley (1970) based on the theory that axially symmetric fluid motions will not provide for stable dynamo action, add credence to our deduction that the abnormally fast increase in l.o.d. recorded for the Late Palaeozoic can be attributed to an abnormally large moment of inertia.

9. Extraterrestrial mechanisms

Crain *et al.* (1969) calculated Fourier power spectra for palaeomagnetic reversal data of the kind shown in Figure 1 compiled by Simpson (1966). They computed two major components in the spectrum in the vicinity of 300 My and 80 My. Later, Crain and Crain (1970) developed the stochastic model for geomagnetic reversals (Cox, 1968) incorporating a variable triggering field. They argued that a terrestrial source for these variations is unlikely and suggested they are connected with the Earth's position in the galaxy, the rotational period of which is 280 My. They associated the shorter period with the vibration of the solar system perpendicular to the galactic plane which has a period of 84 My.

The suggestions must be regarded with some reservation because it appears that the reversal history for the Early Palaeozoic is not shown correctly in Simpson's (1966) curves, possibly because of the ambiguity about the definition of reversed palaeomagnetic directions when the polar wandering curve has crossed the equator (as it had in the Early Palaeozoic). Nevertheless, this work is quoted to stress the possibility that the long-time scale fluctuations of geomagnetic polarity bias and also of the rate of change of l.o.d. might be caused, or at least contributed to, by extraterrestrial sources.

10. Conclusions

The close coincidence in timing, summarized in Table 4, of the grosser features exhibited by the curve illustrating polarity bias of the geomagnetic field (Figure 1) and by the curves illustrating the length of the day (Figures 2 and 3) leads us to suggest that the two phenomena might be related. However, the correlation

TABLE 4. Correlation between the times of discontinuity in geomagnetic field polarity bias and times of discontinuity in the rate of change in the length of the day

Label	Polarity bias (Figure 1) (My)	Palaeontological clock	
		days/month (Figure 2) (My)	days/years (Figure 3) (My)
1	55	50	65
2	220	215	245
3	375	360	360
4	425	420	415

should, as yet, be considered tentative (a) because the palaeomagnetic data for geological subperiods of the Early Palaeozoic are possibly not sufficient in quantity and (b) because the fossil growth ring data can be criticized on two grounds, first the number of shells on which counts have been made is insufficient and it is desirable that a wider variety of genera be used (see Pannella, 1972, for review of the work done) and second, it is often argued (see some of the papers in this volume) that the animals might 'make mistakes' in their 'timekeeping' or, more seriously, that they might possibly not work like 'clocks' at all.

Obviously, more effort by a wider range of observers is required on the palaeontological side. Nevertheless, the overall consistency of the data so far accumulated strongly suggests that the measurements can really be meaningfully interpreted in terms of changes in the length of the day. While future studies may well indicate that the present data do contain appreciable error, their consistency is such as to lead us to expect that the error will be systematic rather than random so that their overall pattern in time through the Phanerozoic is not likely to change substantially.

The quality of the data might be improved if standard statistical methods could be agreed on and adopted by all observers. Also, if the same results could be obtained by different observers studying the same shells, they might carry more conviction. Nevertheless, the consistency of the present data warrants a massive effort in this field in the immediate future.

The result that the length of the day decreased during more than 100 My in the Mezozoic is very important and it is desirable that many new data be obtained for this era either to establish it firmly or to disprove it. Given a time of the order of 100 My in which to operate, it appears that fluctuations in the Earth's moment of inertia of the required magnitude are possible (sections 6 and 7). The possibility that the physical state of the mantle as reflected by, e.g. the depth of transition region C or the difference between the oblateness of the geoid and the hydrostatic figure, may have fluctuated on the 100 My time-scale is an exciting one to investigate because it could show the way to the development of a unified theory of geophysics, i.e. of how the Earth works as a whole, providing a link between processses occurring in the core, the mantle and the crust.

References

Akimoto, S. and Fujisawa, H. (1967). Olivine–spinel solid solution equilibria in the system Mg_2SiO_4—Fe_2SiO_4. *Tech. Rep. of Inst. Solid State Physics*, Tokyo, Series A, No. 273, 41 pp.

Benkova, N. P., Khramov, A. N., Cherevko, T. N. and Adam, N. V. (1973). Spherical harmonic analysis of the palaeomagnetic field. *Earth planet. Sci. Lett.*, **18**, 141–147

Berry, W. B. N. and Barker, R. M. (1968). Fossil bivalve shells indicate longer month and year in Cretaceous than Present. *Nature*, **217**, 938–939

Birch, F. (1968). On the possibility of large changes in the Earth's volume. *Phys. Earth Planet. Interiors*, **1**, 141–147

Bullen, K. E. (1963). *An Introduction to the Theory of Seismology*, Cambridge University Press

Bullen, K. E. and Haddon, R. A. W. (1967). Derivation of an Earth model from free oscillation data. *Proc. U.S. Nat. Acad. Sci.*, **58**, 846–852

Burek, P. J. (1970). Magnetic reversals: their application to stratigraphic problems. *Amer. Assoc. Petrol. Geol. Bull.*, **54**, 1120–1139

Cox, A. (1968). Lengths of geomagnetic polarity epochs. *J. Geophys. Res.*, **73**, 3257–2360

Cox, A. (1969). Geomagnetic reversals, *Science*, **163**, 237–244

Crain, I. K., Crain, P. L. and Plant, M. G. (1969). Long period Fourier Spectrum of geomagnetic reversals. *Nature*, **223**, 283

Crain, I. K. and Crain, P. L. (1970). New stochastic model for geomagnetic reversals. *Nature*, **228**, 39–41

Creer, K. M. (1972). Palaeomagnetic evidence about the assemblage and fragmentation of Pangaea. *Trans. Amer. Geophysics Union*, **53**, 352

Creer, K. M. (1973). A discussion of the arrangement of palaeomagnetic poles on the map of Pangaea for epochs in the Phanerozoic. In *Implications of Continental Drift to the Earth Sciences* (Ed. D. H. Tarling and S. K. Runcorn) Academic Press, London pp. 47–76

Creer, K. M., Georgi, D. T. and Lowrie, W. (1973). On the representation of the Quaternary and Late Tertiary geomagnetic fields in terms of dipoles and quadrupoles. *Geophys. J.*, **33**, 323–345

Doell, R. R. and Cox, A. (1971). Pacific geomagnetic secular variation. *Science*, **171**, 248–254

Fujisawa, H. (1967). Temperature and discontinuities in the transition layer within the Earth's mantle. *Tech. Rep. of Inst. Solid State Physics, Tokyo*, Series A, No. 272, 38 pp.

Fujisawa, H. (1968). Temperature and discontinuities in the transition layer within the Earth's mantle: geophysical applications of the olivine–spinel transition in the Mg_2—SiO_4—Fe_2SiO_4 system. *J. Geophys. Res.*, **23**, 3281–3294

Goldreich, P. and Toomre, A. (1969). Some remarks on polar wandering. *J. Geophys. Res.* **74**, 2555–2567

Harland, W. B., Smith, A. G. and Wilcock, B. (Eds.) (1964). Geological Society Phanerozoic time-scale, 1964. *Quart. J. geol. Soc. Lond.*, **120s**, 260–2

Hays, J. D. and Pitman, W. C. (1973). Lithospheric plate motion, sea level changes and climatic and ecological consequences, *Nature*, **246**, 18–22

Hicken, A., Irving, E., Law, L. K. and Hastie, J. (1972). Catalogue of palaeomagnetic directions and poles, first issue. Publ. Earth Physics Branch, Dept. Energy, Mines and Resources, Ottawa, Vol. 45, No. 1

Hide, R. (1967). Motions of the Earth's core and mantle and variations of the main geomagnetic field. *Science*, **157**, 55–56

Irving, E. and Couillard, R. W. (1973). Cretaceous normal polarity interval. *Nature*, **244**, 10–11

Irving, E. (1966). Palaeomagnetism of some Carboniferous rocks from New South Wales and its relation to geological events. *J. Geophys. Res.*, **71**, 6025–6047

Irving, E. and Parry, L. G. (1963). The magnetism of some Permian rocks from New South Wales. *Geophys. J.*, **7**, 395–441

Johnson, A. H., Nairn, A. E. M. and Peterson, D. N. (1972), Mesozoic reversal stratigraphy. *Nature, Phys. Sci.*, **237**, 9–10

Johnson, L. R. (1967). Array measurements of P velocities in the upper mantle, *J. Geophys. Res.*, **72**, 6309–6325

Kanomori, H. (1967). Upper mantle structure from apparent velocities of P waves recorded at Wakayama microearthquake observatory. *Bull. Earthquake Res. Inst. Tokyo, Univ.*, **45**, 657–678

Khramov, A. N., Rodionov, V. P. and Komissarova, R. A. (1966). *New data on the Palaeozoic History of the Geomagnetic Field in the U.S.S.R.* Translation by E. R. Hope, Defence Research Board, Canada, Publ. T460R, 1–8

Larson, R. and Pitman, W. C. (1972). World-wide correlation of Mesozoic anomalies and its implications. *Bull. Geol. Soc. Amer.*, **83**, 3645–3661

Lilley, F. E. M. (1970). Geomagnetic reversals and the position of the North Magnetic Pole. *Nature*, **227**, 1336–1337

McGugan, A. (1967). In *Abstr. Ann. Meeting Geol. Soc. Amer.*, **145**

Mazzullo, S. J. (1971). Length of the year during the Silurian and Devonian Periods: new values. *Geol. Soc. Amer. Bull.*, **82**, 1085–1086

Miller, G. R. (1966). The flux of tidal energy out of the deep ocean. *J. Geophys. Res.*, **71**, 2485–2489

Munk, W. H. and MacDonald, G. J. F. (1960). *The Rotation of the Earth: a Geophysical Discussion*, Cambridge University Press, London

Newton, R. R. (1968). A satellite determination of tidal parameters and Earth deceleration. *Geophys. J.*, **14**, 505–539

Pannella, G. (1972). Palaeontological evidence on the Earth's rotational history since the Early Precambrian. *Astrophys. and Space Sci.*, **16**, 212–237

Pannella, G., MacClintock, C. and Thompson, M. N. (1968). Palaeontological evidence of variations in the length of the month since the Late Cambrian. *Science*, **162**, 792–796

Pechersky, D. M. and Khramov, A. N. (1973). Mesozoic palaeomagnetic time scale of the U.S.S.R. *Nature*, **244**, 499–501

Runcorn, S. K. (1964). Changes in the Earth's moment of inertia. *Nature*, **204**, 823–825

Scrutton, C. T. and Hipkin, R. G. (1973). Long-term changes in the rotation rate of the Earth. *Earth Sci. Rev.*, **9**, 259–274

Simpson, J. F. (1966). Evolutionary pulsations of geomagnetic polarity. *Bull. Geol. Soc. Amer.*, **77**, 197–203

Stacey, F. (1969). Physics of the Earth. John Wiley, New York, 324 pp.

Valentine, J. W. and Moores, E. M. (1970). Plate tectonic regulation of faunal diversity and sea level: a model. *Nature*, **228**, 657–659

Wells, J. W. (1963). Coral growth and geochronometry. *Nature*, **197**, 948–950

Wells, J. W. (1966). In *The Earth–Moon System* (Eds. G. W. Marsden and A. G. W. Cameron) Plenum Press, New York, pp. 70–81

Wells, J. W. (1970). Problems of annual and daily growth rings in corals. In *Palaeogeophysics* (Ed. S. K. Runcorn) Academic Press, London, pp. 3–9

DISCUSSION

RUNCORN: You have fitted four straight lines to millions of years of palaeomagnetic data, and I think you are reading too much into the data.

MULLER: If the data were accurate, you would get the same answer no matter how the data were fitted.

RUNCORN: This is only if the data were accurate, but in fact we are uncertain how accurate the data are.

CREER: Additional data are certainly needed, but it is instructive to establish a curve as I have done because it will be one means of testing whether the theory in the future is consistent or inconsistent with new data.

O'HORA: Why did you fit the 4th degree polynomial to your data?

MULLER: The higher the order polynomial, the better the fit.

PANNELLA: I fit as high as the 8th degree polynomial to my data. It would seem that reporting the 4th degree by Creer and myself is not extreme.

CREER: I believe palaeontologists have reason to be optimistic about the validity of their data, and I feel it unfortunate that there aren't more workers doing growth increment studies. For geophysicists, I hope palaeontologists can set some ground rules for standards so that if one observer measures a coral he could give it to another person who will get the same answer. It is difficult to count growth increments because they are so irregular, and I urge palaeontologists to adopt some standards for counting and reporting data.

CLARK: There are situations in which one wants to use the mode, some the maximum and others the average number of growth increments counted, depending on how the organism or species forms lines. Sometimes biologists have good reason to choose the way such data is reported. Also, I would like to refer to a 1967 geophysical paper correcting Wells' data. The author assumed that if Wells got 360 lines/365 day modern year, Wells' Devonian year would have to be corrected from 400 to 408 days. I believe such geophysical reasoning was not justified from the palaeontological data.

TIDES AND THE ROTATION OF THE EARTH

R. G. HIPKIN
Department of Geophysics, University of Edinburgh, UK

Abstract

Contrary to the orthodox description, we are unable empirically to detect a net gain in the Moon's orbital angular momentum, or a net loss from the Earth's rotation. The steady transfer of angular momentum from the Earth to the Moon was assumed to occur because tidal torques seemed to be the only physically viable mechanism to explain the observed secular accelerations. To reconcile the momentum equation with these observations, de Sitter proposed changes in the lunar eccentricity while Urey suggested a secular change in the Earth's moment of inertia. These, and other alternatives, are examined in the light of modern tidal data for the solid Earth, the oceans and the atmosphere.

Using recent estimates of the extent of tidal dissipation in different parts of the Earth, together with information about the Earth's modes of free oscillation, the validity of extrapolating the present tidal parameters of the Earth into the remote past will also be assessed.

The equilibrium tide and the tidal distortion of the Earth

Whereas for most people the word 'tide' conjures up a picture of the sea rising and falling across the intertidal shore, its technical use covers the much wider range of phenomena associated with the periodic distortion of the Earth by the Sun and Moon. In the first instance, the word is used for the periodic force itself: there are tidal variations in the acceleration of gravity which can now be measured directly. Secondly, the solid Earth, the oceans and the atmosphere each distort in response to this force. Lastly, there are thermal tides: as well as acting on the Earth gravitationally, the Sun also causes a periodic redistribution of mass, predominantly in the atmosphere but to a lesser extent also in the oceans, due to the daily cycle of thermal radiation.

Of all these phenomena, only the tide-generating force is geometrically simple to describe. If we imagine the Earth covered uniformly by a very deep, frictionless ocean, the tide-generating force would distort its surface into an ellipsoid whose long axis pointed towards the tide-raising body, the Sun or the Moon. This model, called the equilibrium tide, is merely a way of picturing the tide-generating forces and bears absolutely no resemblance to the way in which the real ocean distorts. Because the geometry of the Earth–Moon–Sun system varies with the position of the Earth and the Moon in their orbits, the axis of the bulge changes in length and moves in latitude in a very complicated way. The equilibrium tide is best split up into three separate bulges whose orientation in latitude remains fixed. The most important of these is an ellipsoid whose long axis lies in the plane of the Earth's equator. A point on the Earth's surface is carried under both ends of the bulge in the course of about one day so that as a result it experiences semi-diurnal tides. The second type of bulge has its long axis pointing at an angle of 45° to the equator. Except on the equator itself, where there is no variation, points on the Earth's surface have only one high tide a day, i.e. a diurnal tide, because the two ends of the bulge are in different hemispheres. The last type of tidal bulge is simply a change in the extent to which the Earth is flattened at the poles into an oblate spheroid. The elevation is the same at all longitudes around a particular line of latitude so that there is no variation associated with the rotation of the Earth. Only long-period tides are generated.

The three most important astronomical periods related to the motion of the Earth and the Moon are the rotation period of the Earth, 0·9973 days, the orbital period of the Moon, 27·3217 days, and the orbital period of the Earth, 365·2422 days. These correspond to rotation rates of

$$\omega = 15\cdot0410\ 6864°/\text{hr}$$
$$n_m = 0\cdot5490\ 1653°/\text{hr}$$
$$n_e = 0\cdot0046\ 4184°/\text{hr}$$

There are also less important periods of about 9 years and 18·6 years due to the eccentricity of the Moon's orbit and to the inclination of this orbital plane to the plane of the Earth's orbit about the Sun. Every tidal period can be constructed in a simple way from these five quantities.

Within the broad division of the tide into semi-diurnal, diurnal and long-period parts, there are nearly 300 separate components, called partial tides, each with a measurable elevation and each with its own characteristic period. For most purposes it is only necessary to consider about half-a-dozen of them and they can be referred to by the simple code names given them by Sir George Darwin. Because the particular sequence of high and low tides at any place, together with the record it leaves as a pattern of growth increments, can only be understood in terms of combinations of these partial tides, I shall describe the relation of some of their periods to the three important astronomical periods.

The largest partial tide, by a factor of two, is the semi-diurnal tide raised by the Moon, called M_2. It has a period of 12 hr 24 min and a speed of $2\omega - 2n_m = 28\cdot98410°/\text{hr}$, which is twice the rotation rate of the Earth with respect to the Earth–Moon line.

The second largest partial tide is the diurnal one, K_1, raised by the combined action of the Sun and the Moon. Its period, 23 hr, 56 min, is almost exactly equal to the rotation period of the Earth, not with respect to the Sun or Moon, but with respect to the stars; consequently it is sometimes called the sidereal diurnal tide. Its speed is $\omega = 15\cdot04107°/\text{hr}$.

Next comes the semi-diurnal solar tide, S_2. Its period is exactly 12 hours,

Tides and the Rotation of the Earth

corresponding to the rotation period of the Earth with respect to the Earth–Sun line, which is, of course, the basis of civil timekeeping. Its speed is $2\omega - 2n_e = 30\cdot00000°/\text{hr}$.

The fourth and fifth largest partial tides are the lunar and solar diurnal tides, Q_1 and P_1, with periods of 25 hr, 49 min and 24 hr, 4 min and speeds of $\omega - 2n_m = 13\cdot94304°/\text{hr}$ and $\omega - 2n_e = 14\cdot95893°/\text{hr}$. These figures demonstrate the important points that there are no diurnal partial tides with exactly twice the period of M_2 or S_2. The speed of O_1 is not $\omega - n_m$ but $\omega - 2n_m$ so that its phase gradually gets more and more behind that of M_2. For the same reason, the phase of P_1 gradually falls behind the S_2 component and therefore behind local solar time.

At about 20% of the size of the M_2 tide comes a semi-diurnal component, N_2, which depends upon the lunar eccentricity, and one of the long-period tides, M_f, whose period is about one fortnight. Their speeds are approximately $(2\omega - 3n_m)$ and $2n_m$ respectively.

The list could be extended but the main point is that, in the equilibrium tide, each partial tide has an exactly calculable amplitude and period and its phase is governed only by the relative orientations of the Greenwich Meridian, the Moon and the Sun. In contrast, the real Earth responds to the tide-generating forces, of which the equilibrium tide is a model, in an extremely irregular way. There are two aspects to this irregularity: firstly, the size of the response in a particular ocean basin depends very critically upon frequency, that is the speed of the tide, so that some partial tides are suppressed and others amplified; secondly, the tidal bulge is best modelled by a sphere with warts rather than a simple ellipsoid.

In an analysis of the tidal components recorded on island stations in the North Atlantic, Wunsch (1972) finds that the amplitudes of the diurnal tides are consistently less than those of the equilibrium tide by a factor of about two, while the semi-diurnal components are strongly amplified, preferentially at the slower end. The N_2 component, with a speed of $28\cdot43973°/\text{hr}$, is amplified about six times compared with the equilibrium tide and nearly three times compared with the only slightly faster S_2 component whose speed is exactly $30°/\text{hr}$. This illustrates the tendency of each ocean basin to have certain preferred frequencies of oscillation in the vicinity of which the tidal amplitude is greatly enlarged.

The other aspect of the irregularity of the Earth's response to tides is the shape of the bulge. In the equilibrium tide, there are just two bulges of equal size, each half the Earth's circumference, or 20,000 km, across. In the real tide, bulges with these dimensions are not the best developed; the highest are in the range 13,000–8000 km across and it is only for bulges less than 3000 km across that the elevation drops to less than one-third of the 20,000 km bulge. This is simply one way of picturing the spherical harmonic analysis of the M_2 tide carried out by Hendershott (1972) but the qualitative result has been known for many decades.

The suppression of diurnal tides in the North Atlantic has strongly influenced popular conceptions about tides. On the Atlantic coasts of Europe and North

America most of the tidal variation is described by the semi-diurnal tides M_2 and S_2; these alternately reinforce each other and cancel each other with a period corresponding to their differences in speed:

$$(2\omega - 2n_e) - (2\omega - 2n_m) = 2(n_m - n_e)$$

This corresponds to two sets of higher and lower tides, spring and neap tides, in the period of revolution of the Moon with respect to the Earth–Sun line; in astronomical terms, this period is the synodic month. Because the response of the real ocean may lag or lead the equilibrium tide and the extent to which it does so varies markedly from place to place, the time when M_2 and S_2 reinforce each other, the spring tide, does not usually coincide exactly with full or new Moon. It can be displaced by as much as 8 days. Garrett and Munk (1971) give some statistics on this point.

A tidal cycle related to the synodic month is not a necessary feature: in extensive areas of the Pacific Ocean, the Atlantic situation is reversed and the diurnal tides are amplified at the expense of the semi-diurnal ones. Where this relative amplification exceeds a factor of three or four, for example in the South China Sea, the total tide is predominantly diurnal and it is the interference of the K_1 and O_1 partial tides which govern the observed elevation. The difference in their speed is

$$(\omega) - (\omega - 2n_m) = 2n_m$$

'Spring' and 'neap' tides now occur twice every *sidereal* month, resulting in 26·74 'fortnightly' patterns per year instead of 24·74.

For a lesser degree of amplification, the observed tide will be a rather more complicated 'mixed semi-diurnal–diurnal' tide in which the main pattern may be controlled by M_2 and K_1. In this case, 'spring' and 'neap' tides again occur twice every *sidereal* month.

The growth ridge patterns discovered by Evans (this volume) in clams from the Oregon coast are a beautiful illustration of this phenomenon.

Dissipation and the tidal torque

It is now generally believed that most of the frictional dissipation of tidal energy takes place in continental shelf areas around the major ocean basins. Some friction may also occur internally within the ocean, but the distortion of the solid Earth contributes very little to the energy loss. At least as far as tides are concerned the solid Earth can be modelled quite satisfactorily by an ideal elastic material and therefore the distortion is in phase with the generating forces. From seismological and gravimetric measurements of the spheroidal free-oscillation of the Earth S_2 (Lagus and Anderson, 1968) we should expect a phase lag of about $(1/6)°$ while the detailed analysis of ocean loading effects upon gravimetric Earth tides indicates a lag of less than $(1/4)°$ (Farrell, 1972 and

Tides and the Rotation of the Earth

personal communication). In contrast, if all the dissipation were due to solid Earth friction, we should need a phase lag of about 8·5°. Compared with this, solid Earth friction can be neglected.

The solid Earth response is relatively simple for another reasons: it is only weakly dependent upon frequency. The Earth's core possesses a mode of free-oscillation which is only a few minutes short of the sidereal day and this does produce amplification of the diurnal tides (Melchior, 1966, page 383). For the bulk of the Earth, however, the free periods are all less than 1 hr and do not affect the tidal response.

Nevertheless, measurements of Earth tides, whether variations in Earth strain, the acceleration of gravity or the deflection of the vertical, still cannot be interpreted as a simple elastic bulge analogous to the equilibrium tide because the Earth is distorted not only by the direct attraction of the Sun and Moon but also by the attraction and variable load of the tide in the oceans. Ozawa (1967) has even detected a component of Earth strain due to the pressure variation of the solar thermal tide in the atmosphere. Since we can calculate the elastic response of the solid Earth quite accurately, the importance of Earth tide measurements now lies in their ability to constrain our estimates of tide heights in otherwise inaccessible mid-ocean regions.

Because of the resonant amplification of the tide in the deep ocean, strong tidal currents are generated as the ocean depth decreases abruptly and water is channelled on to the gently sloping continental shelf. These currents are retarded by bottom friction, in which the stress is roughly proportional to the square of the tidal velocity. Both elements must be present before the dissipation is important: the shallow sea and its connection to a large body of deep ocean in which the free periods are favourable to the generation of tides. The second factor makes it difficult to interpret changes in the extent of shallow seas in the past directly in terms of changes in the magnitude of tidal friction. The Baltic, although a substantial shallow sea, dissipates a negligible amount of energy both because it is cut off from tides in the Atlantic and because its own free periods are unfavourable (Defant, 1961).

If enough were known about the elevation and phase of the tide in the world's oceans, the torque exerted by the Moon and Sun on this tidal mass distribution could be calculated and the acceleration of the Earth's rotation and the lunar orbital motion obtained directly. Such a calculation was first attempted by Heiskanen (1921) and later by Groves and Munk (1958). With very few exceptions, tidal parameters are measurable only on the coastline itself and mid-ocean values must be obtained by interpolation. Instead of calculating the direct torque over the whole ocean, alternative energy flux and bottom friction methods, described by Munk and MacDonald (1960), require only tidal parameters over the shelf area; nevertheless, with all three techniques, it is necessary to extrapolate values away from where they have been measured. (Brosche and Sündermann (1971) have in any case shown that the original theory of the bottom friction method was inappropriate to the secular acceleration problem.) It is the improvement in our ability to carry out the necessary interpolation, an improvement which began with Hansen's (1949) work on the North Atlantic, which has made the oceanographic approach so much more valuable.

The tidal charts used by Groves and Munk were constructed by linear interpolation, controlled mainly by the subjective judgment of their authors and, to a lesser extent, by measured tidal parameters on the occasional mid-oceanic island. This interpolation can be replaced by an exact calculation: the bathymetry and shape of the oceans are known with sufficient accuracy and the tide-generating forces are known exactly to that the parameters of any tidal component can be obtained by solving the appropriate equations of hydrodynamics without the need for any observational data at all. Published charts of the M_2 partial tide obtained in this way include those of Pekeris and Accad (1969) and Zahel (1970). Zahel (1973) has recently produced a similar solution for the K_1 tide. A somewhat less demanding calculation uses the hydrodynamic equations to interpolate from measured tidal parameters on the coastlines. Bogdanov and Magarik (1967, 1969) have derived worldwide tidal charts for the M_2, S_2, K_1 and O_1 components in this way. Even though there are still some refinements to be included before any chart can be accepted as definitive, there is already an impressive degree of mutual consistency.

Kuznetsov (1972) has estimated the torque exerted by the Moon on the distribution of mass shown on these calculated charts as the M_2 tidal elevation of sea level. He obtains figures of $-7 \cdot 29$, $-9 \cdot 29$ and $-10 \cdot 13 \times 10^{16}$ Nm from the charts of Bogdanov and Magarik, Pekeris and Accad and Zahel. This is to be compared with a torque of $-3 \cdot 25 \times 10^{16}$ Nm estimated by Groves and Munk from the subjectively interpolated M_2 chart of Dietrich (1944). All of these results assume that the ocean floor is rigid, whereas in reality it is depressed by the weight of the ocean tide above it and elevated by its gravitational attraction. The combination of these two disturbances makes the effective torque less than the results quoted above. Provided that the flexibility of the ocean floor does not alter the overall picture of highs and lows on the tidal chart, it can be allowed for quite simply: the torque must be multiplied by a factor of $(1 + k_2') = 0 \cdot 690$ (k_2' is the load effective Love number described by Munk and MacDonald, 1960). On this basis, the calculations indicate an effective torque by the Moon on the M_2 tidal distribution of about $-5 \cdot 0$, $-6 \cdot 4$ and $-7 \cdot 0 \times 10^{16}$ Nm for the three charts mentioned above.

Hendershott (1972) starts an interative procedure to include the effects on a flexible ocean floor in the initial calculation of the tidal chart but he gives no indication of whether his solution has converged sufficiently. For an earlier solution (Hendershott and Munk, 1970) which was obtained from measured coastal parameters by a hydrodynamic interpolation process, he did allow for some of the consequences of a flexible ocean floor, but excluded all loading and self-gravitational effects. With this chart he obtained a torque of $-4 \cdot 33 \times 10^{16}$

Tides and the Rotation of the Earth

Nm for the M_2 tide (actually he gives the equivalent statistic of a rate of working of $3\cdot03 \times 10^{12}$ W).

There are some reasons for preferring the tidal representation of Zahel, notably the fact that he correctly used a quadratic law of friction and has most accurately modelled the coastlines (even though Pekeris and Accad used a finer grid spacing). The solution is certainly very dependent upon the precision with which the coastlines are modelled and the fineness of the grid spacing: qualitatively, this is well illustrated by the charts in Pekeris' paper.

As an alternative to direct calculation or measurement of the tidal elevation of the oceans, it is possible, in principle, to determine the tidal mass distribution by a suitable global average of its gravitational attraction. Because of the severe loading effects originating on continental shelves, it is difficult to obtain a meaningful average directly from ground-based gravimetry. A much more satisfactory average could be obtained from artificial satellites: the acceleration of the satellite is of course an even more direct means of determining gravitational forces. Nevertheless, many other rather poorly known disturbances, such as radiation pressure, must be eliminated before the very small tidal effect becomes detectable. The most thorough analysis of satellite tracking data to look for the tidal effect is by Newton (1968) but he models the tide by a simple time-lagged ellipsoidal bulge, which should be a particularly poor approximation to the total disturbance as sensed at the altitude of an artificial satellite. Newton's result is a torque of $-4\cdot15 \pm 0\cdot39 \times 10^{16}$ Nm but his quoted uncertainty is unrealistically small.

In addition to the oceanic torque, there is a small torque on the Earth due to the solar thermal tide in the atmosphere. In an ocean tide, kinetic energy derived from the gravitational attraction of the Sun and Moon is converted into heat by frictional forces and the effect is to slow the Earth's rotation. With thermal tides, the process is reversed and becomes a heat engine: heat is supplied by the Sun and generates kinetic energy as atmospheric motion which results in a speeding up of the Earth's rotation. The net torque is small and quite well-determined: from atmospheric pressure data given by Haurwitz (1964), suitably corrected for Earth loading effects, a torque of $+0\cdot22 \times 10^{16}$ Nm can be deduced.

Although this thermal tidal torque is small at the present time, it is potentially important because it is the only tidal effect to produce a positive acceleration of the Earth and also because it may well be sensitive to climatic changes. The mechanism is now quite well understood (see Chapman and Lindzen, 1970) and does not depend upon a simple resonance between the semi-diurnal frequency and the frequency of free oscillation of the atmosphere, but rather upon a matching of the vertical distribution of velocity in one of many possible oscillation modes with the vertical distribution of ozone in the atmosphere. The ozone acts as an energy source for the tidal motion because it is the dominant absorber of solar radiation. Ozone itself is mainly generated by solar radiation so the magnitude of the torque accelerating the Earth's rotation is probably very sensitive to small changes in solar output.

Tidal friction and astronomical data

Munk and MacDonald (1960) used a determination of the tidal acceleration of the Moon by Spencer Jones ($\dot{n} = -22\cdot44''$/cy) to deduce that the tidal torque acting to slow the Earth's rotation is $-3\cdot9 \times 10^{16}$ Nm, equivalent to an angular acceleration of $\dot{\omega} = -4\cdot9 \times 10^{-22}$ rad sec^{-2}. By adopting Jeffreys' (1920) figure for the ratio of the lunar to solar tidal torques, this gives a total tidal increase in the length of the day of 1·8 msec/day/cy. This figure has passed into the literature of growth ridges as 'the astronomically determined rate of increase in the length of the day' even though the assumptions on which it is based are seldom stated and not strongly supported by observational evidence. This criticism is independent of the fact that Spencer Jones' figure for the lunar acceleration is almost certainly an underestimate by a factor of two.

Although positional astronomy can give a very precise measure of the rotation rate of the Earth, it is unable to measure the *secular* change in the length of the day because, for all observational periods from a few days to several centuries, the acceleration is dominated by non-cumulative fluctuations. Because the observed 8000-year variation in the Earth's dipole moment is very probably accompanied by an exchange of angular momentum between the Earth's core and mantle (Yukutake, 1972), the apparent secular acceleration found from ancient eclipse observations over the last $2\frac{1}{2}$ millenia is also likely to be biased by these fluctuations. The orthodox procedure is that used by Munk and MacDonald, which estimates the secular acceleration indirectly.

It is based on the following assumption.

(i) Tidal friction is the dominant cause of secular changes in the lengths of the day and month.

(ii) Except for a theoretically calculable solar tidal torque, angular momentum is conserved within the Earth–Moon system—there are no other external forces acting.

(iii) Except for those predicted by a conservative gravitational theory, no changes take place in any of the dynamical elements describing the size, shape or orientation of the Earth–Moon system, apart from changes in the orbital speed of the Moon and the rotation rate of the Earth caused by tidal friction. (Assumption (ii) then implies that a measurement of the tidal acceleration of the Moon uniquely determines the corresponding tidal torque acting to increase the length of the day.)

(iv) The tidal acceleration so determined can be extrapolated into the remote past because the Earth responds to a tidal disturbance now in a way which is typical of its entire history.

This orthodox procedure is in reality rather more indirect than is usually admitted and the estimate that it gives of the tidal acceleration of the Earth's rotation is probably no less uncertain than the oceanographic techniques.

The attraction of the Sun causes the Earth's equator and the plane of the lunar

Tides and the Rotation of the Earth

orbit to rotate in space. This precessional motion is evidence of angular momentum passing to and fro between the Earth's rotation, its orbital motion about the Sun and the lunar orbit. The precessional torque is about 4 million times greater than the torque due to tidal friction but fortunately the precessional torque acts only in the plane of the Earth's orbit about the Sun. Tidal effects then become observable as changes in momentum *perpendicular* to this plane. In this direction, the angular momentum of the Earth–Moon system consists of a part due to the orbital motion of the Moon:

$$L = \frac{M_e M_m}{(M_e + M_m)} a^2 n(1 - e^2)^{1/2} \cos(i) \tag{1}$$

and parts due to the rotation of the Earth and the Moon. The latter is negligibly small leaving only the Earth's spin momentum:

$$S = C\omega \cos(I) \tag{2}$$

These expressions are often written in the approximation that the Moon's orbit is circular and lies in the plane of the Earth's equator ($e = i = I = 0$) but this exposes the treatment to the probability of serious error: the angular momentum is altered if the orientation or shape of the lunar orbit changes, or if the Earth's axis tilts more in space. These effects will be missed if changes only in the length of the day and month are considered. It may then appear that angular momentum is not conserved. In fact, when only the day and month are treated as variable, the angular momentum gain which is observed as the lunar acceleration is much greater than (about 180% of) that observed to be lost by the secular slowing of the Earth's rotation over the past 2000 years. (It was pointed out above that this latter observation may not be a true measure of the *secular* slowing of the Earth because of core–mantle momentum transfer.)

This discrepancy was first pointed out by de Sitter (1927). He suggested that the observations could be reconciled if the lunar eccentricity also were treated as variable. So far the required rate of increase has been unobservably small: writing $e = \sin \phi$, it is only $d\phi/dt = +0.0088''$/cy. Martin and van Flandern (1970) find $d\phi/dt = +0.10 \pm 0.04''$/cy, really indicating that the observational uncertainty is an order of magnitude greater than de Sitter's predicted change. However, the latter is well within the capabilities of laser range finding to the Moon. In the Earth–Moon distance, there is a term with a period of one anomalistic month, 27.475 days, and an amplitude of 21,103 km, which provides a direct measure of the lunar eccentricity. After observations have been matched to a suitable theory of the lunar motion we should need to detect a residual increase in the amplitude of this term at the rate of 0.24 m/y. The best measurements to date have an uncertainty of ±1.5 m but an improvement by a factor of 5 is expected soon.

Urey (1952) again drew attention to the inconsistency of the astronomical measurements of the changes in the day and month but attributed it to a secular increase in the Earth's moment of inertia. He proposed that the observed increase in the length of the day was only part of the tidal effect: the rest was cancelled by a tendency for the Earth to spin faster as metallic iron settled out of the mantle and caused the core to enlarge. In fact, the continuing differentiation of the crust is a rather more likely process and about three times as effective.

Urey's proposal is not so easily susceptible to observational verification because seismological measurements of the core radius have an uncertainty of several km; however, two recent geophysical techniques now improve the situation slightly. The falling interferometer apparatus developed by Sakuma (1971) measures the absolute acceleration of gravity at the Earth's surface with a basic precision of $\delta g/g = \pm 10^{-9}$ from about one month's observations, while the tracking of artificial satellites yields the shape parameter J_2 with an uncertainty of $\delta J_2 = \pm 2 \times 10^{-9}$ from about four years' data (Kozai, 1970). Although gravity may change locally owing to tectonic effects, on a global scale it is related to the mean radius of the Earth, R, and J_2, by

$$\frac{\delta g}{g} = -(\tfrac{3}{2}\cos^2\theta - \tfrac{1}{2})\delta J_2 - 2\frac{\delta R}{R} \qquad (3)$$

Any change in the moment of inertia of the Earth will change either one or both of these quantities.

If the Earth is secularly increasing or decreasing in size, for example due to an overall secular increase in internal temperature, only R will change and

$$\frac{\delta C}{C} = 2\frac{\delta R}{R} = -\frac{\delta g}{g} \qquad (4)$$

A shape change resulting from incompressible flow within the Earth, a good model for postglacial isostatic adjustment, leaves R unchanged and

$$\frac{\delta C}{C} = 2\delta J_2 \qquad (5)$$

Although it is usually supposed that surface measurements are incapable of detecting changes in the extent to which the Earth's density is concentrated towards the centre, as measured by the moment of inertia factor $z = C/MR^2$, this is only strictly true of a perfectly spherical Earth. (For a homogeneous Earth, $z = 2/5$; for the real Earth, $z = 0.3308..$; if all the mass were concentrated at the centre, $z = 0$.) Provided that the Earth remains close to a state of hydrostatic equilibrium, the factor z is related to J_2 by the Radau equation (see, for example, Kaula, 1968, section 2.1. The parameter m is well determined and is not important to this argument).

$$z = \frac{2}{3}\left\{1 - \frac{2}{5}\left(\frac{4m - 3J_2}{m + 3J_2}\right)^{1/2}\right\} \qquad (6)$$

For any differentiation process which changes the degree of central condensation but leaves the gross state of hydrostatic equilibrium unaltered

$$\delta z = \left(\frac{2m}{m+3J_2}\right)\left\{\frac{1}{(4m-3J_2)(m+3J_2)}\right\}^{1/2} \delta J_2$$

or (7)

$$\frac{\delta C}{C} = \frac{\delta z}{z} \doteqdot \frac{12}{25}\frac{\delta J_2}{J_2}$$

Measurements of moment of inertia changes due to such a process are consequently about 200 times more uncertain.

Apart from changes in the Lunar eccentricity and the Earth's moment of inertia, there could be many other causes of the discrepancy in the observations of increases in the lengths of the day and month. Most are undetectable. Table 1 lists the momentum change due to the currently observed rate of change of each quantity in equations (1) and (2). Where no variation has been detected, the observational uncertainty is given instead, with a rate equivalent to 10 years of data. This makes very clear the weakness of assumption (iii) of the orthodox procedure which treats only the lengths of the day and month as variables. Empirically, it is clearly impossible to eliminate variations in the remaining dynamical elements of the Earth–Moon system.

There is also a serious problem with the second assumption. The common statement in the literature that the solar tidal torque is a known fraction, either 1/5·1 or 1/3·4 according to mechanism, of the lunar tidal torque, and can therefore be calculated, is an illusion derived from a paper by Jeffreys (1920). He showed that, if frictional forces were linear, then the ratio of the solar tidal torque to the lunar one is 1/4·86 (this is obtained with modern constants in Jeffreys' formula; he obtained 1/5·1). Such a linear model is quite unrealistic, although it is the only one which allows a valid calculation of this ratio. Instead, Jeffreys argued in favour of a quadratic law of frictional forces. With this model, *together with the simplifying approximation of the Moon and the Sun moving in a circular orbit in the plane of the Earth's equator*, the ratio is reduced to 1/3·67 (modern constants with Newton's (1968) correction to Jeffreys' arithmetic).

The proviso appeared trivial whereas in reality the calculation is impossible without it: Jeffreys' treatment approximates the whole ocean tide by the two terms M_2 and S_2 and shows that the effect of quadratic friction is to make the lesser of the two terms relatively more important compared with the size of its coefficient in the equilibrium tide. Even if M_2 and S_2 were the largest parts of the tide, it would now be necessary to justify neglecting all the other smaller components; but in fact S_2 is only the third largest tide after K_1, and O_1 is nearly as large. The calculation cannot be extended to include these other terms because when the Moon is allowed to move in an inclined and eccentric orbit it raises not only semi-diurnal tides but also diurnal and long-period ones which depend

TABLE 1. Observed changes in the angular momentum of the Earth-Moon system

Parameter	Observation	Author	Technique	Relative angular momentum change
Rotation rate ω	$\frac{1}{\omega}\frac{d\omega}{dt} = -(2.76\pm0.33)\times10^{-10}/\text{year}$	Newton (1970)	Ancient astronomy	$\frac{1}{\omega}\frac{d\omega}{dt} = -(2.76\pm0.33)\times10^{-10}/\text{year}$
Obliquity of the ecliptic I	$\frac{dI}{dt} = -(0.287\pm0.029)''/\text{cy}$	Duncombe (1958)	Motion of Venus	$-\tan I \frac{dI}{dt} = +(60.4\pm6.1)\times10^{-10}/\text{year}$
	$\frac{dI}{dt} < \pm 0.1''/\text{cy}$	Fricke (1972)	Consistency of stellar proper motions	$-\tan I \frac{dI}{dt} < \pm 21\times10^{-10}/\text{year}$
Moment of inertia C	$\delta g = \pm 1\ \mu\text{gal}$	Sakuma (1971)	Absolute acceleration of gravity	$\frac{1}{C}\frac{d}{dt}(\delta C) = \pm 1\times10^{-10}/\text{year}$ (equation 4)
	$\delta J_2 = \pm 2\times10^{-9}$	Kozai (1970)	Tracking artificial satellites	$\frac{1}{C}\frac{d}{dt}(\delta C) = \pm 4\times10^{-10}/\text{year}$ (equation 5)
				$\frac{1}{C}\frac{d}{dt}(\delta C) = \pm 900\times10^{-10}/\text{year}$ (equation 7)
Lunar mean motion n	$\frac{dn}{dt} = -(42\pm6)''/\text{cy}$	Morrison (1973)	Lunar occultations	$-\frac{L}{3nS}\frac{dn}{dt} = +(3.93\pm0.56)\times10^{-10}/\text{year}$
	$\frac{dn}{dt} = -(41.6\pm4.3)''/\text{cy}$	Newton (1970)	Ancient astronomy	$-\frac{L}{3nS}\frac{dn}{dt} = +(3.89\pm0.40)\times10^{-10}/\text{year}$
Lunar eccentricity $e = \sin\phi$	$\frac{d\phi}{dt} = +(0.10\pm0.04)''/\text{cy}$	Martin and van Flandern (1970)	Lunar occultations	$-\frac{Le}{S(1-e^2)}\frac{de}{dt} = -(12.9\pm5.2)\times10^{-10}/\text{year}$
Lunar inclination i	$\frac{di}{dt} = -(0.04\pm0.09)''/\text{cy}$	Martin and van Flandern (1970)	Lunar occultations	$-\frac{L}{S}\tan i\frac{di}{dt} = +(8\pm19)\times10^{-10}/\text{year}$

Tides and the Rotation of the Earth

upon a different function of latitude. The only way of resolving the problem is to apply the full oceanographic calculation of the response of the world's oceans to the sum of all the different tidal components and to calculate torque exerted by the Sun and Moon on this distribution. Such a calculation is not technically feasible at the present time and is unlikely to become so in the foreseeable future.

From his analysis of artificial satellite data, Newton (1968) finds a lunar tidal torque of $-4\cdot15 \pm 0\cdot39 \times 10^{16}$ Nm and a solar torque of $-0\cdot38 \pm 0\cdot13 \times 10^{16}$ Nm. Because he used the Love number k_2 instead of k_2' to eliminate the effects of the solar *thermal* tidal torque, the solar tidal torque quoted above should be lowered to $-0\cdot11 \pm 0\cdot13 \times 10^{16}$ Nm. The implication is a ratio of only 1/40 but, for the reasons described earlier, the artificial satellite technique must be considered unreliable.

An alternative approach uses the tidal charts of Bogdanov and Magarik (1967). Since the regions of intense quadratic friction are largely outside their computational boundaries it may be a reasonable approximation to superimpose their charts linearly. Pariyskiy et al. (1972) calculate a ratio of the S_2 to M_2 tidal torques of about 1/7. This is probably the best estimate to date but we should really consider the size of the solar tidal torque as an open question.

Dynamical consequences of tidal friction

The literature of this subject is very extensive but, with the benefit of hindsight derived principally from the paper of Gerstenkorn (1955), it could be said that every important contribution to the theory of tidal friction can be found in the papers of Sir George Darwin (1908). Most notably, in Darwin (1880) he derived a method of calculating the dynamical effects of individual partial tides on the Earth–Moon system. The necessary parameters of only M_2, K_1, S_2 and O_1 are available from tidal charts at the present time and it is probable that this parameter is sufficiently reliable for the M_2 component only, so that the technique cannot yet be applied with any rigour. Nevertheless, following an extension of Darwin's approach (Hipkin, 1972), some predictions can be made about the current rates of change of the lunar inclination i and the obliquity of the ecliptic I without such unrealistic approximations as linear friction or a time-lagged equilibrium tide.

Derivation and detailed discussion of this treatment is inappropriate here so that only the results will be quoted:

$$\frac{di}{dt} \doteqdot -\frac{\sin(i/2)}{3}\frac{1}{n}\frac{dn}{dt} \qquad (8)$$

$$\frac{dI}{dt} \doteqdot -\sin(I/2)\frac{1}{\omega}\frac{d\omega}{dt} \qquad (9)$$

These approximations depend upon two assumptions: first, that the long-period tides about which we know very little on a global scale are not strongly amplified or significantly phase-lagged when averaged over the world's oceans; secondly, within the diurnal and semi-diurnal tidal groups the global average of the elevation and phase lag are not *differentially* amplified over one another by more than 50%. With these assumptions, equations (8) and (9) are good to at least ±30%.

These relationships are somewhat surprising: the two diurnal tides O_1 and K_1 each produce a change in I of about half the total, compared with only 4% of the change in the length of the day, but these contributions cancel almost exactly. Similarly, two very insignificant diurnal tides (too insignificant to have been given a name by Darwin, but otherwise known as 145·545 and 165·565) cause about half the total change in i but again they cancel.

The crucial relationship between the lunar tidal acceleration of the Moon and the Earth's rotation is somewhat less certain: if there is no differential amplification at all, then the M_2 tide accounts for more than 85% of the total effect in both cases, but other partial tides contribute to the remaining part of the Earth's accleration differently from the way they contribute to the lunar accleration. Until consistent charts of O_1, K_1 and N_2 become available, we should not assume that the angular momentum change associated with the increasing lengths of the day and of the month can be equated more accurately than ±15%.

$$\frac{1}{\omega}\frac{d\omega}{dt} \doteqdot 1\cdot62\,\frac{1}{n}\frac{dn}{dt} \qquad (10)$$

No general result exists at all for changes in the lunar eccentricity. The partial tides dominating the change of the lunar eccentricity have only a minor effect on any other dynamical variable. Only with very restrictive and probably unrealistic assumptions can any relationship be deduced: with a strict version of the

TABLE 2. Predicted effects of lunar tidal friction

Parameter	Relative angular momentum change	Estimated uncertainty[a]
Rotation rate of the Earth ω	$-3\cdot9 \times 10^{-10}$/y	±15%
Lunar eccentricity e	$-0\cdot061 \times 10^{-10}$/y	1/10 to 10 ×
Lunar inclination i	$-0\cdot016 \times 10^{-10}$/y	±30%
Obliquity of the ecliptic I	$-0\cdot079 \times 10^{-10}$/y	±30%

[a] The estimated uncertainty assumes that the measured lunar accleration of $-42''/cy^2$ is without error.

linear friction model (that described by Darwin (1879) as the small viscosity case) I obtained:

$$\frac{1}{e}\frac{de}{dt} \doteq -1\cdot 74 \frac{1}{n}\frac{dn}{dt} \quad (11)$$

It is quite conceivable that this could be wrong by an order of magnitude.

For the purposes of comparison, Table 2 gives the results of equations (8) to (11) expressed in the same units as the final column in Table 1, that is a rate of change of angular momentum measured as a fraction of the Earth's spin angular momentum. These usually neglected effects therefore appear to account for at least 15% of the discrepancy between the observed and predicted change in the length of the day. Attention should be drawn to the uncertainty in de/dt.

Conclusions

On the assumption that equation (10) is uncertain by no more than 15% then the measured lunar acceleration of $\dot{n}_m = -42''/$cy indicates a lunar tidal torque tending to slow the Earth's rotation of $-7\cdot3 \times 10^{16}$ Nm. The observational uncertainty in the modern astronomical datum is also about 15% according to Morrison, so that the true value of the lunar tidal torque is very unlikely to lie outside the range -5 to -9×10^{16} Nm. Although the extent of this range of uncertainty may reflect excessive caution, it would nevertheless be unwise to treat other torque measurements as inconsistent with the astronomical observations merely because they lay near the extremes of this range.

The other technique which appears reliable is the oceanographic one. The best two determinations so far indicate a torque for the M_2 tide alone of between -6 and -7×10^{16} Nm but it is too early yet to exclude the somewhat smaller value obtained from the Russian chart. With no differential amplification of partial tides, these figures would represent a little over 85% of the total lunar tidal torque. Even with our present indifferent knowledge of the smaller tidal components in the deep oceans, the remaining partial tides are very unlikely to contribute less than an additional 3% or more than a further 25%. On this basis, we can have some confidence that the oceanographic technique determines a torque in the range -5 to -9×10^{16} Nm, which is exactly the range implied by the astronomical data. I shall adopt a torque of $-(7 \pm 2) \times 10^{16}$ Nm.

If we are to use present-day observations of tidal changes in the length of the day to compare with palaeontological measurements, then this is the figure we should use. It is equivalent to a fractional rate of increase in the day length of $(3\cdot9 \pm 1\cdot1) \times 10^{-10}/$y). (The units mean that the length of the day increases by 3·9% in 100 My: msec introduce unnecessary arithmetic.) The famous figure derived from Munk and MacDonald (1960), 1·8 msec/day/cy, is $2\cdot1 \times 10^{-10}/$y and seems in much better agreement with the palaeontological data than the true one.

If one simple conclusion is to be drawn from this discussion, then it is that in this field the predictive role of geophysics and astronomy is no longer viable: with our present understanding of tidal friction in the oceans it is unreasonable to expect the present-day torque, even if we were confident in our ability to measure it now, to be representative of past marine configurations. Now that palaeontology has confirmed that past changes in the lengths of the day and month are of the same order of magnitude as the original geophysical predictions, the role of the geophysicist can only be the reactionary one of interpreting the data given him. Its reliability is a problem for the biologists.

References

Bogdanov, K. T. and Magarik, V. A. (1967). A Numerical solution to the problem of the semidiurnal tidal wave distribution (M_2 and S_2) in the world ocean. *Proc. (Dokl.) Acad. Sci. USSR*, **172** (6), 1315–1317

Bogdanov, K. T. and Magarik, V. A. (1969). A numerical solution of the problem of tidal wave propagation in the world ocean. *Bull. (Izv.) Acad. Sci. USSR, atmos. ocean. Phys.*, **5** (12), 757–761

Brosche, P. and Sündermann, J. (1971). Die gezeiten des meeres und die rotation der erde. *Pure appl. Geophys.*, **86**, 95–117

Chapman, S. and Lindzen, R. S. (1970). *Atmospheric Tides*, Gordon and Breach, New York, 210 pp.

Darwin, Sir. G. H. (1879). On the body tides of viscous and semi-elastic spheroids, and on the ocean tides on a yielding nucleus. *Phil. Trans. roy. Soc. London*, **170**, 1–35

Darwin, Sir G. H. (1880). On the secular changes in the elements of the orbit of a satellite revolving about a tidally distorted planet. *Phil. Trans. roy. Soc. London*, **171**, 713–891

Darwin, Sir G. H. (1908). *Scientific Papers*, Vol. II, *Tidal Friction and Cosmogony*, Cambridge Univ. Press

Defant, A. (1961). *Physical Oceanography*, Vol. II, Pergamon Press, New York, 598 pp.

Dietrich, G. (1944). Die gezeiten des weltmeeres als geographische erscheinung. *Veröffentl. Inst. Meeresk. Univ. Berlin*, **A41**, 1–68

Duncombe, R. L. (1958). The motion of Venus—1750–1949. *Astron. Pap. amer. Ephem. naut. Almanac*, **16** (1), 3–260

Farrell, W. E. (1972). Global calculations of tidal loading. *Nature, phys. Sci.*, **238**, 43–44

Fricke, W. (1972). On the motion of the equator and the ecliptic. p. 196 in *Rotation of the Earth* (Eds. P. Melchior and S. Yumi), D. Reidel, Dordtecht, 244 pp.

Garrett, C. J. R. and Munk, W. H. (1971). The age of the tide and the 'Q' of the oceans. *Deep-Sea Res.*, **18**, 493–504

Gerstenkorn, H. (1955). Über gezeitenreibung beim zweikörperproblem. *Zeit. Astrophys.*, **36**, 245–274

Groves, G. W. and Munk, W. H. (1958). A note on tidal friction. *J. marine Res.*, **17**, 199–214

Hansen, W. (1949). Die halbtägigen gezeiten im nordatlantisches ozean. *Deut. hydrog. Zeit.*, **2**, 44–51

Haurwitz, B. (1964). Atmospheric tides. *Science*, **144**, 1415–1422

Heiskanen, W. (1921). Über den einfluss der gezeiten auf die säkuläre acceleration des mondes. *Suomal. Tiedeakat. Toim.* A, **18** (2), 1–84

Hendershott, M. (1972). The effects of solid Earth deformation on global ocean tides. *Geophys. J. roy. astron. Soc.*, **29**, 389–402
Hendershott, M. and Munk, W. W. (1970). Tides. *A Rev. fluid Mech.*, **2**, 205–224
Hipkin, R. G. (1972). *Some Aspects of the Dynamical History of the Earth–Moon System*, Ph.D. Thesis, University of Newcastle upon Tyne, 220 pp.
Jeffreys, Sir H. (1920). The chief cause of the lunar secular acceleration. *Mon. Not. roy. astron. Soc.*, **80**, 309–319
Kaula, W. (1968). *An Introduction to Planetary Physics: The Terrestrial Planets*, John Wiley, New York, 490 pp.
Kozai, Y. (1970). Seasonal variations of the geopotential. *Smithson. astrophys. Obs. Spec. Rep.*, **312**, 1–6
Kuznetsov, M. V. (1972). Calculation of the secular retardation of the Earth's rotation from up-to-date cotidal charts. *Bull. (Izv.) Acad. Sci. USSR, Phys. solid Earth*, **8** (12), 779–784
Lagus, P. L. and Anderson, D. L. (1968). Tidal dissipation in the Earth and planets. *Phys. Earth planet. Interiors*, **1**, 505–510
Martin, C. F. and van Flandern, T. C. (1970). Secular changes in the lunar ephemeris. *Science*, **168**, 246–247
Melchior, P. (1966). *The Earth Tides*, Pergamon Press, New York, 458 pp.
Morrison, L. V. (1973). Rotation of the Earth from AD 1663–1972 and the constancy of G. *Nature*, **241**, 519–520
Munk, W. H. and MacDonald, G. J. F. (1960). *The Rotation of the Earth*, Cambridge Univ. Press, London, 323 pp.
Newton, R. R. (1968). A satellite determination of tidal parameters and the Earth's deceleration. *Geophys. J. roy. astron. Soc.*, **14**, 505–539
Newton, R. R. (1970). *Ancient Astronomical Observations and the Accelerations of the Earth and Moon*, Johns Hopkins Press, Baltimore, 309 pp.
Ozawa, I. (1967). Observations of the atmospheric tide effects on the Earth's deformation. *Kyoto Univ. Geophys. Inst. Spec. Publ.*, **7**, 133–142
Pariyskiy, N. N., Kuznetsov, M. V. and Kuznetsova, L. V. (1972). The effect of oceanic tides upon the secular deceleration of the Earth's rotation. *Bull. (Izv.) Acad. Sci. USSR, Phys. solid Earth*, **8** (2), 65–70
Pekeris, C. L. and Accad, Y. (1969). Solution of Laplace's equations for the M_2 tide in the world ocean. *Phil. Trans. roy. Soc. London*, **A265**, 413–436
Sakuma, A. (1971). Observations expérimentales de la constance de pesanteur au Bureau International des Poids et Mésures. *Bull géod.*, **100**, 159–163
Sitter, W. de (1927). On the secular accelerations and fluctuations of the longitude of the Moon, the Sun, Mercury and Venus. *Bull. astron. Insts Neth.*, **4**, 21–38
Urey, H. (1952). *The Planets*, Yale Univ. Press, New Haven, Conn., 245 pp.
Wunsch, C. (1972). Bermuda sea level in relation to tides, weather and baroclinic fluctuations. *Rev. Geophys. Space Sci.*, **10**, 1–50
Yukutake, T. (1972). The effect of change in the geomagnetic dipole moment on the rate of the Earth's rotation. *J. Geomag. Geoelec.*, **24**, 19–47
Zahel, W. (1970). Die reproduction gezeitenbedingster bewegungsvorgänge im weltozean mittels des hydrodynamisch-numerischen verfahrens. *Mitt. Inst. Meereskunde Univ. Hamburg*, **17**
Zahel, W. (1973). The diurnal tide K_1 in the world ocean—a numerical investigation. *Pure appl. Geophys.*, **109**, 1819–1825

DISCUSSION

JACOBS: Are you saying it is impossible to postulate a connection between change in length of day and tidal friction, because of differing assumptions?

HIPKIN: I mean one cannot unambiguously say that the change in the length of the month, the secular acceleration of the Moon, uniquely determines the rotation of the Earth, for there are other places angular momentum could go. The best evidence for tidal changes in l.o.d. comes from oceanographic investigations.

RUNCORN: Surely it is possible to isolate the total change in angular momentum of the Earth.

HIPKIN: This could be done only in a co-planar orbit.

MORRISON: It may be that assumptions about lunar orbit are satisfactory for the past 2000 years but changes in eccentricity which the geophysicist needs to consider for the Earth's rotational history are in terms of millions of years.

HIPKIN: On various tidal considerations, one can postulate various changes in eccentricity and rotational changes. We know very little about the different tidal components, and often theoretical extrapolations are based on minute tides, not those examined oceanographically.

MORRISON: Do you think tidal friction changed appreciably in the past 2000 years?

HIPKIN: No. But this problem depends on the validity of ancient eclipse data.

MULLER: I agree on theoretical grounds. Acceleration of the Moon cannot have changed within 2000 years and one should be suspicious of results which claim there was a change.

RUNCORN: Ever since the 1930s there has been a suspicion that when you determine lunar tidal friction for the last few hundred years from accurate telescopic observation you get a value which corresponds to about 1·8 msec/cy variation in length of day, which is only a portion of the value which is obtained from eclipse data which is, of course, for a period of 2000 years. So the question is, what are modern values, those from atomic clocks? The exact value is still not certain. Nothing much is known about tidal friction, but some people believe it couldn't possibly have changed over the past 2000 years. But sea level has changed in the past 2000 years.

TARLING: I do believe change in sea level was significant within the past 2000 years. But if tidal friction is significant in slowing the earth, it is important only within 20–30° of the equator because of the way it exerts its torque. Change in land mass and topography due to volcanicity could be very important relative to changes in sea level.

RUNCORN: From early work of Jeffreys, and others, it seems that tidal friction is effective only in shallow seas, so small changes in sea level might be important. The difficulty is that tidal friction is so poorly understood.

FARROW: Present land–sea configuration could be of little value in extrapolating to the past, because the land–sea configuration has changed so through time.

THE EARTH'S INTERIOR AND THE EARTH'S ROTATION

J. A. JACOBS AND K. D. ALDRIDGE
*Institute of Earth and Planetary Physics, University of Alberta,
Edmonton, Alberta, Canada*

Abstract

The origin of the Earth and its present gross differentiation into crust, mantle and core is briefly reviewed. The effect of a variable rate of rotation on the thermal regime of the core and on the geomagnetic field is discussed in some detail. Observations of the change in the length of the day show that the mantle is decelerating at a very slow rate. The coupling mechanism which causes the liquid outer core to participate in this deceleration should be predominantly viscous rather than electromagnetic in origin because the time-scale of changes in the speed of mantle rotation is extremely long. The importance of viscous coupling is being investigated experimentally in the fluid dynamics laboratory at the University of Alberta. Preliminary results of this and other experiments are briefly presented.

Introduction

As a result of the attraction of the Sun and Moon on the Earth's equatorial bulge and of movements of mass within the Earth, the angular velocity of the Earth is not constant; there are fluctuations, not only in the rate of spin (i.e. changes in the length of the day), but also in the direction of the axis of rotation, i.e. the Earth 'wobbles'. This paper is mainly concerned with changes in the rate of spin. There are a number of peaks in the frequency spectrum of the Earth's rotation, covering a very long time-scale, and these peaks are believed to arise from different causes. Three distinct components have been recognized: a steady increase in the length of the day by about 2×10^{-3} sec/cy (-23×10^{-11}/year), seasonal fluctuations of about 10^{-3} sec ($\pm 2300 \times 10^{-11}$/year), and less regular variations of up to about 5×10^{-3} sec ($\pm 200 \times 10^{-11}$/year) having time-scales of the order of years (the so-called decade fluctuations).

The seasonal variations in the length of the day are chiefly the result of torques on the mantle exerted by oceanic currents and atmospheric winds. However, the rapid irregular variations in the length of the day over a decade cannot be explained by surface phenomena. No transport of mass at the surface could alter the Earth's moment of inertia sufficiently to account for such large changes as are observed. It has been suggested that these 'decade' fluctuations are caused by the transfer of angular momentum between the Earth's solid mantle and liquid core. This in turn implies some form of core–mantle coupling. It is difficult, however, to make quantitative estimates of the horizontal stresses at the core–mantle boundary. Calculations indicate that neither viscous coupling nor electromagnetic coupling is really adequate to account for the decade fluctuations (see e.g. Rochester, 1970, 1973). More recently Hide (1969) has suggested the possibility of topographic coupling as a result of irregular features (bumps) at the core–mantle boundary. It would be interesting to detect any small departures in uniformity in the rate of rotation of other planets such as Mars or Venus and of planetary satellites with rigid surfaces, since any such departures may yield information on a possible liquid convecting core and planetary magnetic fields.

Fluctuations with a time-scale of $\simeq 100$ years (Newcomb's empirical terms) may be due to changes in the Earth's moment of inertia. Changes over geological time are predominantly a constant deceleration as a result of tidal friction. It is difficult to estimate the rate at which this deceleration has taken place since our knowledge of palaeogeography is scant and the result depends critically on a few shallow seas. Urey (1952) has suggested, on the other hand, that, because of differentiation of the materials of the Earth and the growth of the core, the moment of inertia of the Earth about its axis of rotation may have been reduced and as a result the length of the day decreased from about 30–24 hr. The changes caused by a growing core are considerably smaller (and of opposite sign) to those due to tidal friction.

A method of determining the rate of the Earth's rotation in the geological past has more recently been developed by the use of fossil 'clocks'. Wells (1963, 1970) was able to identify both a daily and an annual banding in corals of Middle Devonian age (i.e. about 370 My ago) and calculated that at that time there were about 400 days in the year. Later Scrutton (1964, 1970) found apparent monthly bands in corals of the same age and suggested that the Devonian year consisted of 13 lunar months of 30·5 days each. Using estimates of the length of the day and lunar month in Devonian times, Runcorn (1964, 1970) showed that it is possible to separate the slowing down of the Earth by tidal friction from other processes which affect its rate of rotation. He found that nearly all the change in the length of the day is caused by tidal friction and that any residual effects (e.g. a change in the moment of inertia of the Earth due to the growth of the core) are negligible.

Pannella et al. (1968) obtained values of the length of the synodic month during other geological periods using tidally controlled periodical growth patterns in molluscs and stromatolites. They found that the slowing down of the Earth's rotation has not taken place at a uniform rate—there appear to be two major breaks in the slope of the curve, the slowing down being negligible over a 200 My period from the Pennsylvanian to the Upper Cretaceous. Pannella et al. suggested that these changes in slope may be related to particular events in the Earth's history. Results of laboratory experiments carried out at the University of Alberta with application to changes in the length of the day on a geological time-scale are discussed later in this paper.

The rotation of the Earth and geomagnetism

The magnetic field which we measure at the Earth's surface is predominantly an axial dipole. This dipole is known to reverse its polarity in a more or less random fashion at a rate of a few times per My; on a time-scale of decades the smaller non-dipole part of the Earth's magnetic field is observed to drift westward at a fraction of 1°/year. There is evidence that throughout the history of the geomagnetic field the dipole axis has almost always been nearly aligned with the Earth's rotation axis. The importance of the Earth's rotation is thus revealed both in the dipole direction and the axis around which predominant drift of the non-dipole field occurs.

This relevance of the Earth's rotation to the geomagnetic field is not too surprising if we consider other hydrodynamical processes which take place at the Earth's surface. The balance between pressure gradient and Coriolis forces in the Earth's atmosphere, as evidenced by winds which blow along isobars rather than across them, is a familiar effect. A common feature in the Earth's oceans is the existence of inertial oscillations which owe their existence entirely to the rotation of the Earth. Thus in both the atmosphere and the ocean it is observed that the role of rotation is of fundamental hydrodynamical importance. Such a statement may also be made for the Earth's fluid core with some qualifications.

With regard to these rotational effects we must distinguish between the core and oceanic-atmospheric layers in the following manner. The core should be divided up into thin layers only if it is stably stratified. In such a case dynamo action, though still possible, would be significantly constrained to a predominantly two-dimensional motion in concentric spherical shells. If, on the other hand, the core is not stratified, any perturbation of the otherwise steady rotation would lead to a predominantly two-dimensional motion in planes perpendicular to the rotation axis as predicted by the Proudman–Taylor theorem. In both cases we must amplify these statements to include the effects of the Lorentz force.

In certain special cases the effects of the Lorentz force can be included without too much difficulty. For example, we can consider the possibility of interpreting the westward drift of the non-dipole field as the propagation of hydromagnetic waves in the fluid core around the axis of rotation in the presence of a dominant toroidal magnetic field (Hide, 1966). If the fluid is *not* stably stratified the Lorentz force enters into the equations of motion for hydromagnetic wave solutions in the same manner as the Coriolis force. In fact the flow in this case is described by the Poincaré equation: the same equation which describes the flow in the absence of a magnetic field. But even in this rather simple example we find an added difficulty. We are looking for solutions to a hyperbolic differential equation with certain specified boundary conditions. This is an ill-posed mathematical problem and there are serious difficulties in attempting to find a solution. A major source of the difficulty is the presence of an inner boundary to the fluid—

the solid inner core. It is for these reasons that we have attempted to study rotational effects through laboratory experiments: these will be described later.

Yukutake (1973a) has examined in some detail his earlier (1972) suggestion that fluctuations in the Earth's rate of rotation are related to changes in the dipole moment. He investigated the possibility of such a relationship for three different time periods, 8000, 400 and 65 years, using archeomagnetic data, observations of the variation of the Moon's longitude and recent observatory

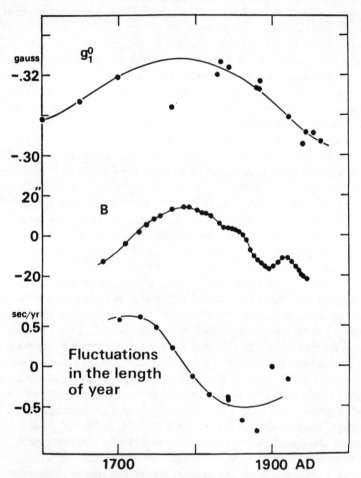

FIGURE 1. Fluctuations in various features of the geomagnetic field and the Earth's rotation. From the top, fluctuations in the gauss coefficient g_1^0 (the dipole term), the Moon's longitude (curve B), and the length of the year. (Reproduced with permission from Yukutake, T. (1973), *J. Geomag. Geoelec.*, **25**, 195)

data. He found that such a relationship does exist, but that it is highly dependent on the period, the magnitude of the change increasing as the period decreases. He was able to account for this dependence of the excitation of the change in rotation rate upon period by electromagnetic coupling between the mantle and core if the conductivity of the lower mantle is as large as 10^{-8} e.m.u. and if the change in the dipole field originates near the surface of the core.

From ancient observations of eclipses, the Earth's rotation is known to have been *accelerated* during the past few millennia in addition to the steady retardation due to tidal friction. Archaeomagnetic studies have shown that the dipole field has also been changing, approximately periodically, with a large amplitude ($\simeq 50\%$ of the present dipole moment) and with a period $\simeq 8000$ years (Bucha, 1967, 1970; Cox, 1968; Kitazawa, 1970). Since the last maximum of the dipole moment (sometime between A.D. 0 and 500), the electromagnetic coupling between mantle and core has been diminishing and acceleration of the Earth's rotation is thus to be expected during the past 2000 years. Observation confirms the theoretical prediction (Yukutake, 1972) of a phase difference of about π between rotation and dipole moment change, for periods $\simeq 8000$ years.

Yukutake (1971) has also shown that the magnitude of the gauss coefficient g_1^0 increased during the 17th and 18th centuries and then began to decrease after the early 19th century; this variation being superimposed on the general trend discussed above (the gradual decrease since about 2000 years ago). During this period there was also a large fluctuation in the observed longitude of the Moon which has been ascribed to a change in the Earth's rate of rotation (See Figure 1). The length of the year curve leads the dipole curve by about $\pi/2$ for these periods $\simeq 400$ years. Variations in the Earth's rate of rotation on a decade time-scale are considered in the next section.

Decade fluctuations in the Earth's rate of rotation

The non-tidal variations in the length of the day must conserve the total angular momentum of the Earth: for variations of the order of a decade, the core of the Earth appears to be the only reasonable source. Yukutake (1973a, see Figure 2) found a variation in the dipole field in phase with the length of the year fluctuations ($\simeq 65$-year period). Variations in the westward drift of the secular variation are, however, very different from those obtained earlier by Vestine (1953). Figure 3 (by Vestine and coworkers) shows a comparison between the motion of the eccentric dipole and the deviation in the length of the day. Vestine and Kahle (1968) have interpreted these curves as showing that the angular velocity of the eccentric dipole is related to changes in the angular momentum of the outer $\simeq 200$ km of the core. Figure 3 also shows that there is an apparent phase lag between variations in the rate of rotation of the mantle and in the velocity of the eccentric dipole. Ball *et al.* (1969) suggested that this is the effect of diffusion of the magnetic signal through the conducting mantle. Yukutake (1973b) has questioned, however, whether the movement of the eccentric dipole really represents that of the geomagnetic field as a whole. He showed that the westward movement of the eccentric dipole over the last 150 years is determined almost entirely by the westward drift arising from only one term ($n = 2$, $m = 1$) of the magnetic potential.

Kahle et al. (1967) attempted to determine the fluid velocity v near the surface

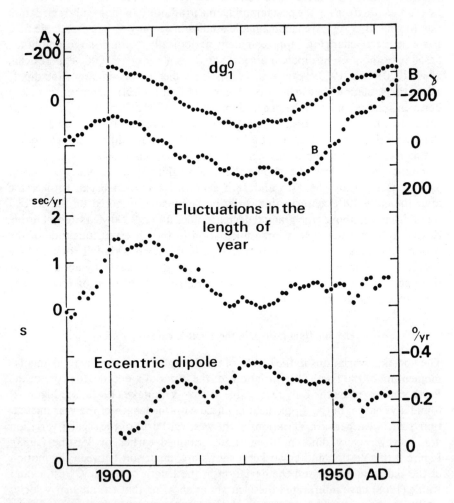

FIGURE 2. Comparison of fluctuations in the length of the year with those of various features of the geomagnetic field. From the top, variations in the dipole term (dg_1^0)—curve A from an analysis of 21 observatories, curve B from an analysis of 6 observatories—the length of the year, and the drift rate of the eccentric dipole. (Reproduced with permission from Yukutake, T. (1973), *J. Geomag. Geoelec.*, **25**, 195)

The Earth's Interior and the Earth's Rotation

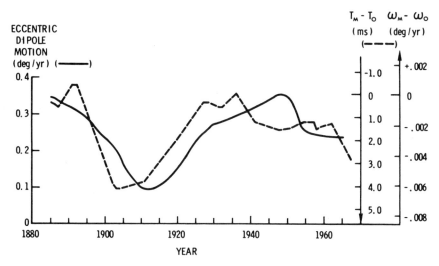

FIGURE 3. Comparison of astronomical and geomagnetic data on the speed of the Earth's rotation. The solid line represents the westward motion of the eccentric dipole relative to the mantle, the dashed line the eastward angular velocity of the mantle (with scales appropriate to deviations from both standard angular velocity and standard length of day). (Reproduced with permission from Ball, R. H., Kahle, A. B. and Vestine, E. H. (1968), *Rand Corp. Rept.* RN-5717-PR, The Rand Corporaporation, Santa Monica California)

of the core from a knowledge of the magnetic field and secular variation at the top of the core, which information may be obtained by extrapolation from observations made at the surface of the Earth. Backus (1968) has shown that the goal which Kahle *et al.* set themselves is in principle unattainable and that there is no unique solution to the problem. This applies not only to fine-scale features, but also to arbitrary functions whose scales may be of the order of the core's circumference. Kahle *et al.* obtained a unique solution only because they truncated the spherical harmonic expansion of v. Their truncated expansion of v does not fully reproduce the secular variation, i.e. reduce the r.m.s. difference between the measured and computed values to zero.

If magnetic diffusion is neglected (i.e. if the electrical conductivity is considered infinite) in the free stream in the core, the external geomagnetic field is completely determined by the fluid motion at the top of the free stream. Backus (1968) has shown that it is thus possible to test whether there is *any* motion of a perfectly conducting core which will produce the observed secular variation. If the observed secular variation passes this test, Backus shows how to obtain explicitly all 'eligible' velocity fields, i.e. all velocity fields at the top of the free stream in the core which are capable of producing exactly the observed secular variation. Because of this multiplicity of eligible velocity fields, it is not possible to determine whether there is westward drift of core fluid, or latitude dependence of the westward drift of the core fluid, at the top of the free stream. (That the *magnetic* field itself shows a westward drift was established by Bullard *et al.* in 1950.) Backus showed how it is possible to select from among all eligible velocity fields those which are of particular geophysical interest, such as the one which, in the least-squares sense, is most nearly a uniform rigid rotation (westward drift) about the geographic polar axis of the Earth or the one which is most nearly a latitude-dependent westward drift. Booker (1969) has used the methods outlined by Backus to obtain velocity components which are compatible with primarily latitude-dependent westward drift, but not with the velocity field obtained by Kahle *et al.* (1967).

It is generally assumed that the westward drift of the non-dipole part of the

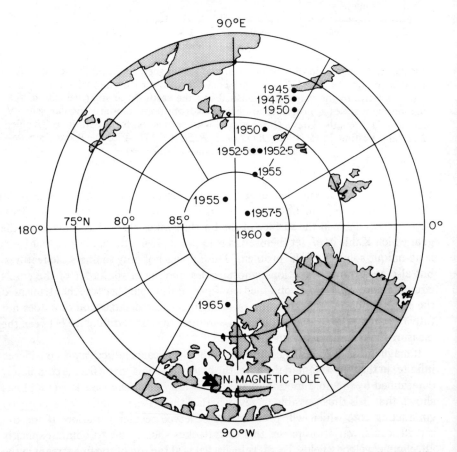

FIGURE 4. Position of the pole of rotation of the non-dipole part of the Earth's magnetic field. (Reproduced with permission from Malin, S. R. C. and Saunders, I. (1973), *Nature*, **245**, 25–26)

Earth's magnetic field is a rotation about the geographical axis. Malin and Saunders (1973) have recently investigated the possibility that the secular change in the magnetic field might be more closely represented by rotation about another axis. They estimated pole positions and rotation rates which would give maximum correlation between magnetic field models for different epochs. They found that the pole of rotation has apparently moved from near Novaya Zemlya in 1945 to near the north magnetic pole in Canada in 1965 at a fairly steady rate (See Figure 4). During this time, the drift rate remained near $0.18°$/year, approximately westwards, but showed a maximum amplitude when the rotation pole was near the geographical pole.

Malin and Saunders concluded that in general the pole of rotation differs significantly from the geographical pole: had the true pole of rotation coincided with the geographical pole, it would be expected that the points in Figure 4 would be randomly distributed about the geographical pole. If the movements of the field reflect corresponding movements of the outer layers of the core, their results imply that the absolute rotation of the outer core is about an axis slightly different from that of the mantle, the rotation poles of the core and the mantle being on opposite sides of the total angular momentum pole (the direction of which is fixed in space).

Laboratory experiments

1. *Time varying*

Certain aspects of the origin of planetary magnetic fields may be investigated in the laboratory. Although direct experimentation of hydromagnetic phenomena is not practical for laboratory work because of the size and speed limitations dictated by necessarily high magnetic Reynolds numbers, it is possible to study the constraints produced by rotation on a contained fluid mass such as the Earth's core.

As an example, the possibility of the existence of wave solutions to the problem of the westward drift can be investigated by the following procedure: a fluid contained between rigid spherical boundaries is set into uniform rotation about a fixed vertical axis. The fluid is perturbed by an axisymmetric oscillation of the boundary. The response of the fluid as observed quantitatively by differential pressure measurements reveals certain preferred frequencies of oscillation. These frequencies lie within the so-called inertial range—less than twice the frequency of the uniform rotation—thus confirming that oscillations of inertial type do indeed exist in a thick shell of rotating fluid (Aldridge, 1972). Related experiments on non-axisymmetric inertial oscillations in a thick spherical shell are at present under way in our laboratory (Aldridge, 1974). Since these waves are completely analogous to their hydromagnetic counterparts, interpretation of these experiments provides an essential step in the understanding of the role of rotation in the dynamo process.

Higgins and Kennedy (1971) have argued that the radial density gradient in the core is very stable (i.e. highly subadiabatic or 'bottom heavy'), in which case purely toroidal hydromagnetic waves would be much easier to excite than those with a radial particle velocity component. However, both Jacobs (1971) and Birch (1972) have stressed the uncertainties in the estimates of the melting-point and adiabatic gradients in the core and suggested that the density gradient in the core is only weakly stable or even unstable. In such a case the 'non-toroidal' modes would not be suppressed and could possibly be excited by buoyancy forces (see e.g. Braginskii, 1967), whereas purely toroidal modes would be hard to excite.

The assymmetric oscillations of a rotating finite circular cylinder of fluid have been investigated both theoretically and experimentally by Stewartson (1959) and Wedemeyer (1966) with good agreement (as far as the principal resonances are concerned) at values of the Ekman* number $E \simeq 10^{-5}$. For the Earth, the value of $E \ll 10^{-5}$, which would appear to indicate that the effect of viscosity on the principal modes of oscillation in the Earth may be neglected. Two other effects, however, may be significant: turbulence in the core and ohmic dissipation in the core and mantle. There is some evidence that the effects of turbulence may be neglected, except possibly in the boundary layer near the mantle. Dissipation effects are also probably of secondary importance in their influence on hydromagnetic waves in the Earth's core: viscous effects dominate in the boundary layer at the core–mantle boundary, but not elsewhere.

2. *Steady state*

As already mentioned, studies of the growth rings in corals have shown that the Earth's mantle has been decelerating at a very slow rate over geological time. Whatever the source of this deceleration, it can hardly be doubted that the Earth's fluid core must participate in this slowing down. The necessary coupling between core and mantle could be either conservative or dissipative. Although non-axially symmetric irregularities of the core–mantle boundary might, in principle, contribute to the former type, we consider here only those dissipative mechanisms which are either electromagnetic or viscous in origin. On relatively short time-scales of variation in mantle rotation speed we would expect that even a relatively poor conducting mantle would be coupled to a highly conducting core by Maxwell stresses. On the other hand, changes in mantle rotation speed on a geological time-scale would be predominantly communicated to the fluid core by the action of viscous stresses. We have thus considered the role of viscosity in the very long-term retardation of the Earth's fluid core.

The problem of coupling between a slowly decelerating rigid spherical boundary and an incompressible, constant density fluid which fills the spherical cavity

* The Ekman number $E = \nu/\omega a^2$, where ν is the kinematic viscosity, ω the angular velocity of rotation and a a typical length parallel to the axis of rotation. The Ekman number is thus a gross measure of the ratio of the viscous force to the Coriolis force.

has been considered by Bondi and Lyttleton (1948). They showed that the final settled drift of the fluid near the axis in a spherical cavity rotating at a rate ω rad/sec and decelerated at a rate $\dot\omega$ is scaled as $\dot\omega a/(v\omega)^{1/2}$ rad/sec where a is the radius of the container and v the kinematic viscosity. Thus the fluid near the axis rotates relative to the shell at a rate equal to the angular velocity attained by the shell at a time $a/(v\omega)^{1/2}$ earlier. More precisely, this relative angular velocity of the fluid, independent of the coordinate parallel to the axis of rotation as required by the Proudman–Taylor theorem, can be written, except for terms smaller by a factor $0(E^{1/2})$, as

$$\alpha(\theta) = \frac{\dot\omega a}{(v)^{1/2}} \cos^{3/2}\theta,$$

where θ is the colatitude of the circle on the boundary which is at polar radius r from the rotation axis.

Bondi and Lyttleton were concerned with the steady state behaviour of the fluid as the boundary decelerated with time, and placed the restriction

$$\frac{\omega}{\dot\omega} \gg \frac{a^2}{v}$$

on their results. For the long-term deceleration of the mantle, $\dot\omega \simeq 6 \times 10^{-22}$ rad/sec², and for a steady state to exist it would be necessary that $v \gg 3$ cgs units. Recent estimates of the kinematic viscosity of the fluid core by Gans (1972), however, have yielded a value of $v \leqslant 6 \times 10^{-3}$ cgs units. Experiments have been undertaken to establish conditions for the steady state and the dependence of drift rate on position in the fluid within a decelerating spherical boundary.

The apparatus consists of a spherical cavity of diameter 20 cm in a highly polished upright Perspex cylinder filled with water. This container is rotated about a fixed vertical axis at a rate $\omega(t) = \omega + \dot\omega t$ where ω and $\dot\omega$ are fixed positive and negative constants respectively. This time-varying rotation speed is maintained by a closed-loop feedback control system. Rotating with the container and mounted vertically above it is a 35 mm camera with a motorized back. The shutter for the camera may be triggered either manually or automatically at some fixed rate. The water in the cavity contains a small amount of starch and potassium iodide which produces a dark blue dye at the anode when a potential difference is applied between two electrodes in the fluid. This dye-releasing electrode is a 0·010″ wire embedded in the container wall at a colatitude of 49°. The time history of the dye release is recorded on film along with a digital clock running at 60 × normal rate.

In a typical experimental run the container is set rotating at a steady rate until the fluid in the spherical cavity has come to solid body rotation with the container. After dye is released to confirm this condition deceleration at a constant rate $\dot\omega$ is begun. Convergence of the dye in the boundary layer towards the pole leads to a concentration near the pole followed by a very slow sinking of the dye into the fluid interior. Photographs are taken at sufficiently long intervals for a measurable drift of the dyed fluid to occur. Since each photograph includes the clock, the position of a particular dye configuration may be plotted against time. Drift rate relative to the container for the final settled motion of the fluid is calculated directly from the final straight line portion of the graph of dye position against time.

Some preliminary experimental results for the steady drift rate $\alpha(\theta)$ around the axis of rotation for each of three values of $\bar{\omega}/\dot{\omega}$ where $\bar{\omega}$ is an average rotation speed of the container (typically about 10^{-2} rad/sec) are shown in Figure 5. The

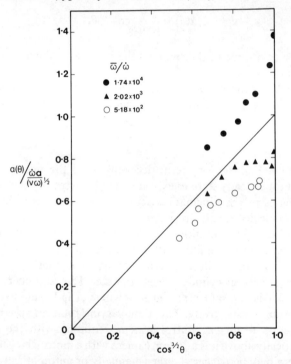

FIGURE 5. Angular velocity of the fluid $\alpha(\theta)$ relative to the decelerating container. $\theta = \sin^{-1} r/a$. The straight line is from Bondi and Lyttleton (1948)

drift rate is plotted in units of $\dot{\omega}a/(\nu\omega)^{1/2}$; radial position r of the dye is expressed in terms of $\cos^{3/2}\theta$, where $\theta = \sin^{-1} r/a$. The straight line shown in the figure is taken from the work of Bondi and Lyttleton with the omission of terms $O(E^{1/2})$. The plotted points appear to be in qualitative agreement with the straight line, though it is noted that only the points above the line were obtained for the condition

$$\frac{\bar{\omega}}{\dot{\omega}} > \frac{a^2}{\nu}.$$

Thus the restrictions placed by Bondi and Lyttleton on their analysis can apparently be relaxed somewhat, probably to

$$\frac{\bar{\omega}}{\dot{\omega}} \gg \frac{a}{(v\bar{\omega})^{1/2}}.$$

This would correspond to the condition that the time-scale of the boundary motion be long compared to the spin-up time for steady motion of the fluid to result. This implies that even if the viscosity of the core were as low as 6×10^{-3} cgs units, a steady drift would be expected in the fluid core due to the deceleration of the boundary at a rate of 6×10^{-22} rad/sec².

It must be emphasized that the experimental results shown are of a preliminary nature. The scatter in the data points, particularly near the axis of rotation, is under investigation. The effect of a filling hole near the 'north pole' of the spherical cavity may be responsible for part of this somewhat irregular behaviour. However, even though the points shown in the Figure are barely adequate for the range of values of $\bar{\omega}/\dot{\omega}$ shown, there does appear to be a progression of increasing drift rate with $\bar{\omega}/\dot{\omega}$. This effect is also being studied in more detail.

It is of interest to consider the magnitude of the steady azimuthal motion in the fluid core due to a deceleration of the core–mantle boundary at a rate $\dot{\omega} = 6 \times 10^{-22}$ rad/sec². Taking a value of the radius $r = a/2$, we can estimate the velocity to be

$$\frac{a}{2} \frac{\dot{\omega}a}{(v\omega)^{1/2}} \cdot \cos^{3/2}\frac{\pi}{6} = \tfrac{1}{2} \text{ mm/sec}$$

for $v = 6 \times 10^{-3}$ cgs units. This implies that long-term changes in rotation speed, communicated to the fluid core by viscous coupling, lead to drift speeds of the same order of magnitude as those generally associated with the westward drift of the Earth's magnetic field.

References

Aldridge, K. D. (1972). Axisymmetric inertial oscillations of a fluid in a rotating spherical shell. *Mathematika*, **19**, 163–168

Aldridge, K. D. (1974). Inertial waves in a rotating spherical container. In preparation

Backus, G. E. (1968). Kinematics of geomagnetic secular variation in a perfectly conducting core. *Phil. Trans. Roy. Soc. A*, **263**, 239–266

Ball, R. H., Kahle, A. B. and Vestine, E. H. (1968). Variations in the geomagnetic field and in the rate of the Earth's rotation. *Rand Corp. Rept.* RN-5717-PR, Santa Monica, California

Ball, R. H., Kahle, A. B. and Vestine, E. H. (1969). Determination of surface motions of the Earth's core. *J. Geophys. Res.*, **74**, 3659–3680

Birch, F. (1972). The melting relations of iron and temperatures in the Earth's core. *Geophys. J.*, **29**, 373–387

Booker, J. R. (1969). Geomagnetic data and core motions. *Proc. Roy. Soc. A.*, **309**, 27–40

Bondi, H. and Lyttleton, R. A. (1948). On the dynamical theory of the rotation of the Earth. *Proc. Camb. Phil. Soc.* **44**, 345–359

Braginskii, S. I. (1967). Magnetic waves in the Earth's core. *Geomag. Aeron.*, **7**, 851–859

Bucha, V. (1967). Archaeomagnetic and palaeomagnetic study of the magnetic field of the Earth in the past 600,000 years. *Nature*, **213**, 1005–1007

Bucha, V. (1970). Changes in the Earth's magnetic field during the archeological past. *Comments Earth Sci. Geophys.*, **1**, 20–27

Bullard, E. C., Freedman, C., Gellman, H. and Nixon, J. (1950). The westward drift of the Earth's magnetic field. *Phil. Trans. Roy. Soc. A*, **243**, 67–92

Cox, A. (1968). Length of geomagnetic polarity intervals. *J. Geophys. Res.*, **73**, 3247–3260

Gans, R. F. (1972). Viscosity of the Earth's core. *J. Geophys. Res.* **77**, 360–366

Hide, R. (1966). Free hydromagnetic oscillations of the Earth's core and the theory of the geomagnetic secular variation. *Phil. Trans. Roy. Soc. London A*, **259**, 615–650

Hide, R. (1969). Interaction between the Earth's liquid core and solid mantle. *Nature*, **222**, 1055–1056

Higgins, G. and Kennedy, G. C. (1971). The adiabatic gradient and the melting-point gradient in the core of the Earth. *J. Geophys. Res.*, **76**, 1870–1878

Jacobs, J. A. (1971). Boundaries of the Earth's core. *Nature Phys. Sci.*, **231**, 170–171

Kahle, A. B., Ball, R. H. and Vestine, E. H. (1967). Comparison of estimates of fluid motions at the surface of the Earth's core for various epochs. *J. Geophys. Res.*, **72**, 4917–4925

Kitazawa, K. (1970). Intensity of the geomagnetic field in Japan for the past 10,000 years. *J. Geophys. Res.*, **75**, 7403–7411

Malin, S. R. C. and Saunders, I. (1973). Rotation of the Earth's magnetic field, *Nature*, **245**, 25–26

Pannella, G., MacClintock, C. and Thompson, M. N. (1968). Palaeontological evidence of variations in length of synodic month since late Cambrian. *Science*, **162**, 792–796

Rochester, M. G. (1970). Core–mantle interactions: geophysical and astronomical consequence. In *Earthquake Displacement Fields and the Rotating of the Earth* (Eds. L. Mansinha, D. E. Smylie and A. E. Beck), D. Reidel Publ. Co., Dordrecht-Holland, pp. 136–148

Rochester, M. G. (1973). The Earth's rotation. *Trans. Amer. Geophys. Union EθS*, **54**, 769–780

Runcorn, S. K. (1964). Changes in the Earth's moment of inertia. *Nature*, **204**, 823–825

Runcorn, S. K. (1970). Palaeontological measurements of the changes in the rotation rates of Earth and Moon and of the rate of retreat of the Moon from the Earth. In *Palaeogeophysics* (Ed. S. K. Runcorn), Academic Press, pp. 17–23

Scrutton, C. T. (1964). Periodicity in Devonian coral growth. *Palaeontology*, **7**, 552–558

Scrutton, C. T. (1970). Evidence for a monthly periodicity in the growth of some corals. In *Palaeogeophysics* (Ed. S. K. Runcorn), Academic Press, pp. 11–16

Stewartson, K. (1959). On the stability of a spinning top containing fluid. *J. Fluid Mech.*, **5**, 577–592

Urey, H. C. (1952). *The Planets, Their Origin and Development*, Yale Univ. Press, 1952

Vestine, E. H. (1953). On variations of the geomagnetic field, fluid motions, and the rate of the Earth's rotation. *J. Geophys. Res.*, **58**, 127–145

Vestine, E. H. and Kahle, A. B. (1968). The westward drift and geomagnetic secular change. *Geophys. J.*, **15**, 29–37

Wedemeyer, E. H. (1966). Viscous corrections to Stewartson's stability criterion. BRL Rept. 1325 AD 489 687, Ballistics Res. Lab., Aberdeen Proving Ground, Maryland

Wells, J. W. (1963). Coral growth and geochronometry. *Nature*, **197**, 948–950
Wells, J. W. (1970). Problems of annual and daily growth-rings in corals. In *Palaeogeophysics* (Ed. S. K. Runcorn), Academic Press, pp. 3–9
Yukutake, T. (1971). Spherical harmonic analysis of the Earth's magnetic field for the 17th and 18th centuries. *J. Geomag. Geoelec.*, **23**, 39–59
Yukutake, T. (1972). The effect of change in the geomagnetic dipole moment on the rate of the Earth's rotation. *J. Geomag. Geoelec.*, **24**, 19–47
Yukutake, T. (1973a). Fluctuations in the Earth's rate of rotation related to changes in the geomagnetic dipole field. *J. Geomag. Geoelec.*, **25**, 195–212
Yukutake, T. (1973b). The eccentric dipole, an inadequate representation of movement of the geomagnetic field as a whole. *J. Geomag. Geoelec.*, **25**, 231–235

DISCUSSION

RUNCORN: Why have you stressed viscous coupling over electromagnetic coupling?
JACOBS: While viscous effects are not necessarily the most important effects to be considered, relative to electromagnetism, viscous effects can't be neglected.

GRAVITY AND THE EARTH'S ROTATION

PAUL S. WESSON

St. John's College, Cambridge, England

Abstract

Of the seven cosmological theories that have notable geophysical consequences, the most sound are those of Jordan–Brans–Dicke (following Dirac) and Hoyle–Narlikar: they predict $\dot{G}/G = -2 \times 10^{-11}$/year and $\dot{G}/G = -1 \times 10^{-10}$/year respectively, with planetary expansions over the history of the Earth of amounts of 180 km and 500 km. Both of these amounts of expansion are much smaller than that of ~ 3000 km indicated by data in favour of the expanding Earth hypothesis (about 20 determinations), which latter can be given a firm foundation in general relativity with no need for G to vary. If G does vary, it is expected that other constants like α, h, e, c, etc. may have varied too, the whole field of possibilities being interrelated. The best limits on $\dot{G}$ are $|\dot{G}/G| \lesssim 4 \times 10^{-10}$/year (direct determination) and $|\dot{G}/G| \lesssim 2 \times 10^{-11}$/year (indirect determination). A decreasing G would lead to an expansion of the Earth (of the amounts noted above) with a concomitant increase of its moment of inertia and the appearance of a cosmological term in the planet's rate of spin-down: such a term is not indicated by direct observations of the rotation of the Earth or by data on the ancient balance of angular momentum in the Earth–Moon system; the latter implies periods in which the planet has expanded and (if a variation is allowed) periods in which G has increased. It can be stated that the accumulation of evidence points to G being a true constant in time, but with a non-negligible expansion of the Earth due to some other cause. In contradistinction, the hypothesis of G (and other constants in physics) being space-variable is still largely an open question. The constants involved are G (Newtonian constant of gravity), g (local acceleration due to gravity on the Earth), h (Planck's constant), α (the fine-structure constant), m_e (the electron mass), m_p (the proton mass) and c (the velocity of light). Other definitions and variational limits are given at the end of section 2.

1. Introduction

There are two main aspects to the problem of gravitational theory as it applies to the Earth: the first concerns a direct influence of a possible change in the parameter G on the state of compression of the globe, the effects of a modified lunar orbit on the tides etc. and other phenomena such as palaeographical and palaeoclimatic consequences of a differing and time-dependent force of gravity. The second aspect of the problem is slightly less well-known and is concerned with a direct expansion of the Earth; if G decreases, the Earth must expand, of course (or possibly contract if G increases as for an interpretation of Milne's cosmology); but it is important to remember that in one sense Einstein's general relativity can be thought of as giving a direct expansion of the Earth at a rate deducible from measurements of the recession of the nebulae.

I have considered both these aspects of the problem at some length (Wesson, 1973), having treated expansion of the globe and those cosmologies which are expected to have salient effects on the Earth. With regard to the former subject, there now exist about 20 determinations, based on different methods, of the rate of increase of the Earth's radius. The palaeomagnetic determinations are now known to be spurious, and on disregarding them there is left a group of 13 estimates which, with the exception of one wildly differing point, form a closely spaced group in dr/dt when a plot is made of rate of increase of the radius (r) versus time or age of the data (Wesson, 1973, p. 41). The deduced rate of increase is $dr/dt \simeq 0.5$ mm/year, agreeing remarkably closely with the equivalent Hubble expansion velocity calculated from general relativity ($= 0.6$–0.7 mm/year) and with the rate needed (0.66 mm/year) if the Earth began expanding from an initially sial-covered state 4.5×10^9 years ago. Although it is easy to think of objections to the methods used to calculate dr/dt, I do not think these are sufficient to warrant ignoring the conclusion arrived at, and I am personally of the opinion that the planet may be expanding at 0.6–0.7 mm/year in radius. This expansion, if true, probably has nothing whatsoever to do with the parameter G, being a result of expansion of space as inferred from conventional general relativity.

Cosmologies that predict notable geophysical effects number seven: (i) Milne (1935); (ii) Dirac (1937); (iii) Jordan (1949); (iv) Kapp (1960); (v) Brans–Dicke (1961); (vi) Prokhovnik (1970); (vii) Hoyle–Narlikar (1971). These cosmologies have been examined consecutively (Wesson, 1973, pp. 17–40), and I concluded that the theories of Hoyle–Narlikar and Jordan–Brans–Dicke are theoretically the most well-founded. Both the amounts of planetary expansion given by Hoyle–Narlikar and Brans–Dicke are considerably smaller than the 3000 km implied by evidence in favour of the expanding Earth hypothesis.

Both of the most acceptable theories with time-dependent G are consistent with the best direct determination to date of $\dot{G}$, namely that of Shapiro et al. (1971) which is $|\dot{G}/G| \lesssim 4 \times 10^{-10}$/year. It will be seen in section 2 that an indirect method of estimating $\dot{G}$, using the history of the Earth's rotation, gives $|\dot{G}/G| \lesssim 2 \times 10^{-11}$/year, which, if upheld, would tend to support the Brans–Dicke theory against that of Hoyle–Narlikar.

The subject of a possible variation in time or space of the parameter G and other constants of physics has attendant to it a large bibliography. In as far as various theories predict such changes, I have already dealt with one aspect of the subject (Wesson, 1973). The present discussion is devoted more to the aspects of variable G theory as it affects the Earth, and not as to the *a priori* justification of the cosmologies involved. I have included references of all papers I know of in this subject which have appeared since my previous article (Wesson, 1973), and the present discussion represents, I hope, an up-to-date account of the topic. I have tried to keep the mathematics involved to a minimum as far as the text is concerned, banishing most of it to the Appendix: this may give the impression in some places that I have pulled results out of the hat, so to speak; but the

Gravity and the Earth's Rotation

back-up material is available, to those interested, in the paper cited. For those who are not conversant with the variable G theories, let me sum up the situation briefly with the following primer.

Newton's law of gravitation says that two bodies of masses m_1 and m_2 are attracted towards each other with a force that varies as the product of the masses and the inverse square of the distance, r. The law is thus of the form $F \propto m_1 m_2/r^2$. To enable the proportionality sign here to be replaced by an equals sign, it is necessary to define a system of units: choosing these as the centimetre, gram and second, it is found that $F = Gm_1 m_2/r^2$. The parameter G is a universal constant that can be determined by experiment, and is found to be about $6 \cdot 6 \times 10^{-8}$ in the units adopted. Ordinarily, no question arises of G being variable in any way at all, since it is somewhat perverse to start complicating the subject with nonconstancy at this stage of the game. However, for several reasons workers in the field of gravitation have suggested that G may not be a strict constant: in particular, G might vary with time or with position. If I were able to get into a Tardis-like machine and return to (say) the Carboniferous period with a Cavendish balance, I might find, on getting there and carrying out the experiment, that G was not equal to $6 \cdot 6 \times 10^{-8}$. The implication would be that G varies with time. Alternatively, if I removed myself to Jupiter (say), I might again find that G was not equal to $6 \cdot 6 \times 10^{-8}$, and the implication would be that G is variable in space.

It is known from experiments on the Earth within historical times that any variation of G in these two ways must be very small: the limits set and expected have been quoted above, and are indeed very small. I would emphasize that there is no compelling reason to believe that G *does* vary: the hypothesis that G is variable is one that must be rigorously tested to determine its plausibility. One way of doing this, which is of interest to biologists, is to look for signs of any change (probably with time) of the local acceleration of gravity, g. This parameter should not be confused with G; G determines g in the following way: if I drop a solid body of mass m_b from the Leaning Tower of Pisa, it accelerates towards the ground with an acceleration governed by a force $F = m_b g$. This force is caused by the Earth's gravity acting on the body and by the body acting on the Earth: it is given by the previous law ($F = Gm_1 m_2/r^2$) with m_1 taken as the mass of the Earth (M_E) and m_2 as the mass of the body (m_b), the distance r being about equal to the radius of the Earth (r_E). It is seen that $m_b g = F = GM_E m_b/r_E^2$. From this expression, m_b goes out (a profound coincidence on which Einstein founded general relativity) and g is left as being $g = GM_E/r_E^2$. Thus, if G were numerically larger in the past, then g would be, too: a larger G in the Carboniferous would mean that g was larger and the pull of the Earth (assumed unchanging as regards M_E, r_E) on animals, rocks etc. would have been larger: pterodactyls for instance, would have found it harder to fly than they would do if they lived at the present time.

The time-variability of G just considered must not be confused with space-variability. The latter is much harder to look for and has not been discussed as much as the time-variability hypothesis. In what follows, I shall endeavour to make the difference clear. Section 2, which follows this Introduction, is a short, concise account of the present situation as regards limits on the variation of G and other constants of physics (if G varies with time at a rate $\dot{G}$, it is expected that other constants will also vary, and *vice versa*). Section 3 is on rotation. Section 4 is a Conclusion, and an Appendix contains calculations I have made on the variation of G as it affects the shape of the Earth and stresses in the lithosphere.

2. Gravitational theory

The laws of physics give rise to certain constants that must be evaluated numerically before the form of the equation of the law and the figures calculated from experiment can be brought into correspondence. Such numbers are G, e, h, m_e, m_p etc., and while certain of these constants can be set equal to unity by a suitable choice of units, the dimensionless numbers that can be formed from them are independent of whether one measures the component constants in mks cgs or fps or whatever. One can therefore take the view that the dimensionless numbers, such as the coupling constants and the ratios of elementary particle masses to the fundamental gravitational mass $(\hbar c/G)^{1/2}$, contain the essence of the laws of physics. It is obviously of prime importance to see if these dimensionless constants really are constants, independent of space and time, as assumed in the promulgation of the Strong Equivalence Principle (Dicke, 1964a, b) of general relativity.

Arguments concerning the space or time variation of G (Wheeler, 1962, pp. 19, 110–120) depend on the dimensionless numbers mentioned in the last paragraph. There are four main dimensionless constants in Nature (Bertotti, Brill and Krotkov, 1962, p. 375) which have quite distinctive magnitudes.

(i) The coupling constant for strong interactions is of order of size $\approx 10^0$, the fine-structure constant being $\alpha = 1/137$.

(ii) The weak coupling constant $(g' m_\pi^2 c/\hbar^3)^2 = 3 \times 10^{-14}$ (where g' is the beta-decay coupling constant, $g' = 1\cdot 4 \times 10^{-49}$ erg/cm^3), and so is of overall magnitude $\approx 10^{-20}$.

(iii) The ratio of gravitational to electrical force in the hydrogen atom is $Gm_e m_p/e^2 \simeq 5 \times 10^{40}$, approximately equal to the ratio of an atomic period to the age of the Universe, $e^2/m_e c^3 T \simeq 0\cdot 2 \times 10^{-40}$. Both are thus $\approx 10^{-40}$.

(iv) There is some reason for thinking that the number of particles in the Universe is $4\pi R^3 \rho/3m_p \simeq 0\cdot 3 \times 10^{80}$ (R = Hubble radius of the Universe = cT = 10^{28} cm; $\rho \approx 10^{-30}$ gm/cm^3), so that the reciprocal of this number is $\approx 10^{-80}$.

These numbers, often called Eddington numbers after the first person to realize the significance of the progression in sizes involved in them, have been used extensively to propound possible cosmologies lying outside general relativity (although an explanation of their sizes has been given by Klein (1958) based on general relativity and without violating the Strong Principle of Equivalence).

Dirac (1937) based his hypothesis concerning $G \propto t^{-1}$ on the coincidence between some of the sizes of these numbers, and so started the investigation, which is still in progress today, of the possible time variation of the constants of physics.

The basis of the formation by Dicke of Dirac's hypothesis has some interesting consequences (Dicke, 1961). Denoting quantities by the same symbols as used in discussing the numbers (i)–(iv) previously, Dirac noted that the dimensionless gravitational coupling constant

$$\frac{Gm_p^2}{\hbar c} = 5 \times 10^{-39} \left[\text{or} \ \frac{Gm_e m_p}{c^2} = 5 \times 10^{-40} \right] \quad (1)$$

and the number

$$\frac{Tm_p c^2}{\hbar} \simeq 10^{42} \left[\text{or} \ \frac{Tm_e c^3}{e^2} \simeq 10^{40} \right] \quad (2)$$

are related in that the reciprocal of (1) is approximately equal to (2). Since (2) varies with time, the implication is that (1), which includes G, also varies with time. (The first numbers in (1) and (2) are those used for illustration of the coincidence by Dicke (1961) in discussing Dirac's hypothesis; the second numbers are those mentioned, e.g. by Bertotti (1962, p. 178) and originally used by Dirac. Note that m_p here denotes the mass of the proton, and is not to be confused with the similar symbol for passive gravitational mass used elsewhere in this section.) If M is the mass of the Universe, there is also the number

$$\frac{M}{m_p} \approx 10^{80} = (10^{40})^2 \quad (3)$$

to be considered, so that it might be thought that (1) varies as t^{-1}, (2) as t^{+1} and (3) and t^{+2}. To do this consistently, notes Dicke (1961), the presumed interconnection between the numbers must itself be independent of time, and the coincidence in sizes only works out if T is the *present* age of the Universe. Dicke's philosophy is that (1), (3) could well be constant, and that (2) varies with the cosmological epoch such that physicists just happen to be alive at the present time to note the relationships between the magnitudes of the numbers.

While G is variable in time in the Brans–Dicke theory it is to be recalled that Sciama's analysis of the origin of inertia gave rise to the formulation of a space-variable G (Sciama, 1953), this having been interpreted by many (see, e.g. Bertotti, 1964) in the form

$$G = G_0 \left(1 + \frac{aV}{c^2} \right) \quad (4)$$

where a is a dimensionless constant of the order of unity and V is the gravitational potential in the region of the body or bodies being considered. The relation (4) has been commented on in passing by several workers (Einstein, 1950; Jordan,

1955, 1959; Brans and Dicke, 1961; Dicke, 1962a, 1962b), but the most notable attempt to interpret it practically has been made by Finzi (1962), who investigated the variation of the gravitational self-energy of a body as G varies. This gives rise to a force on white dwarfs which is non-negligibly in need of correction as the white dwarf pursues its path in the Galaxy, this path being expected to be slightly different from that of a normal star. In particular, white dwarfs in very weakly bound clusters will escape, while those in more strongly bound clusters will occupy an anomalous position. While the proposed tests of (4) due to Finzi (1962) still await application, Steiner (1967) has made some rather wild observational speculations concerning orogenies and palaeomagnetism etc. (he believes these to occur with a 350-My period), which he justifies by appealing to the variation in G as the Earth pursues its elliptical orbit around the Sun, and the Sun its orbit about the Galactic centre. The existence of gravitational anisotropies in the Solar System has been investigated by Will (1971a) in so far as metric theories of gravitation (like the Brans–Dicke theory, with general relativity as a limiting null case) predict anisotropies in the passive gravitational mass of bodies like the Earth and Sun.

Regarding anew the content of physical laws as they are summed up in the constants of physics, I would like to think of the dimensionless numbers of Nature as proper invariants, in space at least: the component parameters making up these dimensionless numbers (e.g. $Gm_e^2/\hbar c = [0]$), could vary, however. The possibility of such a conspiracy between the dimensional physical constants was apprehended by Dirac (see Harwit, 1971), and the subject is one that needs careful handling because the values of constants without zero dimension depend on the system of units we use to measure them, and changing our metre rules for yardsticks in measuring a length is certainly not a deeply significant act that needs to be incorporated into physical theory. On the other hand, if I imagine myself carrying out (say) Cavendish's experiment and discovering that G as given by the experiment is not the expected value of $G \simeq 6 \times 10^{-8}$ cgs, I have just cause to be surprised, if not amazed: assuming that my apparatus is in good working order, I would be justified in regarding the result of the experiment as physically significant.

The meaning of variability of dimensional constants is an important one for theories in which G is taken to be variable in either time or space. Experiments now in progress along the lines of that of Shapiro *et al.* (1971) will, within a decade, attain an accuracy sufficient to validate or disprove changes in G of the sizes claimed by the Jordan–Brans–Dicke and Hoyle–Narlikar theories. This and other possible tests have been discussed by Wesson (1973), and I also proposed (*loc cit*, p. 45) a new test based on the trajectories and distribution on the ground of ejecta that originated in ancient volcanoes: this test now appears to be one which would be slightly difficult of application, as pointed out by Molyneux (private communication): (a) meteorological conditions would have altered many times during the eruption of a Palaeozoic volcano and would also have been

Gravity and the Earth's Rotation

different at various altitudes; (b) the estimation of the ejection velocity would be problematical, as would (c) the measure of the horizontal range to sufficient accuracy and (d) corrections for local variations in g, the gravitational acceleration. These objections to the proposed test, while serious, might be circumvented if a statistical method could be employed to find the parameters necessary to estimte G from the formula as proposed:

$$\text{Range} = \frac{v^2 \sin 2\beta}{g} \quad (5)$$

where v is the velocity of projection, β the angle of projection from the horizontal, and g the local gravitational acceleration, from which G could be calculated. Whether the use of (5) would enable $\dot{G}$ to be found to an accuracy good enough to evaluate the validity of the Jordan–Brans–Dicke or Hoyle–Narlikar theories is, of course, not known.

To conclude: the limits of possible variations in the physical constants, especially in G, are summarized in what follows. The symbols are defined thus: G (constant of gravitation), $\dot{G}$ (rate of change with time of G); m_p, m_1 (passive and active gravitational mass of electron, e^-, or positron, e^+); α (fine structure constant), $\dot{\alpha}$ (rate of change with time of α); $\Delta m_1/m_1$ (anisotropy of inertial mass in the Universe); h, c (Planck's constant, velocity of light); g_s (strong coupling constant). For the purposes of geophysics, the most important limit to note is that G, if changing with time, is doing so at a rate definitely less than 4 parts in 10^{10}/year, and probably less than 2 parts in 10^{11}/year.

To sum up the limits arrived at, in the order of the symbols just defined: the Eötvös experiment is upheld by numerous data, with a direct determination (section 1) of $|\dot{G}/G| \lesssim 4 \times 10^{-10}$/year, and with an indirect determination of $|\dot{G}/G| \lesssim 2 \times 10^{11}$/year (see the next section); also $m_p(e^-) \simeq m_1(e^-)$ to one part in 10^4, $m_p(e^-) = m_p(e^+)$ to one part in 100, $m_1(e^+) \simeq m_1(e^-)$; $|\dot{\alpha}/\alpha| < 3 \times 10^{-13}$/year $\Rightarrow |\dot{G}/G| \lesssim 2 \times 10^{-11}$/year; $\Delta m_1/m_1 < 10^{-20}$ for inertial anisotropy; $|\dot{G}/G| = 10^{-10}$/year $\Rightarrow |\dot{\alpha}/\alpha| = 10^{-12}$/year; $|\dot{G}/G| = 10^{-8}$/year (annual) $\Rightarrow |\dot{\alpha}/\alpha| = 10^{-10}$/year to 10^{-13}/year (undetectable directly); $d[\log_e(hc)]/dz < 3 \times 10^{-4}$, $d[\log_e(hc)]/dt < 10^{-10}$/year; $d[\log_e(\alpha^2)]/dt < 10^{-11}$/year; $e^{-2} d(e^2)/dt < 10^{-13}$/year; $g_s^{-2} d(g_s^2)/dt < 10^{-11}$/year; $\alpha^{-1} d\alpha/dt < 2\cdot 3 \times 10^{-11}$/year.

3. Rotation of the Earth

Cosmological theories in which G decreases with time lead, as noted in section 1, to the inevitable result that the Earth should be expanding. There is also independent evidence in favour of expansion, unrelated to the variable G theories (Wesson, 1973). As the Earth expands, its moment of inertia increases and its spin decreases if no other forces are acting: in general, the actual situation would be complicated because neither $I\Omega^2$ or $I\Omega$ (Runcorn, 1968) is constant for the Earth (Ω = angular velocity, I = moment of inertia). The expansion effect is likely in any case to be masked by the slowing due to tidal friction (Jeffreys, 1970, pp. 308–313), the total spin-down rate being about 2 sec in 10^5 years.

The theory of the rotation of the Earth is a complicated subject, having ramifications in many attendant fields of study. The spin axis has an inertial orientation that changes with periods of 26,000 years (the steady precession, amplitude $=23°$; general relativistic effects contribute $\simeq 2''$ to the precession), 18·6 years (the principal nutation, amplitude $=9''$), 9·3 years/1 year/6 months/2 weeks (parts of the nutation in obliquity and longitude, amplitudes $\lesssim 1''$), and possibly other changes due to ill-understood, perhaps general-relativistic effects (secular decrease in the obliquity with a discrepant rate $\approx 0\cdot 1''$/cy). The spin axis moves with respect to the surface of the Earth also, having components with periods of 20–40 years (a perturbation of amplitude $\approx 0\cdot 01$ arc sec, sometimes referrred to as the Markowitz wobble), 440 days (the Chandler wobble, amplitude $\simeq 0\cdot 15$ arc sec, but variable, damping out in ≈ 30 years), 1 year/6 months (seasonal perturbations of amplitudes $=0\cdot 10$, $0\cdot 01$ arc sec respectively), 1 month/2 weeks (amplitude expected to be $\approx 0\cdot 001$ arc sec), 1 sidereal day (diurnal, period $\lesssim 0\cdot 01$ arc sec), and secular motions (irregular, rate $\simeq 0\cdot 1$ arc sec in 50 years), and changes with the periods of the nutations (but with amplitudes $\approx 0\cdot 01$ arc sec). The instantaneous axis of spin defines the angular velocity, Ω, of the planet, and this changes over periods of 10^{10} years (secular acceleration, $\dot{\Omega}/\Omega \simeq -5 \times 10^{-10}$/year), 100 years ($\dot{\Omega}/\Omega \lesssim \pm 5 \times 10^{-10}$/year), 1–10 years ($\dot{\Omega}/\Omega \lesssim \pm 100 \times 10^{-10}$/year), months ($\dot{\Omega}/\Omega \lesssim \pm 500 \times 10^{-10}$/year), and other short-period changes, some of which are of doubtful recognizement (2 years, $\Delta\Omega \simeq 10$ msec; 1 year, $\Delta\Omega \simeq 20$ msec; 6 months, $\Delta\Omega \simeq 10$ msec; month/2 weeks, $\Delta\Omega \simeq 1$ msec). These movements and changes to do with the spin can be measured at present to an accuracy of $\pm 0\cdot 01$ arc sec over periods $\lesssim 6$ months, but a potential accuracy of order $0\cdot 001$ arc sec is offered by the technique of very long baseline interferometry using radio sources. As noted by Rochester (1973), any divergences that turn up from now on between ephemeris time and atomic time will indicate inadequacies in the theory of the Moon and Newtonian celestial mechanics.

Recently, Weinstein and Keeney (1973) have re-examined the data of Pannella *et al.* with a view of estimating the apparent loss of angular momentum in the Earth–Moon system: they find that there has been an overall 5% loss, with the Moon having lost angular momentum over the Precambrian to Silurian, been constant in the interval Silurian to Pennsylvanian, and finally gained angular momentum from Pennsylvanian to Recent. The first two phases are somewhat anomalous, since tidal friction can only increase the lunar momentum; since the length of the sideral year has been approximately a constant over the last 10^9 years at least (Hipkin, 1970), the implication is either that G is changing or else I (the moment of inertia of the planet) is changing. If $\dot{G} \neq 0$, the data require that G has *increased* secularly; this harks back to Milne, but is hardly satisfactory. The alternative is that I has increased by 25–30% over the last $1\cdot 7 \times 10^9$ years, a large part of the increase having occurred since the Devonian. I take this as in support of an expanding Earth model but, as noted by Weinstein and Keeney, the anomalous behaviour of the lunar momentum over the period Precambrian to Pennsylvanian cannot be entirely due to changes in I. The conclusion, therefore, is that the Earth is probably expanding but that some other (possibly cosmological) effect is also operative, in combination with tidal friction.

Tidal friction theory is only accurate to a factor of two, so the spin-down could, alternatively, be due entirely to tidal friction, the variabilities that are encountered in eclipse observations, telescope observations and palaeontological data being results of changes in the configuration of shallow seas. This is ignoring the possibility that there could be processes tending to accelerate instead of decelerate the Earth's rotation. On this interpretation, there is no need for an expansion term; a substantial contribution due to expansion *may* well be present, and this is acceptable since the theory and data are subject to such uncertainties as discussed earlier. (Dooley, 1973, has reviewed the expansion argument, while Birch, 1967, estimates that a change $2G \to G$ on its own would cause an expansion of 370 km or so.) The view that all the spin-down is due to expansion is obviously wrong, since tidal slowing is definitely occurring to some extent, even if small.

The secular acceleration of the Earth's rotation gradually slows down its spin rate: as this process continues, the shape of the planet must alter. Newton (1968) has noted that data on the Devonian length of day, satellite orbits and astronomical latitude discrepancies as given by ancient eclipses, allow a possible variation in G of 1 part in 10^8/cy, this leading, in turn, to an increase in the polar moment of inertia (C) of several parts per 100. Most of the change in C is due to an overall expansion of the globe, this rate being compatible with the extrusion of matter taking place at the mid-oceanic ridges. (Newton takes the solar-induced tide in the atmosphere as contributing $2 \cdot 7 \times 10^{-9}$ parts a cycle to $\dot{\Omega}/\Omega$, and mentions that Dicke has used ancient eclipse data to obtain $\dot{\Omega}/\Omega = -15 \cdot 9$ $(\pm 0 \cdot 7) \times 10^{-9}$/cycle, which Dicke would like to attribute to $\dot{G}$; Newton considers $\dot{\Omega}/\Omega \simeq -18 \times 10^{-9}$/cycle to be a more reliable figure obtained from eclipse records, and is not convinced that any contribution to $\dot{\Omega}$ from a $\dot{G}$ term is required.) If G changes, then $\dot{\Omega}/\Omega = -2\dot{G}/G$, and as $|\dot{G}|$ is considered as larger in size so the consequent change in C is expected to have progressed from a smaller ancient value. In fact, Newton (1968, p. 3769) is able to show that $C/M_E v_E^2$ has not changed in the past, to an order of $e \times$ (change in $\Omega^2/G\rho_E e$), as a result of changes in Ω, G, ρ_E or e (M_E = mass of the Earth, ρ_E = its mean density, e = its ellipticity). If material is being extruded at the ridges and not resubducted, then the volume of the Earth is increasing at a rate $\dot{V}/V = 5 \cdot 5 \times 10^{-9}$/cycle. If C changes slowly due to this, then $\dot{C}/C = 3 \cdot 7 \times 10^{-9}$/cycle leads to an evaluation of $C(\text{now})/C(\text{Devonian}) = 0 \cdot 987$. The concomitant expansion of the planet would have increased the temperature of the interior by about 1000 K since the Devonian. The conclusion arrived at by Newton is that C is indeed increasing, with a probability of 5:1, but that the rate of increase is extremely uncertain.

If the $\dot{C}$ term is due to $\dot{G} \neq 0$, then $\dot{G} \neq 0$ is upheld also with a probability of 5:1. This does not, of course, mean that $\dot{G}$ *is* non-zero, since there are quite independent grounds for thinking that the Earth might be expanding.

One of the main consequences of a slowing in the Earth's rotation is that the excess equatorial bulge is probably in a state of slow relaxation. As pointed out by McKenzie (1966), the energy stored in the excess equatorial bulge is $\simeq 2 \times 10^{30}$ erg (gravitational) plus 2×10^{29} erg (elastic). Ignoring the latter, the energy in the J_2 harmonic is seen to be greater than that in any other harmonic: this strongly suggests that it is due to an anelastic response to the spin-down with an equivalent lower mantle viscosity of $v_m \approx 10^{26}$ cgs. The bulge is not due to convection currents (McKenzie, 1966, p. 3998); its presence produces maximum shear stresses of $\approx 10^8$ dyne cm^{-2} in the mantle, and as the Earth undergoes spin-down, the moments of inertia A, A, C (C = polar axis moment of inertia) changes secularly. MacDonald (1964) has also considered the question of the relaxation of the bulge, and he gives an interpretation of the coefficients J_n in terms of surfaces of equal density: the deviation from hydrostatic equilibrium is largest in J_2 (MacDonald, 1964, p. 220), leading to the evaluation of a viscosity of the mantle that is $\approx 10^4$ times that of the upper mantle as calculated from the uplift of Lake Bonneville. The excess bulge results in stresses in the mantle of $\simeq 14$ bars according to MacDonald, this figure being in approximate agreement with that derived by McKenzie. The energy stored in the crust due to J_2 is $\approx 10^{29}$ erg, and MacDonald derives the total energy associated with the bulge as being about 2×10^{34} erg.

Dicke's work on the secular deceleration of the Earth has been influenced by a wish to detect a term due to $\dot{G}$ as opposed to one due primarily to expansion. The attempt to isolate such a term from historical data (Dicke, 1966) is suspect because there are irregular fluctuations in the rotation that are greater than the secular decrease and may be connected with slight changes in the tilt of the axis (Dicke, 1966, p. 111, also Guinot, 1970, has examined short-period terms in Universal Time, these being abrupt changes in spin rate of <1 msec and of 12–30-day period, tending to change the vertical and the shape of the geoid). Averaging, over a period of order 1000 years, is necessary to remove these fluctuations, and even when this is done there remain numerous effects other than a change in G that can alter the spin rate: (i) lunar, solar and atmospheric tides; (ii) angular momentum transfer to and from the solar wind; (iii) angular momentum transfer from the solid Earth to the atmosphere and oceans; (iv) electromagnetic coupling across the CMB; (v) changes in sea-level that alter I; (vi) tectonic processes changing I; (vii) a possible continuing growth of the core causing a change in I; (viii) a change in I due to expansion or contraction of the Earth unconnected with $\dot{G}$, e.g. thermal or cosmological in origin. Despite the smallness noted previously of any substantial term not explicable by tidal friction, Dicke (1966) has sought to explain the small anomaly ($6 \cdot 8 \times 10^{-11}$/year) between his calculation of $\dot{\Omega}$ by eclipses and the expected $\dot{\Omega}$ arising from all the previously considered processes (i)–(viii). Since the total $\dot{\Omega}$ being dealt with is

Gravity and the Earth's Rotation

about 20×10^{-11}/year, I do not think there is much justification for the subsequent calculation of $\dot{G}/G \simeq -4 \times 10^{-11}$/year, since tidal friction alone is uncertain by a factor of two and can explain nearly all of the observed $\dot{\Omega}$. There is, of course, no contradiction in postulating such a term due to $\dot{G}$, just as there is nothing wrong with invoking the expanding Earth hypothesis to explain other geophysical data, but it must be clearly recognized that while there is no clear-cut evidence against either postulate there is also no convincing evidence in favour of it.

Concluding the whole field of the Earth's rotation, it is manifestly clear that more data of increased accuracy are needed in which the number of days/year and not merely the number of days/synodic month are calculated from fossil specimens. A definitive account of eclipse observations is vital to a knowledge of the rotation over the last 10–20 cy, and theoretical models are needed to explain the short-term fluctuations in the spin-down rate. Until these things are available it will not be possible to say with certainty whether or not the Earth is expanding. The present situation is that expansion is not contradicted, especially if G is variable, but that the angular momentum loss from the Earth–Moon system implies that $\dot{G}$ is either positive or zero. In view of this problem and the low limit ($|\dot{G}/G| < 2 \times 10^{-11}$/year) set on changes in G by studies of the rotation of the Earth, it is best tentatively to accept Earth expansion without changes in G as a working hypothesis as far as the secular rate of change of the planet's rotation is concerned.

4. Conclusion

Having considered, in the previous three sections, the large amount of information available on the possible variation of G and the rotation of the Earth, it is a relief to be able to sum up the situation succinctly: there is at the moment no evidence that such a tendency is having any effect on the rotation of the Earth. The direct limit set by Shapiro et al. (1971) of $|\dot{G}/G| < 4 \times 10^{-10}$/year is close to the value of $\dot{G}/G \simeq -1 \times 10^{-10}$/year predicted by the Hoyle–Narlikar cosmology but comfortably remote from the value of $\dot{G}/G = -2 \times 10^{-11}$/year predicted by the Brans–Dicke cosmology. In the Appendix the effects that the spin-down and a change in G would have on the excess equatorial bulge of the Earth are tentatively examined: I find that there is a serious plausibility gap in adopting the hypothesis that G changes at all, the implication being that $\dot{G}$ is smaller than 10^{-12}–10^{-13}/year and/or global expansion of order 1000 km has occurred. This limit is an indirect one, derived from studying the gravity harmonic J_2, and is ambiguous in not deciding between expansion of the globe and a change in G. The best indirect method of estimating $\dot{G}$ is that of Morrison (1973), who, using data on the rotation of the Earth, has derived $|\dot{G}/G| < 2 \times 10^{-11}$/year. The angular momentum of the Earth–Moon system has been varying in a somewhat puzzling manner since the Precambrian (Weinstein and Keeney, 1973), and does not support the idea of any negative variation in G, even of the small size given by Brans–Dicke and Morrison's limit. The most sensible interpretation of the angular-momentum results is that the Earth is undergoing a secular expansion, with G constant, but that some other unknown effect is recorded in the ancient rotation data. What this effect could be is still obscure, and the further elucidation of this problem will probably prove an absorbingly interesting topic for further investigation.

APPENDIX

Stresses set up by changes in Ω and G due to the equatorial bulge

A1. *Preamble*

The equatorial bulge of the Earth due to its spin rate represents a notable store of energy that is released gradually as Ω is decreased by tidal friction, and by expansion and decreases in G (if they occur). McKenzie (1966) gives 2×10^{30} erg as the gravitational energy of the bulge (with 2×10^{29} erg as elastic energy) while MacDonald (1964) gives 10^{29} erg as the energy stored in the lithosphere and 10^{34} erg overall. Since the characteristic relaxation time of the bulge is 10^7 years (if its excess part is taken as being the result of anelasticity in the mantle), it is releasing energy at a present rate of about 2×10^{27} erg/year, which is approximately 0·2 of the geothermal flux, and much larger than the 10^{25} erg/year released as seismic energy (MacDonald, 1964). The energy budget of the upper mantle and crust is therefore seen to be seriously influenced by the state of the equatorial bulge.

To obtain a grip on the problem, let it be assumed that there is $U_b = 10^{32}$ erg in the whole bulge and $U_l = 10^{30}$ erg in the lithosphere at the present time (these being mean figures from the above-quoted results). These data should be compared to the total self-gravitational energy of the planet, which is $GM_E^2/a \simeq 4 \times 10^{38}$ erg ($M_E \simeq 6 \times 10^{27}$ gm, $v_E \equiv a \simeq 6380$ km). The forces in the Earth due to the bulge, acting in a tangential sense, are $\sigma_T \approx U_b/2\pi a$ and $U_l/2\pi a$ respectively, to order of magnitude. The length-scale $2\pi a$ ($= 4 \times 10^9$ cm) is approximately equal to the wavelength of the J_2 harmonic (20,000 km). The total force is thus $\simeq 3 \times 10^{22}$ dyne, while the force in the crust is $\simeq 3 \times 10^{22}$ dyne. If the whole mantle supports the force, the area of the interior of the planet over which it is distributed is roughly $2\pi a \times 3000$ (km) $\simeq 10^{18}$ cm² and the stress is $\sim 3 \times 10^4$ dyne cm⁻². If the lithosphere supports its part of the force, it is distributed over an area of roughly $2\pi a \times 30$ (km) $\simeq 10^{16}$ cm² and the stress is $\sim 3 \times 10^4$ dyne cm⁻². The tangential stresses set up by the bulge are thus of order 10^5 dyne cm⁻².

This should be compared to the direct-loading shear stress of 10^8 dyne cm⁻² in the mantle due to the bulge (McKenzie, 1966) and the expected stresses of

Gravity and the Earth's Rotation 365

10^7–10^9 dyne cm^{-2} caused by convection currents (Jeffreys, 1970, p. 428; Wesson, 1972b). In actual fact the stresses calculated for the bulge above are extreme minima, because there are several agencies that act to increase the effective stress.

(i) Stresses do not distribute themselves uniformly in the Earth, but tend to nucleate in certain regions and around already existing faults (Jeffreys, 1970).

(ii) Part of the mantle is probably stirred by convection currents and so cannot support long-term ($\gtrsim 10^7$ year) stresses; this almost certainly applies to the upper mantle, and so the stress has to be supported by the lower mantle and the crust (Wesson, 1972b). If part of the region below 700 km is also weak, this argument is extended, implying that a large part of the stress is supported in the crust.

(iii) The effective radius for supporting the mantle stress is r, where $r < r_E$ (see equation A3 below), this causing an increase in the effective stress by a factor of about three.

(iv) The bulge was larger in the past, as will be seen below more exactly; stresses tend to accumulate until fracture in rigid materials, and a change in J_2 of (say) $\Delta J_2 = 100 \times 10^{-6}$ (J_2 was about 1500×10^{-6}, 1·6 Æ ago) would produce a total force, by equation (A1) below, of 1×10^{27} dynes, which would imply stresses in the Earth of $10^{27}/10^{18} = 10^9$ dyne cm^{-2} (mantle) and roughly $10^{25}/10^{26} = 10^9$ dyne cm^{-2} (lithosphere), if an analogy is made with the present bulge, which has about 1/100 of the total energy stored in the crust. This stress would build up in a time of $\sim 100/500$ (Æ) $= 2 \times 10^8$ years, since J_2 (now) $\simeq 1000 \times 10^{-6}$ and J_2 (1·6Æ) $\simeq 1500 \times 10^{-6}$. For comparison, the breaking strains of rocks in the crust (Jeffreys, 1970, pp. 99, 25) are in the range 10^8–10^9 dyne cm^{-2}.

(v) The stresses are preferentially concentrated in midlatititude belts, as can be seen from equation (A3) below.

It can be seen from these figures that the stresses set up by the bulge over geological time are in the range 10^5–10^9 dyne cm^{-2}. McKenzie (1972, p. 327) is of the opinion that the movement of crustal plates over the non-spherical shape of the bulge will produce strains of 1 part in 100. Since the time-scale for extensive continental drift is of the order of 10^8 years, movements of the plates are not expected to nullify the build-up of stresses in the lithosphere of the order of the breaking strains of its component rocks. Thus, tectonic consequences of the Earth's spin-down, acting via the bulge, might be discernible: it is certainly to be expected that the geothermal flux and seismicity will be affected by the changing shape of the bulge (these last two points are not open to the criticism that the mantle or crust might adapt like a plastic to the bulge stresses: this last point does not seem very certain to me in view of the relative time-scales involved). I will now proceed to examine more closely the distribution of stresses produced by the scular change in the bulge of the Earth, connecting it up later with possible changes in G and large planetary expansions. N.B.: The energy U employed below is the potential energy per unit mass, and similarly for σ_T etc. To find the total energy or stress, M_E must be replaced by M_E^2.

A2. Dynamical deformation of the Earth

Consider a dynamically flattened globe of mean radius R, and semi-major and semi-minor axes, a, b. If the colatitude is θ, the gravitational potential of the spheroid at distance r from its centre is given (Wesson, 1972, p. 2144), by the usual expression:

$$U = \frac{GM_E}{r}\left[1 - \sum_{n=2}^{\infty} J_n \left(\frac{a}{r}\right)^n P_n(\cos\theta)\right] \quad (A1)$$

where the P_ns are polynomials in the solution of Laplace's equation. The term in J_2 is connected with the excess bulge due to the spin, so that, neglecting terms for which $n > 2$,

$$U \simeq \frac{GM_E}{r}\left[1 - J_2 \left(\frac{a}{r}\right)^2 \tfrac{1}{2}(3\cos^2\theta - 1)\right] \quad (A2)$$

If the bulge adjusts to the spin-down of the planet, J_2 continuously decreases with time. After a certain lapse, J_2 has changed by an amount ΔJ_2, causing a change in the potential energy, ΔU. The existence of ΔU causes stresses to be set up tangential to the surface and directed along lines of longitude. The tangential stress is

$$\sigma_T = \frac{1}{r}\frac{\partial}{\partial \theta}(\Delta U)$$

which from (A2) is

$$\sigma_T = \frac{GM_E}{r^2} \Delta J_2 \left(\frac{a}{r}\right)^2 3\sin\theta \cos\theta \quad (A3)$$

The stress is a maximum at $\theta = \pm\pi/4$, so that tensional tectonic features are expected primarily at about latitude 45° in both hemispheres, along with enhanced heat-flow and seismicity. On examining maps of the major mountain ranges, midoceanic ridges and rift systems, etc. I find the expected effects are not observed. Their non-existence needs explaining.

Five possibilities of explanation occur: (a) the Earth is not undergoing spin-down; (b) the material of the lithosphere can adjust itself to the stresses σ_T without fracture; (c) the gravitational constant G may be altering with time in a way such as to compensate the expected release of gravitational potential energy; (d) some other process, notably continental drift, sea-floor spreading or large changes in the planet's radius, may have masked the effect; (e) the bulge may not be the result of spin-down of the Earth.

Of these five hypotheses, (a) was discussed by Kelvin (Jeffreys, 1970, p. 307), who was concerned with a thermal atmospheric tide that tries to accelerate the Earth's spin. There was some disagreement as to whether this tide could maintain the planet's rotation, but this is now generally held to be implausible (Wesson,

Gravity and the Earth's Rotation

1973). Hypothesis (b) does not seem likely in view of the stresses and time-scales noted previously, and in any case does not explain the absence of heat-flow and seismic anomalies around $|\theta| = \pi/4$. Possibility (d) also seems unlikely as regards drift, on the time-scale argument and by the existence of some areas of the globe that have not been crossed or shifted by drift since the early Palaeozoic; the midoceanic ridge system in several major oceans (like the Atlantic) is orientated North–South and not East–West as expected, therefore sea-floor spreading also seems unlikely as an explanation; a large-scale expansion of the Earth, however, could indeed swamp the effect being looked for (see section A3 below). Hypothesis (e) is held to be true by Dicke (1966), and Goldreich and Toomre (see McKenzie, 1972): the viewpoint of the latter authors has been effectively criticized by McKenzie (1972, p. 355); and Dicke's explanation depends on whole-mantle convection, which is not feasible on several grounds (Wesson, 1970, 1972a, 1972b). Hypothesis (c), involving $\dot{G} \neq 0$, is examined (see section A4) below.

The latter can be estimated by noting that the radius of the globe in colatitude θ is

$$r = R[1 + e(\tfrac{1}{3} - \cos^2 \theta)] \tag{A4}$$

which contains a time-dependent part $e/3$. This changes as the axes of the planet alter to accommodate the spin-down and maintain hydrostatic equilibrium:

$$b = a(1 - e)$$

At an epoch 1·6 Æ ago (1 Æ = 10^9 years), e was 1/210 (Jeffreys, 1970, p. 409), compared to the present value of 1/297. If the Earth condensed out of a solar nebula, it probably rotated once every 5 hours or so, and e (4·5 Æ ago) = 1/13. The Earth accommodates itself to the spin-down by decreasing the radius of curvature of its surface near the equator and increasing it near the poles. The crust is compressed over the equator and is stretched over the poles. Overall (Jeffreys, 1970, p. 409), there is a symmetrical contraction of 10 km or so in radius, with an additional shortening of the equatorial circumference by 18 km (over the last 1·6 Æ) to 1000 km (over the last 4·5 Æ). Clearly, the expected effects are large: the readjustment of the lithosphere would tend, unless nullified by one of the hypotheses (c) or (d), to produce mountain ranges along lines of longitude and rifts along lines of latitude, these effects being most pronounced in region near $|\theta| = \pi/4$.

A3. *No-strain conditions on the radius*

To see what magnitude of size is concerned in the problem, I will derive conditions on the radii of the globe at two different epochs subject to the non-existence of extensive straining at $\theta = \pm 45°$ and elsewhere. To do this, I take variations of equation (A2):

$$\delta U = \frac{GM}{r}\left[-\frac{\delta r}{r} + \frac{3}{2}\left(\frac{a}{r}\right)^2 J_2(3\cos^2\theta - 1)\frac{\delta r}{r} - \frac{1}{2}\left(\frac{a}{r}\right)^2 \delta J_2(3\cos^2\theta - 1)\right] \tag{A5}$$

Roughly speaking, one finds that for $\delta U \equiv 0$ (as needed to avoid any change of potential),

$$\frac{\delta R}{R} = -\gamma \delta J_2 \qquad (A6)$$

where γ is a factor of order unity ($\gamma = -1/2$ for $\theta = \pi/2$; $\gamma = 1/4$ for $\theta = \pi/4$; $\gamma = 1$ for $\theta = 0$). This, even when there is no change in the potential U, there is still a finite change in the radius. This change, by equation (A6), would still be present even if the upper layers of the Earth could adjust to a null energy-change scheme. Unlike the general effects noted before (equation A2: $\delta U \neq 0$ gives circumferential lengthening of 18–1000 km, corresponding to radius changes of about 6–170 km), the changes in radius are now small:

$$\delta b \simeq +3 \cdot 2 \text{ km} \quad \text{(a polar expansion, i.e.)}$$
$$\delta a \simeq -1 \cdot 6 \text{ km} \quad \text{(an equatorial contraction)} \qquad (A7)$$
$$\delta R(\theta = 45°) \simeq +0 \cdot 8 \text{ km} \quad \text{(in regions where } \delta U \equiv 0)$$

since an epoch 1·6 Æ ago; while

$$\delta b \simeq +164 \text{ km}$$
$$\delta a \simeq -82 \text{ km} \qquad (A8)$$
$$\delta R(\theta = 45°) \simeq +41 \text{ km}$$

since an epoch 4·5 Æ ago.

I do not wish to investigate here the mechanism by which gravitational energy from the bulge might be released without producing strains at $\theta = \pm\pi/4$. Changes in radius of the sizes given by equations (A7), (A8) are geophysically non-negligible, and some straining must have taken place in some parts of the globe. A similar argument, which would apply if δU of equation (A5) were extremized, leads to even more drastic changes: the extrema occur at $\theta = 0$ and $\theta = \pi/2$, so that tectonic processes would absorb or release work energy at these places.

To explain the non-observation of the expected rifts requires a no-strain condition at $|\theta| \simeq \pi/4$. The two ways left to accomplish this (A2) are (c) a variation in G, and (d) large changes in the Earth's radius sufficient to swamp out smaller-scale changes of ≈ 1–100 km.

A4. *Secular changes in* G

To take into account both expansion and a change with time of G, from equation (A2) I form

$$\delta U = \frac{M_E}{r}\left[1 - \frac{J_2}{2}\left(\frac{a}{r}\right)^2 (3\cos^2\theta - 1)\right]\delta G - \frac{GM_E}{r^2}\delta r$$
$$+ \frac{3GMJ_2}{2r^2}\left(\frac{a}{r}\right)^2 (3\cos^2\theta - 1)\delta r - \frac{GM_E}{2r}\left(\frac{a}{r}\right)^2 (3\cos^2\theta - 1)\delta J_2 \quad (A9)$$

Gravity and the Earth's Rotation

If the expansion is due entirely to changing G,

$$\frac{\delta r}{r} = -\alpha \frac{\delta G}{G} \qquad (A10)$$

where α is a constant. Defining

$$\beta = \left(\frac{a}{r}\right)^2 (3\cos^2\theta - 1) \qquad (A11)$$

and putting (A10) into (A9) with $\delta U \equiv 0$ gives

$$\frac{\beta}{2}\langle G \rangle \delta J_2 = \left(1 - \frac{\beta}{2}\langle J_2 \rangle + \alpha - \frac{3}{2}\alpha\beta < J_2\right)\delta G \qquad (A12)$$

In the last equation, $\langle\;\rangle$ denotes a mean value. At the present time and to the desired accuracy, for all past epochs, $\langle J_2 \rangle \approx 10^{-3}$ and is a pure number. To a good approximation, therefore,

$$\delta J_2 = (1 + \alpha)\frac{2}{\beta}\frac{\delta G}{\langle G \rangle} \qquad (A13)$$

It is immediately obvious from equation (A13) that a monotonic secular change in G cannot produce a uniform effect on J_2 and continue to satisfy the no-strain criterion, since β changes sign at some point near $|\theta| = 55$ degrees (by (A11), β is such that $-1 \leq \beta \leq 2$). What this means is that higher terms of J_n ($n > 2$) need to be taken into account if a no-strain condition is to hold. But it can be verified that the inclusion of higher harmonics is of little consequence as regards the energy budget involved. The general form of the result equation (A13) still holds when a variational approximation is made which includes a term expressing small changes in planetary mean radius of non-gravitational origin; I have verified this by a somewhat tedious calculation up to terms of order $(J_2)^2$ following the method of equation (A20) below. However, by arbitrarily picking the expected regions of maximum strain ($\theta = \pm\pi/4$) and imposing no-strain conditions at these areas, it is possible to make a first approximation to the desired goal of explaining a tectonic null observation.

It is necessary, as it was implicitly in deriving equations (A7), (A8) from (A6), to know J_2 in terms of e and Ω (the angular velocity of ration; $e \propto \Omega^2$). In fact

$$J_2 = \tfrac{2}{3}J \qquad (A14)$$

where

$$J = e[\text{1st order}] - \tfrac{1}{2}m \qquad (A15)$$

and

$$m \equiv \frac{\Omega^2 a^3(1-e)}{GM_E} \qquad (A16)$$

In equation (A16), M_E is the mass of the Earth, which is assumed to be fixed ($M_E = 5\cdot977 \times 10^{27}$ gm). Substituting for numerical values (a[now] = 6378·39 km, Ω[now] = $7\cdot29 \times 10^{-5}$/sec) and collecting together equations (A14), (A15), (A16) gives

$$J_2(1\cdot6 \, \text{Æ ago}) = 1560 \times 10^{-6}$$

at an epoch 1·6 Æ ago, when the planet rotated in 0·84 of our present days (this is derived by extrapolating the lunar secular acceleration; see Jeffreys, 1970, pp. 170, 409). Similarly, at formation, when the Earth rotated in about 5 hours,

$$J_2(4\cdot5 \, \text{Æ ago}) = 26600 \times 10^{-6}$$

The present theoretical value of J_2 based on hydrostatic theory is $1072\cdot1 \times 10^{-6}$, while the observed J_2 is slightly larger at $1082\cdot6 \times 10^{-6}$. The difference, which is probably connected with relaxation in the mantle, and not, as was once thought, with ice ages (McKenzie, 1966), is negligible for the accuracy being aimed at here. (The result given below is exact to within a factor of about 2.) From the above figures one gets

$$\delta J_2(1\cdot6 \, \text{Æ}) = -488 \times 10^{-6} \tag{A17a}$$

$$\delta J_2(4\cdot5 \, \text{Æ}) = -25{,}500 \times 10^{-6} \tag{A17b}$$

To evaluate $\delta G/\langle G \rangle$, I shall use $\alpha = 0\cdot1$ in (A13), following Dicke (1962). At the latitudes of maximum strain ($\theta = \pm\pi/4$) this gives

$$\frac{\delta G}{\langle G \rangle} = 0\cdot23 \, \delta J_2 \tag{A18}$$

Employing equations (A17a, b) in (A18) and dividing by $1\cdot6 \times 10^9$ and $4\cdot5 \times 10^9$ respectively gives

$$\frac{\dot{G}}{\langle G \rangle} \simeq 0\cdot07 \times 10^{-12}/\text{year} \quad (\text{over } 1\cdot6 \text{ Æ}) \tag{A19a}$$

$$\frac{\dot{G}}{\langle G \rangle} \simeq 1\cdot30 \times 10^{-12}/\text{year} \quad (\text{over } 4\cdot5 \text{ Æ}) \tag{A19b}$$

These are the rates of change of G needed to satisfy the no-strain condition at $\theta = \pm\pi/4$.

Indirect geophysical rotation arguments (Morrison, 1973) have given the limit $|\dot{G}/G| \lesssim 2 \times 10^{-11}$/year. The previous best direct upper limit of the rate of change of G with time is that of Shapiro *et al.* (1971), who used planetary tracking data to obtain $|\dot{G}/G| \lesssim 4 \times 10^{-10}$/year. The values of equation (A19) are considerably smaller than this, and it is an effect of the size of (A19a) which is needed to give no-strain at $\theta = \pm\pi/4$. The difference between (A19a) and (A19b) reflects ignorance about the origin of the Earth (A19b) and about the rate of tidal dissipation of energy by the Moon in the remote past (A19a). I would expect that (A19a) is the more firmly based of the two estimates. Both (A19a) and (A19b)

Gravity and the Earth's Rotation

are weaker changes of G with time than those deduced from the theories of Brans–Dicke (Dicke, 1962), and Hoyle–Narlikar (1971), which are of order 10^{-11}/year and 10^{-10}/year respectively. However, it is likely that (A19a) and (A19b) represent *lower* bounds if one is determined to invoke only changes of G to be consistent with tectonic observations. A very strong objection to adopting this philosophy of explanation (c) is to be found in the change of sign of β in (A13) which cannot be removed from the theory of the deformation if produced solely by changes in G and by adjustment of J_2 only. I think this forms a strong motive for considering explanation (d) of section A2, i.e. large changes in radius, to be, by default, the most likely hypothesis to adopt.

A5. *Secular global expansion*

Thus, having ruled out alternatives (a), (b) and (c), (e), I am left with (d) as a possible way of explaining the non-observation of the features expected, following spin-down. I recently made a comprehensive study of global expansion and the hypothesis that $\dot{G} \neq 0$, finding (Wesson, 1973) that there is definite evidence in favour of a growth of the radius of the Earth at about 0·6–0·7 mm/year. This conclusion rests on about 20 determinations of the rate of increase of the Earth's radius, by different workers and spread over all of geological history. This growth, I would emphasize, probably has nothing whatsoever to do with a varying G: the needed overall expansion in 4·5 Æ from a planet that was originally *half* its present size is much too large an effect to be predicted from any of the $\dot{G} \neq 0$ theories (e.g. Brans–Dicke (Dicke, 1962) predict an expansion of 180 km, while Hoyle–Narlikar (1971) predict 500 km. Both are small compared to the indicated expansion of $\simeq 3000$ km). The conclusion, that the Earth is expanding at $\simeq 0.66$ mm/year, is statistically significant, on the basis of empirical estimates, and implies some cosmological connexion, probably the general relativistic expansion of space expected from Hubble's law (see MacDougal *et al.*, 1963). This interpretation of the data has since been verified at least in part by Weinstein and Keeney (1973), using a geochronological method.

The question now is: does G vary *too*, or is the expansion an unrelated phenomenon? Any considerable increase in the Earth's radius would account for the lack of spin-down effects, but certain theories, as mentioned, deduce the expansion as being caused by a secular decrease of G. In as far as these theories predict values for $\dot{G}/G$ which adequately satisfy the no-strain lower limits of (A19), and which also produce some global expansion, they are acceptable. An optimist or a friend of the Jordan–Brans–Dicke–Hoyle–Narlikar school might, indeed, look upon equation (A19) as evidence in support of these theories. On the other hand, I feel inclined to use Occam's razor and veto this attitude on the grounds that it introduces an unnecessary hypothesis: viz, the variation of G. All that is required is a large expansion, sufficient to swamp the spin-down effects: this can be obtained from general relativity (MacDougal *et al.*, 1963), and recent data (Weinstein and Keeney, 1973; Wesson, 1973) strongly support it.

The necessity of (d) can be seen by varying equation (A2) with account taken of a change in the mean radius:

$$\delta U \simeq (1+\alpha)\frac{\delta G}{G} - \frac{\delta R}{R} - \frac{\beta}{2}\delta J_2 \qquad (A20)$$

The changes in R must be bigger than the 180 km given by Dicke's theory and preferably larger than the 500 km of the Hoyle–Narlikar theory.

A6. Conclusion

To recap on the above calculations: (i) the Earth is undergoing spin-down, leading to (ii) secular change in the ellipticity of the planet which shows up as (iii) a secular change in J_2. This secular change causes (iv) changes in the potential U which should (v) produce tension along lines of longitude and compression along lines of latitude, leading to the expectation (vi) of finding longitude-aligned fold mounts and latitude-aligned rifts, with enhancement of the geothermal flux and seismic energy release near midlatitudes ($\theta = \pm\pi/4$). (vii) These are not preferentially present over tectonic features tending in other directions, so that either (c) G is changing with time (at $|\dot{G}/G| \gtrsim 10^{-13} - 10^{-12}/\text{year}^{-1}$) and/or (d) there have been very large changes ($\gtrsim 1000$ km) in the Earth's radius in the past. Of the last group of hypotheses, that of global expansion with no change of G is intrinsically the more sound. The expansion may be an effect of general relativity (MacDougal et al., 1963) or the result (Dicke, 1962; Hoyle and Narlikar, 1971; Weinstein and Keeney, 1973) of cosmology.

References

Bertotti, B. (1962). The theory of measurement in general relativity, pp. 174–201. In *Proceedings of the XX International School of Physics (Enrico Fermi)*, Academic Press, 264 pp.

Bertotti, B. (1964). Discussion, p. 91. In *Relativistic Theories of Gravitation* (Ed. L. Infeld), Pergamon Press, 379 pp.

Bertotti, B., Brill, D. and Krotkov, R. (1962). The theory of measurements in general relativity, pp. 1–48. In *Gravitation: An Introduction to Current Research* (Ed. L. Witten), Wiley Interscience, 481 pp.

Birch, F. (1967). On the possibility of large changes in the Earth's volume. *Phys. E. Planet. Ints.*, 1, 141–147

Brans, C. and Dicke, R. H. (1961). Mach's Principle and a relativistic theory of gravitation. *Phys. Rev.* 124, 925–935

Dicke, R. H. (1961). Dirac's cosmology and Mach's Principle. *Nature*, 192, 440–441

Dicke, R. H. (1962a). Mach's Principle and invariance under transformation of units. *Phys. Rev.*, 125 (6), 2163–2167

Dicke, R. H. (1962b). Gravitation without a principle of equivalence. *Rev. Mod.Phys.*, 29, 363–376

Dicke, R. H. (1964a). Experimental relativity, pp. 165–314. In *Relativity, Groups and Topology* (Ed. C. and B. DeWitt), Gordon and Breach, London, 929 pp.

Dicke, R. H. (1964b). *The Theoretical Significance of Experimental Relativity*, Blackie and Sons, London, 153 pp.

Dicke, R. H. (1966). The secular acceleration of the Earth's rotation and cosmology, pp. 98–164. In *The Earth–Moon System* (Ed. B. G. Marsden and A. G. W. Cameron), Plenum Press, New York, 288 pp.

Dirac, P. A. M. (1937). The cosmological constants. *Nature*, **139**, 323

Dooley, J. C. (1973). Is the Earth expanding? *Search*, **4**, 9–15

Einstein, A. (1950). *The Meaning of Relativity*, p. 102, Princeton University Press, Princeton, New Jersey, U.S.A.

Finzi, A. (1962). Tests of possible variations in the gravitational constants by the observation of white dwarfs within galactic clusters. *Phys. Rev.*, **128** (5), 2012–2015

Guinot, B. (1970). Short period terms in universal time. *Astron. and Astrophys.*, **8**, 26–28

Harwit, M. (1971). A possible observational test of determine variations of Planck's constant on a cosmic scale. *Bull. Ast. Inst. Czech.*, **22**, 22–29

Hipkin, R. G. (1970). P. 54 in *Palaeogeophysics* (Ed. S. K. Runcorn), Academic Press Ltd., New York

Hoyle, F. and Narlikar, J. V. (1971). On the nature of mass. *Nature*, **233**, 41–44

Jeffreys, Sir H. (1970). *The Earth*, 5th ed., C.U.P. Cambridge, 542 pp.

Jordan, P. (1949). Formation of the stars and development of the universe. *Nature*, **164**, 637–640

Jordan, P. (1955). *Schwerkraft und Weltall*, 2 Aufl., Braunschweig, F. Vieweg

Kapp, R. O. (1960). *Towards a Unified Cosmology*, Hutchinson, London

Klein, O. (1958). Some considerations regarding the earlier development of the system of galaxies. In *Institut International de Physique Solway*, 11ᵉ Congres de Physique, R. Stoops, Brussels

MacDonald, G. J. F. (1964). The mechanical properties of the Earth, pp. 203–245. In *Advances in Earth Science* (Ed. P. M. Hurley), M.I.T. Press, Cambridge, Mass. U.S.A., 326 pp.

MacDougal, J., Butler, R., Kronberg, P. and Sandquist, A. (1963). A comparison of terrestrial and universal expansion, *Nature*, **199**, 1080

McKenzie, D. P. (1966). The viscosity of the lower mantle. *Journ. Geophys. Res.*, **71**, 3995–4010

KcKenzie, D. P. (1972). Plate tectonics, pp. 323–360. In *The Nature of the Solid Earth* (Ed. E. C. Robertson), McGraw-Hill, London, 677 pp.

Milne, E. A. (1935). *Relativity, Gravitation and World Structure*, Oxford Univ. Press, 364 pp.

Morrison, L. V. (1973). Rotation of the Earth from AD 1663–1972 and the constancy of G. *Nature*, **241**, 519–520

Newton, R. R. (1968). A satellite determination of tidal parameters and Earth deceleration. *Geophys. J. R. astr. Soc.* **14**, 505–539

Prokhovnik, S. J. (1970). Cosmological theory of gravitation. *Nature*, **225**, 359–361

Rochester, M. G. (1973). The Earth's rotation, in 2nd GEOP Research Conference on the Rotation of the Earth and Polar Motion (Ohio State University, Columbus, Feb. 8–9, 1973). Trans. Americ. Geophys. Union. Aug. 1973

Runcorn, S. K. (1968). Fossil bivalve shells and the length of month and year in the Cretaceous. *Nature*, **218**, 459

Sciama, D. W. (1953). On the origin of inertia. *M.N.R.A. S.*, **113**, 34–42

Shapiro, I. I., Smith, W. B., Ash, M. B., Ingalls, R. E. and Pettingill, G. H. (1971). Gravitational constant: experimental bound on its time variation. *Phys. Rev. Lett.*, **26**, 273–80

Steiner, J. (1967). The sequence of geological events and the dynamics of the milky way galaxy. *J. Geol. Soc. Australia*, **14**, 99–131

Weinstein, D. H. and Keeney, J. (1973). Apparent loss of angular momentum in the Earth–Moon system. *Nature*, **244**, 83–84

Wesson, P. S. (1970). The position against continental drift. *Q. Jl. R. astr. Soc.*, **11**, 312–340

Wesson, P. S. (1972a). Objections to continental drift and plate tectonics. *J. Geol.*, **80**, 185–197

Wesson, P. S. (1972b). Modified lomnitz vs. elasticoviscous law and problems of the new global tectonics. *B.A.A.P.G.*, **56** (11), 2127–49

Wesson, P. S. (1973). The implications for geophysics of modern cosmologies in which G is variable. *Q. Jl. R. astr. Soc.*, **14**, 9–64

Wheeler, J. A. (1962). *Geometrodynamics*, Academic Press, New York, 334 pp.

Will, C. M. (1971a). Relativistic gravity in the solar system. I. Effect of an anisotropic gravitational mass on the Earth–Moon distance. *Ap. J.*, **165**, 409–412

Will, C. M. (1971b). Relativistic gravity in the solar system. II. Anisotropy in the Newtonian gravitational constant. *Ap. J.*, **169**, 141–155

DISCUSSION

(Paper delivered *in absentia* by Professor Runcorn)

HIPKIN: How small must $\dot{G}/G$ be before it is irrelevant to the problem of changes in l.o.d. and month?

RUNCORN: It is a difficult question to answer and it depends on the age of the universe. Assuming the age was 10,000–20,000 My, then since some time in Mid Precambrian, say 2000 My ago, a change in G of 5–10% would be in order.

KEENEY: The present value of $\dot{G}/G = 10^{-11}$/year.

RUNCORN: We need to determine $\dot{G}/G$ more accurately. The question is, does 5–10% change in G from mid Precambrian have any geological or palaeontological effect?

PANNELLA: Stromatolites might contain information on a value of G significantly different from the modern value. Perhaps these changes would be recognized as a temperature effect.

MULLER: As for changes in the earth's radius accompanying changing G, satellite data rules out changes in earth radius of 50 cm/year.

HIPKIN: Satellite gravity measurements are able to determine radius changes only if the change exceeds about 1 mm/year.

PALAEONTOLOGY AND THE DYNAMIC HISTORY OF THE SUN–EARTH–MOON SYSTEM

D. H. WEINSTEIN AND J. KEENEY

The Superior Oil Company, Houston, Texas

Abstract

We have recently suggested that Hubble's Law in a generalized sense implies the existence of a cosmological drag force of magnitude Hp on any moving particle of momentum p. Here H is Hubble's constant ($\sim 10^{-18}$/sec). We have also indicated that the dynamic behaviour of the Sun–Earth–Moon system may be noticeably influenced by this force.

However, other parameters, namely the gravitational constant, the Earth's moment of inertia and the length of the year all play a part in determining the apparent motion of the system. Our purpose here is to review the effects of variations in these constants in the light of the palaeontological evidence, Hubble's Law and the emerging results of Shapiro's radar measurements of the period of Mercury.

The possibility of perceiving cosmological influences on the Earth has been advanced from time to time (Dicke, 1962; Runcorn, 1969; Wesson, 1973). In this connection, particular attention has been paid to Dirac's idea (Dirac, 1938) that the gravitational constant G should decrease with time because of cosmological expansion.

Recently, another type of cosmological influence has been suggested (Weinstein and Keeney, 1973b) which may have common roots with the origin of inertia (Mach's Principle). It was inferred by generalizing Hubble's law (Weinstein and Keeney, 1974). This expression, which can be written as (Geller and Peebles, 1972)

$$\frac{dv}{dl} = -H\frac{v}{c} \qquad (1)$$

relates the observed change of photon frequency dv with the change in distance dl from astronomical sources, where H is Hubble's constant, v the photon frequency, and c the velocity of light.

Transformation of equation (1) reveals a photon-particle symmetry which suggests that every particle possessing momentum $\bar{p}$ experiences a force $\bar{F}_H$ opposing its motion given by

$$\bar{F}_H = -H\bar{p} \qquad (2)$$

In an effort to determine if such a force does exist, we have turned to the Sun–Earth–Moon system and the planets, particularly Mercury, as perhaps being capable of yielding some information on this matter. Runcorn (1969) has already introduced the palaeontological method in considering the response of the Sun–Earth–Moon system over the last half aeon to changes in G and the Earth's moment of inertia, but changes in the orbital eccentricities of the Earth and Moon were considered negligible. We include here these eccentricities, which appear to be important over the longer time intervals which have more recently become accessible. In this case the equations of motion are

$$\frac{d}{dt}[M_E(GM_s)^{2/3} \omega_E^{-1/3}(1-e_E^2)^{1/2}] = \tau\beta - H[M_E(GM_s)^{2/3} \omega_E^{-1/3}(1-e_E^2)^{1/2}] \quad (3)$$

$$\frac{d}{dt}(I_E \Omega_E) = -(1+\beta)\tau - HI_E \Omega_E \quad (4)$$

$$\frac{d}{dt}[M_L(GM_E)^{2/3} \omega_L^{-1/3}(1-e_L^2)^{1/2}] = \tau - H[M_L(GM_E)^{2/3} \omega_L^{-1/3}(1-e_L^2)^{1/2}] \quad (5)$$

where

M_s = solar mass
M_E = Earth mass
M_L = lunar mass
I_E = Earth's moment of inertia
ω_E = mean Earth orbital angular velocity
Ω_E = mean Earth spin angular velocity
ω_L = mean lunar orbital angular velocity

G = gravitational constant
H = Hubble's constant
τ = lunar tidal couple
β = ratio of solar to lunar tidal couples
e_E = Earth orbit eccentricity
e_L = lunar orbit eccentricity.

Equations (3) and (5) refer, respectively, to the motion of the Earth about the Sun and the motion of the Moon about the Earth, while equation (4) describes the Earth's spin about its axis of rotation. Kepler's third law has been incorporated in equations (3) and (5). The small effect of rotation of the Earth–Moon about its centre of mass is neglected, as is the lunar spin.

The motions of the system are reflected in the tidal and seasonal variations of the daily growth increments of certain fossil organisms. Accordingly, in terms of the number N of solar days/year and n/synodic month, we write

$$\Omega_E = \omega_E(N+1) \quad (6)$$

$$\omega_L = \omega_E \left(\frac{N}{n} + 1\right) \quad (7)$$

To gain some insight into the behaviour of the system in terms of the data set (N, n) we neglect, for the moment, H, β, e_E, and e_L, and assume G and I_E constant.

If we now add equations (4) and (5), incorporate the relations (6) and (7), and integrate, we have

$$I_E(N - N_0) + M_L(GM_E)^{2/3} \omega_E^{-4/3} \left[\left(\frac{N}{n} + 1 \right)^{-1/3} - \left(\frac{N_0}{n_0} + 1 \right)^{-1/3} \right] = 0 \quad (8)$$

which expresses the conservation of angular momentum under the assumptions mentioned, but with dissipative Earth–Moon tidal interaction. Subscript zero refers to an arbitrary zero of time.

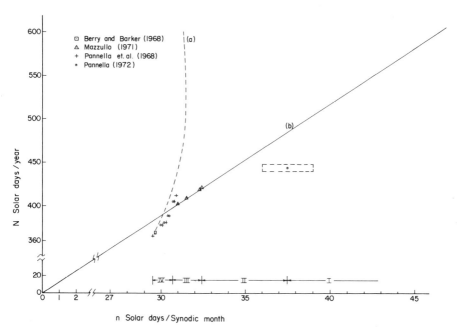

FIGURE 1. The fossil record of the number of solar days N/year and n/month over the past 1·8 aeons. Average values of n and N, corrected upward by 1%, are shown. Limits of error are given for the Precambrian stromatolite (1·8 aeons in age, $n = 36$–39) since it appears to be somewhat more uncertain than the other data. Curve (a) is a graph of equation (8) and represents the conservation of angular momentum, with Earth–Moon tidal interaction, but with I_E and G constant and neglecting solar tides, orbit eccentricities and Hubble's constant. Line (b) is a curve of constant N/n and is a graph of equation (11) under the conditions of no tidal interaction and constant lunar eccentricity.

Apparent trends in the data suggested the division of the graph into intervals I–IV. Since N and n decrease with the passage of time, a crude time-scale is provided as follows:

II—early Precambrian (age 1·8 aeons) to Silurian (0·4 aeons);
III—Silurian to Pennsylvanian (0·3 aeons);
IV—Pennsylvanian to recent.

Interval I contains no data

For comparison, the available fossil data (Pannella, 1972; Pannella et al., 1968; Mazzullo, 1971; Berry and Barker, 1968) are plotted in Figure 1 along with a graph of equation (8). The data cover several geological ages ranging from Precambrian (Gunflint, age 1·8 aeons) to recent. In general, N and n both decrease with the passage of time, the smallest values thus occurring at present. On the graph of equation (8), n reaches a maximum (Miller, 1969; Weinstein and Keeney, 1973a) of 31·4 when $N \simeq 500$–550, a feature not exhibited by the data. Instead, n seems to depart increasingly from this curve as we go back in time. The assumptions in equation (8) are evidently not all satisfied and so we now proceed to examine the system in more detail.

We consider the lunar motion first. From equation (5), we find

$$\frac{\dot{u}}{u} = \frac{2\dot{G}}{G} + 3H - \frac{\dot{\omega}_E}{\omega_E} - \frac{3e_L \dot{e}_L}{1 - e_L^2} - \frac{3\tau}{L_L} \tag{9}$$

where $u = N/n + 1$ and L_L is the lunar angular momentum.

To simplify the expression, we make use of equation (1) for the Earth. From equation (1) then

$$\frac{\dot{\omega}_E}{\omega_E} = \frac{2\dot{G}}{G} + 3H - \frac{3e_E \dot{e}_E}{1 - e_E^2} - \frac{3\tau\beta}{L_E} \tag{10}$$

where L_E is the Earth's orbital angular momentum.

By substitution of equation (10) into equation (9), we find

$$\frac{\dot{u}}{u} = \frac{3e_E \dot{e}_E}{1 - e_E^2} - \frac{3e_L \dot{e}_L}{1 - e_L^2} - 3\tau\left(1 - \frac{\beta L_L}{L_E}\right) \tag{11}$$

which is independent of G, H and I_E. As $\beta L_L/L_E \lesssim 10^{-6}$, we neglect this term.

We now consider the data plotted in Figure 1 in terms of equation (11). Since the data of interval IV do not deviate substantially from curve (a), we turn to interval III. A portion of the data in this interval appears to lie on a straight line through the origin, with the result that $\dot{u}/u$ is zero. If $e_E \dot{e}_E$ and $e_L \dot{e}_L$ are essentially zero, as seems likely in this relatively recent era, then it follows that τ is also zero and that, hence, there was no tidal dissipation. This situation is illustrated in Figure 1 by the straight line (b).

Here an interesting consequence emerges. The conditions indicated in region III allow a solution of equation (4) for the Earth. Taking into account equation (10), we find

$$-\frac{\overline{(N+1)}}{(N+1)} = \frac{2\dot{G}}{G} + \frac{\dot{I}_E}{I_E} + 4H \tag{12}$$

Runcorn (1964) has found little change in I_E since the Devonian epoch. We therefore neglect $\dot{I}_E$ over interval III which ranges from Silurian to Pennsylvanian, a span of roughly 0·1 aeon. Also neglecting $\dot{G}/G$, which we will discuss shortly, we find from equation (12) that $H \simeq 4 \times 10^{-18}/\text{sec}$. This is approximately two to three times the astronomical value ($1\cdot3^{+0\cdot8}_{-0\cdot4} \times 10^{-18}/\text{sec}$) (Branch and Patchett, 1973) but the scarcity of data and the attendant uncertainties must be born in mind.

In interval II we note that the Gunflint point lies below the extension of the line (b) defined in interval III. This seems anomalous. Here $\dot{u}/u$ is positive. The present small size of e_E suggests that it was either zero or decreasing in this interval. Consequently, equation (11) can only be satisfied by a decrease in e_L with passing time. This implies, in the absence of torque, an eccentricity e_{L0} of about 1/4 at the beginning of the interval (Gunflint). If the Precambrian torque had its estimated present value ($3\cdot9 \times 10^{23}$ dyne cm) (Munk and MacDonald, 1960), then e_L would have been about 4/10, or more for a larger torque. Taking $e_{L0} = 1/4$ and ignoring all other parameters, calculation of H by means of equations (3)–(5) over the 1·8 aeon interval to the present gives $H \simeq 1/2 \times 10^{-18}/\text{sec}$.

To accomplish the reduction in e_L some mechanism is required. Goldreich (1963) and Urey et al. (1959) have discussed changes in orbital eccentricity resulting from dissipative tidal forces. On the basis of their consideration, decreases in e_L cannot be ruled out, but neither do they appear to be calculable at present.

On the other hand, a value of $N \simeq 470$, some 5% greater than the given value of 440–448, would put the Gunflint point on the line (b). In this case, e_{L0} could be zero and calculation would give $H \simeq 1 \times 10^{-18}/\text{sec}$. However, we cannot conclude from equation (11) that e_{L0} is necessarily zero.

Thus, while the dynamic behaviour of the Sun–Earth–Moon system does not conclusively demonstrate the existence of F_H, it is not inconsistent with it. We have already shown elsewhere (Weinstein and Keeney, 1973a) that, if H (as well as e_L) is neglected, then the conservation of momentum requires that, since the Gunflint epoch, either I_E has increased by about 30% or that G has increased correspondingly. An increase in G is not expected (Dirac, 1938). Taking account of e_L we find from equations (3)–(5) that a decrease from about 1/3 in the Gunflint to the present zero would also satisfy the conservation of momentum. However, these observations are based on one Precambrian datum point. Clearly, more data would be helpful.

Some climatic consequences of F_H are to be expected. Dicke (1962) has already considered temperature changes in response to changes in G. He finds that with $\dot{G}/G \simeq -3 \times 10^{-11}/\text{year}$, the Earth might have been about 60 K warmer 3 aeons in the past than it is now. On the other hand, with the same assumptions relating the Earth's orbital radius to the surface temperature T, we find $\dot{T}/T = H$ if G is neglected. Consequently, 3 aeons in the past the temperature could have been some 30 K lower than now. Again we see the opposing effects of G and H.

Shapiro's radar measurements (Shapiro, 1972) of the rate of change of the orbital period of Mercury are of interest here. Equation (3) applies to Mercury as well as to the Earth. If ω_M is the mean orbital angular velocity of Mercury, M_M the mass, and e_M the eccentricity,

$$\frac{\dot{\omega}_M}{\omega_M} + \frac{3 e_M \dot{e}_M}{1 - e_M^2} = \frac{2\dot{G}}{G} + 3H - \frac{3\tau_M}{L_M} \tag{13}$$

where τ_M represents a possible solar tidal torque and L_M the orbital angular momentum. Thus, measurement of $\dot{\omega}_M$ and $\dot{e}_M$ for Mercury gives a measure of the combined effects of H and $\dot{G}/G$ with some uncertainty introduced by the possibility of a solar tidal torque.

The measurements of $\dot{\omega}_M$ are in progress, but Shapiro presently indicates that $\dot{\omega}_M \sim 0$, and he finds 'no evidence for a time variation of the gravitational constant, the magnitude of the estimate of $\dot{G}/G$ being only a small fraction of the formal standard error, 1.5×10^{-10}/year'. This determination, however, does not appear to include a measurement of $\dot{e}_M$ of sufficient precision (Shapiro, personal communication), nor was F_H taken into account.

Shapiro (Counselman and Shapiro, 1968) has pointed out that e_M is an adiabatic invariant under changes in G, but this does not prevent changes in e_M due to other causes. The importance of $\dot{e}_M$ may be seen in equation (13). On the right-hand side $\dot{G}$ and $-\tau_M/L_M$ are expected to both be negative or zero, leaving only H positive. Therefore, taking $\dot{\omega}_M = 0$, a positive value of $\dot{e}_M$ would imply the existence of H. If $\dot{e}_M$ is negative or zero, no conclusion can be drawn. We would expect at most $\dot{e} \simeq H/e_M \simeq 2.5 \times 10^{-10}$/year, and this would be reduced by $\dot{G}/G$ or τ_M.

In this connection, Morrison (1973) has concluded from lunar occultations of stars, that $|\dot{G}/\dot{G}| \leq 2 \times 10^{-11}$/year. Also we note that Barnothy and Tinsley (1973) have presented an argument against decreases in $\dot{G}$ of the same order on the basis of observations of the magnitudes and temperatures of distant galaxies.

References

Barnothy, J. M. and Tinsley, B. M. (1973). *Astrophys. J.*, **182**, 343–349
Berry, W. B. N. and Barker, R. M. (1968). *Nature*, **217**, 938
Branch, D. and Patchett, B. (1973). *Monthly Notices Roy. Astronom. Soc.*, **161**, 71–83
Counselman, C. C., III and Shapiro, I. I. (1968). *Science*, **162**, 352
Dicke, R. H. (1962). *Science*, **138**, 3451, 653–664
Dirac, P. A. M. (1938). *Proc. Roy. Soc.* (Lond.), Ser. A, **165**, 199–208
Geller, M. J. and Peebles, P. J. E. (1972). *Astrophys. J.*, **174**, 1
Goldreich, P. (1963). *Monthly Notices Roy. Astronom. Soc.*, **126**, 3, 257–268
Mazzullo, S. J. (1971). *Geol. Soc. Amer. Bull.*, **82**, 1085
Miller, R. H. (1969). *Science*, **164**, 67–68
Morrison, L. V. (1973). *Nature*, **241**, 519–520

Munk, W. H. and MacDonald, G. J. F. (1960). *The Rotation of the Earth*, Cambridge University Press, London
Pannella, G. (1972). *Astrophysics and Space Science*, **16**, 212–237
Pannella, G., MacClintock, C. and Thompson, M. N. (1968). *Science*, **162**, 792
Runcorn, S. K. (1964). *Nature*, **204**, 823–825
Runcorn, S. K. (1969). A palaeontological method of testing the hypothesis of a varying gravitational constant. In *The Application of Modern Physics to the Earth and Planetary Interiors* (Ed. S. K. Runcorn), Wiley, London
Shapiro, I. I. (1967). Private communication. See also, *Astronom. J.*, **72**, 3, 338–350 (1967)
Shapiro, I. I. (1972). *General Relativity and Gravitation*, **3** (2), 135–148
Urey, H. C., Elsasser, W. M. and Rochester, M. G. (1959). *Astrophys. J.*, **129**, 842
Weinstein, D. H. and Keeney, J. (1973a). *Nature*, **244**, 83–84
Weinstein, D. H. and Keeney, J. (1973b). *Lettere al Nuovo Cimento*, **8** (5), 299–302
Weinstein, D. H. and Keeney, J. (1974). Generalization of Hubble's law, *Nature*, **247**, 140
Wesson, P. S. (1973). *Quart. J. Roy. Astronom. Soc.*, **14**, 9–64

DISCUSSION

MOHR: What degree of eccentricity does the Moon's orbit need in order for Pannella's data to fall on the curve of conservation of angular momentum, assuming a nearly circular orbit?

KEENEY: An eccentricity of 0·25 and eccentricity increases as the lunar torque increased.

ASTRONOMICAL EVIDENCE OF CHANGE IN THE RATE OF THE EARTH'S ROTATION AND CONTINENTAL MOTION

E. PROVERBIO AND A. POMA
International Astronomical Latitude Station, Cagliari, Italy

Abstract

The existence of secular movements in the Earth's crust is emphasized by geological and astronomical measurements. The hypothesis of the displacement of large plates of the Earth's surface could, however, also be considered valid for periodical movements (for example annual) to which these same large plates seem subjected. Evidence has recently been derived from time observations that indicate relative periodical motion of continental blocks (Pavlov, 1968). The astronomical time-scales of sixteen Astronomical Observatories related to the BIH time-scale and subdivided in six groups lying on the same plate have been considered by the authors. The comparisons between the averaged time-scales corresponding to these six groups during the same year and the comparisons of the variations in the rate of the Earth's rotation for different years are carried out. The results suggest the existence of a connection between variations in the Earth's rotation speed and longitude variations caused by deformations in the Earth's crust; both variations can be attributed to atmospheric circulation and to tidal forces.

It is well known that the rate of the Earth's rotation is subject to secular and periodical variations due to tidal forces and geophysical and meteorological change in angular momentum.

Together with these variations in the Earth's rotation as a whole there is much evidence that the Earth's crust is wandering under the action of different forces. On the basis of the existence of various plates in the Earth's crust according to Le Pichon's theory and emphasizing that elastic and/or plastic deformation is possible between different plates and the underside of the lithosphere, one cannot disregard the possibility that small relative movements due either to tidal forces or to mechanical stresses on the surface of the Earth may take place between different plates. According to the equation of the Earth's motion about the axis of rotation given by Munk and MacDonald (1960), the excitation function causing variation in the rate of the Earth's rotation (excluding the atmosphere and disregarding changes in the moment of inertia) is defined substantially by torques depending on wind stress and tidal forces acting upon the surface of the solid Earth. According to the theory of plate tectonics, there is no reason for denying that the same forces may cause periodical and irregular movements in the relatively movable lithospheric fragments of which we suppose the Earth's crust to be composed. Mintz (1951) finds that the magnitude of wind stress on the surface of the Earth between 35° N and 90° N is of the order of 1 dyne/cm². This pressure is not sufficient to produce elastic deformation of the Earth's crust (Ljustih, 1962) large enough to be revealed by astronomical observations. But if the Earth's surface is considered as subdivided into single plates, one moving another on account of elastic-viscous deformations owing to continuous stresses, then in this case these movements can be observed. Very recently, Knopoff and Leeds (1972) calculated the relative lithospheric motion of ten plates depending on the dynamical effect of deceleration in the Earth's rotation relative to an inertial frame of reference. The mean axial velocity component of this lithospheric drift was found to be 2·3 cm/year. This drift is not negligible as regards the spreading secular angular velocity of the Earth's plates, the existence of which was pointed out by geological and geophysical data as well as by the analysis of latitude and longitude astronomical observations (Proverbio and Quesada, 1973a,b, 1974). The secular deceleration in the Earth's rotation is about $5·5 \times 10^{-8}$ sec/day², while the seasonal acceleration and deceleration deduced from the variations in the Earth's rotation given by the Bureau International de l'Heure by means of the standard relation

$$\Delta T = UT2 - UT1 = 0^s \cdot 025 \sin \omega(t + 336) + 0^s \cdot 009 \sin 2\omega(t + 66)$$

can be expressed in units of sec/day² by the equation

$$d^2 \Delta T/dt^2 = (-7\cdot3 \sin \omega(t + 336) - 11\cdot1 \sin 2\omega(t + 66)) \times 10^{-6}$$

where $\omega = 2\pi/365$, we can see that this last effect is 10^2 times larger than the secular deceleration; consequently the torque responsible for the periodical variations in the Earth's rotation applied to irregularly fragmented regions of the Earth might produce periodical and irregular motions in the lithosphere corresponding to the mechanical deformations caused by wind stress on the Earth's surface. Evidence of a relation between seasonal variations in the Earth's rotation and longitude variations was emphasized by Stoyko (1931, 1932, 1942, 1962). More recently Pavlov (1968) pointed out the existence of relative motions of continental blocks derived from time observations and Bostrom (1932) found irregular and westward displacements varying regionally in the lithosphere. These results show that the hypothesis of the displacement of large plates of the Earth's surface correlative with periodical and irregular change in the Earth's rotation cannot be disregarded. In order to study the longitudinal movement of continents relative to one another, we have considered the astronomical time-scales of sixteen Astronomical Observatories related to the BIH time-scale and have subdivided them into six groups according to geographical position. The distribution of the sixteen stations is reported in Table 1. This table gives the geographical coordinates (longitude and latitude) of each observatory, the type of astronomical instrument utilized in time determination: visual transit instru-

TABLE 1

Station	Longitude	Latitude	Instrument	EQM	Weight
GROUP I					
S. Fernando	6 12	36 28	VTI	15·8	4
Greenwich	−0 20	50 52	PZT	5·7	31
Paris	−2 20	48 50	AST	8·1	15
Milan	−9 11	45 28	VTI	10·9	8
Hamburg	−10 14	53 29	PZT	6·9	21
GROUP II					
Riga	−24 07	56 57	PTI	7·9	16
Pulkovo	−30 20	59 46	PTI	5·8	29
Nikolaev	−31 58	46 58	PTI	9·8	10
Moscow	−37 34	55 45	PTI	5·7	31
GROUP III					
Tashkent	−69 18	41 20	VTI	8·0	16
Novosibirsk	−83 00	55 02	AST	7·7	17
Irkutsk	−104 17	52 16	AST	7·9	16
GROUP IV					
Tokyo	−139 32	35 40	PZT	6·0	28
GROUP V					
Washington	77 04	38 55	PZT	8·7	13
GROUP VI					
La Plata	57 56	−34 55	VTI	15·5	4
Santiago	70 33	−33 24	AST	6·6	23

ment (VTI), photoelectric transit instrument (PTI), astrolabe (AST), photographic zenith tube (PZT), the typical mean square error (EQM) and the weight attributed at each time series. In particular we have examined the mean monthly values of the quantity for the years 1962–1967:

$$u_i = UT1(0)_i - UTC_{BIH}, \quad (i = 1, 2, \ldots, 6)$$

for each group i where

$UT1(0)_i$ = rotational observed time-scale corrected by polar variations;
UTC_{BIH} = coordinated time-scale of the Bureau International de l'Heure.

These are calculated from the single daily values u_n ($n = 1, 2, \ldots, 16$) of the same quantities recently published by the authors (Poma and Proverbio, 1973). Putting

$$u_n^* = u_n - \frac{1}{16} \sum_{n=1}^{16} u_n$$

and averaging successively the values u_n^* for each of the six groups given in

Table 1 we obtained the quantities u_i, which are consequently independent of the Earth's rotation. Apart from secular variations, the quantities u_i contain:

(a) seasonal systematic errors due to observational catalogues and other causes;
(b) periodic and irregular variations caused by deformations in the Earth's crust.

Therefore, we can write:

$$u_i(r) = \sum_s A_{i,s} \sin(\omega_s t + B_{i,s}) + f_i$$

where s indicates the terms depending on crust motion or apparent causes and the term f_i represents residual non-periodical variations.

Starting from u_i we calculated the differences

$$\Delta u_{i,j} = u_i - u_j \quad (i \neq j = 1, 2, \ldots, 6)$$

which depend on terms due to both causes (a) and (b), and the differences

$$\Delta u_i^k = (u^{k+1} - u^k)_i \quad (k = 1962, 1963, \ldots, 1966)$$

which, on the contrary, depend chiefly on non-annual variations in the motion of the crust. The differences $\Delta u_{i,j}$ minus the mean value $\Delta \bar{u}_{i,j}$ averaged over the period 1962–1967 are plotted in Figure 1. The accuracy of E_{ij} of each point on

TABLE 2

	$E_{i,j}$ (msec)	$A_{i,j}$ (msec)
I–II	±4·6	19
I–III	±6·0	19
I–IV	±2·5	5
I–V	±2·5	13
I–VI	±6·4	28

the curves of Figure 1 is given in Table 2 together with the amplitudes A_{ij} of the fluctuations of the quantities $\Delta u_{i,j} - \Delta \bar{u}_{i,j}$. Comparing the amounts of the m.s.e E_{ij} with the amplitude A_{ij} we see that the fluctuations plotted in Figure 1 cannot be attributed to accidental error. Only the variations between Europe and South America and between Europe and Asia are apparently less significant, because of the smaller amount of data. In Figure 2, by means of a more suitable scale, the annual relative fluctuations between Europe, North America and Japan are shown. The relative stability of Europe and Japan is to be noted, at least as regards periodical variations.

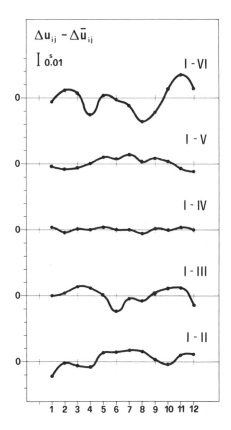

FIGURE 1. Relative variations of timescales observed between different pairs of groups of stations. The progressive number of the different groups is explained in Table 1. These variations correspond to relative variation of opposite signs in the longitude of each group of stations

Table 3 shows the differences of longitudes $\Delta\lambda_{i,j}$ between two different plates calculated from the data represented in Figure 1. In analysing these longitude differences it is very interesting to note that the greatest negative differences are associated with the Asiatic plate. This agrees with the larger eastward displacement of the Asiatic continent in correspondence to the epoch in which the acceleration of Earth rotation shows the greater value. Considering the monthly variations in the rate of the Earth's rotation ($d_2 \Delta T_s$) given in the last column of

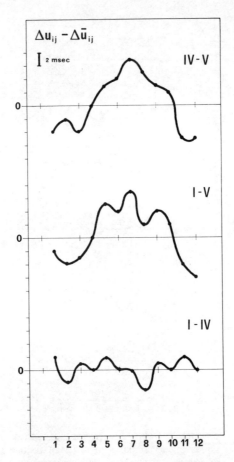

FIGURE 2. Evidence of periodical relative variations in the longitude of the American plate (group V) and Europe and Japan (groups I and IV)

Table 3 it is evident that there is good agreement between the longitude differences associated with the Asiatic plate and the latter values of the Earth's acceleration. This result seems to confirm, on the one hand, the existence of an elastic deformation in the Earth's crust caused by the same torque which is responsible for the annual and semi-annual variation in the Earth's rotation and, on the other, the hypothesis formulated by Pavlov (1968) that the Asiatic plate has greater mobility because of the stronger mechanical forces acting on its surface.

Although in the quantities $\Delta u_{i,j}$ catalogue errors will appear only as differential terms, it is probable that the longitude differences derived from them will be distorted by errors depending on the accuracy of the different star catalogues utilized in time observations. These errors are, on the contrary, eliminated from the differences Δu_i^k plotted in Figure 3 in which every seasonal fluctuation in longitude is also eliminated. Because the uncertainty of each point of the

TABLE 3

Month	$\Delta\lambda_{\text{I-V}}$ (msec)	$\Delta\lambda_{\text{II-V}}$ (msec)	$\Delta\lambda_{\text{III-V}}$ (msec)	$\Delta\lambda_{\text{III-IV}}$ (msec)	$\Delta\lambda_{\text{III-I}}$ (msec)	$d_2 \Delta T_s$ (msec)
1	+2	−9	+2	−2	0	−3
2	+4	+3	+6	+4	+2	+2
3	+3	0	+10	+6	+7	+5
4	0	−4	+6	+6	+6	0
5	−5	+2	−4	−1	+1	−7
6	−4	+3	−16	−12	−12	−13
7	−7	+2	−9	−2	−2	−11
8	−2	+6	−6	−1	−4	0
9	−4	−2	−2	+2	+2	+17
10	−2	−4	+2	+4	+4	+11
11	+4	+9	+10	+4	+6	+4
12	+6	+12	−1	−7	−7	−3
Amplitude	13	21	26	18	19	30

graphs is of the order of 2 ÷ 3 msec, the existence of irregular fluctuations in the motion of different plates cannot be neglected. Apart from the irregular variations, it is possible to trace from Figure 3 the existence of systematic variations in longitude from year to year which are nearly common to different plates. We have calculated the mean values of the quantities Δu_i^k for the intermediate epoch 1962·5, 1963·5, etc., the negative sign denoting westward movement of the Earth's crust. If these movements were even only partly connected with the Earth's rotation we would notice similar change in the rate of the Earth's rotation. Table 4 gives the mean values of displacements Δu_i^k for the different plates and for the Earth's surface as a whole, together with the values of the rate of change of the length of day for the same period. The good agreement between the Earth's acceleration and crust movements seems to confirm the hypothesis that we have a westward movement of the Earth's crust corresponding to a deceleration in the Earth's rotation.

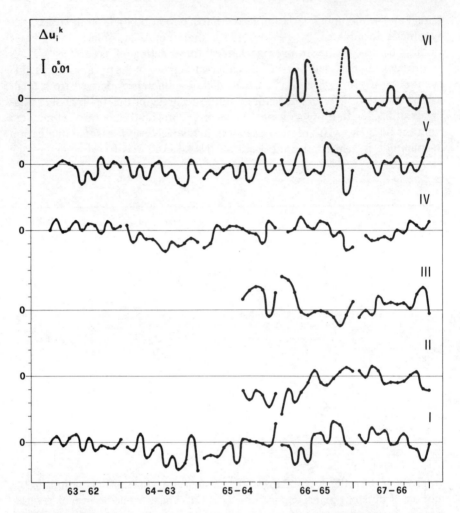

FIGURE 3. Irregular and long-term variations in the longitude of different groups from year to year. These variations seem to correspond to variations in the rate of the Earth's rotation as shown in Table 4 and to confirm the hypothesis that the torque responsible for the variation in the Earth's rotation may produce relative movements between the different plates of the Earth's crust

Finally, in Figure 4, the quantities

$$\Delta u_{i,j}^k = (u^{k+1} - u^k)_i - (u^{k+1} - u^k)_j$$

which represent relative drift and irregularities in the motion of different plates, are plotted. The arrows emphasize the existence of fairly similar variations for the Euro-Asiatic plates, though the annual drift of the European and Asiatic stations presents a different behaviour. It is very difficult to give the reason for the mutual relations between such different plates; nevertheless the same remarkable regularities can be evinced from the analysis of the quantities $\Delta u_{i,j}^k$.

TABLE 4

Period	I (msec)	II (msec)	III (msec)	IV (msec)	V (msec)	VI (msec)	Earth (msec)	Δ_2 l.o.d. (msec/day/year)
1963–62	−0·8	—	—	2·7	2·9	—	−0·4	−0·15
1964–63	−8·9	—	—	−7·5	−5·2	—	−7·2	−0·48
1965–64	−3·4	—	—	−2·0	−3·1	—	−2·8	−0·22
1966–65	−7·0	−6·2	2·9	−0·7	−0·6	(8·0)	−2·3	−0·28
1967–66	0·6	−2·0	3·4	−1·2	2·2	−3·3	−0·1	−0·10

In particular, the relative stability of the European station, at least for the periods 1966–65 and 1967–66, is evident, whereas the greatest mobility appears to be attributable to the Asiatic stations.

A confirmation of this is implicitly given by analysing the values

$$*\Delta u_{i,j}^k = \sum_{1}^{12} (\Delta u_{i,j}^k - \Delta \bar{u}_{i,j}^k)^2$$

$\Delta \bar{u}_{i,j}^k$ indicating the annual mean of $\Delta u_{i,j}^k$, calculated by the authors and reported in Table 5. In spite of the lack of data in the period 1962–65 for the USSR

TABLE 5

	$*\Delta u_{i,j}^k$ in 10^{-6} sec			
	1–2	1–3	1–4	1–5
1963–1962	—	—	114	227
1964–1963	—	—	690	534
1965–1964	—	—	541	605
1966–1965	809	3390	1717	1627
1967–1966	466	1640	1139	1371

stations, it appears clear enough that the larger variations in longitude must be attributed to the Asiatic plate, upon which the greater exchanges of energy between the atmosphere and the Earth's surface take place, and which shows the higher value of $*\Delta u_{i,j}^k$ (second column of Table 5). These data seem once again to confirm the results of the work carried out by Pavlov (1968) and the hypothesis of the relative mobility of the Earth's crust.

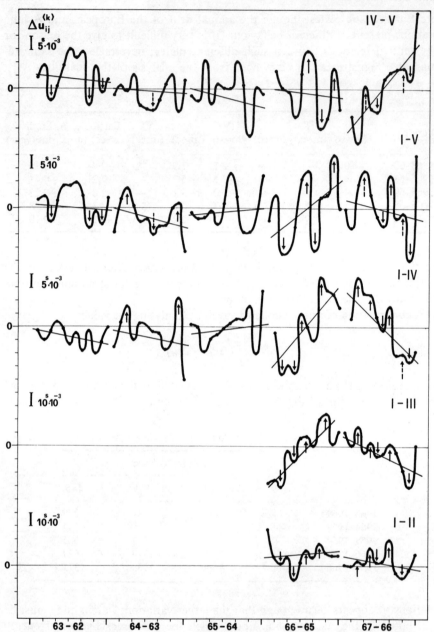

FIGURE 4. Relative drift and irregularities in the motion of different groups of stations. Diagrams I–IV, I–III and I–II show the existence of very similar short-period terms attributed to the European group of stations, the Asiatic stations on the contrary appear provided with very large drift

Astronomical evidence hence suggests that the lithosphere is subject to secular, periodical and irregular displacement in agreement with geological and geophysical observations. With regard to the annual and small-term fluctuations of the Earth's crust, according to the segmentation of its surface, mechanical torque, caused by wind stresses and tidal forces, cannot be disregarded. The global energy resulting from this event is probably larger than the energy dissipated by plate movement.

References

Bostrom, R. C. (1972). *The Measurement and Interpretation of Change of Strain in the Earth*, Meeting of the Roy. Soc., London

Knopoff, L. and Leeds, A. (1972). Lithospheric momenta and the deceleration of the Earth. *Nature*, **237**, 93

Ljustih, J. N. (1962). *Izv. AN SSSR*, Geol. Ser., No. 1

Mintz, T. (1951). The geostrophic poleward flux of angular momentum. *Tellus*, **3**, 195

Munk, W. H. and MacDonald, G. J. F. (1960). *The Rotation of the Earth*, Cambridge Univ. Press, London

Pavlov, N. (1968). Variations of the Earth's rotation, deformations in the Earth's crust and solar activity. *Izv. glav. Astron. Obs. Pulkove*, **183**, 1

Poma, A. and Proverbio, E. (1973). Observed and smoothed values of time scales. *Circ. Staz. Astron. Intern. Latit.*, Carloforte-Cagliari, A (1), No. 6

Proverbio, E. and Quesada, V. (1973a). Analysis of secular polar motion and continental drift. *Bull. Geod.*, **109**, 281

Proverbio, E. and Quesada, V. (1973b). The variation of longitudes and plate tectonic motion. *Bull. Geod.*

Proverbio, E. and Quesada, V. (1974). Secular variations in latitudes and longitudes and continental drift. *J. Geophys. Res.*, in press

Stoyko, N. (1931). Thèses de la Fac. de Sc. de Paris, Gauthier-Villars, Paris.

Stoyko, N. (1932). Sur les déplacements periodiques des continents. *C. R. Acad. Sc.*, **204**, 2225

Stoyko, N. (1942). Sur les variations séculaires et periodiques des longitudes. *C. R. Acad. Sc.*, **214**, 558

Stoyko, A. (1952). Sur l'irregularitè de la rotation de la Terre dans les deux hémisphères. *C. R. Acad. Sc.*, **234**, 2258

GEOLOGICAL PROCESSES AND THE EARTH'S ROTATION IN THE PAST

D. H. TARLING
*Department of Geophysics and Planetary Physics,
University of Newcastle upon Tyne, U.K.*

Abstract

The only factor which affects past rates of the Earth's rotation and also the Moon's rate of recession is the lunar tidal torque. Most geological considerations indicate that this torque would probably be greater and at least comparable to the present, in contrast to the slower rates indicated by the palaeontological and past tidal evidence. It seems that this conflict can only be resolved if deep oceans did not exist in the past, and there is some geological evidence for this supposition. Changes in the coupling between the Earth's core, mantle and atmosphere have changed through geological time, but not by amounts which would significantly affect the Earth's net surface rotation. Similarly, movements of the continents and convective motions appear to have little effect on the Earth's moment of inertia, although there is some evidence that the Earth may have been slightly decreasing its total mass by degassing, but this also depends on the supposition that volatile release has affected the past depth of the World's oceans.

Introduction

Studies of the length of the day and estimates of the Earth–Moon distance in the geological past based on geological and palaeontological observations differ significantly from those expected from extrapolation of astronomical observations made during historical times. Extremely precise measurements made during the last 25 years (O'Hora, 1975) and very precise measurements over the last 250 years (Morrison, 1975) show a secular deceleration of the Earth's rotation on which accelerations of varying periodicity and magnitude are superimposed. Extending over the last 2500 years, total eclipse observations indicate similar deceleration rates for the Earth's rotation and also a uniform recession of the Moon (Muller and Stephenson, 1975). When these recession rates for the Moon are extended back through geological time, assuming no change in either the obliquity or eccentricity of the Moon's orbit, then the Moon would have been some 20,000 km from the Earth's surface, much less than 1000 My ago. Such a near approach would have multiplied the present amount of tidal energy to be dissipated by about 8000 so that not only would huge oceanic tides have been created, but at least the outer layers of the Earth would have become molten and possibly the whole Earth. In contrast to these extrapolations, the tidal amplitude in Precambrian seas was not excessive, and only some

0·5–1·0 m in extensive shallow seas 2000 My ago (Truswell and Eriksson, 1975). Furthermore, Precambrian geology shows no global melting could have occurred during the last 2700 My and probably much earlier than that. However, if the Moon's obliquity was much greater in the past (Hipkin, 1975), then the magnitude of the lunar tidal torque would be much less and therefore the secular deceleration of the Earth and the recession of the Moon would be less. Such a situation should be testable by comparison of daily and monthly growth increments as only very weak solar tides would exist when the Moon was in polar or subpolar positions so that the lunar monthly cycle would be very strongly emphasized. It is uncertain if this was the situation in the Precambrian, as the number of reliable observations of rhythmic growth increments in stromatolites are few, but the distinctions between daily and monthly growth increments that have so far been reported do not require a near polar lunar orbit at these times, although this may reflect interpretations of different rhythmic bands rather than eliminate such explanation. The palaeontological evidence for the length of the day in the geological past (Berry, 1975; Johnson and Nudds, 1975; Mohr, 1975; Pannella, 1975) also indicates much lower deceleration rates for the Earth's surface rotation in the past than today and that different rates operated at different times, although persisting over very extended time spans. The number of observations and the reliability of some interpretations of the palaeontological measurements make it difficult, at the moment, to assess either the timing or abruptness of these changes and such clarification would almost certainly define the mechanism involved. In this chapter, therefore, certain geological factors are considered that may offer explanations for these long-term changes, but many interrelated factors are involved and it is emphasized that these are only speculations that indicate lines of enquiry, rather than affording definitive answers.

The present secular deceleration is attributed almost entirely to the action of lunar and solar torques operating on the Earth's tidal bulges (Munk and MacDonald, 1960; Jeffreys, 1970). This arises because the solid Earth is not perfectly elastic and the oceans are not perfect fluids, thus the Earth's rotation carries the tidal bulges out of alignment with the maximum gravitational attraction of the Moon and Sun, so that these operate decelerating torques on the Earth's rotation. The solar torque is much less than that of the Moon (estimates of the ratio vary between 1:24–1:5) and has no effect on the rate of recession of the Moon as the change it causes in the Earth's rate of rotation has negligible effect on the lunar tidal phase lag. The lunar torque on the Earth has an equal and opposite effect on the Moon's orbital velocity, causing the Moon's orbital velocity to increase and thus the Moon recedes from the Earth. The lunar torque is therefore the most critical factor to be considered as this affects both the rotation of the Earth and the Moon's recession, but other factors that affect the Earth's rotation do not affect the rate of recession of the Moon. Studies of historical irregular fluctuations in the length of the day that last for only a few weeks show that these are associated with persistent wind patterns, and fluctuations persisting for several years appear to be related mainly to changes in the strength of the electromagnetic coupling between the Earth's core and mantle (Rochester, 1973).

Geological factors that may alter the effectiveness of tidally controlled secular deceleration are considered in the next section. As the Earth's core, crust and atmosphere are rotating at different rates, geological factors affecting the coupling between these regions are discussed in the third section, although these are obviously secondary to the processes causing the different zones to rotate at different rates. In the fourth section, geological influences on the Earth's moment of inertia are outlined as changes in this parameter would affect the Earth's rate of rotation and Earth mass losses would also alter the magnitude of the lunar tidal torque. The direct effects of hypothetical changes in physical 'constants', such as the gravitational and Hubble 'constants', on the Earth's rate of rotation are considered in detail elsewhere (Weinstein and Keeney, 1975; Wesson, 1975) and the possible effects of these changes on geological processes are not pursued here as, despite considerations by Egyed (1956) and Creer (1965), there is no evidence requiring significant changes in these 'constants' during the last 3000 My, even if it is eventually found that changes may have occurred in the very early history of the Universe, some 12–15,000 My ago.

Secular tidal deceleration

As the Moon's present orbital plane is close to the ecliptic and the natures of the lunar and solar tidal torques are similar, most of the following discussion is restricted to consideration of the lunar torque. The moment that the lunar torque exerts on the Earth's rotation is obviously only that component of the torque that operates parallel to lines of latitude and as the moment is strongly dependent on latitude (it is a function of $\cos^2 \psi$, where ψ is the latitude, and ignoring the additional effect of the equatorial bulge), it is only the moment that is exerted in low latitudes, within 40° of the Equator, that is important in its effect on the Earth's rotation and the recession of the Moon.

The locus at which this decelerating torque operates is critical. Kant (1754) speculated that oceanic tidal friction was important and this was pursued more rigorously by Delaunay (1865), but there was at that time opposition to such explanations for the orbital acceleration of the Moon since discrepancies in the Sun's longitudinal position were not confirmed until the work of Cowell (1905) and particularly that of Spencer Jones (1939). Thomson and Tait (1879) suggested that oceanic tides would be significant but Darwin (1887), although conceding that they may be important, attributed any tidal deceleration of the Earth to body tides, even though this required an extremely fluid Earth.

The technique usually used to determine the locus is by comparison of the tidal energy dissipated in different regions with that expected by the action of the full lunar tidal torque, as derived from the rate of change of the Moon's orbital acceleration. Munk and MacDonald (1960) estimated the lunar tidal torque to be $3·9 \times 10^{23}$ dyne cm, which did work of $2·7 \times 10^9$ erg/sec, but more recent determinations of the secular change in the Moon's longitude give a torque between 4 and 6×10^{23} dyne cm and therefore energy dissipations between 3 and 5×10^{19} erg/sec. Oceanic tides are obvious features and therefore probable sources for such energy dissipation. The viscosity of sea-water is too low for significant dissipation by laminar flow (Jeffreys, 1920), the energy dissipated by degradation into internal modes (Munk, 1968) accounts for less than 5×10^{18} erg/sec (Rochester, 1973) and skin-friction forces are even less adequate, being of the order of 10^{16} erg/sec (Jeffreys, 1970). This means that the bulk of lunar tidal energy dissipation in the oceans must take place as turbulent frictional losses from bottom tidal currents. As the frictional stress between a current and the sea floor is proportional to the current velocity squared the energy dissipation is a function of the cube of the velocity. Using Swallow's (1955) determination of the average bottom tidal currents in the deep oceans as about 1 cm/sec, the total energy dissipation in the deep ocean is some 10^{16} erg/sec (Jeffreys, 1970), suggesting that this locus is insignificant compared with regions of shallow seas where shelving action magnifies tidal current velocities, particularly where the sea-floor geometry leads to resonance. Taylor (1919) originally calculated the tidal energy dissipation within the Irish Sea, and his calculations were extended by Jeffreys (1920) and Heiskanen (1921, revised by Lambert, 1928) to most shallow seas and showed that the total tidal energy dissipation in shallow seas was comparable with contemporary estimates of the total lunar tidal energy. Subsequently, both estimates have increased as more data became available, but they have remained comparable, suggesting that the shallow seas of the World account for most of the tidal energy dissipation and therefore are the locus for the operation of the lunar tidal torque. However, while there is general agreement that very little tidal energy is now being dissipated within the solid Earth (Munk, 1968; Lagus and Anderson, 1968), it is still not clear exactly how much oceanic tidal energy is, in fact, taken up within shallow seas, even though these may predominate. Detailed studies in the North Sea show that the regional lunar tides actually accelerate the Earth (Brosche and Sündermann, 1971) and, on a larger scale, the net tidal effects in the Bay of Bengal also exert an accelerating torque even though the total oceanic tidal moment (Pariiskii et al., 1972) is comparable with the total lunar tidal decelerating torque. Nonetheless, it is likely that deep oceanic tidal currents contribute more to the deceleration than is generally allowed, and tidal current activity along the continental slopes may also be significant. The uncertainties in present estimates of tidal currents in the deep ocean are clearly shown by comparison of different co-tidal charts, as these afford the basic data for tidal amplitudes in the deep oceans. These charts are based on tidal measurements on continental coastlines and oceanic islands, and major features, such as amphidromic points, vary significantly from one chart to another. Furthermore, even if the tidal amplitudes are well defined, the determination of bottom tidal currents is not simple and the corrections are not always applied uniformly. Studies of electric currents induced in abandoned telegraph cables (Runcorn, 1964) offer one method by which regional tidal bottom velocities can be established. Nonetheless, it appears that low-latitude shallow seas are the major locus in which the lunar semi-diurnal tidal moment is currently operating, but it is necessary to consider geological changes that could affect the magnitude of the torque in the oceans, the solid Earth and its atmosphere.

1. *Geological changes in the oceans*

The extent of shallow waters throughout the world depends on the shape of the hypsometric curve (Figure 1), the shape of the oceanic basins and the volume of sea-water (Wise, 1972). Any change in the degree of mountainousness or in the number and depth of oceanic trenches has little effect on the total amount of sea-water compared with changes in the number and extent of oceanic rises or changes in the total volume of sea-water.

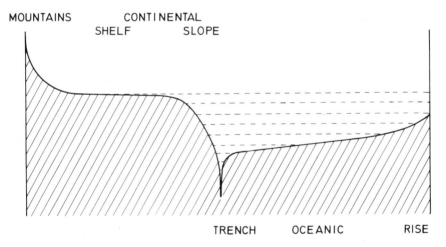

FIGURE 1. The Hypsometric Curve, modified to illustrate the present oceanic and continental topographic features

The elevation of the oceanic ridges seems to reflect mainly the upper mantle temperatures, and pulses of rapid sea-floor spreading correlate with eustatic changes in sea-level during the Tertiary and Cretaceous (Rona, 1973; Hays and Pitman, 1973; Fleming and Roberts, 1973) and presumably reflect both higher standing ridges and greater rates of volatile release. The simple existence of shallow seas in low latitudes, however, only provides a location in which tidal currents could be amplified. The establishment of even non-tidal current circulation patterns in past oceans is extremely difficult. The early Central Atlantic appears to have been stagnant on its western flanks, but aerobic currents circulated in the east until about 100 My ago (Vogt *et al.*, 1971) but even when the proto-Gulf Stream began to flow in the western Atlantic, it is unclear if this formed a true gyre until much later (Hart and Tarling, 1974). Clearly it is almost impossible to determine the actual relationship of past tidal currents to shallow seas but, for much of the last few hundred My, the continental configurations suggest that the lunar semi-diurnal deep ocean tidal currents, if present, would be carried onto shallow seas and strongly magnified.

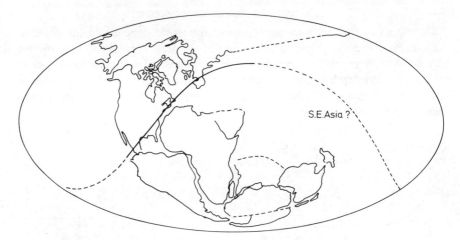

FIGURE 2. Pangaea and the Tethys Ocean some 300 My ago. The equator is based on palaeomagnetic observations for Europe and North America. There are major uncertainties about the position of south-east Asia, and probable movements in north-eastern Asia and Alaska have not been shown

There is general agreement about the relative position of the different continents some 300 My ago, with the northern continents (North America, Europe and most of Asia) forming one block and the southern continents (South America, Africa, Antarctica, India and Australia) forming another, even though there may be some disagreement on the precise fit of the individual continents (Smith and Hallam, 1970; Tarling, 1972). These blocks were joined along N.W. Africa and western North America but elsewhere were separated by the Tethys Ocean which, at this time, was largely equatorial (Figure 2). This configuration of the Tethys would probably constrict tidal currents onto the flanking shallow oceanic areas, and would be particularly effective when access from the Tethys to the proto-Pacific was restricted. On the other hand, the equatorial position of what is now south-eastern Asia may have acted as a barrier, preventing substantial tidal flow into the Tethys. In extremely general terms, therefore, the palaeogeography of the last few hundred My suggests that secular deceleration by tidal torques would have been highly effective, even though the palaeontological evidence suggests lower rates. This could mean that other factors may have been accelerating the Earth's rotation much more in the past than today, but such explanations would not account for the low lunar recession rate which depends exclusively on the magnitude of the lunar tidal torque. This suggests that either the palaeogeographic or the palaeotidal model is invalid. One possibility is that the total volume of sea-water could have varied substantially and therefore there were no deep oceans in the past from which tidal currents could flow to be magnified on the shallow-sea regions.

The volume of sea-water, over geological time, depends on two main factors: the ambient temperature and the rate of production of juvenile water. (The amount of water trapped on the continents or within sediments will probably average out over prolonged periods.) An increased temperature would cause an expansion in sea-water (a rise of 60 cm/°C for the present oceans), increased evaporation rates, and also more water in the atmosphere. The study of oxygen isotopes in a variety of marine organisms suggests that World average temperatures were some 8°C higher during 'normal' geological time than during the present interglacial period, although equatorial temperatures were only some 1–2°C higher and high latitudes were some 10–15°C higher. It is probable that the increased volume of water held in the atmosphere would more than compensate for the thermal expansion of sea-water and also for the 70 m rise in sea-level if no ice-caps existed. Irrespective of these balancing effects, it is still the production rate of juvenile water that is likely to be the dominant influence over very long periods. Until the acceptance of Hess's concept of ocean-floor generation (presented in 1960, published 1962), the total amount of water added to the Earth's surface by volcanic activity was thought to be small, since only a few hundred volcanoes were known, but Hess's model required volcanic activity along the entire oceanic ridge system and corresponds to a formation of more than 10 km^3 of oceanic crust each year, and this process must be accompanied by the release of volatiles, of which water is the dominant constituent. Unfortunately, the nature of ocean-floor generation is obscure and the volume of volatiles released must depend largely on whether dyke intrusion or lava extrusion is dominant at oceanic ridge crusts. It seems probable that lavas are very common and that volatile release will, in fact, be considerable and could account for the production of the World's oceanic waters during only the last 200–300 My. Studies of deep-oceanic sediments do suggest that the World's oceans were much shallower some 100 My ago than they are today (Hart and Tarling, 1974) and this is also supported by determinations of the salt balance of the oceans, which again suggests an age for the present oceanic waters of only some 200 My (Kennedy and Bowden, 1970). Such a radical hypothesis does not mean that oceanic basins containing sea-water did not exist in the past, as marine sediments are present throughout the geological column and the movement of plates during at least the last 2700 My (Tarling, 1971) requires the generation and subduction of oceanic crust, with volatile release, unless an entirely different continental drift mechanism is invoked. The suggestion is that ancient oceans may have been much shallower than those of today and that the magnification of tidal currents could have been minimal. Unfortunately, much of the evidence for the actual depth of such past oceans has either been lost by subduction or has been metamorphosed during the addition of oceanic remnants onto the continental crust.

2. *Solid Earth tides*

The tidal energy dissipated within the solid Earth is considered to be small at the moment (Munk and MacDonald, 1960), and this is generally supported by seismic determinations of the damping coefficient, Q, of the mantle. However, the Q may be of the order of 50 (Helmberger, 1973) within the lithosphere and this value is likely to have changed over geological time. In particular, lower Q values presumably existed locally immediately prior to, and during, the eruption of the extensive flood basalts some 160 My ago in Africa, India, Australia and Antarctica and 120 My ago in South Africa and South America when some 10^7 km^3 of basalts were released. However, these areas were in fairly high palaeo-latitudes at that time and any increased lunar tidal torque on the upper mantle in these regions would cause little change in the Earth's rotation rate, but the Siberian flood basalts of some 200 My ago were erupted in much lower palaeo-latitudes and the associated low Q upper mantle could have increased the effectiveness of the lunar solid–Earth tidal torque. However, any long-term changes in the Q of the upper mantle are likely to have been small compared with the possible oceanic changes and would probably average out over long time periods.

3. *Atmosphere*

The present thermal tide in the atmosphere has peaks at 10 a.m. and 10 p.m., local time, and thus results in an accelerating torque (Thomson and Tait, 1879) and a corresponding energy dissipation of some $2 \cdot 2 \times 10^{18}$ erg/sec (Munk and MacDonald, 1960), i.e. some 30 times less than the decelerating oceanic torque. Geological effects on this torque are difficult to evaluate as the generally higher temperatures of the past would initially decrease the atmospheric density, but increased evaporation would probably lead to an overall increase in its mass (water vapour currently forms 10% of atmospheric pressure). Furthermore, the atmosphere is an evolving system which depends on the various ways in which the original composition gradually changes (e.g. Berkner and Marshall, 1965) as well as on the quantity and composition of volatiles being released at different times. The problem of volatile release from oceanic ridge systems is therefore of major importance. In general terms, it seems likely that the magnitude of the accelerating tidal torque on the Earth's atmosphere may have been greater in the past, but even if it were doubled, this would still be insignificant compared with the present lunar oceanic tidal deceleration.

4. Mass losses from the Earth and Moon

The effectiveness of the lunar tidal torque on the total system depends on the masses of the two planetary bodies. The Earth is likely to have increased its total mass from meteoritic particle accumulations at a faster rate than the Moon, but as the accumulation rates on the Earth today are only some 10–20 kg/year and the Moon's surface shows only very low accretion rates for the last 3000 My, such mass increases can be ignored on both bodies. Both planets are also losing mass to outer space by degassing, but changes in the Moon's total mass for at least the last 3000 My are likely to have been insignificant as no substantial volcanic activity has occurred during this time. The loss of volatiles from the Earth to its atmosphere, and eventually to outer space, is obviously much greater than for the Moon, but the total mass lost depends critically on the past rates of volatile release at oceanic ridges, as discussed earlier. If the present oceanic waters have only been released in the last 200 My or so, then the solid Earth must have lost some 1.4×10^{24} gm, i.e. some 0.003% of the mass of the mantle, during 2×10^8 years. Such a rate of mass loss is unlikely to have been sustained throughout geological time ($>3 \times 10^9$ years) and, if oceans of the past were really shallower, then past rates of volatile release could have been less. Nonetheless, even if the World's present oceanic waters are, in fact, extremely old, there has still been some mass loss from the Earth's interior, much of which will have eventually been lost to outer space, and this would mean that the magnitude of the lunar torque would be somewhat higher in the past than today. This mass loss would, in addition, change the Earth's moment of inertia and thereby change its rate of rotation (see below). However, despite major uncertainties about the present and past rates of volatile loss, it seems unlikely that changes in the mass of the Earth–Moon system are major contributors to changes in the magnitude of the lunar tidal torque and, if anything, they increased its effectiveness in the past.

Coupling between the atmosphere, crust, mantle and core

If the coupling between different zones is poor, then changes in their individual rotation could occur which did not affect the Earth's surface for which the rotational rates are determined. However, changes in the strength of the coupling would mean that the Earth's surface would be accelerated or decelerated according to the relative rotation rates involved. At the moment the Earth's core is rotating some 0.003% slower than the Earth's surface, shown by the westward drift of the geomagnetic field (Bullard et al., 1950; Vestine, 1952), and parts of the upper atmosphere are rotating as much as 20% faster than the Earth's surface (Rishbeth, 1973).

The geomagnetic field provides an electromagnetic coupling between the Earth's core and its ionosphere so that changes in the strength of the Earth's magnetic field could be relevant. Long-term variations in geomagnetic field strength, for example when it decreases to 20% of its usual value during a polarity transition lasting some 3–5000 years (Tarling, 1971), are likely to be more important in the effect such changes would have had on the height of the ionosphere and thus the temperature of the upper atmosphere and associated climatic change. The nature of the ionosphere, and therefore of the coupling, must also have changed with time as some elements are preferentially lost to outer space and as the injection of new volatiles proceeds. Again, the effects of such changes are likely to be more significant in terms of changes in mass distribution than on the electromagnetic coupling. As the masses of the core and mantle are so great, changes in electromagnetic coupling between them are likely to be more significant than the coupling with the atmosphere and this is discussed more fully elsewhere (Creer, 1975; Jacobs, 1975), but even this effect appears to have been small compared with the acceleration and deceleration rates required.

The frictional coupling between the atmosphere and the Earth's solid surface would have changed through geological time with the appearance of land plants some 400 My ago and with changes in the extent and magnitude of mountain areas. However, such changes would, at the most, have doubled the frictional coefficient and again this effect is negligible. It appears that although changes in the coupling between different zones have been quite large, their effect on the relative rotation rates can have been only negligible and, in any case, such changes would still depend on a primary mechanism causing and maintaining different rotation rates in the different zones.

Non-tidal torques and moment of inertia

The distribution of either volcanic or seismic activity along the edges of plates is unlikely to be radially symmetric but can be considered simply as sudden mass displacements in either horizontal or vertical directions (Mansinha and Smylie, 1967). This means that all such activity can be considered as part of convective motion and therefore as affecting the Earth's moment of inertia. Any net downward motion of mass would decrease the Earth's moment of inertia and therefore increase its rate of rotation and, similarly, any net equatorial mass movement decreases the moment of inertia and increases the Earth's rotation.

Possibly the most important problem in geophysics and geology is the nature and extent of convective motions within the Earth. Although the existence of convection is no longer in serious doubt following the evidence for sea-floor spreading, it is not yet clear if the total convecting layer is confined within the low-velocity layer, i.e. within the top 200–250 km, or extends throughout the mantle. In the latter case, convective motions should result in the continued growth of the core. Urey (1952) suggested that the growth of the core could account for the observed accelerations of the Earth during the last 2500 years but, as pointed out by Munk and MacDonald (1960), this growth of some 10^{11} g/sec would mean that the entire core would form in about 0.6×10^9 years. Runcorn (1966) suggested the initial fast growth would decline so that the core's radius may have only increased by some 10% during the last 10^9 years and the corresponding rate of acceleration would therefore be reducing gradually throughout geological time. Such a model has an additional advantage that, if the core–mantle coupling is weak, the 'rain' of iron droplets from the mantle could result in an acceleration of the mantle and retardation of the core, explaining the present difference in their angular velocities. The total core–mantle coupling is obviously critical, but the dominant consideration is the rate of growth of the Earth's core at different times and therefore the depth of convection. Most recent theories, based on the geochemical and thermal evolution of the Moon, suggest that the Earth's core formed extremely rapidly during differentiation which took place during the first few My after the accretion of the Earth and Moon. If such concepts are valid, then the Earth's core has probably not changed its size significantly during the last few 10^8 years and this seems to be supported by observations that the geomagnetic field, generated at the core–mantle boundary, has remained of similar strength and that the amplitude of this secular variation has remained similar during most of the last 2×10^9 years (Tarling, 1971). However, such evidence is merely consistent with the evidence for the early rapid formation of the Earth's core and it is certainly conceivable that mantle-wide motions are still involved and therefore that the core may still be growing at a rate that is significant in its effect on the Earth's rate of rotation.

Even if it is assumed that convection is restricted to the upper mantle, vertical and lateral motions are involved, but the motions alone have no effect on the Earth's rotation (Munk and MacDonald, 1960), unless a net mass redistribution is involved. The ascent of less dense material at oceanic rises should, at first sight, result in an increase in the moment of inertia and decrease in the rate of rotation, but determinations of gravity do not indicate any significant mass redistribution. Conversely, at subduction zones, the descent of colder, denser rocks should mean that the Earth's moment of inertia is decreased and its rate of rotation increased, but the mass redistributions are very localized as their gravitational effects are not discernible from satellite observations and the positive and negative gravity anomalies observed at the Earth's surface must therefore balance each other when considered over larger areas of the Earth's surface. These observations suggest, therefore, that there is little or no net vertical mass redistribution actually occurring at either oceanic trenches or ridges, and the present and past distribution of such features has only minimal effect on the Earth's rotation. If vertical mass redistribution is really negligible at the present time, then it should follow that the total mass of continental rocks should not be increasing and that there should be no change in the composition of the oceanic crust formed by upper-mantle differentiation throughout geological time. Most orogenic belts appear to comprise reworked older crustal rocks and there seems to be little or no difference between major and minor element abundances in modern and Precambrian ophiolites (although such evaluations are difficult because of the metamorphic changes which often occurred when old ophiolites were incorporated into continental crust). This compositional consistency suggests that there has been little differentiation during recycling of mantle rocks whether convection is restricted to shallow or deep convective cells.

Lateral motions of the Earth's surface would appear to be highly significant as the Himalayas alone comprise some 3×10^{22} g lying above the normal continental level (Munk and MacDonald, 1960), and small changes in the distribution of atmospheric pressures have easily detectable effects on the Earth's moment of inertia and its rate of rotation, yet the total mass of the atmosphere is only $5 \cdot 1 \times 10^{21}$ g. However, the motions of lithospheric plates of the Earth's surface extend to depths of some 200–250 km and density differences between the oceanic and continental parts of the lithosphere extend to at least 120–150 km. Examination of these density differences shows that, above a depth of some 30 km, the moment of inertia of the continental plate is some 16% higher than that of an oceanic plate, but above a depth of 120–150 km, the differences are extremely small and within the errors of present density determinations. Petrological studies based on rocks erupted on the continents, but derived from the upper mantle, suggest that the continental lithosphere moves as a unit some 150–200 km thick, and therefore any changes in the Earth's moment of inertia as a result of plate tectonic movements can only be caused by local differences in the density distribution within and beneath the continents and oceans and thus are likely to be small in their total effect.

The fact that all the above considerations indicate, as first-order approximations, only minor changes in the Earth's moment of inertia as a result of convective motion can be tested by the fact that only small asymmetric changes in the moment of inertia would cause major changes in the position of the axis of principal inertia (Goldreich and Toomre, 1969). As the Earth's rotational axis must be closely linked to an inertial axis, the rate of motion of the rotational pole can be used to define the order of magnitude of changes in the positions of the Earth's inertial axes. Palaeomagnetic evidence extending over the last 20 My confirms that the average geomagnetic pole position coincides with that of the Earth's rotational pole within better than $5°$ and the axial geocentric dipole model for the average geomagnetic field is supported by both theoretical and palaeoclimatic arguments (Tarling, 1971). The motion of the average geomagnetic pole position during the last 300 My has averaged some $0 \cdot 3°/\text{My}$, and the maximum rate of movement appears to be of the order of $2–3°/\text{My}$. This rate of motion is comparable with known rates of plate motion determined from the spacing of dated oceanic magnetic anomalies; the maximum defined rate is some $2°/\text{My}$ for the motion of the Indian plate some 75 My ago (McKenzie and Sclater, 1971). This means that the apparent movements of the rotational pole can be explained entirely by plate motions, without recourse to 'true' polar wandering, although the asymmetric distribution of subduction zones, as today, will mean that the summation of motions for all plates (Mueller and Schwarz, 1972) will show a net apparent polar motion. The observed rate of polar motion therefore rules out major asymmetric changes in the Earth's moment of inertia, but minor asymmetric changes and large symmetrical changes cannot be excluded at the moment.

The release of volatiles at the oceanic ridges and subduction zones has been discussed in terms of the possibility of its changes affecting the total mass of the Earth and the depth of the oceans. As the volatiles released would be poorly coupled to the mantle, the process could cause a net acceleration of the Earth's surface and a deceleration of the atmosphere and ocean. Dicke (1969) considered that the effect of extra water loading on the oceans would be counterbalanced if flowage of material were possible within the mantle. As the greatest volatile loss is likely to occur when the upper mantle is hottest, it seems probable that such compensation will take place. The release of volatiles as part of an upper or complete mantle circulation system could therefore have a small symmetrical effect on the Earth's moment of inertia that could cause an acceleration of the surface with little or no change in the location of the axes of inertia. The total effect of this acceleration is impossible to evaluate until the past and present rates of volatile release have been established.

Conclusions

In order to account for a reduction in both the recession of the Moon and the rate of rotation of the Earth in the past, major changes in the magnitude of the lunar torque must have occurred through geological time. Most geological factors during the last few hundred years would appear to operate in the opposite direction to that required unless major changes have occurred in the depths of the oceans. Changes in the Earth's rate of rotation can be expected from a variety of processes but, surprisingly, the changing distribution of the continents appears to have had little or no effect on the Earth's rate of rotation, and most other geological processes have had only small effects compared with those predicted from changes in the magnitude of the lunar tidal torque. The previous depths of the World's oceans are, therefore, absolutely critical and it seems difficult to escape from the remarkable conclusion that the Earth's oceans, in the past, were generally very much shallower than those of today. Such a conclusion is difficult to reconcile with either the existence of extensive, shallow seas in the Precambrian, which imply that oceanic basins were filled, or with the quantity of water locked up in ice caps at different periods of the Earth's history, but clearly such a possibility must be considered in view of the astronomical and palaeontological observations reported elsewhere in this volume.

References

Berkner, L. V. and Marshal, L. C. (1965). On the origin and rise of oxygen concentration in the Earth's atmosphere. *J. Atmos. Sci.*, **22**, 225–261

Berry, W. B. and Barker, R. M. (1975). This volume

Brosche, P. and Sündermann, J. (1971). Die gezeiten des meeres und die rotation der Erde. *Pure Appl. Geophys.*, **86**, 95–117

Bullard, E. C., Freedman, C., Gellmann, H. and Nixon, J. (1950). The westward drift of the Earth's magnetic field. *Phil. Trans. Roy. Soc.* A, **243**, 67–92

Cowell, P. (1905). Lunar theory from observation. *Nature*, **73**, 80–81

Creer, K. M. (1965). An expanding Earth. *Nature*, **205**, 539–544

Creer, K. M. (1975). This volume

Darwin, G. (1887). On the influence of geological changes on the Earth's axes of rotation. *Phil. Trans. Roy. Soc.*, **A162**, 27

Delaunay, C. (1865). Sur l'existence dune cause nouvelle ayant une influence sensible sur la valeur de l'equation seculaire de la lune. *C. R. Acad. Sci., Paris*, **61**, 1023–1032

Dicke, R. H. (1969). Average acceleration of the Earth's rotation and the viscosity of the deep mantle. *J. Geophys. Res.*, **74**, 5895–5902

Egyed, L. (1956). A new theory on the internal constitution of the Earth and its geological–geophysical consequences. *Acta Geol. Acad. Sci., Hung.*, **IV**, 43–83

Fleming, N. C. and Roberts, D. G. (1973). Tectono-eustatic changes in sea level and seafloor spreading. *Nature*, **243**, 19–22

Goldreich, P. and Toomre, A. (1969). Some remarks on polar wandering. *J. Geophys. Res.*, **74**, 2555–2567

Hart, M. B. and Tarling, D. H. (1974). Cenomanian palaeogeography of the North Atlantic and possible Mid-Cenomanian eustatic movements and their implications. *Palaeogeog. Palaeoclimat., Palaeocol.*, **15**, 95–108

Hays, J. D. and Pitman, III, W. C. (1973). Lithospheric plate motion, sea level changes and climatic ecological consequences. *Nature*, **246**, 18–22

Heiskanen, W. (1921). Uber den einfluss der gezeiten anf die sakulare acceleration des mondes. *Ann. Acad. Acient. Fennicae A*, **18**, 1–84

Helmberger, D. V. (1973). On the structure of the low velocity zone. *Geophys. J. Roy. Astr., Soc.*, **34**, 251–263

Hess, H. H. (1962). History of the ocean basins, pp. 599–620. In *Petrologic Studies—Buddington Memorial Volume*, Geol. Soc. Amer., N.Y.

Hipkin, R. G. (1975). This volume

Jacobs, J. C. (1975). This volume

Jeffreys, H. (1920). Tidal friction in shallow seas. *Phil. Trans. Roy. Soc. A*, **221**, 239–264

Jeffreys, H. (1970). *The Earth*, Cambridge Univ. Press, 5th edit., 525 pp.

Jeffreys, H. (1973). Tidal friction. *Nature*, **246**, 346

Johnson, G. A. L. and Nudds, J. R. (1975). This volume

Kant, I. (1754). Untersuchung der frage, of die Erde in ihrer umdrehung um die ackse wodurch sie abwechselung des tages und der nacht hervorbringt wochenl frag- und anzeigung-nachrichten, 23 and 24 konginsberg

Kennedy, W. Q. and Bowden, P. (1970). Aspects of oceanic development. *J. Earth Sci., Leeds*, **8**, 1–14

Lagus, P. L. and Anderson, D. L. (1968). Tidal dissipation in the Earth and planets. *Phys. Earth Planet. Interiors*, **1**, 505–510

Lambert, W. (1928). The importance from a geophysical point of view of a knowledge of the tides in the open sea. Bull. ii, Section D'Oceanog., Union Geodesique Geophysique Internat., 52

Mansinha, L. and Smylie, D. E. (1967). Effect of earthquakes on the chandler wobble and secular polar shift. *J. Geophys. Res.*, **72**, 4731–4743

McKenzie, D. P. and Sclater, J. G. (1971). The evolution of the Indian Ocean since Late Cretaceous. *Geophys. J. Roy. Astr. Soc.*, **24**, 437–528

Miller, G. R. (1966). The flux of tidal energy out of the deep oceans. *J. Geophys. Res.*, **71**, 2485–2489

Mohr, R. (1975). This volume

Morrison, L. V. (1973). Rotation of the Earth from A.D. 1663–1972 and the constancy of G. *Nature*, **241**, 519–520

Morrison, L. V. (1975). This volume

Mueller, I. I. and Schwarz, C. R. (1972). Separating the secular motion of the pole from continental drift. Where and what to observe, pp. 68–77. In *Rotation of the Earth* (Eds. P. Melchior and S. Yumi), Reidel, Netherlands

Muller, P. and Stephenson, R. (1975). This volume

Munk, W. H. (1968). Once again—tidal friction. *Quart. J. Roy. Astr. Soc.*, **9**, 352–375

Munk, W. H. and MacDonald, G. J. F. (1960). *The Rotation of the Earth*, Cambridge Univ. Press, pp. 232

O'Hora, N. P. J. (1975). This volume

Pannella, G. (1975). This volume

Pariiskii, N. N., Kuznetsov, M. V. and Kuznetsova, L. V. (1972). The effect of oceanic tides on the secular deceleration of the Earth's rotation. *Izv., Earth Phys.*, **8** (2), 3–12

Rishbeth, H. (1973). Superrotation of the upper atmosphere. *Revs. Geophys. Space Phys.*, **10** (3), 799–819

Rochester, M. G. (1973). The Earth's rotation. *Eos*, **54** (8), 769–780

Rona, P. A. (1973). Worldwide unconformities in marine sediments related to eustatic changes of sea level. *Nature Phys. Sci.*, **244**, 25–26

Runcorn, S. K. (1964). Changes in the Earth's moment of inertia. *Nature*, **304**, 823–825

Runcorn, S. K. (1966). Changes in the moment of inertia of the Earth as a result of a growing core, pp. 82–92. In *The Earth–Moon System* (Ed. A. G. W. Cameron and B. G. Marsden), Plenum Press, New York

Smith, A. G. and Hallam, A. (1970). The fit of the southern continents. *Nature*, **225**, 139–144

Spencer Jones, H. (1939). The rotation of the Earth and the secular accelerations of the Sun, Moon and Planets. *Mon. Not. Roy. Astron. Soc.*, **99**, 541–548

Swallow, J. (1955). A neutral-buoyancy float for measuring deep currents. *Deep-Sea Res.*, **3**, 74–81

Tarling, D. H. (1971). *Principles and Applications of Palaeomagnetism*, Chapman & Hall, London, p. 164

Tarling, D. H. (1972). Another gondwanaland. *Nature*, **238**, 92–93

Taylor, G. I. (1919). Tidal friction in the Irish Sea. *Phil. Trans. Roy. Soc.*, **A220**, 3–33

Thomson, W and Tait, P. (1879). *Treatise on Natural Philosophy*, Vol. 1, pt. 1, Cambridge

Truswell, J. F. and Eriksson, K. (1975). This volume

Urey, H. C. (1952). *The Planets*, Yale Univ. Press, 275 pp.

Vestine, E. H. (1952). On variations of the geomagnetic field, fluid motions, and the rate of the Earth's rotation. *Proc. Nat. Acad. Sci.*, **38**, 1030–1038

Vogt, P. R., Anderson, C. N. and Bracey, D. R. (1971). Mesozoic magnetic anomalies, sea-floor spreading and geomagnetic reversals in the SW North Atlantic. *J. Geophys. Res.*, **76**, 4796–4823

Weinstein, D. H. and Keeney, J. (1975). This volume

Wesson, P. (1975). This volume

Wise, D. U. (1972). Freeboards of continents through time. In *Studies in Earth and Space Studies* (Ed. R. Shagam et al.), Mem. 132, Geol. Soc. Amer., pp. 87–100

DISCUSSION

BROSCHE: It is by no means certain that tidal dissipation occurs predominantly in shallow seas. If one considers energy dissipation in an inertial frame, calculations using a 10° model of oceans with nearly no shallow seas can, surprisingly, account for half of the astronomically variable tidal friction. Perhaps the direction of current is more regular in deep oceans and although perhaps the amount of friction at one surface remains small, the total amount of dissipation in deep oceans is nevertheless large.

CLIMATE, THE EARTH'S ROTATION AND SOLAR VARIATIONS

J. GRIBBIN

Nature, 4 Little Essex Street, London, UK

Abstract

Variations in solar activity over the 11-year sunspot cycle cause variations in the Earth's rotation rate (Challinor, 1971). During an investigation of geophysical implications of this, Plagemann (NASA, Goddard) and I sought evidence of direct links between specific solar flare activity and changes in length of day. We think we found this, others disagree, but Challinor's work is beyond doubt. Plagemann continues to study geophysical implications; in seeking the mechanism linking solar changes with changes in Earth's rotation I looked at the way the atmosphere reacts to changes in solar activity, specifically how changes in solar output affect climate and weather.

The three variations are closely linked, and there is no doubt of the importance of climatic change for biological processes. There is a great body of little-known work on the relation between solar and climatic changes. King (1973) has reviewed and discussed some evidence and presented a model involving injection of solar cosmic rays into the atmosphere at high latitudes. This ties in well with observations by Roberts and Olsen (1973) and with evidence that stratospheric air is injected into the lower troposphere after solar flares (Reiter, 1973). Climatic change is even more interesting. We are on the fringe of understanding how climatic changes relate to changes in the Sun and other gross properties of the Earth; biological data, such as tree rings, oxygen isotopes in marine shells, etc., will play a key role. If the solar cycle does affect weather, it seems possible to predict meteorological change by studying variations in solar activity.

At first sight, it may seem a little odd to find what purports to be a learned paper being contributed to such a conference as this by a journalist. But this conference has been specifically advertised as being 'interdisciplinary', and in these days, when the information explosion makes it difficult to keep up with any one specialist field, the scientific journalist is in a unique position to notice developments as they occur in what seem at first sight to be totally different fields, but which can in fact be pieced together to give a new insight into processes affecting the Earth. I do not claim that this insight has any great depth, or even any particular intellectual merit, but I do hope that my superficial assessment of how the pieces of geophysics, geodesy, meteorology, astronomy, and even perhaps astrology, might be fitted together will encourage others to carry out more thorough investigation of some remarkable effects which seem not to have been generally realized before.

Solar activity and the Earth's spin

The story began for me nearly 3 years ago, when Challinor (1971) published the results of an investigation into the changing length of day and its relation to geophysical and astronomical phenomena. He showed in particular that there is a very clear link between variations in the overall activity of the Sun, revealed by the number of sunspots visible on the Sun's disc and in other ways, and the rate of rotation of the Earth. Of course, there are many other processes affecting the Earth's spin, including seasonal effects and the Chandler Wobble, and this direct solar effect is small. But with atomic clocks now available to provide very accurate measurements of the changing length of day Challinor could identify a positive correlation between the solar activity and the rotational acceleration of the Earth.

This kind of effect produces changes in the length of day measured in msec, and the cycle of solar activity has a mean period of about 11·1 years (at present, we are just at the minimum of such a cycle). Therefore we are not talking about something which is likely to have a direct effect on the habits of biological organisms here on Earth (except to set some of them puzzling about the cause of these changes).* But several possibilites arise from Challinor's work. First, it lends considerable weight to the rather less conclusive work by earlier investigators, who lacked the accurate timing facilities available to Challinor but who found hints of a link between the level of solar activity, measured by cosmic ray activity, for example (Danjon, 1962a,b), and the length of day. Since the activity of the Sun is far from uniform from one cycle to the next, it seems likely that changes in the average level of solar activity will have affected the Earth's spin over a period of hundreds or thousands of years. That possibility must remain largely an interesting speculation, for lack of observational data extending over sufficiently long periods of time. But two other developments from this discovery of a link between solar activity and changes in the length of day—a link now confirmed by work at the Royal Greenwich Observatory (O'Hora, 1973)—are immediately testable, and could be of great importance to biological organisms on Earth.

Effects attributed to the solar storm of August 1972

Since I lack the patience of the true astronomer, the thought of waiting for a couple of decades to see how changes in the length of day could be tied to mean levels of solar activity did not appeal to me. The alternative, however, seemed equally unpromising, since the only other way I could see to tie changes in the

* Rosenberg (personal communication) indicates that there is, in fact, evidence of an 11-year 'rhythm' in population size of animals such as rabbits and deer. Perhaps this might arise through a direct influence of solar activity; I think it more likely, however, that such a rhythm reflects the changing climatic factors, notably rainfall, which are influenced by the solar cycle of activity as discussed below.

length of day to changes in the activity of the Sun was to find a dramatic event on the Sun which could be linked conclusively to a dramatic change in the Earth's spin. The kind of effect needed, given the accuracy of present-day measuring techniques, would have to be a solar storm at least as big as the great storm of 1959, which had indeed been investigated for an effect on the Earth's spin by Danjon. Danjon suggested that there was an increase in the length of day when the nucleonic component of solar cosmic rays increased; but other observers disputed this and the issue remained unclear (see discussion by Schatzman, 1966). Two things made it seem particularly unlikely that another storm like that of 1959 might occur in the early 1970s: first, the last maximum in the solar activity cycle had just been passed; second, the 1959 solar storm was the biggest such event ever to be recorded in more than 250 years of continuous monitoring of sunspot activity.

But, as it turned out, we did not have to wait for 250 years, or even for 11 years. To the bafflement of solar astronomers, in August 1972, right at the minimum of its cycle of activity, the Sun suddenly produced a storm even bigger than the 1959 event. The spot group which was associated with the all-important flare activity built up from 29 July and reached a maximum size on 4 August; the cause of this activity at such a time remains a mystery, but with the various satellites and space probes operating at that time available to turn their detectors on the phenomenon, astronomers gained more data about sunspots in 1 week than they had in the previous 250 years. The obvious thing to do was to look for an effect produced by this storm on the changing length of day. That would have proved Danjon's ideas correct (if further proof were required after Challinor's work) and would perhaps encourage investigations of the mechanism linking the Earth's rotation with solar activity. Or so I hoped.

After waiting until the end of 1972, in order to have data concerning changes in the length of day for a reasonable period of time on either side of the August solar storm, Gribbin and Plagemann (1973) analysed the variations in AT–UT2 shown by observations made at the US Naval Observatory in Washington. These data show a very clear disturbance on 8 August 1973; this takes the form of a sudden increase in AT–UT2, that is, an increase in the length of day, just as we expected if Danjon were correct. Although the effect was very small, it was the largest 'glitch' in the USNO data covering 3 months on either side of the great solar storm, and we felt sure that the two events were causally related, particularly since the best explanation of the phenomenon would seem to be through a solar influence on the circulation of the atmosphere, and the 4-day interval between solar storm and change in the length of day looked about right for such a mechanism.

But, unfortunately for this happy-sounding argument, the days of the gentleman amateur in science seem to be over. It seems that we should not have relied on data from just one observatory, because there are inevitably local errors in the measurement of UT. A real astronomer would have used an appropriate average of data supplied by many observatories around the world, such as the data used by the Bureau International de l'Heure in Paris. Using this correct approach to the problem, O'Hora and Penny (1973) could find no evidence of the 'glitch' which is so clearly present in the Washington data.

There is an important distinction to be made here, however, between the possible effect of one solar storm on the measurably changing spin of the Earth, and the cumulative effect of solar activity over the 11-year solar cycle. The first effect is controversial; we cannot point to one discrete change in the length of day and say, to the satisfaction of all astronomers, that this change was produced by one specific event on the Sun. What we can do, without any reasonable doubt at all, is point to the more or less continuous changes in the length of day over one or more solar cycles and say, to the satisfaction of everyone who studies such variations, that these changes are related to changes in the average level of solar activity over that cycle or those cycles.

Whichever way the argument about the 1972 solar storm is eventually resolved —and it certainly looks as if the correlation with the Earth's spin cannot be proved for that particular event—it is certainly a remarkable coincidence, to say the least, that when searching for such an effect we just happened to obtain our data from the only observatory in the World where there is a fluctuation of exactly the right size in those data, occurring at just the time we expected. But, that aside, it would certainly have been futile to press our luck by developing this specific line of research further, since that would have meant that we should await yet more solar storms of the size needed to check the idea that solar activity affects the length of day. Therefore, I have turned to the other side of the problem: if the link between solar activity and length of day exists, what is the mechanism through which it operates?

Solar activity and climatic change

The question of a relation between solar activity and meteorological phenomena has become fashionable in recent months, and was the subject of a symposium at the NASA Goddard Space Flight Center in November last year. Quite apart from the possibility that this might provide a mechanism through which events on the Sun could affect the rotation of the Earth, the subject is particularly appropriate to the theme of this conference, both because it would be an unusual biological organism which could not be affected by climatic changes, and because the techniques used to investigate how the habits of those organisms have changed over the millenia might, in this case, provide a clue to how the climate of the Earth has changed over the same periods of time. But although this may now be a fashionable area of research, and the recent tragedies in Africa and India emphasize the need for urgent investigation of any idea which might provide a clue to the causes of climatic change, the basic concept is far from new. Approximately 100 years ago, on 8 January 1874, an article in *Nature* reported work by 'a German physicist Dr. Küppen' relating to temperature cycles on the Earth and possible relationships between these temperature variations and the sunspot cycle. 'Dr Küppen proposes', our counterparts of 100 years ago were told, 'to examine the influence of periodic weather changes . . . on some phenomena of organic nature' (Anonymous, 1874).

Following in Küppen's footsteps, then, what is the evidence for a link between small-scale variations in solar activity, particularly over the 11-year cycle, and meteorological and climatic events on Earth? The prime source for any student of climatic change is Lamb's book *Climate: Present, Past and Future* (1972), and Lamb does summarize attempts which have been made to relate solar cycles to climatic change. The remarkable thing about this work is that it has surfaced time and again over the past 100 years or so, but has never until now become firmly accepted as a basis for further work. Perhaps we can do something to rectify that. The latest re-emergence of the particular idea of a link between climatic changes and the 11-year solar cycle comes as a result of King's (1973) study; the evidence comes from many sources, including conventional meteorological records and interpretation of the statistics in Wisden to obtain an indication of the occurrence of 'good' and 'bad' summers during the 19th century. There remains no room for doubt; variations and trends in the Earth's climate over decades can indeed be associated with solar variations over the 11-year sunspot cycle.

But King does more than 'merely' suggest that climatic variations can be related to solar changes over any one solar cycle. The average activity of the Sun over a whole cycle, or a series of cycles, also seems to be related to important, practical meteorological effects on Earth. The length of the 'growing season' is a well-defined climatic parameter, and depends on the number of days in the year for which the air temperature 1·25 m above the ground at an observing station exceeds 5·6°C. King points out that 'the season [at Eskdalemuir] is on average about 25 days longer near sunspot maximum than near sunspot minimum'. Further, the growing season tends to be longest approximately a year after sunspot maximum, and if the sunspot activity at the time of maximum solar activity is unusually high, then there is a correspondingly high peak in the plot of the length of the growing season.

This has obvious biological implications. Between 1920 and 1970, the average activity of the Sun in successive cycles has shown marked changes which are reflected in the length of the growing season. The first two sunspot maxima in this century were not particularly impressive, and the length of the growing season remained around 200 days (apart from the 11-year variation associated with the solar cycle) until the mid-1930s. Subsequently, three solar cycles followed in which each sunspot maximum was more pronounced than the one before; the effect on the growing season was to produce peaks at around 220 days, 225 days, and, in 1959, 235 days. The latest peaks, in both sunspot activity and length of growing season, show a decline from this 'maximum of maximums'.

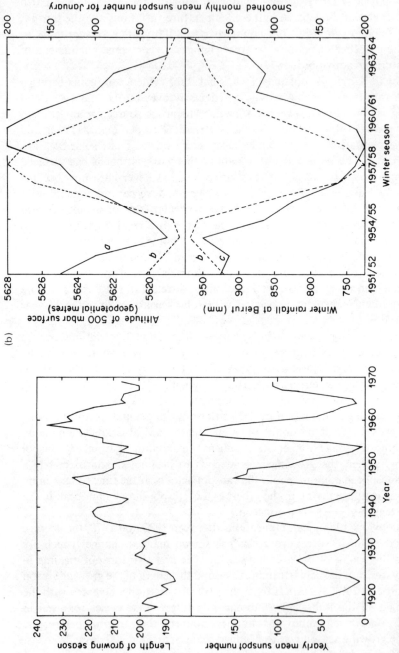

FIGURE 1a. Top: the length of the 'growing season' (that is, the number of days in the year on which the averaged temperature exceeds 5·6°C) at Eskdalemuir (55° N, 3° W); bottom: yearly mean sunspot numbers. 1b. Relation between atmospheric parameters and solar activity. The altitude of the 500 mbar surface at 40° N between 110° W and 70° E (a) and the rainfall at Beirut, 34° N, 36° E (c) (both for October–May periods) are compared with sunspot number (b). The sunspot number scale is inverted in the lower figure to make the comparison easier. (Reproduced with permission from King, J. W. (1973), *Nature*, **245**, 443)

Variations in many other terrestrial parameters can now be linked to variations in the Sun's activity (see King, 1973, for a recent summary); but we still have not found a detailed mechanism to explain the phenomena. Nevertheless, in this part of the jigsaw puzzle at least, the pieces are beginning to fit together coherently. When solar cosmic rays arrive at the Earth, they produce auroral activity which can be monitored with no more sophisticated equipment than the human eye, and over the past decade Roberts and colleagues (see, for example, MacDonald and Roberts, 1960; Roberts and Olsen, 1973) have used this indicator of solar activity to find a link with the weather at high latitudes. Shortly after the arrival of solar cosmic ray bursts at the Earth (indicated by spectacular auroral displays) low-pressure systems forming in the Gulf of Alaska are larger than average; this ties in with still earlier work (Duell and Duell, 1948) which showed that under certain conditions geomagnetically disturbed days were followed by pressure rises of about 3 mbar in the Greenland–Iceland area; as King (1973) points out, 'it seems reasonable to suggest . . . that this behaviour may be a consequence of the incursion of energetic particles', which are particularly likely to influence the atmosphere in that area because of the guiding influence on these charged particles of the Earth's magnetic field.

But it is not just at high latitudes that the arrival of solar particles in strong bursts affects the distribution of air in the Earth's atmosphere. Observations carried out at Zugspitze, 3 km above sea level, in the Bavarian Alps, show that shortly after solar cosmic and x-rays arrive at the Earth, there is a sharp increase in the levels of the radionuclides beryllium-7 and phosphorus-32 in the troposphere. These increases cannot be explained directly, in terms of cosmic-ray particles themselves penetrating into the troposphere, but rather correspond to the injection of stratospheric air, under the influence of the solar activity, down into the lower troposphere (Reiter, 1973). Such effects are of crucial importance to the general circulation of the atmosphere, and to the climate on Earth.

Changes in the general circulation

The fundamental facet of the climatic puzzle which must be resolved before the picture of climatic change can become clear is the problem of atmospheric circulation and changes in the general circulation. There is no doubt that the circulation pattern as a whole does move; because of the higher proportion of land, as opposed to ocean, in the northern hemisphere, the 'atmospheric equator' lies north of the geographical equator. That is, the Intertropical Convergence Zone (ITCZ), where the circulation patterns of the northern and southern hemispheres meet, lies in the northern hemisphere. But the exact position of the ITCZ has changed, certainly over thousands of years and probably within historical times.

This convergence of winds in the ITCZ produces a rainy belt which today lies across the Galapagos Islands; geological evidence shows that between 30,000 and 10,000 years ago, the Galapagos were much drier, and clearly at that time the ITCZ did not lie across the islands. The question of whether it then lay further to the north or further to the south than at present is unresolved, and I do not wish to enter that controversy here; the point is simply that there are significant changes in the circulation pattern.

That kind of change has recently been the subject of discussion in the context of the droughts affecting the region south of the Sahara, Ethiopia and parts of India (Winstanley, 1973a,b). Winstanley's evidence concerns three related effects in particular: since the mid-1950s, rainfall in the Middle East and Africa

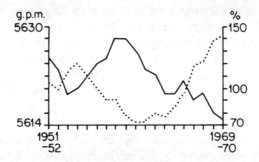

FIGURE 2. ...., 5-year running means of the mean percentage of normal seasonal rainfall (October–May) at Bikaner (India), Shiraz (Iran), Mosul (Iraq), Beirut (Lebanon), Jerusalem (Israel), Jiddah and Riyadh (Saudi Arabia), Nalut (Libya), Marrakech (Morocco) and Atar (Mauritania); ——, 5-year running means of the mean altitude of the 500 mbar surface (geopotential metres) along 40° N, between 110° W and 70° E, from October–May. (Reproduced with permission from Winstanley, D. (1973), *Nature*, **243**, 414)

north of the Sahara during the rainy season (October to May) has been increasing; to the south of the desert, there has been a corresponding decrease in rainfall, building up to the present drought situation; and there is a very clear anti-correlation between the changing rainfall in the belt north of the Sahara and the height of the 500 mbar surface along 40° N, between 110° W and 70° E.

The third effect provides the clue to how changes in the atmospheric circulation are causing the climatic changes. The decrease in the altitude of the 500 mbar surface during the 1960s signifies, says Winstanley, 'an increase in the southward extent of the troughs in the circumpolar westerlies' which are a dominant feature of the weather of the northern hemisphere. In less technical terms, the whole pattern of climatic zones in the northern hemisphere is shifting south. Droughts along the southern edges of northern hemisphere deserts, increased rainfall along

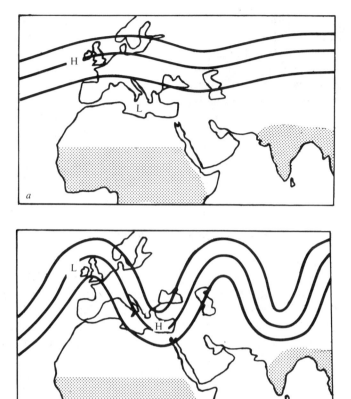

FIGURE 3. Diagrammatic representation of extreme circulation types, according to Winstanley (1973b), and their associated rainfall patterns. Top: Strong zonal circulation gives high rainfall over England and allows monsoon rains to extend well into India. Bottom: weak zonal circulation allows 'waves' to develop in the circumpolar westerlies; gives low rainfall over England and suppresses the northward extent of the monsoon rains (H, High rainfall; L, Low rainfall). According to Winstanley, we are now moving from a situation like that of the upper figure to one more accurately represented by the lower figure. (Reproduced with permission from Winstanley, D. (1973), *Nature*, **245**, 190)

the northern margins of those same deserts, and even the establishment of drier conditions over Britain are all part of the same pattern, and all have their cause in a change in the general circulation. The fact that this change in circulation is not at all pronounced, and has only been noticed by analysis of records covering 20-odd years, only serves to emphasize how sensitive biological organisms are to the slightest change in climate. Of course, this very sensitivity of biological organisms to climatic change is proving invaluable to climatologists trying to unravel patterns of climatic change over the the millenia. Tree rings and periodic changes in the growth patterns of invertebrates are two obvious examples, and it is easy to see how a great deal of the work described at this conference could be used for climatic studies. But if we are to tackle the problem from the other direction, in an attempt to explain observed changes in the growth patterns of these organisms in terms of changes in the Earth's climate, it is necessary to look outside the Earth for the cause of the changes in the atmospheric circulation. In particular, if cycles or patterns can be found in the variations of the external influences, then these 'rhythms' could provide the means to extrapolate backwards in time, 'predicting' the state of the climate hundreds or thousands of years ago.

Planetary alignments and solar activity

We have already seen that one of the influences affecting the Earth's climate, through its influence on the atmospheric circulation, is the activity of the Sun. There is no doubt that there are other influences which affect the climate, and studies of the variation of the Earth's overall magnetic field look particularly promising (King, 1974 and personal communication); but if it is possible to find out why the Sun's activity varies then we will have at least a grasp of the problem of why the Earth's climate changes.

Wood (1972) has pointed out a remarkable correlation between the activity of the Sun, indicated by sunspot number, and the (height of) the tide raised on the Sun by the four 'tidal planets', Venus, Earth, Mercury and Jupiter. These planets produce between them a tide which varies with a period of roughly 11 years; the exact period varies just as the sunspot cycle varies, with the maximum of the solar cycle occurring at the time of greatest tide. This agreement is so good that it seems difficult not to believe that the tidal influence of these planets is the cause of the observed sunspot variations, modulating the sunspot activity of the Sun. But there is more to this work than the correlation of the '11-year' solar cycle with the '11-year' tidal variations. The height of the tide raised by the four tidal planets can also be correlated with the number of sunspots occurring at any particular maximum of the solar cycle. If the maximum smoothed monthly number of sunspots is examined, there is some evidence of a pattern which repeats after about 170 to 180 years. This is particularly intriguing because, although Wood's work is concerned only with the four planets which produce the greatest tidal effect on the Sun, it seems likely that the outer planets will also produce some effect; and there is a period of 179 years between the alignments of these outer planets (Jupiter, Saturn, Neptune and Uranus). If these planets do indeed have a tidal effect on the Sun which produces a variation in sunspot activity, then the effect should indeed be seen as a repetition in the solar cycle pattern with a period of 'about 170 to 180 years' (Gribbin, 1973).

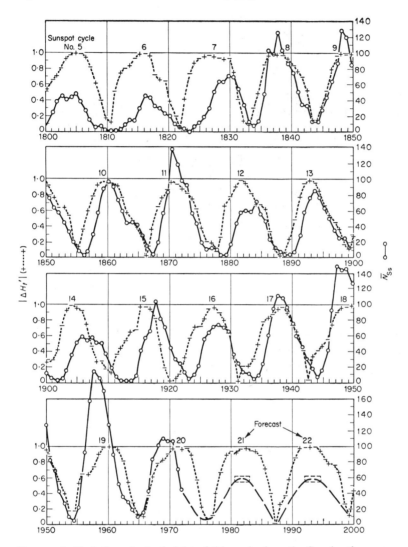

FIGURE 4. Variations in the height of tides raised on the Sun by the combined effects of Venus, Earth, Mercury and Jupiter (+) compared with variations in sunspot number (○) since 1800. $\Delta H_t'$ is a dimensionless measure of how tidal height (H_t) varies relative to the maximum value. (Reproduced with permission from Wood, K. D. (1972), *Nature*, **240**, 91)

Can this influence be detected in climatic records? According to Winstanley (1973b), the present variations in rainfall north and south of the Sahara are part of a cycle which has a period of about 200 years. Lamb (1972, p. 443) has reviewed attempts to relate climatic changes to long solar cycles, and one of these relations links a roughly 200-year cycle of climatic change with a more or less periodic variation in radiocarbon levels, which in turn are presumably affected by changing levels of solar cosmic rays. This link between solar fluctuation and associated climatic change with a period of about 200 years has been found in data going back for 4400 years.

Perhaps the similarity between the approximately 200-year variation of climate and radiocarbon levels, and the '170 to 180 year' variation in solar activity can both be tied to the influence of the outer planets on the Sun, boosting the tides produced by the four 'tidal planets', when they align together on one side of the Sun, every 179 years. The theory can easily be tested, since fortunately the next such alignment is due in 1982; according to these ideas, it should produce a marked effect on the Sun's activity, which in turn should affect the Earth's atmosphere and climate to a small, but perhaps noticeable, degree. I have mentioned elsewhere (Gribbin, 1973) that meteorologists and others concerned with predicting climatic change could do worse than look at Wood's (1972) predictions of sunspot activity, based on the movements of the planets over the next few decades; I would now add that anyone interested in the way that biological clocks have varied during the course of the Earth's history would also be well advised to consider how planetary alignments have affected tides on the Sun, and possibly sunspot activity, both because of the resulting effect on the climate (which must affect biological organisms) and because of the direct effect of solar activity on the length of day.

References

Anonymous (1874). On temperature cycles. *Nature*, **9**, 184–185
Challinor, R. A. (1971). Variations in the rate of rotation of the Earth. *Science*, **172**, 1022–1024
Danjon, A. (1962a). Sur la variation continue de la rotation de la Terre. *C. R. Acad. Sci., Paris*, **254**, 2479–2482
Danjon, A. (1962b). La rotation de la Terre et le soleil calme. *C. R. Acad. Sci., Paris*, **254**, 3058–3061
Duell, B. and Duell, G. (1948). The behaviour of barometric pressure during and after solar particle invasions and solar ultraviolet invasions. *Smithsonian misc. Collection*, **110**, No. 8 (Washington)
Gribbin, J. (1973). Planetary alignments, solar activity and climatic change. *Nature*, **246**, 453–454
Gribbin, J. and Plagemann, S. H. (1973). Discontinuous change in Earth's spin rate following great solar storm of August 1972. *Nature*, **243**, 26–27
King, J. W. (1973). Solar radiation changes and the weather. *Nature*, **245**, 443–446
King, J. W. (1974). Weather and the Earth's magnetic field. *Nature*, **247**, 131–134
Lamb, H. H. (1972). *Climate: Present Past and Future*, Methuen, London, 613 pp.
MacDonald, N. J. and Roberts, W. O. (1960). Further evidence of a solar corpuscular influence on large-scale circulation at 300 mb. *J. geophys. Res.*, **65**, 529–534
O'Hora, N. P. J., unpublished work mentioned by King, J. W. in a personal communication

O'Hora, N. P. J. and Penny, C. J. A. (1973). Rotation of the Earth during the 1972 solar event. *Nature*, **244**, 426–427

Reiter, R. (1973). Increased influx of stratospheric air into the lower troposphere after solar Hα and x-ray flares. *J. geophys. Res.*, **78**, 6167–6172

Roberts, W. O. and Olsen, R. M. (1973). Geomagnetic storms and wintertime 300 mb trough development in the North Pacific–North America area. *J. atmos. Sci.*, **30**, 135–140

Schatzman, E. (1966). Interplanetary Torques, pp. 12–25. In *The Earth–Moon System*, (Ed. A. G. W. Cameron and B. G. Marsden), Plenum, New York

Winstanley, D. (1973a). Recent rainfall trends in Africa, the Middle East and India. *Nature*, **243**, 464–465

Winstanley, D. (1973b). Rainfall patterns and general atmospheric circulation. *Nature*, **245**, 190–194

Wood, K. D. (1972). Sunspots and planets. *Nature*, **240**, 91–93

THE DETECTION OF RECENT CHANGES IN THE EARTH'S ROTATION

N. P. J. O'HORA

Royal Greenwich Observatory, Herstmonceux, Sussex, UK

Abstract

Universal Time (UT) is briefly explained and improvements achieved at the Royal Greenwich Observatory (RGO) in the determination of UT by the use of the Photographic Zenith Tube (PZT) are demonstrated. The principles underlying the operation of atomic clocks and the methods employed in the derivation of uniform time-scales from these clocks are described. Comparisons between the PZT observations of UT and the AT scale of the RGO indicate how irregularities in the rate of rotation of the Earth may be detected, and illustrate some of the practical difficulties in the work. The nature of the changes in rate of rotation that occur is explained and some research in the geophysical causes of the changes is described.

Universal Time

The measurement of time has always been closely connected with the rotation of the Earth. For obvious reasons the Sun is the natural reference body for measuring the rotation but in precise work it is never used directly for this purpose, mainly because the period of the Earth's rotation with respect to the Sun varies throughout the year due to the orbital motion of the Earth and also because it is difficult to observe, with sufficient accuracy, transits of meridian by the Sun. The true period of the rotation is the interval between successive meridian transits of the same fixed star and, since stellar transits can be observed with comparatively high accuracy, the rotation of the Earth is measured by timing with a clock the transits of selected stars whose directions relative to the centre of the Earth are specified in a reference system of co-ordinates. Such observations give directly Sidereal Time (ST) which, despite its advantages for many purposes, is not a convenient time-scale for ordinary usage because it is out of phase with the diurnal motion of the Sun which governs so much of the activity of everyday life. A solar time-scale, closely related to the diurnal motion of the Sun and free from the annual variations due to the orbital motion of the Earth, is achieved by the use of a fictitious body known as the Mean Sun. The period of revolution with respect to the Mean Sun is nominally constant and is the same as the period measured with respect to the true Sun averaged over a year. The time-scale based on the Mean Sun is mean solar time and this, when referred to the meridian of Greenwich, is entitled Greenwich Mean Time (GMT). For use in international time-keeping GMT is formally designated Universal Time (UT).

Observed values of UT, generally referred to as UT observations, are computed from ST observations by the use of a numerical formula. Basically, this formula calculates the instantaneous co-ordinates of the Mean Sun in the same reference system as that of the stars. The ST and UT scales are thus rigorously connected and so nearly equivalent that the choice of either is frequently one of convenience. The two scales are strictly proportional: any interval such as a day, hour, minute or second in the UT scale is 1·002738 times the duration of the length of the corresponding interval in the ST scale (the exact ratio is known to higher accuracy and contains a small linear term). It follows that, in general, the properties of UT are also characterstic of ST.

Until about the beginning of this century the rate of rotation of the Earth was generally assumed to be constant, so that time-scales based on the rotation were deemed to be uniform and this assumption was justified in so far as the performance of the Earth as a clock had always been superior to contemporary mechanical clocks. Early in the century, however, evidence of irregularities in the rate of rotation was obtained by comparisons of the rotational motion with the orbital motions of some solar system bodies. It is now recognized that three types of departure from uniformity affect the rate of rotation of the Earth:

(a) a secular retardation leading to a progressive increase in the length of the day;
(b) irregular and seemingly random changes in speed of rotation;
(c) a fairly regular and comparatively small seasonal variation that can conveniently be represented by annual and semi-annual components that are quite repetitive, with the rotation faster in spring and slower in autumn.

Physical explanations for (a) and (c) have been derived and research is now mainly centred on the irregular changes in speed. The three types of variation may easily be detected by comparing UT observations with the time kept by a clock with a constant rate. It is of interest that all three were discovered before such clocks became available, but the introduction of uniform scales of atomic time nearly 20 years ago has greatly improved the quality and ease of such comparisons.

The fundamental interval in the scale of UT is the length of the day (l.o.d.) which is determined by the period of rotation of the Earth with respect to the Mean Sun. Because of variations in the rate of rotation of the Earth the UT scale is not uniform and consequently it is inappropriate for some scientific purposes. The scale is sufficiently uniform for most applications and, because it is a measure of the instantaneous orientation of the Earth with respect to the stars, it is necessary for use in navigation, geodetic surveying, radio astronomy and space science. New techniques are now being developed for the determination of UT but it seems likely that for many years it will continue to be determined by measuring the clock times of stellar transits.

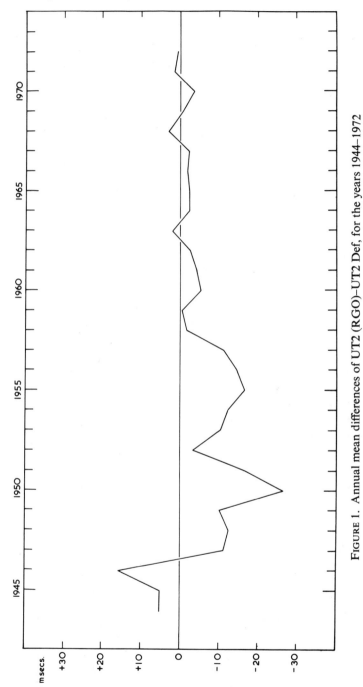

FIGURE 1. Annual mean differences of UT2 (RGO)–UT2 Def, for the years 1944–1972

In the United Kingdom, UT is determined with a Photographic Zenith Tube (PZT) at the Royal Greenwich Observatory (RGO), Herstmonceux, Sussex. This instrument was designed by the late D. S. Perfect in the RGO laboratories and was built by Grubb Parsons. It was installed at Herstmonceux in 1955 and, after 2 years of preliminary observations, it replaced the small transit telescope. Descriptions of the instrument (Perfect, 1959) and of the methods employed in the reductions (Thomas, 1963) have been published. The UT observations of individual observatories are combined in order to establish an improved scale known as Temps Universel Définitif (TU Def) and differences between TU Def and the individual scales are published. Figure 1 shows the mean annual difference between UT determined at the RGO and TU Def: it illustrates the substantial improvement in the long-term stability of the RGO system since 1957 and, to a lesser extent, the contemporaneous enhancement in the precision of TU Def. There has also been a great reduction in the day-to-day scatter of the observations. In 1936, the last complete year of small transit operation in the RGO, the probable error of observation of a single star exceeded $\pm 0^s \cdot 04$; in 1972 this quantity for a PZT observation amounted to $\pm 0^s \cdot 014$. A further advantage of the PZT is its ability to measure the latitude of the observatory, which varies because of a motion of the pole of rotation of the Earth about the pole of figure; this polar motion is obtained by analysis of such latitude observations. The high degree of international co-operation this work requires is organized by the Bureau International de l'Heure (BIH), the international agency which is also responsible for the computation of values of TU Def and of the co-ordinates of the instantaneous pole of rotation with respect to the mean pole. The co-ordinates are needed for correcting UT observations for the differential variations in the longitudes of observatories caused by the polar motion, and corrected in this way the observations are designated UT1. This reduction to standard co-ordinates for each observatory enables their time-scales to be compared with each other; these standard co-ordinates are also necessary for the definition of a common geodetic reference system.

The UT1 observations provide the most accurate measures currently available (on a routine basis) of the rate of rotation of the Earth. Apart from their use in position finding, they are of great interest in several fields of geophysical research. Both the time and latitude observations are affected by regular deflexions of the vertical caused by changes in luni-solar gravitational attraction and the measurement of these can be used for evaluating the elastic properties of the Earth. Systematic deflexions have been measured with the Herstmonceux PZT with an accuracy of $\pm 7 \times 10^{-9}$ (O'Hora, 1973). Geophysical research in the Chandler Wobble, the free polar motion, depends on the data furnished by latitude observations. Perhaps the most intense and widespread interest in the results is concerned with the changes in the rate of rotation of the Earth because these involve so many different disciplines. The scope of this paper is limited to consideration of some geophysical phenomena that might influence the rotation but a wide-ranging review of the geophysical interest both in changes in the rate of rotation and in the polar motion was published recently (Rochester, 1973).

Atomic time-scales

Quartz crystal clocks can maintain the same rate, with a precision of the order 1×10^{-10} over a period of a few weeks, or even months, but because they are susceptible to spontaneous changes in rate, or in rate of change of rate, their accuracy over longer periods falls off rapidly. A caesium beam resonator was brought into service at the NPL in 1955 (Essen and Parry, 1957): other scientific establishments also developed atomic standards and commercial standards became generally available. The long-term uniform time-scales which came into general use about 1958 are based mainly on commercial caesium-beam atomic clocks.

An atom can exist in different states depending on the configuration of its electrons; each state corresponds to a particular energy value and changes from one state to another are accompanied by quantized emission or absorption of an electromagnetic wave with frequency f where

$$f = (E_1 - E_2)h$$

In this equation, since h is Planck's constant and E_1 and E_2 are the two energy levels corresponding to the state of the atom, it follows that f is a constant for transitions between these levels. The value of f is directly proportional to the difference in the energy levels and, in order to obtain a frequency low enough for comparison with that of a quartz crystal oscillator, suitable values of E_1 and E_2 must be close to each other. The caesium atom proved particularly suitable for use in this work when a convenient method was devised for triggering and maintaining transitions between closely spaced energy levels corresponding to a frequency of approximately 9200 MHz. The original NPL standard could only function for short periods; it was operated two or three times a week in order to monitor the frequency of quartz crystal oscillators. In principle the output frequency of the crystal oscillator was tuned to match the resonant frequency of the transitions so that the number of transition cycles in the period of one vibration of the oscillator could be determined. The crystal oscillator was then used to monitor the caesium frequency in terms of an astronomical time-scale. The frequency of the NPL caesium standard was calibrated in this way by comparisons with the UT observations of the RGO in 1955 and later it was calibrated in terms of Ephemeris Time by co-operation with the US Naval Observatory. Ephemeris Time (ET) is a uniform time-scale based on the orbital motion of the Earth and, for practical reasons, it is very little used outside astronomy.

In principle, the commercial atomic clocks are similar to the NPL prototype, but it is now possible to maintain them in continuous operation. The clock time is obtained from the quartz crystal oscillator that is incorporated in the clock, with its frequency locked in a fixed ratio to that of the atomic transitions which it triggers. The locking is achieved by the use of a servo-mechanism which adjusts the frequency of the crystal whenever the frequency of the alternating magnetic field, which induces the transitions, diverges from the value that gives maximum response. Correcting the crystal frequency restores the alternating field to the invariant resonant frequency of the transitions, and the self-adjusting system thus generates, through the quartz crystal, a frequency that is defined by the atomic transitions.

The value obtained for the caesium frequency in terms of the ET second (Markowitz *et al.*, 1958) was confirmed by later analyses and was subsequently used for defining the second; the SI unit is now defined as follows:

The second is the duration 9 192 631 770 periods of the radiation corresponding to the transition between the two hyperfine levels of the ground state of the caesium atom 133.

No absolute reference standard that can be employed for assessing the performance of individual atomic clocks has yet been produced and, in practice, each clock is checked by comparisons with others. In the RGO, the 5 clocks upon which GA2, the U.K. scale of atomic time is based, are monitored by continuous recording and regular digital registration of timing differences, from which the relative performance of each clock is deduced. Timing of the pulses of the Loran-C system of navigational radio transmissions enables comparisons to be carried out between the standard in laboratories in different countries. The results of these international comparisons are reported regularly to the BIH where the information is utilized in the formation of the scale of International Atomic Time (IAT). In turn, data published by the BIH can be used to evaluate the performance, relative to IAT, of all the individual standards upon which the international scale is based. Even though the statistical methods employed in the BIH are designed to ensure that effects in the scale, due to changes in the number or in the performance of standards upon which it is based, are kept to a minimum, it is clear that the lack of an absolute reference standard renders this improvised procedure for the formation of IAT far from ideal. On the other hand, it is right to point out that it is most unlikely that existing methods and standards would allow the accumulated error of IAT, over a decade, to exceed the value of, say, 50 μsec. In terms of the range of accuracy of UT at any epoch—of the order of a few msec—IAT is extremely uniform. An issue of the Proceedings of the IEE devoted to time and frequency measurement contains an up-to-date comprehensive review of the development of the different time-scales (Smith, 1972).

Research on the rotation of the Earth

Information on the rotation of the Earth in modern times has mainly been derived from studies of the motions of the Sun, Moon, Mercury and Venus and

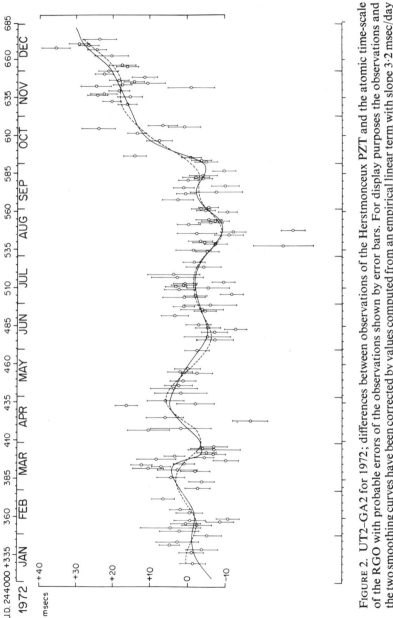

FIGURE 2. UT2−GA2 for 1972; differences between observations of the Herstmonceux PZT and the atomic time-scale of the RGO with probable errors of the observations shown by error bars. For display purposes the observations and the two smoothing curves have been corrected by values computed from an empirical linear term with slope 3·2 msec/day

also by means of comparisons of RT with clocks. This chapter is chiefly concerned with clock measures of the rotation.

It may be of interest to recall that one of the first tasks undertaken by Flamsteed at Greenwich, following his appointment as first Astronomer Royal, was an investigation of the regularity of the Earth's rotation, as measured by the Observatory clock (Howse, 1970). The accuracy of the clock was of the order 1×10^{-4} and it is hardly surprising that it failed to reveal irregularities in the rotation. The next recorded investigation of this type in the RGO was made in 1929 when comparisons of the free pendulum clocks with UT were analysed for confirmatory evidence of variations in the rotation of the Earth that had been provided by lunar and planetary observations. The short-term accuracy of these clocks was, at best, 1×10^{-8} but the long-term performance was inferior to that of the Earth and proved incapable of checking it (Jackson, 1929). Only a few years later, by combining the results of selected free pendulum clocks of a number of observatories, reliable evidence of an annual variation in the rate of rotation was discovered (Stoyko, 1936). Considering the quality of the clocks and of the observations used in the analysis, this was a remarkable result. Confirmatory evidence could not be obtained until 14 years later; this was provided by an analysis of comparisons of the quartz crystal clocks of the RGO with the UT results of the observatory (Finch, 1950). Both these analyses and also a third investigation in Germany (Schiebe and Adelsberger, 1950) yielded much larger amplitudes for the seasonal variation than the currently accepted value; the first determination of the phase and amplitude in agreement with present-day values was also based on RGO data (Smith and Tucker, 1953). Comparisons of UT with AT furnish the most refined information available on the rotation of the Earth since 1955. In examining the long-term changes that have occurred, it is convenient to remove the effects of seasonal variation and UT1 corrected in this way is known as UT2. Figure 2 shows the UT2 results of the Herstmonceux PZT compared with GA2. The plotted points have different weights, depending on the number of stars observed in a night. Two smoothing curves are shown, one obtained by a numerical method based on a least-square solution and the other by the traditional method of manual curve fitting. This diagram was originally prepared for an evaluation of the two methods; it is presented here in order to illustrate the quality of the results and to demonstrate the numerous small changes in trend in the observations. Any of these changes could be attributed to any one of a number of different effects such as a change in the rate of rotation of the Earth, systematic errors in the adopted positions of the reference stars, an instrumental change or even an improbable association of purely random errors. Considerable reduction in ambiguity can be achieved by the use of TU Def results to represent the behaviour of the Earth; these are largely free from the systematic errors of a single observatory. The ability to ascertain whether a change in trend occurred abruptly or developed gradually would obviously be advantageous in tracing the cause of the change. Coupled with the

The Detection of Recent Changes in the Earth's Rotation

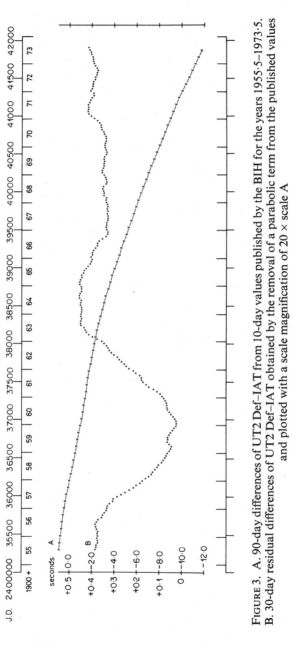

FIGURE 3. A. 90-day differences of UT2 Def–IAT from 10-day values published by the BIH for the years 1955·5–1973·5. B. 30-day residual differences of UT2 Def–IAT obtained by the removal of a parabolic term from the published values and plotted with a scale magnification of 20 × scale A

importance of deciphering the nature of a change is the need for timing its onset and, in general, even with the use of the BIH data, the scatter of the observations is too large to allow such information to be obtained with sufficient accuracy. The large increase in accuracy claimed for new techniques, such as very long base-line interferometry (VLBI), for the determination of UT should greatly facilitate the identification of geophysical activities responsible for changes in trend.

Because the l.o.d. is inversely proportional to the rate of rotation of the Earth it is a convenient parameter for monitoring the rotation. The change in UT2-AT in a day gives directly the departure of the l.o.d. from its nominal length of 86,400 sec (SI). From the first and second differences of a table of values of UT2-AT it is possible to calculate changes and rates of change in the l.o.d. In an investigation of the UT results of the US Naval Observatory for the years 1956–69 Challinor obtained evidence of a correlation between the rate of change of the l.o.d. and solar activity as defined by sunspot numbers (Challinor, 1971). Other attempts to establish a relationship between the l.o.d. and solar activity of one kind or another, and excluding the normal heating effects of the Sun, have been challenged and Challinor was careful to indicate how his own results differed from those of other investigators. It is known that weather affects the rotation of the Earth; the annual term in the seasonal variation has long been explained in terms of regular atmospheric changes that are meteorological in origin and recent work has established that the intermittent biennial term in UT can be explained in terms of variations in global zonal wind circulation at high altitudes (Lambeck and Cazenave, 1973). Challinor pointed out that the validity of the correlation he had established would depend on positive evidence of the influence of solar activity on weather. The convincing evidence of such influence recently published (King, 1973) has led to a re-examination of the evidence of a connection between sunspot activity and changes in the rate of rotation of the Earth, using different data over a longer period.

The data analysed are 10-day values of TU2 Def-IAT, abstracted from BIH publications, for the period 1955·5–73·5. The general trend of the data is shown in Figure 3, where values are plotted for 90-day (approximately quarterly) intervals. In order to present the measures in greater detail a parabolic ephemeris was removed from the data and 30-day values of the residual values are also included in the Figure. The square term of the parabola equals $-1·993 \times 10^{-7}$ sec/day^2 which represents an average increase of $14·6 \times 10^{-3}$ sec/cy in the l.o.d. This is much greater than the measures derived for the past two millenia from eclipse records (Newton, 1970) and for the past three centuries from lunar observations (Morrison, 1973); these amounted to $2·3 \times 10^{-3}$ and $1·5 \times 10^{-3}$ sec/cy, respectively. Recent lunar observations confirm that the current retardation of the rotation of the Earth greatly exceeds its long-term average value.

Data used in the analysis are summarized in Table 1, in which mid-yearly (1 July, 0^h) values of TU2 Def-IAT are tabulated in column 2. The first differences

The Detection of Recent Changes in the Earth's Rotation

TABLE 1

(1) Date	(2) UT2-IAT sec	(3) Excess l.o.d. msec	(4) Rate of change of l.o.d. msec/year	(5) Zurich sunspot numbers	(6) Seismic strain release erg$^{0.5}$
1955·5	+0·962				$10^{13}\times$
		−0·726			
56·5	+0·696		−0·421	141·7	0·89
		−1·147			
57·5	+0·277		−0·285	190·2	1·59
		−1·432			
58·5	−0·246		+0·154	184·8	1·11
		−1·278			
59·5	−0·713		−0·013	159·0	1·06
		−1·291			
60·5	−1·186		+0·163	112·3	1·08
		−1·128			
61·5	−1·598		−0·071	53·9	0·88
		−1·199			
62·5	−2·036		−0·145	37·5	0·75
		−1·344			
63·5	−2·527		−0·483	27·9	1·09
		−1·827			
64·5	−3·196		−0·215	10·2	1·23
		−2·042			
65·5	−3·942		−0·282	15·1	1·42
		−2·324			
66·5	−4·791		−0·096	47·0	0·30
		−2·420			
67·5	−5·675		+0·031	93·7	0·33
		−2·389			
68·5	−6·550		−0·220	105·9	0·56
		−2·609			
69·5	−7·503		−0·126	105·5	0·68
		−2·735			
70·5	−8·502		+0·052	104·5	0·72
		−2·683			
71·5	−9·482		−0·427	66·6	0·79
		−3·110			
72·5	−10·621		+0·003	68·9	0·50
		−3·107			
73·5	−11·756				

of these measures reduced to units of msec/day are given in column 3. This column thus contains values of the nominal minus the actual length of the day, averaged over periods of a year and tabulated for the beginning of calendar years, corresponding to mid-period dates. The differences of column 3, which

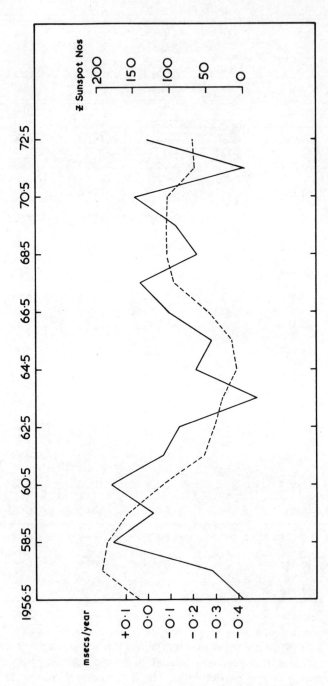

FIGURE 4. Values of annual rates of change in the length of the day (joined by full lines) and Zurich Sunspot Numbers (joined by hatched lines), for the years 1956–1972

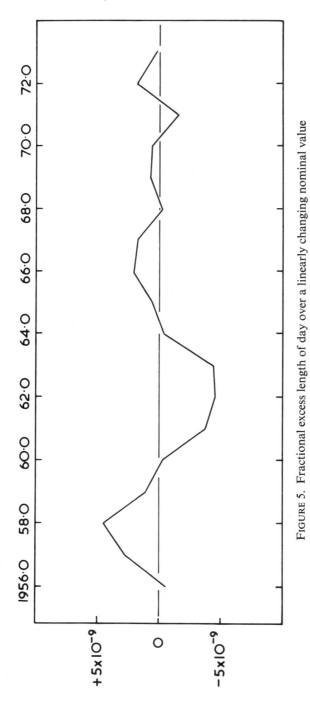

FIGURE 5. Fractional excess length of day over a linearly changing nominal value

are tabulated in column 4, are therefore measures of the rate of change of the l.o.d. in msec/year for the mid-year dates. Zurich sunspot numbers for each year are listed in column 5 of the table.

The annual sunspot numbers together with the annual changes in the l.o.d. are shown in Figure 4, from which it may be seen that although there are occasional similarities in trends, the overall correlation is fairly weak. The correlation factor computed for the deviations of annual values of the two variates from their mean annual values over the period amounts to 0·31. The diagram also indicates that, assuming both are periodic, a phase-adjustment of one curve relative to the other should not significantly increase this factor because the most striking difference between the two is in period. King has shown that variations in the altitude of the 500 mbar pressure surface are nearly synchronized with changes in sunspot numbers, and it is relevant that 14 years of the 18-year period covered in his study are within the range 1955–73. His results indicate that the altitude of the atmosphere appears to have been least about 1954 and 1964 and greatest in 1959 and also (by extrapolation) about 1970.

If the moment of inertia of the Earth varies significantly with the altitude of the atmosphere and, bearing in mind angular momentum, is conserved in the rotation, it might be expected that the l.o.d. should therefore be maximum about 1959 and 1970 and least about 1964. The deviations of the l.o.d. from a linearly varying length, expressed as fractions of a day, are plotted in Figure 5, which shows no turning points for those years. It is not to be supposed that this incongruity is regarded as conclusive evidence but it does indicate that, of the meteorological phenomena that have been linked with solar activity, the one most likely to directly influence the rate of rotation of the Earth has no appreciable effect.

As Challinor took care to point out, there are good grounds for regarding with suspicion the similarities of such variations, especially when they suggest a dubious connection, as in his own results between sunspot numbers and earthquakes. His correlations are not quantified so it is not possible to determine whether or not his UT data suggested a closer connection with solar activity than have the BIH results. In comparing his results with those obtained here, account should be taken of the 4 years extra data now available for use. It is unlikely that his use of the USNO observations could seriously affect his conclusions.

Such evidence in Figure 4 of regular fluctuations in the deceleration of the Earth also indicates that the period is shorter than that of the sunspot cycle and it is possible that there are terms of about 9 years duration that influence the rotation of the Earth. From 1956–66 it was possible to represent the variations in UT2–AT by a polynomial together with periodic terms with argument 2Ω (Ω is the longitude of the ascending node of the Moon's orbit) (Stoyko, 1966). The cyclic motion of Ω in 18·6 years is geophysically pertinent in this work because associated with it there is a tidal effect that affects the rotation of the

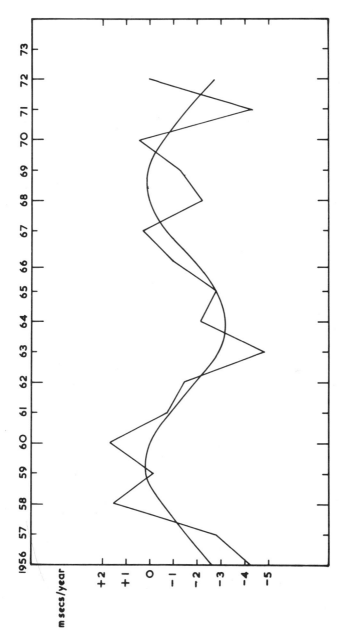

FIGURE 6. Annual rates of change in l.o.d. (joined by full lines) and the harmonic representation of the variation, with argument 2Ω, computed from a least-square solution of the annual values

Earth (Munk and MacDonald, 1960). The correlation factor between annual values for the deceleration of the Earth, i.e. rate of change of the l.o.d., and values of $\cos 2\Omega$, computed for the tabular dates, amounts to 0·58, almost twice the value obtained in the similar solution for sunspot numbers. A solution to determine how closely such a term could represent the data furnishes the following expression for the rate of change of the l.o.d.

$$\text{Rate of change of l.o.d.} = -0\cdot155 \pm 0\cdot028 + (0\cdot167 \pm 0\cdot041)$$
$$\times \cos(2\Omega - 4°\cdot4 \pm 12°\cdot7) \text{ msec/year}$$

where the statistical quantities are probable errors.

The data and the variation computed from this expression are shown in Figure 6, where it can be seen that the harmonic representation of the data is just convincing enough to justify further study. If the deceleration is really cyclic the period is certainly close to 9 years because a solution with argument Γ' (the mean longitude of perigee which has a period of 8·8 years) gives almost the same standard error for the phase and amplitude as those obtained in the 2Ω solution.

A suitably large effect that is not symmetrical about the axis of rotation could change the axial tilt relative to the body of the Earth and thus it is plausible to argue that large earthquakes influence the Chandler Wobble (Smylie and Mansinha, 1968). Until a relationship is discovered between changes in the direction of the axis and variations in the rate of rotation it may be inferred that the agencies responsible for the latter act symmetrically about the axis of rotation (Spencer Jones, 1955). Connections between seismic activity and changes in the rate of rotation have, however, been put forward and the BIH data was examined for evidence of this. Annual values of the total strain release for seismicity in the years 1956–72 are given in column 6 of Table 1. Deviations of the tabulated values from the mean annual value were computed and compared with the deviations of annual values of the rate of change of the l.o.d. over the same period. No significant evidence of a relationship could be discerned and the correlation factor is only −0·24. The dates and magnitudes of recent changes, i.e. over the past 18 years, in Earth spin have, by the use of atomic timing, been ascertained with much greater precision than was previously possible. But earlier measures, mainly derived from lunar observations and extending over the past 3 centuries, must be taken into consideration in a discussion of the rotation of the Earth. Three aspects of the information furnished by the earlier measures that are particularly relevant are: (i) the changes in rotation were irregular, (ii) occasionally they were relatively large; around 1900 and 1918 there were quite sudden changes of 4 or 5 msec in the l.o.d. (Spencer Jones, 1955), (iii) there were periods of more than 20 or 30 years of uniform deceleration, free from fluctuations. Current lunar observations show that the rotation in recent years exhibits type (iii) activity. It is difficult to accept that over the centuries effects of a recurring term with a period of the order of a decade have not been disclosed by the observations. The preponderance of evidence suggests that the periodicity exhibited in recent measures is accidental and without physical significance and, until further evidence can be adduced, it should be regarded with scepticism.

References

Challinor, R. A. (1971). Variations in the rate of rotation of the Earth. *Science*, **172**, 1022–1024

Essen, L. and Parry, J. V. L. (1957). The caesium resonator as a standard of frequency and time. *Phil. Trans.*, **250**, 45–69

Finch, H. F. (1950). On a periodic fluctuation in the length of the day. *Mon. Not. R. astr. Soc.*, **110**, 3–14

Howse, D. (1970). The Tompion clocks at Greenwich and the dead-beat escapement. *Antiquarian Horology*, (December 1970), 18–34

Jackson, H. (1929). Shortt clocks and the Earth's rotation. *Mon. Not. R. astr. Soc.*, **89**, 239–250

King, J. W. (1973). Solar radiation and the weather. *Nature*, **245**, 443–446

Lambeck, K. and Cazenave, A. (1973). The Earth's rotation and atmospheric circulation. *Geophys. J. R. astr. Soc.*, **32**, 79–93

Markowitz, W., Hall, R. G., Essen, L. and Parry, J. V. L. (1958). Frequency of caesium in terms of Ephemeris Time. *Phys. Rev. Lett.*, **1**, 105–106

Morrison, L. V. (1973). Rotation of the Earth from A.D. 1663–1972 and the constancy of G. *Nature*, **241**, 519–520

Munk, W. H. and MacDonald, G. J. F. (1960). *The Rotation of the Earth*, Cambridge University Press, London and New York, 313 pp.

Newton, R. R. (1970). *Ancient Astronomical Observations and the Acceleration of the Earth and Moon*, John Hopkins Press, London, 309 pp.

O'Hora, N. P. J. (1973). Semi-diurnal tidal effects in PZT observations. *Phys. Earth Planet. Inter.*, **7**, 92–96

Perfect, D. S. (1959). The PZT of the Royal Greenwich Observatory. *Occ. Notes R. astr. Soc.*, **3**, 223–233

Rochester, M. G. (1973). The Earth's rotation. *Trans. Am. Geophys. Union*, **54**, 769–781

Scheibe, A. and Adelsberger, U. (1950). Die gangleistungen der PTR-quarzuhren und die järliche schwankung der astronomischen tageslänge. *Z. Physik*, **127**, 416–428

Smith, H. M. (1972). International time and frequency coordination. *Proc. Inst. EEE.*, **60**, 479–487

Smith, H. M. and Tucker, R. H. (1953). The annual fluctuation in the rate of rotation of the Earth. *Mon. Not. R. astr. Soc.*, **113**, 251–257

Smylie, D. E. and Mansinha, L. (1968). Earthquakes and the observed motion of the rotation pole. *J. Geophys. Res.*, **73**, 7661–7673

Spencer Jones, H. (1955). The rotation of the Earth. *Handb. der Physik.*, **47**, 1–33

Stoyko, A. (1966). Variation de la vitesse de rotation de la Terre et sa representation analytique. *Bull. Classe des Sciences, Academie Royale de Belgique*, **52**, 1462–1468

Stoyko, N. (1936). Sur l'irrégularité de la rotation de la Terre. *Compt. Rend. Acad. Sci.*, **203**, 39–40

Thomas, D. V. (1963). PZT instrument and method of reductions. *Rot. Observ. Bull.*, No. 81

DISCUSSION

MULLER: Is the 9-year oscillation in the length of the day which you showed related to half the nodal period of the Moon?

O'HORA: The data can be well-represented by a periodic term, cosine half-nodal period. However, the data can be equally well represented by a periodic term varying with the longitude of perigee.

CHANGES IN THE EARTH'S ROTATION FROM ASTRONOMICAL OBSERVATIONS

L. V. MORRISON

H.M. Nautical Almanac Office, Royal Greenwich Observatory, Herstmonceux, Sussex, U.K.

Abstract

Changes in the Earth's spin and the magnitudes of the associated torques over the past 2000 years are deduced from astronomical observations. The secular deceleration due to the luni-solar tidal torque of $-5 \cdot 10^{23}$ dyne cm acting on the oceans and body of the Earth has superimposed on it torques of greater magnitude, but shorter duration, which act to accelerate as well as decelerate the spinning Earth. Over decades, torques as great as $\pm 4 \cdot 10^{24}$ dyne cm have occurred which are attributed to interactions at the core–mantle interface. The regular annual fluctuation in the Earth's spin implies torques of up to $\pm 3 \cdot 10^{25}$ dyne cm, and these could be mainly due to the variations in zonal wind patterns during the year, but significant departures from this seasonal pattern have occurred as, for example, in 1971. These irregular fluctuations, extending over periods of weeks to months, imply torques of the order of 10^{26} dyne cm: higher frequency fluctuations of comparable magnitude caused by meteorological disturbances are expected to be found when high resolution data from Lunar Laser Ranging or Very Long Base-line Interferometry are analysed.

Introduction

The purpose of this note is to explain (with an interdisciplinary readership in mind) how variations in the rate of rotation of the Earth are determined from astronomical observations made during the period -700 to the present, and to consider briefly some of the geophysical implications of the results.

Astronomers do not directly observe changes in the Earth's rate of rotation. They use observations of the transits of stars to set up a scale of mean solar time, known as universal time (UT), whose natural period is a fixed fraction of the period of rotation of the Earth on its axis, and then they compare this time-scale with a standard time-scale which is regarded as being more uniform. Any non-linearity in the difference between the time-scales is ascribed to variations in the rotation of the Earth or, possibly, to non-uniformity in the standard scale. An atomic time-scale, denoted by TAI, whose precision and uniformity is much better than that of universal time, is available as a standard of comparison from 1955·5. Prior to this we have to rely on a dynamical time-scale, known as ephemeris time (ET), which is based on observations of the motions of the Sun,

Moon and planets with respect to the stars, to provide the standard time-scale, but such a scale is much less precise and there is doubt about its uniformity.

A convenient interval of time on the comparison scales is the day of 86,400 SI* sec. The duration of the SI sec was arbitrarily chosen so that the reference day of 86,400 SI sec was equal to the universal day near the epoch, 1900·0. The period of rotation of the Earth on its axis is the sidereal day: it is a fixed fraction of the length of the universal day. Therefore, a fractional change in the length of the universal day over the reference day of 86,400 SI sec is the same as a fractional change in the period of rotation on that day compared with the period (or rate) of rotation near 1900·0. The fractional change in the length of the universal day can be expressed as the excess length in msec of the universal day over the reference day by multiplying the fractional change by $86,400 \times 10^3$. I shall denote by ΔT the *cumulative* excess of the length of the universal day over the reference day since the arbitrary epoch near 1900·0: this is the quantity which the astronomer measures directly.

In order to interpret the changes in ΔT as arising from geophysical or other causes, it is necessary to form the first and second derivatives of ΔT at particular epochs with respect to uniform time, t. The first derivative is the excess length of the (universal) day at those particular epochs (usually in units of msec) over the length of the reference day of 86,400 SI sec. If we divide the value of the derivative expressed in msec by $86,400 \times 10^3$ we obtain the fractional increase in the length of the day (l.o.d.), which is the same as the fractional decrease in the rate of rotation. The second derivative is the *rate* of change in the l.o.d., and hence it represents the instantaneous acceleration in rotation. Most investigations are concerned with the *average* acceleration in rotation in some time interval, t_1, t_2; it is given by

$$\frac{d^2(\Delta T)}{dt^2} = -\frac{2(\Delta T_1 - \Delta T_2)}{(t_1 - t_2)^2}$$

If ΔT and t are both expressed in units of years, the above expression gives the fractional change in the rate of rotation (or l.o.d.) per year; I shall denote it by $\dot{d}$. Alternatively, if ΔT is measured in seconds and t in centuries, we can express the result as the rate of change in l.o.d. per century (msec/cy) by multiplying the above expression by 10^3 and dividing by 36,525. Further definitions of notation will be given later as the need arises.

The span in time of the astronomical observations can be conveniently subdivided into three ranges according to the different techniques used to determine the accumulated time difference, ΔT: (1) *atomic clocks*, 1955·5 to the present; (2) *telescopic observations*, 1660–1955; (3) *pre-telescopic observations*, medieval around 1000, and ancient around A.D. 1.

---

* Systéme International d'Unités.

Changes in the Earth's Rotation from Astronomical Observations

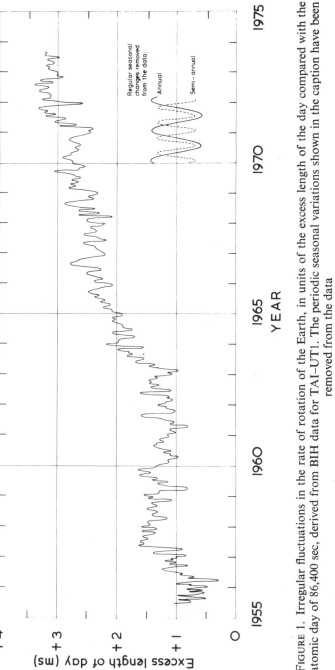

FIGURE 1. Irregular fluctuations in the rate of rotation of the Earth, in units of the excess length of the day compared with the atomic day of 86,400 sec, derived from BIH data for TAI–UT1. The periodic seasonal variations shown in the caption have been removed from the data

Observations

1. *Atomic clocks*

In the present context, the TAI time-scale can be considered as being uniform. The UT scale is presently measured to a precision of about 0·001 sec using a photographic zenith tube (PZT). The results of individual PZTs are combined and published centrally by the Bureau International de l'Heure (BIH) in Paris. Figure 1 shows 10-day values of the first derivative of ΔT with respect to t calculated from the values of TAI-UT1 published by the BIH: UT1 is the version of UT which has been corrected for changes in direction of the Earth's axis of rotation (polar motion). In Figure 1 the following regular 'seasonal' terms have been removed:

$+0·43 \cos(2\pi y - 0·50)$ msec annual term
$-0·32 \cos(4\pi y - 0·86)$ msec semi-annual term

where y is the fraction of a year. Figure 1 therefore shows the irregular changes in the rate of rotation (or excess length of the day) with the highest resolution in time available to date. The acceleration in rotation, $\ddot{d}$, averaged over several weeks, ranges between $\pm 3 \times 10^{-8}/y$ in the course of any one year; but there are instances where the average acceleration reaches twice these values. Markowitz (1972) has pointed out that the data under discussion to not permit changes in acceleration to be detected with a resolution finer than about 2 weeks. At this level there is no evidence for impulsive or abrupt changes in acceleration. Future high-resolution data from lunar laser ranging (Bender *et al.*, 1973), laser ranging to artificial satellites (Smith *et al.*, 1972) and very long base-line radio interferometry (Hinteregger *et al.*, 1972) will probably reveal changes at diurnal level.

2. *Telescopic observations*

Before 1955·5, a comparison time-scale is determined from observations of the regular motions of the Sun, Moon and planets. The principle of the method is that the mean longitude, L, of the Sun or Moon, following Newton's law for motion under point-gravitational attraction, is given by an expression of the form

$$L = c + n\tau + \tfrac{1}{2}\dot{n}\tau^2$$

where c is a constant, n is the mean motion at some arbitrary epoch, and τ is the time elapsed since that epoch. Here, time is the implicit argument in Newtonian dynamics which is taken to be uniform; it is usually denoted ephemeris time (ET). The acceleration in mean longitude, $\dot{n}$, is measured with respect to this time-scale. For the motion of the Sun (the reflected orbital motion of the Earth) $\dot{n}$ arises solely from the gravitational attractions of the planets, treated as point masses, on the Earth. Given the masses of the planets, $\dot{n}$ can be calculated accurately. We note, in passing, that if the gravitational constant G is variable, there will be an additional contribution to the value of $\dot{n}$. However, for the present, I shall assume that the value of G is constant.

With the above expression for L we can compute the longitude of the Sun and Moon and, likewise, other solar system bodies, for a given instant of ET (after including the periodic perturbations arising from their mutual attractions). If we observe the longitude at an instant of UT, we thus find

$$\Delta T = \text{ET} - \text{UT}$$

Because of its comparatively rapid angular motion, observations of the longitude of the Moon provide the most accurate values of ΔT, but there is one major problem: the torque exerted on the Moon by the tidal deformation of the Earth makes a large contribution to the value of $\dot{n}$. The magnitude of this torque cannot be predicted theoretically and we have to rely on astronomical observations to determine its value. How then can we separate the contribution due to $\dot{n}$ (tidal) from the acceleration of the Earth's rotation, $\dot{d}$, in the values ΔT deduced from lunar observations?

Spencer Jones (1939) solved this problem by (essentially) deriving values of ΔT from observations of the longitudes of the Sun and Mercury made between 1700 and 1930: here the values of $\dot{n}$ are only dependent on the planetary masses. He compared these values of ΔT with those deduced from the observed longitude of the Moon and attributed the difference in the values of ΔT, varying as the square of time, to the 'tidal' part of $\dot{n}$ in the expression for the mean longitude of the Moon. He found

$$\dot{n} = -22 \pm 1''/\text{cy}^2$$

Since the introduction of the atomic time-scale, another solution to the problem of separating the two contributions to ΔT has been possible. We adopt an ephemeris time-scale, ET*, since 1955·5 using the relation,

$$\text{ET*} = \text{TAI} + \text{constant}$$

where the constant allows for the difference between the origins of the two timescales. The observed times of lunar occultations are reduced using this reference time-scale to give residuals in longitude, ΔL, which are analysed for a term varying as t^2. This term is interpreted (assuming G is constant) as a correction to the adopted value of $\dot{n}$ (tidal) in the expression for L in the lunar ephemeris. Using this procedure, Van Flandern (1970) found

$$\dot{n} = -52 \pm 16''/\text{cy}^2$$

from about 6000 occultation observations in the period 1955·5–1969·0. From about 40,000 occultations in the period 1955·5–1972·0 I (Morrison, 1973) found

$$\dot{n} = -42 \pm 6''/\text{cy}^2$$

Oesterwinter and Cohen (1972) found

$$\dot{n} = -38 \pm 8''/\text{cy}^2$$

from an analysis of meridian circle observations of the Sun, Moon and planets made between 1912 and 1968.

I used my solution for $\dot{n}$ to revise previous values of the accumulated time difference ΔT for the period 1660–1972 which had been obtained by Brouwer (1952) and Martin (1969) using $\dot{n} = -22''/\text{cy}^2$. It should be noted that this procedure for deriving values of ΔT relies on the (reasonable) assumption that $\dot{n}$ (tidal) has been effectively constant since around 1600. The revised values of ΔT are plotted in Figure 2(a), with the first derivative with respect to t plotted in (b).

Figure 2(b) shows a gradual increase in the l.o.d. of 1·5 msec/cy (equal to a fractional decrease in the rate of rotation of 17×10^{-11}/year), but there are changes of ± 17 msec/cy ($\pm 200 \times 10^{-11}$/year) in a few decades around 1900.

3. Pre-telescopic observations

Before the invention of the telescope around 1600, we have to make deductions about the behaviour of the Earth's rotation from eye-witness accounts in historical records of astronomical phenomena such as solar and lunar eclipses. I illustrate the method of deduction using the so-called eclipse of Hipparchus in the year -128.

We compute the positions of the Sun and Moon for the date of the eclipse using the appropriate value of τ for the time elapsed before 1900·0 on the ET scale. We then compute where the narrow belt of totality would have occurred on the Earth's surface, assuming the Earth kept a constant rate of rotation over the time interval. The computed path (for which I am indebted to F. R. Stephenson) of totality, from sunrise in the west to sunset in the east, is shown in Figure 3 crossing the North American continent; yet we have accounts of the eclipse having been total in the vicinity of the Hellespont in Greece. The displacement in longitude is about 80°, which is equivalent to a value of $+5\cdot3$ hr for ΔT. Inserting this value in the appropriate relation given in the introduction, we find

$$\dot{d} = -30 \times 10^{-11}/\text{year} \qquad (+2\cdot6 \text{ msec/cy})$$

In this illustration I have made a number of simplifying assumptions, not least of which is that the value of the 'tidal' deceleration ($\dot{n}$) of the Moon in mean longitude is $42''/\text{cy}^2$. In fact, this is still uncertain by a factor of two. The values of both $\dot{n}$ and $\dot{d}$ have to be considered as unknowns in the observational equation for ΔT, and since their coefficients are highly correlated some investigators doubt whether they can be separated in a solution which only uses solar eclipse observations, even though these provide the most precise data available. Observations of lunar eclipses give $\dot{d}$ directly, but the problem here is obtaining sufficient numbers of reliable observations. There is also some dispute as to whether the

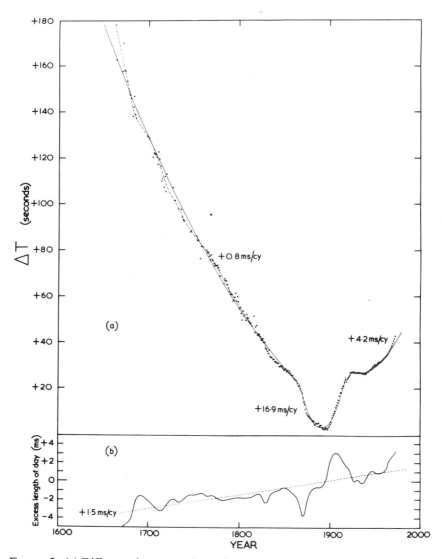

FIGURE 2. (a) Difference between ephemeris and universal time (ΔT) for the period 1663–1972. It is equal to the cumulative excess in the variable length of the universal day over the ephemeris day which has a fixed length close to that of the universal day near 1900

(b) First derivative with respect to time of the curve in (a), showing irregular fluctuations in the rate of rotation, expressed as the excess length of the universal day in msec over the ephemeris day. The slopes of lines fitted to sections of (b) show variations in the length of the day in the range ± 17 msec/cy. Since the curve in (b) is the first derivative of that in (a), changes in the length of the day are represented by arcs of a curve of degree two (parabola) in (a). The three representative arcs in (a) show increases in the length of the day of the amounts indicated. Both figures reproduced from Morrison, L. V., *Nature*, **241**, 519–520 (1973), with permission from Macmillan (Journals) Ltd.

Eclipse of Hipparchus 129 B.C.

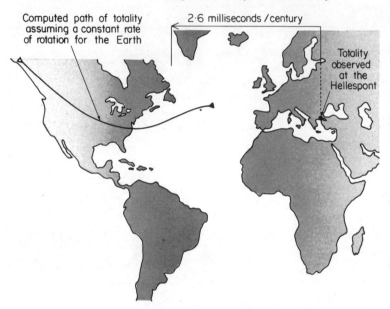

FIGURE 3. Illustration of how the long-term deceleration in the rotation of the Earth is deduced from the observed occurrence of a total solar eclipse

eclipse of Hipparchus was indeed that of the year −128, and not some other eclipse around that date. These doubts apply to many other eclipse identifications and as a result all the ancient observations between −700 and 500 are usually taken together and mean values $\dot{n}$ and $\dot{d}$ for an epoch around the year 1 are derived. The most recent results are given by Newton (1970) and Stephenson (1972), respectively

$$\dot{d} = -27\cdot 7 \pm 3\cdot 4 \times 10^{-11}/\text{year}; \quad \dot{n} = -42 \pm 4''/\text{cy}^2$$
$$\dot{d} = -25\cdot 6 \pm 1\cdot 2 \times 10^{-11}/\text{year}; \quad \dot{n} = -34 \pm 2''/\text{cy}^2$$

for an epoch around 1000, Newton (1972) finds the value

$$\dot{d} = -22 \pm 4 \times 10^{-11}/\text{year}$$

if we fix the value of $\dot{n}$ at $-42''/\text{cy}^2$. This implies a change of $+6 \times 10^{-11}/\text{year}$ in the value of $\dot{d}$ between the years 1 and 1000. Newton (1972) prefers not to draw this conclusion but rather to question the assumption that $\dot{n}$ has remained constant. Fractional changes in l.o.d. of $\pm 4 \times 10^{-8}$ occurred during the period of a few decades around 1900. Similar changes, but of longer duration, around 1000 would contribute a change of $\pm 4 \times 10^{-11}$/year to the average value of $\dot{d}$.

This concludes my summary of the observational results: I will now summarize the geophysical and other implications of these results.

Geophysical and other considerations

Long-term angular momentum balance

A change of 1 part in 10^{11} in the Earth's rate of rotation means a change of $5 \cdot 8 \times 10^{22}$ m² kg/sec in angular momentum. Adopting the value -27×10^{-11}/year (+2·3 msec/cy) for the average deceleration in rotation over the past 2000 years, the corresponding annual decrease in angular momentum is $1 \cdot 6 \times 10^{24}$ m² kg/sec. An observed angular deceleration of $42''$/cy² in the mean longitude of the Moon corresponds to an annual increase in orbital angular momentum of $2 \cdot 3 \times 10^{24}$ m² kg/sec. This implies a net gain of angular momentum in the Earth–Moon system of $0 \cdot 7 \times 10^{24}$ m² kg/sec: either the observed values for $\dot{d}$ and $\dot{n}$ are in error, or there are other variables to be considered.

A fractional decrease of 2×10^{-10}/year in the Earth's principal moment of inertia would account for the apparent imbalance of angular momentum. Runcorn (1966) has suggested that this decrease could be explained by the hypothesis that the core is still growing.

Dicke (1966), and later Hoyle (1972), proposed an alternative solution that the gravitational constant G is decreasing with time. This would account for about half of the observed value of $-42''$/cy² for $\dot{n}$. But Morrison (1973) concluded from the close agreement of the values of $\dot{n}$ measured with respect to TAI and ET that G is effectively constant. If G is decreasing with time, then this would imply an expansion of the Earth which would have important geophysical consequences as discussed, for example, by Hoyle (1972). However, more extended analyses of lunar occultations or new and more precise observations from, say, lunar laser ranging are required before the value of $\dot{n}$ can be measured with confidence.

Tidal torque and energy dissipation

The observed long-term deceleration in the Earth's rotation of 27×10^{-11}/year corresponds to a torque of $5 \cdot 0 \times 10^{16}$ Nm. If this torque arises solely from tidal forces, the rate of dissipation of energy in the oceans and body of the Earth would be $3 \cdot 5 \times 10^{12}$ W. Unless the principal moment of inertia or G is decreasing with time, as discussed above, we must regard the value of $3 \cdot 5 \times 10^{12}$ W as being a minimum, because the observed orbital deceleration of $42''$/cy² for the Moon implies a greater value for the torque. Adding a contribution for the solar tidal torque, we find that the total rate of dissipation of energy could be nearer 6×10^{12} W.

Estimates of the rate of dissipation of energy have been made independently of astronomical observations by studying the tidal flow in the seas and oceans and computing the amount of friction required along the sea-bed and coastline to give the observed tides. Munk and MacDonald (1960, Table 11.3) have estimated the total dissipation due to lunar and solar tides to be $4 \cdot 2 \times 10^{12}$ W, whereas more recent work by Pekeris and Accad (1969) and Pariiskii et al. (1972) find a higher value around 6×10^{12} W. We note that the astronomical and geophysical estimates of the rate of dissipation of energy cover the same range of values, but neither method can yet fix the value within a factor of about two.

Core–mantle interactions

The changes in acceleration in rotation around 1900 of $\pm 200 \times 10^{-11}$/year, imply torques of about $\pm 40 \times 10^{16}$ Nm, which are eight times greater than the long-term tidal torque. Because of their magnitude, duration and reversibility, these torques have been attributed to differential motions between the core and mantle. Rochester (1970) has discussed various forms of core–mantle interactions, among which inertial, topographic and electromagnetic coupling are the most likely; but there are problems in explaining the magnitude of the observed changes completely. Vestine and Kahle (1968) have further investigated the apparent correlation between the changes in the rate of rotation (such as those shown in Figure 2b) and changes in the rate of westward drift of the Earth's magnetic eccentric dipole.

Atmosphere–mantle interaction

The amplitude (0·43 msec) of the annual term given in the section of atomic clocks implies an amplitude of about 3×10^{25} m² kg/sec for the change in angular momentum of the Earth. Munk and MacDonald (1960) have shown that the seasonal imbalance of the angular momenta of the zonal winds between the Earth's hemispheres has an amplitude of 2×10^{25} m² kg/sec, so this is thought to be the main cause of the annual term. But they predicted wide variations from this regular pattern, as exhibited in Figure 1. Sidorenkov (1969) has calculated for the years 1957–1965 the expected monthly mean deviation in the rate of rotation of the Earth due to the recorded atmospheric pressure and wind speeds. He not only found a strong correlation with the observed fluctuations in the rate of rotation (such as those shown in Figure 1), but also evidence for longer period trends. But he stressed the need for more meteorological data on wind speeds and atmospheric pressure in order to carry out a more rigorous analysis. Lambeck and Cazenave (1973) have used more comprehensive meteorological data on zonal wind circulation to calculate the angular momentum transferred to the mantle and they find fluctuations which are in very good agreement in phase and amplitude with the observed changes in the rate of rotation.

Challinor (1971) has tentatively pointed to the apparent correlation between the changes in the rate of rotation (as exhibited by the approximately 9-year, oscillation in Figure 1) and the number of sunspots over the period 1956 to 1969, but as little more than one cycle has been completed it would be premature to conclude that this correlation is real.

The astronomical data do not permit us to detect changes in acceleration and hence torque with a resolution finer than about 2 weeks: at this level there is no evidence for impulsive changes. Future high-resolution data may make it possible to separate short-term changes in the rate of rotation due to internal mechanisms associated with the core–mantle interface from those due to atmospheric circulation.

Body tides

There are pseudo-rotational changes in the observed l.o.d. due to the tidal deformation of the Earth by the Sun and Moon and also to the varying load of the atmosphere. Part of the semi-annual term shown in Figure 1 is probably due to the bodily tide induced by the Sun (Munk and MacDonald, 1960). Monthly and fortnightly tides have been resolved in the data by Markowitz (1959) and Guinot (1970), who have then deduced values of the Love number k, which quantifies some of the Earth's elastic properties. Semi-diurnal tides have also been detected by Guinot (1970) and O'Hora (1973).

Conclusion

For a more comprehensive survey of the geophysical consequences, I refer the reader to a review paper by Rochester (1973) in which he considers the effects of

TABLE 1. Observed fractional changes in the rate of rotation (acceleration) and their causes

Time span year	Acceleration year$^{-1} \times 10^{-11}$	Torque Nm $\times 10^{16}$	Principal cause
2000	−27	5·0	Tidal torque (change in moment of inertia?)
300	−17	3·1	Tidal torque and/or long-period core–mantle interaction?
30	±200	40	Core–mantle interaction
1	±3000	600	Seasonal winds
0·1	±7000	1300	Irregular wind patterns

the changes in direction of the Earth's axis of rotation as well as the changes in the rate of rotation. In conclusion, I summarize in Table 1 the observed accelerations in rotation and the current explanations of their causes.

Acknowledgment

I am grateful to Dr. G. A. Wilkins for his helpful criticism of the draft of this note.

References

Bender, P. L. et al. (1973). The Lunar Laser Ranging Experiment. *Science*, **182**, 229–238
Brouwer, D. (1952). A study of the changes in the rate of rotation of the Earth. *Astr. J.*, **57**, 125–146
Challinor, R. A. (1971). Variations in the rate of rotation of the Earth. *Science*, **172**, 1022–1025
Dicke, R. H. (1966). The secular acceleration of the Earth's rotation and cosmology. *The Earth–Moon System*, Plenum Press, New York, pp. 98–164
Guinot, B. (1970). Short-period terms in universal time. *Astron. & Astrophys.*, **8**, 26–28
Hinteregger, H. F., Shapiro, I. I., Robertson, D. S., Knight, C. A., Ergas, R. A., Whitney, A. R., Rogers, A. E. E., Moran, J. M., Clark, T. A. and Burke, B. F. (1972). Precision geodesy via radio-interferometry. *Science*, **178**, 396–398
Hoyle, Sir F. (1972). The history of the Earth. *Q. Jl. R. astr. Soc.*, **13**, 328–345
Lambeck, K. and Cazenave, A. (1973). The Earth's rotation and atmospheric circulation—I. Seasonal variations. *Geophys. J. R. astr. Soc.*, **32**, 79–93
Markowitz, W. (1959). Variations in rotation of the Earth, results obtained with the dual-rate Moon camera and photographic zenith tubes, *Astr. J.*, **64**, 106–113
Markowitz, W. (1972). Rotational accelerations, pp. 162–164. In *Rotation of the Earth* (Eds. P. Melchior and S. Yumi), D. Reidel, Holland
Martin, C. F. (1969). A study of the rotation of the Earth from occultations of stars by the Moon, 1627–1860. Ph.D. Thesis, Yale University. 83 pp. and appendix
Morrison, L. V. (1973). Rotation of the Earth A.D. 1663–1972 and the constancy of G. *Nature*, **241**, 519–520
Munk, W. H. and MacDonald, G. J. F. (1960). *The Rotation of the Earth*, Cambridge University Press, London. 323 pp.
Newton, R. R. (1970.) *Ancient Astronomical Observations and the Accelerations of the Earth and Moon*, Johns Hopkins Press, London. 309 pp.
Newton, R. R. (1972). Astronomical evidence concerning non-gravitational forces in the Earth–Moon system. *Astrophys. & Space Sci.*, **16**, 179–200
Oesterwinter, C. and Cohen, C. J. (1972). New orbital elements for the Moon and planets. *Celestial Mech.*, **5**, 317–395
O'Hora, N. P. H. (1973). Semi-diurnal tidal effects in PZT observations. *Phys. of Earth & Plan. Int.*, **7**, 92–96
Pariiskii, N. N., Kuznetzov, M. V. and Kuznetsova, L. V. (1972). The effect of oceanic tides on the secular decelerations of the Earth's rotation. *Izv. Acad. Sci. USSR Phys. Solid Earth*, **10**, 65–70
Pekeris, C. L. and Accad, Y. (1969). Solution of Laplace's equations for the $M2$ tide in the world oceans. *Phil. Trans. R. Soc.*, A, **265**, 413–436
Rochester, M. G. (1970). Core–mantle interactions: geophysical and astronomical consequences, pp. 136–148. In *Earthquake Displacement Fields and the Rotation of the Earth*, D. Reidel, Holland
Rochester, M. G. (1973). The Earth's rotation. Address to the second GEOP Research Conference on the Rotation of the Earth and Polar Motion, Ohio State University, Columbus

Rochester, M. G. (1973). The Earth's rotation. *Trans. Am. Geophys. Union*, **54**, 769–781

Runcorn, S. K. (1966). Change in the moment of inertia of the Earth as a result of a growing core, pp. 82–92. In *The Earth–Moon System*, Plenum Press, New York

Sidorenkov, N. S. (1969). The influence of atmospheric circulations on the Earth's rotational velocity, 1956·8–1964·8, *Soviet Astr. J.*, **12**, 706–714

Smith, D. E., Kolenkiewicz, R., Dunn, P. J., Plotkin, H. H. and Johnson, T. S. (1972). Polar motion from laser tracking of artificial satellites. *Science*, **178**, 405–406

Spencer Jones, Sir H. (1939). The rotation of the Earth, and the secular accelerations of the Sun, Moon and planets. *Mon. Not. R. astr. Soc.*, **99**, 451–558

Stephenson, F. R. (1972). Some geophysical, astrophysical, and chronological deductions from early astronomical records. Ph.D. Thesis, Univ. of Newcastle upon Tyne. 204 pp.

Van Flandern, T. C. (1970). The secular acceleration of the Moon. *Astr. J.*, **75**, 657–658

Vestine, E. H. and Kahle, A. (1968). The westward drift and geomagnetic secular change. *Geophys. J.*, **15**, 29–37

Wesson, P. S. (1973). The implications for geophysics of modern cosmologies in which G is variable. *Q. Jl. R. astr. Soc.*, **14**, 9–64

THE ACCELERATIONS OF THE EARTH AND MOON FROM EARLY ASTRONOMICAL OBSERVATIONS

P. M. MULLER

Jet Propulsion Laboratory, Pasadena, California, USA

AND

F. R. STEPHENSON

School of Physics, University of Newcastle upon Tyne, England

Abstract

The data analysed in this paper consist of the observations of central solar eclipses as recorded in history by astronomers and laymen alike, commencing in 1375 B.C. We define a method for sifting the many diverse eclipse records, to find those few of extremely high reliability in terms of magnitude observed and date. Arguments are presented to justify confidence in these accepted records as scientifically usable data.

The method of statistical analysis is shown to be stable against errors in the third requisite observational detail, the place where the eclipse was seen. The resolution achieved in determining the function ΔT is much better than that previously available. We also obtain estimates of the earth acceleration, $\dot{e}$, and the lunar acceleration, $\dot{n}$, considered as independent parameters, having accuracies comparable with other recent determinations.

The primary conclusion of this paper is that the lunar acceleration, and both tidal and non-tidal earth accelerations, have been sensibly constant during the historical period. This is contrary to the findings of some other recent papers, but reconsideration of their analyses confirms our conclusion. Numerical results include: lunar acceleration in longitude, considered as an independent parameter, $\dot{n} = -37 \cdot 5 \pm 5$ arc sec/cy^2; earth rotational acceleration, as an independent parameter, $\dot{e} = -91 \cdot 6 \pm 10$ sec of time/cy^2 or $\dot{\omega}/\omega_e = -29 \cdot 0$ cy^{-1}; and ΔT (for the above $\dot{n}$) = $66 \cdot 0 + 120 \cdot 38T + 45 \cdot 78T^2$ (± 100 sec of time, or $\pm 1 \cdot 0$ sec/cy^2). There is some evidence, rather speculative, that there is a fluctuation in ΔT of the form, $\Delta T = +80 \sin((\pi/6)(T+7))$ where T is in centuries from 1900.

> 'On the day of the new moon, in the month of Ḥiyar, the Sun was put to shame, and went down in the daytime, with Mars in attendance.'

Tablet recovered from the city of Ugarit (Ras Shamra), and is an observation of the Total Solar Eclipse of 3 May 1375 B.C. This is the oldest viable record of a solar eclipse known at the present time.

Introduction

The analysis of large solar eclipses in antiquity has had a long and chequered career. Ancient eclipses were first used astronomically, so far as we know, by Hipparchus (ca. 135 B.C.) to determine the lunar parallax. For the purposes of the present paper, the first use of such data was by Sir Edmund Halley (1695), who correctly concluded that there must be a secular acceleration of the Moon to account for the observations. Further work made it possible to produce Canons of ancient eclipses in support of historical and astronomical research such as those of Oppolzer (1887), Ginzel (1899), Schroeter (1923) and Neugebauer (1932). Fotheringham (1920) and De Sitter (1927) analysed eclipses, occultations and equinoxes. Spencer Jones (1939) considered the modern data (since 1672) and concluded that there were secular accelerations both of the Moon in longitude, and of the Earth's rotation. Into more recent times, Van der Waerden (1961), Curott (1966), Oesterwinter and Cohen (1972), Newton (1970, 1972), Van Flandern (1970), and Morrison (1973) have all considered this problem, including the use of very modern data and atomic clocks in the last two papers.

We attempt to present our analysis in a manner convenient for the interdisciplinary audience of the conference. Biology, history, geophysics, astronomy and many other diverse studies and disciplines play their role in this research. If, at times, the matters discussed appear to be trivial, we have attempted to err in the direction of going back to fundamentals, rather than the reverse.

Part I investigates the observable we deal with, namely central or near central solar eclipses, and we derive a set of criteria for data selection. Part II applies these criteria to all of the available historically preserved data, and determines which observations are of the highest reliability. Part III undertakes the numerical analysis of the selected observations, produces the astronomical and geophysical solutions, and comparatively analyses the results of other authors.

PART I: The analysis of naked-eye observations of large solar eclipses

1. *Nature of a solar eclipse*

All educated people of the 20th century understand that a solar eclipse takes place when the Moon comes directly between the Earth and the Sun, so that the lunar shadow falls upon the Earth. Since the Sun is larger than the Moon, the lunar umbral (total) shadow is a cone, with its apex at the point in space where the Sun and Moon appear to have the same angular size. The lunar penumbral shadow, in which part of the Sun is cut off, expands as one moves away from the Moon. The cone of 'umbral' shadow, Figure 1, has two parts. The one nearest the Moon is the region of complete 'total' shadow, and the one beyond the apex is the region in which the Moon is wholly within the solar disk, but fails to cover the Sun completely. An eclipse seen from within the penumbra is termed partial, and the exposed portion of the Sun forms a crescent shape

The Accelerations of the Earth and Moon

ranging from a full circle with only the slighest dent, to a very thin line. An eclipse seen from within the umbral region is termed total, because no portion of the Sun's photosphere is visible. An eclipse seen from beyond the apex of the shadow cone is termed 'annular' because the Moon fails to cover a ring-shaped rim of the Sun.

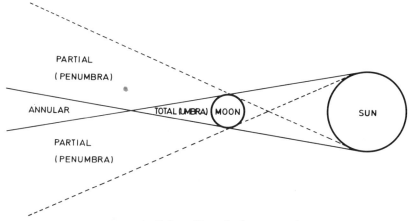

FIGURE 1. Solar eclipse shadow geometry

It so happens that at the Moon's mean distance from the Earth, the apex of the umbral cone is just able to reach the point on the Earth nearest the Moon. Because of the Moon's elliptical orbit, this means that almost exactly half of all central solar eclipses are 'total' at some point on the Earth, and the other half are annular. We will not discuss the non-central 'partial' eclipse except to say that it is obviously possible for the Earth to enter the penumbra without entering, at any point, the umbral cones. A central eclipse is, by definition, a period of time during which some part of the Earth is passing through any part of either umbral cone. When this happens, the elliptical intersection of the umbral cone and the Earth's surface sweeps out a narrow path across the Earth as the Moon moves in its orbit. In all but high latitudes, the shadow must therefore sweep eastwards across the Earth in the same direction as the lunar orbital motion. Viewed from the Earth, the Moon covers first the western limb of the Sun, then crosses and leaves the Sun on the eastern limb. If the observer on the Earth is at a location which passes through the umbral cone of total shadow, there will come a time when the Sun is completely covered by the Moon, and the Sun will remain covered for the time this observer remains within the cone (typically a few minutes). If the observer is within the umbral-annular cone, there will come a time when the Moon is immersed in the Sun completely, but a ring of the intensely bright solar disk will remain visible.

We can now appreciate why the regions of the Earth within which a total or annular eclipse is visible consist of a narrow band, running approximately eastward across the Earth. Widths of these bands, in temperate latitudes, can run from zero (at the shadow cone apex) to approximately 300 km in width for 'ideal' eclipse geometries. Total eclipses occur approximately twice every three years on the Earth as a whole. They are very rare for a given location, occurring approximately every 360 years on the average.

2. *The observer's impression of large solar eclipses*

As we have seen, observers within a total eclipse shadow track on the Earth will witness a complete disappearance of the Sun. The darkness resulting from this event is not quite complete, however, for two reasons. First, the Sun has a tenuous upper atmosphere called the corona, with a brightness of about 10^{-7} that of the Sun itself. This fact was known by Plutarch, who wrote, 'Even if the Moon, however, does sometimes cover the Sun entirely, ... a kind of light is visible about the rim which keeps the shadow from being profound and absolute.' (Plutarch, ca. A.D. 80). The second cause of light within the total eclipse shadow is scattering in the atmosphere, the intensity of which ranges between about 10^{-6} and 10^{-7} that of the Sun. This scattering arises from dust and clouds in the atmosphere outside the region of total eclipse. The subjective impression is similar to that of the deep twilight night sky with a half to full Moon visible.

One of the authors (Muller) has personally witnessed four total eclipses, twice in clear conditions on the ground, once from an aircraft in clear conditions, and once from an aircraft clouded out. Both authors have had considerable experience in reading the accounts of others, both historical and scientific, and Stephenson

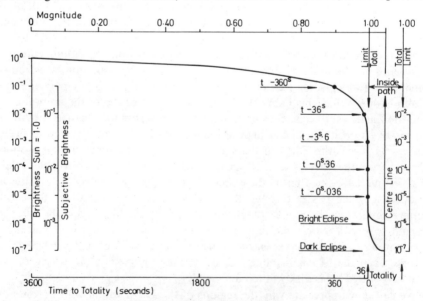

FIGURE 2. Total eclipse light curve

The Accelerations of the Earth and Moon

has read hundreds of such records in the native languages. Since the subjective impression of an eclipse on the part of a human observer is the data to be used in this investigation, there is no definable and repeatable set of experiments which will *prove* the conclusions we are about to give in this matter. In that sense, this investigation must necessarily be somewhat 'unscientific'. We, and others before us, have been convinced that at least some of the admittedly imprecise and often emotional records retained through history by eye-witnesses to large eclipses can, nevertheless, constitute highly reliable and usable scientific observations.

Figure 2 is an approximate light-curve for a typical solar eclipse. A large solar eclipse will have a duration of roughly 2 hr overall, with perhaps 250 sec of total eclipse under ideal circumstances (it can be anything between zero and 480 sec). We create a fictious eclipse of duration 2 hr in the partial phases, 1 hr each side of totality, and 360 sec of totality, and compute the data for Figure 2. Since the two halves of an eclipse light-curve are symmetrical, only the first half is shown in the Figure. The changing of light intensity has the subjective feel to human observers according to something approximating the square root of the brightness. We therefore plot the solar intensity (as a fraction of the whole) on the log scale for compression. The magnitude of an eclipse is defined to be the fraction of the solar *diameter* covered by the Moon, and we ignore the slight difference between this and the area function. We plot the time remaining to totality at each order-of-magnitude drop in the light intensity.

The progress of an eclipse is as follows. When the magnitude reaches 0·90, only 360 of the total 3600 sec remain before totality. The light level has been reduced to 0·10 of the original amount, and it seems *subjectively* reduced to about 0·3 of the original. This is the darkness produced by average overcast, and without atmospheric dimming or optical aid such an eclipse will frequently pass unnoticed by those not expecting it. In 320 sec, with about 36 sec remaining to totality, the eclipse is 0·99 complete. Light is reduced by a factor of 100, perhaps a subjective impression of 10 to 1. This is quite noticeable, and an observer not expecting the eclipse would presumably look up into the sky for the cause of the darkening. It is at this level that the darkening equals or exceeds the greatest which can be produced by clouds alone. This is also approximately the earliest phase at which the solar crescent can be reliably distinguished. In the experience of Muller, at 0·98 eclipse the Sun's crescent shape cannot be reliably discerned, but by 0·99 it can be without great difficulty. It can be appreciated from these considerations, that even a 0·99 partial eclipse might pass unnoticed, though experience indicates that 0·98 is the largest magnitude likely to actually pass unnoticed. Direct observation of the 0·99 eclipsed Sun is possible only for a second or two without permanent injury to the eyes, and temporary 'bright-blindness' seeing conditions are produced by the briefest glances. At 0·99 magnitude, the observer is approximately 35 km (on the normal) removed from the path edge.

Things progress quite rapidly after this point, and the attention of those now awaiting the eclipse is brought repeatedly back to the Sun, again and again, and for progressively longer views as the brightness change accelerates. About 4 sec before totality, the brightness is down by 3 orders of magnitude, and the crescent has changed into a short straight line of intensely bright light at the extreme edge of the black lunar disk. Subjectively, the reduction is about 30 to 1, and the rate of darkening is quite apparent. The lunar shadow can be seen racing in from the west (if there is raised ground there) at a speed of 2000–3500 km/hour.

In the last 4 sec before totality, the further light reduction is the same or more as in the 1st hr of eclipse. The thin line of 'crescent' suddenly breaks up into discrete beads of intense light shining through the deepest lunar valleys along the limb. Each seems like the brightness of a welder's arc-point, and does *not* have the smeared appearance reproduced in photographs. To be sure, there is irradiation in the eye just as on the photographic plate, but the brain interprets it as an intensely bright *point*, with 'heavenly' streamers spreading out around it much like the sunbeams we often see shining down through a hole in heavy clouds. For an observer situated 4 km outside a path of totality, the bright beads remain in view for a minute or more, and it is possible that he might report 'the complete disappearance' of the Sun. We should perhaps allow this narrow region of uncertainty in the distinguishing of 'totality'. Such an uncertainty is, however, quite negligible for the purposes of this study. An observer further outside the path will see the thin crescent progress around the Sun from the east limb to the west, via the south limb if the observer is south of the path, otherwise via the north. We find reports of this phenomenon as can be noted below and, in such cases, we may be certain of partiality (denial of totality). During this final 4 sec, the light declines by a further factor of 1000, 30 subjectively, and when the last bead winks out, the solar corona with about the brightness of a gibbous Moon, bursts forth around the Sun and the brighter stars and planets appear. Muller characterizes this as 'having a blanket thrown over one's head'. This is an indescribably impressive experience even for professional astronomers who know exactly what is happening, and have expectantly travelled across a continent to witness the eclipse.

During totality, the brighter stars and planets can always be seen. Though Muller has yet to witness what would qualify as a very dark eclipse, he has always seen at least two stars/planets without advance preparation regarding where to look. For example, on 30 June 1973 in Kenya, members of his party nearly all saw exactly four objects, Venus, Mercury, Saturn, and one bright star (unidentified). This was not a particularly dark eclipse despite the very wide path and long duration because of considerable scattering from dust in the desert atmosphere.

Eclipses can be so dark that a desk clock cannot be read, a fact which some experienced astronomers overlooked, to the detriment of their experiments, during the eclipse of 1918 seen in the north-western United States. In such an

The Accelerations of the Earth and Moon

eclipse, a sky full of stars is indeed visible. Most eclipses, at least in this century, have been brighter, and produce the equivalent of deep twilight at about the time when the brighter stars begin to show themselves. In such brighter eclipses, the print in this book could be read, though perhaps with some difficulty.

The difference between bright and dark eclipses is shown, approximately, in Figure 2. Near, but within the path edge, the darkening probably does not go below 10^{-6} because of nearby clouds and landscapes outside the path. In a bright eclipse, the darkening will be about the same on the centre line. In a dark eclipse, it can decline another order of magnitude to about 10^{-7}. It appears that many of the eclipses reported historically were darker, or the reports subject to exaggeration, since 'skies full of stars' is not an uncommon observation. Perhaps the industrial revolution has polluted the air on a global scale, rendering modern eclipses brighter on average than in earlier times.

The report of 'stars seen' has been used by some interpreters of these records as tantamount to a report of totality. Since Venus can be seen by day, and certainly could be seen in a 0·99 eclipse, perhaps even in company with other stars, we do not trust this single observational consideration as implying totality. Newton (1970) also gives a detailed analysis of these matters and reaches similar conclusions.

We have taken note of the subjective impressions which arise from total, and near-total, eclipses. In historical records, the kinds of statements most frequently encountered are made in general terms. 'The sun suddenly disappeared'; 'night fell suddenly'; 'the sun was completely hidden'; 'stars were seen', or 'the crescent moved around the sun', are typical comments. All of these except the last two would imply totality, and the last would specifically deny it. Basically, this is the nature of the reports we shall be analysing in detail below.

3. *The question of observation*

Colleagues have questioned our reliance on the ability of untrained, or even terrified, observers to distinguish between totality and 0·99 partiality. There are several factors which answer this question. One is the light curve, Figure 2, and consideration of the fact that the last 4 sec to totality are the subjectively impressive ones. Another is the voice of experience. We have not found any astronomer who has witnessed a total solar eclipse who denies the premise. The observations compiled by Halley (1715) are very revealing, as was an informal experiment carried out in New York City on 24 January 1925. The south limit of this eclipse crossed the centre of New York City near 86th Street and Broadway. We know this because observers, untrained in eclipse observation, were set out at the rate of one per block along Broadway with a questionnaire to complete after the eclipse. In answer to the crucial question regarding totality, every observer north of 86th Street reported affirmative, and every other one reported negative. There was not a single exception, and the resolution was approximately 300 m! Halley's (1715) observers also easily distinguished the path edge, and the resolution achieved by his small group was a km or two.

We believe that there is adequate evidence to support the contention that untrained observers can, in fact, distinguish between totality and partiality. Some colleagues have suggested that this is a central problem to our analytic approach. We respectfully suggest that such is not the case, and would add that those who have not personally witnessed a total eclipse are in a position to judge only from the conclusions of those who have done so. In this connection we recommend the New York experience, the light curve in Figure 2, Halley (1715), and many other commentaries on the nature of totality. We add here the firm opinion, based upon personal experience, that making this distinction is probably the most reliable observation that a layman can make which satisfies the requirements of science.

While this study, in our view, does not suffer from uncertainty in the ability of the observers themselves to distinguish the phase, it most certainly does depend upon *our* ability to determine which records prove that the observer *intended* to say that totality occurred. In other words, the problem is in the linguistic, personal and historical evaluations which go into the detailed understanding of what was meant by a given record of a man's impressions.

It is for this reason, of course, that we have talked about what a person perceives, subjectively, at the onset of totality. We must be able to distinguish reports of totality from those which are ambiguous, or which deny it. This, and no other, is the central difficulty, challenge and characteristic hallmark of the analysis presented here.

As will become apparent below, we depend heavily upon admitting to the study only those records which unambiguously report the phase of eclipse achieved. In general, we require either the specific statement that the Sun disappeared, or the use of the term *chi* by a Chinese professional astronomer, before totality is accepted as the observation of magnitude. For denial of totality, statements describing the movement of the crescent around the disk, or equivalent detail, are necessary. In the case of annular eclipses, we trust the Chinese astronomer to be precise, but all others are allowed a 1% magnitude uncertainty because of the difficulties in distinguishing phase in such conditions. Our confidence in the Chinese astronomers regarding annular eclipses is from indirect reading. We know they kept watch for eclipses, and correctly observed many hundreds of relatively small partial phases. They probably used optical aid of some kind to dim the Sun though, unfortunately, we have no direct information from them on the subject. We do know, however, that the Romans used pitch in a vessel as a mirror for solar observation (Seneca, ca. 1st century B.C.), and that many ancients utilized water in reflecting pools for the same purpose. It is a reasonable presumption that the Chinese used some method of dimming the solar disk to permit safe and reliable observations.

We now provide a summary of typical observational details versus the magnitude achieved in the eclipse. We limit the records admitted to this study to those

The Accelerations of the Earth and Moon 467

in the first two, and the last, categories, so that the implied magnitude is known precisely. Other analyses (such as Newton, 1970) have used large bodies of data from other categories. Note the comments under magnitude = 0·95 or less to appreciate the possibilities inherent in these data for the erroneous admission of small eclipses with a presumed large magnitude. This error source does not affect us, since such data are automatically excluded.

OBSERVATION DETAIL VERSUS PROBABLE MAGNITUDE

Magnitude	*Commentary*
Total	Sun disappeared; Sun suddenly disappeared; Darkness suddenly fell; No part of the Sun remained to our view; Sun completely covered; Ring seen around the Sun (in an eclipse Total on the Earth's surface); Luminous bag seen around the Sun (etc., the corona); Chinese astronomer's *chi*.
Annular	Very rarely observed or reported because of similarity to partial eclipse. Exceptions: Chinese use *chi* to mean annular, and have reported, 'Sun was like a golden ring'. The latter was probably seen through clouds, or with an optical aid of some kind. Always difficult to observe accurately, we allow 0·01 magnitude error in reports by laymen.
0·98–0·999	Stars seen probably implies this large a magnitude, but there are many exceptions to this. Crescent is frequently described as 'like a two- (to five-) day Moon'. Such descriptions do not imply that the uneclipsed portion of the Sun was the same as this level of lunar crescent, because of irradiation. We find the majority of such records fall in this magnitude class, though here again, there are many exceptions (optical aid?). 'Great eclipse' and similar statements, including practically all detailed, impressive records probably fall in this range.
0·95–0·98	Simple mention of an eclipse is frequently in this category.
0·95 or less	Frequently missed entirely unless a regular eclipse watch is being kept as it was in China. Eclipses of even very small magnitude can be observed through 'ideal' cloud cover, or when such occur near the horizon, or if optical aid is being used to view by reflection in reduced light. It is risky to assume even magnitude 0·90 if any of these conditions might apply.
Totality denied	To deny totality unambiguously, we need a statement to the effect that 'the solar crescent proceeded from one limb to the other, and revolved about the lunar disk'. There is a small, but acceptable risk in the admission of records giving detailed

accounting of the crescent, and not mentioning anything about 'total disappearance' or the like.

4. Requirements on the admission of data to this study

To be of use astronomically in this study, we must be able to determine three elements from the record: (i) the date or unambiguous identification of which eclipse was observed, (ii) the *precise* magnitude observed, (iii) the place of observation.

Our approach is to take the absolute minimum risk, hopefully negligible, in the direction of establishing elements (i) and (ii). Newton (1970) and others have utilized many observations for which the magnitude can only be presumed to be statistically greater than some arbitrary value such as 0·95, and depend upon averaging to cancel the variations. There is the real danger of admitting small eclipses as large in such an approach, thereby risking bias in the results. As noted above, we shall adopt the requirement of unambiguous magnitude determination, and thereby throw out the vast majority of all data available from history. For reasons already noted, and others which will become clear later, we believe that the quality of the results can be maintained despite this overwhelming deletion of most 'observations'. In summary, we emphasize that criteria for the observations (i) and (ii) above are intended to be very specific and restrictive in this study, and that our approach differs fundamentally from some of the previous studies.

When we turn to the establishment of the place of observation, the issues are not as clear-cut. First, many records fail to make any mention whatsoever of the place of observation. This is understandable, as the observer either feels it is implicit, unnecessary, or the mention of it simply fails to occur to him. This problem is shared in common by all studies based upon these data. Newton (1970) was the first to pioneer the large-scale research into the historical and sociological contexts so critical to determining which sources are independent and, therefore, which presumptions of observer location are probably correct and which are too risky. This is the approach used in the present analysis, and we are indebted to Newton for his previous careful analysis and descriptions of methodology in this connection. Stephenson has extended these techniques, and others, to the consideration of the Chinese records as well as the European. Part II of this paper details the analysis of each individual observation, not only to satisfy criteria (i) and (ii) as first priority, but also the determination of (iii) with adequate reliability for the purposes at hand. We summarize by emphasizing that sensible risks *must* be taken in the establishment of the places of observation, and difficulties with this one criterion are the most probable source of error in the data considered here.

The strategy of analysis can now be defined. We shall admit only a small fraction of the 'available' data to actual analysis. Criteria (i) and (ii) will be quite explicitly satisfied, and errors in these categories should be very unlikely or non-existent. Criterion (iii), the place of observation, must be presumed or

The Accelerations of the Earth and Moon

determined indirectly in the majority of cases. Errors in this determination are not only possible, but probable. The whole question of uncertainties in the solutions to be obtained from these data will be considered in detail later.

The data will be set into two classes, A and B. The first is reserved for observations which satisfy all three criteria to a very high degree of confidence. The second is for the remainder, about which there is some doubt regarding the place of observation. The class A records are to be more heavily weighted, or be more surely relied upon, in the final analysis than will class B data. Both classes constitute very probable observations, however, and the assignment of class labels is solely for convenience in considerations which may arise in the analysis.

We shall, almost certainly, have a small minority of erroneous observations admitted to the study. The criteria and historical analysis are not perfect. We emphasize that the most probable error is in the place of observation and, in fact, the criteria have been established, practically speaking, to rule out any other errors. In the iterative procedure employed in providing astronomical solutions from the observations, we may find a few inconsistent cases. These may be presumed to be errors. If any such are identified by this means, attempts will be made to find out why the wrong observation location was assigned. If it can be rectified, it will be, and the observation assigned to class B. If not, the data will be deleted in the process of analysis. Any and all such cases, if they occur, will be clearly documented in this paper. Furthermore, it will turn out that some perfectly reliable data actually have no sensitivity in determining the solutions. This arises from the path geometry (parallel to line of latitude is a good example). If such data are consistent with the final solution, then all is well. If not, we have an unresolved inconsistency and, should this happen, the details will be clearly spelled out.

These steps are necessary to proper exposition of the analysis and results, since the question of consistency within the data set adopted for analysis ultimately determines the success or failure of the method employed.

We now provide a summary of the data admission requirements, and the implications resulting from adoption of this particular set of criteria.

DATA ADMISSION REQUIREMENTS AND STRATEGY OF ANALYSIS

(1) Identification of the eclipse (dating)

This must be unambiguous and be provided either from the record itself or from direct information relating to the record.

(2) Establishing the magnitude

Very high reliability is required here, and this results in the dismissal from analysis of a considerable amount of data. In return, we are left with precise observational constraints instead of large uncertainties in magnitude. Totality, or specific denial of the same, is required (in general).

(3) The place of observation

This must be assessed from complex historical and linguistic considerations, and the methods employed in this study do not differ markedly from those used (of necessity) by other recent analysts (i.e. Newton, 1970, 1972). This is the most probable source of error in the data admitted to analysis.

(4) Data weighting and erroneous records

The data are placed into two arbitrary classes, A and B. Account is taken of any and all records found to be inconsistent whether they are used in the final solutions or not. Correctable interpretations are reinstituted in class B with notation to that effect. It is noted that some perfectly consistent and reliable observations may have little or no sensitivity in the determination of the solution.

5. *Summary and conclusion to Part I*

The above rationale regarding the nature of naked-eye observations of large solar eclipses is a mixture of semi-scientific experiment, observation, personal experience, and (hopefully) sound judgments. To make use of these data as scientific observations requires that we calibrate and understand the observers as human beings, and see the events in the context of what the impact would be on these observers. This kind of analysis cannot completely satisfy the usual scientific requirement of repeatable and definable experimental testing. The diverse disciplines of history, psychology, physiology, common sense, personal experience and professional judgment enter in complex ways.

We have tried to follow the procedure of setting clearly defined criteria bearing on the admissability of data, and to remain as objective in their application as possible. Further, it was thought more profitable to spend our time in seeking to uncover further reliable observations from the historical records, rather than in attempting to rationalize second-best choices from the known data. The system of data certification was designed to restrict probable errors to the single matter of defining the place of observation, leaving magnitude and identification as relative certainties. This was not an arbitrary choice. It will become apparent below that many benefits in the elimination or reduction of errors result from this scheme. While the vast majority of observations are eliminated through application of the criteria, this does not necessarily imply that the solutions obtained from the restricted data set are inferior in any way to those secured by the more traditional 'bulk data' methods of analysis.

PART II: Discussion and selection of historical solar eclipse observations

1. *Introduction*

With a single exception (a solar eclipse observed to begin 12 min before sunset at Babylon in 322 B.C.) this paper is confined to the analysis of total or near total solar eclipse observations. These require no measurements, either of time or

The Accelerations of the Earth and Moon

position, and as such are free from the random and systematic errors inherent in other types of data. In particular, the accurate measurement of time presented considerable problems in the ancient and medieval world. To give an example, discussed further below, Stephenson (1974) analysed a Late Babylonian text (323–320 B.C.) containing about 100 observations of what Sachs has called 'lunar sixes'. These are measurements of the interval of time between the rising or setting of the Sun and Moon at new and full Moon. Analysis of this data provides little information of geophysical interest, but shows clearly that the water clock used was subject to systematic drifts varying from 4 to 27%!

Only two kinds of geophysically useful observation do not require measurements. These are:

(i) occultations of stars and planets by the Moon;
(ii) large solar eclipses.

In each case a mere description of what actually took place together with an approximate date is all that is required.

The present investigation is restricted to the second of these categories but it is appropriate to make a few remarks about the first category since an analysis is planned in the near future.

In order to prove of value, an *untimed* observation of an approach of the Moon to a star or planet must relate to a graze, in which the limb of the Moon passed within, say, 1 arc min of the object. Where only a small number of conjunctions are recorded (as in Greek history) the probability of finding a critical observation is correspondingly low. By far the most prolific (and as yet untapped) source of such material is the Chinese dynastic histories. From a general survey of the astronomical chapters of the histories of the principal dynasties (Han, Chin, T'ang and Sung), we would estimate that up to A.D. 1400 there are at least 500 occultations of planets and stars in Chinese records. These observations are the work of the imperial astronomers and in most cases we are told precisely which planet or star was occulted, but without further details (i.e. no measurements of time or position). There should be a significant number of grazes which will allow us to obtain a restricted set of values of $\dot{n}$ (the lunar acceleration in arcsec/cy^2) and $\dot{e}$ (the Earth's rotational acceleration in sec of time/cy^2) at a few selected epochs. Scarcely any of this material has ever been translated and we propose to begin translating and analysing it fairly soon.

2. *Selection of observations of large solar eclipses*

Well over 1000 solar eclipses are recorded in world history, principally in the following sources:

(i) the Greek and Roman classics;
(ii) the town and monastic chronicles of medieval Europe;
(iii) the Chinese annals and astronomical treatises;
(iv) the Late Babylonian astronomical texts.

Eclipses in the first two sources are recorded mainly on account of their spectacular nature. The observations in sources (iii) and (iv) are much more scientific. Before commenting on these four sources we propose to outline the reasons why we have preferred to confine our attention only to those records which either (a) mention the complete disappearance of the Sun or the formation of the annular phase or (b) specifically deny that the central phase was witnessed. We have thus rejected all but a minute proportion (about 1%) of the available material.

We consider a statistical approach to large solar eclipses in medieval times as of doubtful validity on account of the non-random distribution of medieval centres of population. The existence of 'population bias' is readily demonstrated by consideration of the total solar eclipse of 3 June 1239.

Figure 3 shows the zone of totality over S. France and N. Italy as computed on standard elements. Because the track was almost parallel to the equator in this region, the positions of the northern and southern limits would be much the same (within a few km) on any reasonable choice of $\dot{n}$ and $\dot{e}$. The eclipse was extensively observed in Italy but there appear to be few records of it from elsewhere. On the map shaded circles denote a mere reference to an eclipse of unspecified magnitude while asterisks indicate that stars or darkness (or both) were reported. A least-squares fit to the data points would place the central line rather close to the northern limit shown in the figure, but this is physically impossible: the northern edge of the track did not reach further north than latitude $45°\cdot 2$ on any part of the Earth's surface. The explanation of this apparent anomaly is that in 1239, just as at the present day, the density of population in N. Italy was

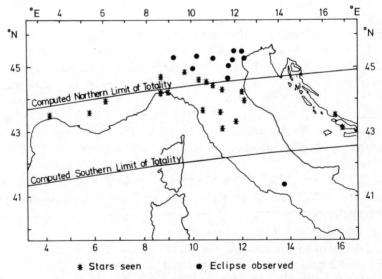

FIGURE 3. The total solar eclipse of 3 June A.D. 1239

much greater than that in the central and southern parts of the country. As a consequence almost all of the observations were made in places lying to the north of the central line.

In the example just given no choice of values of $\dot{n}$ and $\dot{e}$ would satisfy the least-squares fit but there are several examples where finite but unacceptable values are indicated. Figure 4 is a plot similar to Figure 3 for the eclipse of 22 August 1039. The belt of annularity was here only a few km wide. Shaded circles again denote mere mentions of the eclipse; there are no more definite observations. Of the two tracks shown, 'A' represents that computed using the solution of this paper and 'B' the best least-squares fit to the observations, assuming equal weights. The

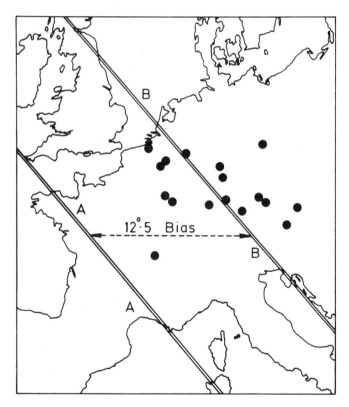

● − Eclipse observations
A − Actual path.
B − Path implied by Least Squares Fit to observations.

FIGURE 4. Population bias effects for the annular solar eclipse of 22 August A.D. 1039

difference in longitude between the two zones is some 12·5°, corresponding to a rate of lengthening of the day of about three times the present rate. A similar analysis for the eclipses of 1093, 1187, 1191 and 1263, all of which were extensively observed in Europe, actually indicates a speeding up of the Earth in each case. Here again the explanation is population bias, resulting from the heavy concentration of monasteries in what is now Germany.

Newton (1970, p. 88) accepted the existence of population bias in 1239 but he later (pp. 89–90) argued that for the eclipse of 1241 (the records of which he took almost entirely from a paper by Celoria, 1877b) there was no significant bias, giving as his reasons:

(1) The places where this eclipse was recorded lay at the time of Celoria's work in either the German Empire or the Austro-Hungarian Empire. Since Celoria had access to the records which he discussed he should also have had access to the records of other places in both empires.

(2) If records of the eclipse were made at points much to the east of any listed by Celoria they would have been in Poland or Hungary. Both kingdoms were as likely to be sources of eclipse reports as the Holy Roman Empire.

Both of these arguments are invalid. Celoria took almost all of his material relating to the eclipse of 1241 from *Monumenta Germaniae Historica* (Pertz, 1826→). This work is very largely a compilation of monastic annals. The distribution of monasteries in the 13th century A.D. was in no way affected by the political situation in 1877 (when Celoria wrote his paper). Additionally, the historical maps of Menke (1880) show a very sharp fall-off in the concentration of monasteries to the east of the present borders of Germany and Austria. The presence of a geographical bias is thus probable.

The only other solar eclipses which were extensively observed took place in 1133 and 1147. Once again it is possible to argue that population bias is present to a small but indeterminable extent. If we had access to a very large number of well-observed eclipses then it might prove possible to reduce the effects of geographical bias to an acceptable level, but as things are, the signal-to-noise ratio is uncomfortably low.

It would appear that the only satisfactory solution to the problem is to follow Fotheringham's (1920) method of approach, suitably adapted. The object of this is to reject all those records which only mention the occurrence of an eclipse or describe it merely in general terms and concentrate on the few select observations which mention either the complete disappearance of the Sun or the formation of the annular phase or alternatively deny that the central phase was witnessed. Similar remarks will apply to the occultation observations when they are eventually analysed.

To be of value an eclipse record must satisfy three observational, and one geometric, requirements.

(1) The date must be accurately known, or at least sufficiently accurately to allow an unambiguous date to be deduced astronomically.

The Accelerations of the Earth and Moon

(2) The place of observation must be able to be identified with a high degree of confidence.

(3) Only a simple description of the magnitude which the eclipse attained is required, but it must make it clear whether or not the central phase is witnessed.

(4) The track must make an appreciable angle with the equator in the vicinity of the place of observation, otherwise the magnitude will be insensitive to changes in $\dot{n}$ and $\dot{e}$.

The last point is well illustrated by the eclipse of 1239. Detailed descriptions of totality are given in the chronicles of Montpellier, Florence, Siena, Arezzo, Cesena and Salona but for any reasonable value of $\dot{n}$ and $\dot{e}$ each place lay well within the central zone (Figure 3).

3. Sources of reliable records

Deductions which we propose to make concerning $\dot{n}$ and $\dot{e}$ are obtained from about 30 observations. Selection of early records of high reliability requires a careful examination of the original texts. At no time have we relied upon secondary sources and this has meant much time engaged on purely historical research.

Let us consider the four major sources in turn.

a. The Greek and Roman classics

The classics contain numerous references to solar eclipses, but in far from ideal form. In many instances all three fundamental details, date, place and magnitude, are uncertain.

Fotheringham, in his well-known 1920 paper, took most of his material from this source. However, Newton (1970) has severely criticized his choice of observations. Fotheringham based his final deductions on three eclipses (Eponyn canon, Hipparchus and Plutarch) for which he gave the dates 763 B.C., 129 B.C. and A.D. 71. The record in 763 B.C. tells us nothing about the magnitude of the eclipse, and Fotheringham's assertion that because this is the only eclipse mentioned in the Assyrian Chronicle (which covers the periods 858–704 B.C.) it must have been total is without foundation. His assumption gives the line in his famous diagram corresponding to the southern limit of totality an exactness which is quite artificial. We had formerly regarded the eclipse of Hipparchus as a reliable observation but we are now of the opinion that the date is not satisfactorily determinable. Cleomedes, Pappus and Ptolemy all refer to Hipparchus' use of this eclipse in calculating the lunar distance but all omit the date. Historically the date 129 B.C. seems most probable, but two earlier dates, 310 and 190 B.C., are also acceptable.

Fotheringham may have been right in identifying the total eclipse mentioned by Plutarch (*De Facie*, 931D) as that of A.D. 71, but Newton (1970, pp. 115–18) gives some valid arguments for questioning the reliability of this eclipse. See Part III, section 9, below.

The remaining observations which Fotheringham used in his paper are, in our view, also of low reliability. In each case at least two of the three critical details are doubtful. Although out of context here, it is perhaps worth commenting on the so-called 'Eclipse of Babylon'. The date of the phenomenon 'the day was turned to night and fire in the midst of heaven' given by Fotheringham as 1063 B.C. is certainly incorrect. The event occurred during the 7th year of a king who reigned at least 17 years. This puts the date at within a few years of 1020 or 1000 B.C. (Brinkman, 1968). Rowton (1946) suggests the eclipse of 9 May 1012 but this reached a magnitude of about 0·8 at sunset. Possibly the phenomenon was a violent thunderstorm: an eclipse seems problematical.

In only two examples from the classics do we know both the date and place of observation of a very large eclipse, but in neither case is the magnitude certain. These eclipses occurred at Rome in 188 B.C. and Athens in A.D. 484. Although interesting historically, we must omit them from the present study since they make no direct reference to the degree of obscuration of the Sun. In 310 B.C. (the eclipse of Agathocles) we have the magnitude and date but lack a location.

b. *The town and monastic chronicles of medieval Europe*

These in the main cover the period from A.D. 900 to 1300. The various chronicles are principally concerned with domestic events and occasionally make reference to earthquakes and the more spectacular astronomical events such as unusually large eclipses of the Sun and bright comets. Frequently an eclipse is regarded as the major event of the year. Occasionally annalists show a considerable interest in astronomical matters recording aurorae, lunar eclipses, meteors and even occultations, as well as the more striking phenomena, but this attitude is rare.

Dating of events recorded in the medieval chronicles seldom presents problems since the correct year is frequently given, and large errors are rare. On the Continent of Europe, as distinct from the British Isles, the chronicles are to a considerable extent independent of one another and are mainly concerned with local events. Thus, except in rare cases, we can be confident that the place of observation was where the chronicle was compiled. This conclusion is not true for the various annals of the British Isles. Several of these are the work of individuals who are known to have travelled throughout the country. Others bear the names of monasteries, but at no time did they attach special significance to the affairs of the monastery. Each set out to be a chronicle of the whole country. Under these conditions it would be appropriate to include material from almost any source and it is clear that the eclipse records come into this category.

A very large number of the chronicles of continental Europe have been published, notably in *Monumenta Germaniae Historica* (Pertz, 1826→) and in *Rerum Italicarum Scriptores* (Muratori, 1723→). In their investigations of the acceleration of the Moon—before the discovery of the apparent solar acceleration —Celoria (1877a and 1877b) and Ginzel (1884a, 1884b and 1918) used these compilations as the basis of their investigations, but they also consulted a number of relatively inaccessible sources (including unpublished chronicles). We have made a careful search of the published chronicles but have been unable to make much advance on the list of observations given by Celoria and Ginzel, and reviewed by Newton (1972).

c. *The Chinese annals and astronomical treatises*

The earliest eclipse observations from China date from the 13th or 14th centuries B.C. These are recorded on the oracle bones of the Shang dynasty which were discovered at An-yang at the end of the last century (cf. Liu chao-yang, 1945; Tung tso-pin, 1945; Dubs, 1951). With one exception (see below), it is too early to consider using these few scattered observations in a geophysical investigation but it may prove possible to deduce some results of chronological importance. An analysis of these observations is planned in the near future.

Between the Shang dynasty and the Unification in 221 B.C. (i.e. during the Chou dynasty and Warring states period) there are less than 50 records of solar eclipses, nearly all confined to the annals of a single small state: Lu, the home of Confucius. That this situation exists is almost certainly the result of the famous burning of the books in 213 B.C. under Shih Huang-ti, the first Emperor of China (Fitzgerald, 1942). Eclipses were probably mentioned in the annals of other states as well as Lu, but these records may well have perished for ever.

Three total solar eclipses are mentioned in the Lu annals under dates which correspond to 709, 607 and 549 B.C.

After the Unification (more accurately from the beginning of the Former Han dynasty in 202 B.C.) we have a very extensive series of observations. Each dynastic history (except in the case of a few short-lived dynasties) contains an astronomical section compiled largely from the records of the imperial astronomers. Observations of solar eclipses form only a small part of this material. Other data include sunspots, sightings of Venus in daylight, lunar eclipses, occultations of stars and planets by the Moon, planetary conjunctions, close approaches of planets to stars, and sightings of meteors, comets and novae, etc.

That we have access to such an exhaustive list of astronomical observations results from the attitude to celestial events prevalent in China from very early times. Events occurring in the heavens, no matter how trivial, were regarded as forerunners of events occurring on Earth or warnings to the ruler. The following extract from Chapter 12 of the *Chin Shu* (*History of the Chin* (*Tsin*) *Dynasty*, dealing with the period A.D. 220–420) typifies the attitude of the emperor and his court to an eclipse.

'At the beginning of the T'ai-ho reign period (A.D. 277/232) of Ming Ti, the Astronomer-Royal Hsü Chih submitted to the Throne (a warning of) a solar eclipse and requested that he and the Prime Minister should be granted leave to sacrifice at the Imperial Observatory (Ling T'ai) in order to avert the impending calamity. The Emperor replied, "I have heard that when the Emperor's actions are faulty the heavens are alarmed and manifest portents as warnings to enable him to repent. Thus when the Sun and Moon are veiled or eclipsed one can infer irregularities in the administration".'

(Trans. Ho, 1966, p. 154)

This preoccupation with astrology led to the appointment of professional skywatchers at the imperial court from at least the beginning of the Han dynasty. One of the Jesuit missionaries to the Ch'ing (Manchu) court at Peking gives the following eyewitness account of the continual watch of the sky which had been going on for some two millennia (Lecomte, 1696).

'They still continue their Observations. Five Mathematicians spend every Night on the Tower in watching what passes over head; one is gazing towards the Zenith, another to the East, a third to the West, the fourth turns his eyes Southwards, and a fifth Northwards, that nothing of what happens in the four Corners of the World may scape their diligent Observation. They take notice of the Winds, the Rain, the Air, of unusual Phenomena's such as are Eclipses, the Conjunction or Opposition of Planets, Fires, Meteors, and all that may be useful. This they keep a strict Accompt of, which they bring in every Morning to the Surveyor of the Mathematicks, to be registered in his Office.'

A daily, as well as nightly, watch was also maintained, for we find frequent records of sunspots, solar haloes and sightings of Venus and occasional novae in daylight apart from solar eclipses.

The observations of solar eclipses, as of other astronomical phenomena recorded in the astronomical treatises of the various dynastic histories, should be treated as having been made at the appropriate capital except when it is expressly stated to the contrary.

In confirmation of this conclusion there is no evidence of systematic observation in the provinces. Dubs (1938, p. 338) conveys a false impression of the proficiency of provincial observers when he considers a solar eclipse recorded in 147 B.C. No eclipse occurred on the stated date and Dubs arbitrarily selected the nearest eclipse visible in China, which was two years later. As this was only observable at sunrise at the tip of the Shan-tung peninsula, Dubs rather rashly proposed hypothetical officials who were capable of making such an observation. Regrettably, Needham (1959, p. 420) followed this suggestion. During the former and later Han dynasties the eclipses of 188 B.C., 28 B.C. and A.D. 120 (among others) were reported as 'almost complete'. In each case the wide zone of totality crossed the entire country over heavily populated areas, but if reports of totality reached the astronomer royal, they have not found their way into the records from which the *astronomical* sections of the histories were compiled.

Especially during the Later Han dynasty there are occasional announcements stating that a particular eclipse was not observed by the imperial astronomers, presumably on account of cloudy weather. The place of observation when specified is invariably at a considerable distance (frequently more than 1000 km) from the capital. Over a period of some 200 years there are only about 10 such announcements and with only a single exception each eclipse was observed at a different location. During this time several total eclipses were visible as such over extensive areas of China, but in each case we have only a bare record of the occurrence of an eclipse. This suggests that eclipse reports came to the capital in an extremely haphazard manner, i.e. there were no systematic observations in the provinces.

The evidence is certainly strong that if an eclipse is recorded in the astronomical treatise of a dynastic history, and the place of observation is not named, the observation can be taken as having been made at the capital. However, during the Former Han (202 B.C.–A.D. 9), Later Han (A.D. 23–220) and T'ang (618–906) dynasties we can be more specific than this. During these periods, a measurement of the right ascension of the Sun, usually expressed to the nearest degree, normally accompanies the observation. When this is the case it is a positive proof that the report comes direct from the records of the imperial astronomers. Towards the close of both Han dynasties these details are almost always omitted, which suggests that perhaps the original astronomical records for these periods were lost and replaced by less reliable material from some other source. Because a similar situation could occur in any other dynasty and go undetected (owing to the omission of the celestial position of the Sun) we have decided to restrict our investigation almost exclusively to those Han and T'ang records contained in the astronomical chapters of the histories which are accompanied by a measurement of the right ascension of the Sun. We would estimate that under these circumstances the capital can be assumed to be the place of observation with at least 95% confidence.

It should be noted that the imperial annals, which always commence a dynastic history, contain eclipse and occasional cometary records but it is clear that these are in the main abbreviated from the records of the court astronomers with additions from elsewhere. Material from this source is of lower reliability.

A translation of the Former Han solar eclipse records has been given by Dubs (1938, 1944 and 1955). Wylie (1897) gives a summary of the Chinese eclipse observations from earliest times until 1785 but his translations are frequently unreliable. We have found it desirable to consult original sources for every observation we use.

d. The Late Babylonian astronomical texts

These texts were discovered in the 1880s and 1890s and are now mainly housed in the British Museum. They cover the period from about 450–50 B.C. A large number of the inscriptions have been copied, notably by T. G. Pinches, and many of these have been published by Sachs (1955). The tablets were collected in a haphazard manner and for many years the literature repeated the statement that the provenance was Sippar. However, Sachs (1948) has proved conclusively that the texts come from Babylon, giving the following reasons.

(a) The provenance expressly mentioned in the published non-astronomical texts of the same museum collections is Babylon.

(b) The improved reading of the colophon of a lunar table results in the elimination of the only alleged mention of Sippar in the astronomical texts themselves.

(c) The character of the few personal names which occur in the colophons and the deities mentioned in the introductory invocations both point to Babylon.

In view of the extreme rarity of total solar eclipses at any one place (three in a typical millenium) we consider ourselves extremely fortunate to have obtained two independent descriptions of totality in 136 B.C. among such fragmentary material. As far as we are aware, the Babylonian texts contain no other reference to a total solar eclipse.

4. The critical observations

In this section the individual observations are discussed in chronological order, having first separated them into two groups: Europe (including the Middle East) and China. Each observation is classified as 'A' or 'B' depending on its reliability. Class A data give us no cause for doubting that a given eclipse was total (or in certain instances non-total) at a known place and on a known date. Class B records contain an element of uncertainty, usually in the place. Two annular eclipses, seen in 197 B.C. and A.D. 1153, although otherwise seemingly of high reliability, are included in this latter category because near the edge of a zone of annularity the narrow portion of the ring may be overpowered by the bright portion, making it difficult to decide whether or not the ring phase is being witnessed (see Part I).

The language of the European texts is principally Latin, while that of the Chinese texts is invariably Classical Chinese (apart from the very earliest record). Except where stated to the contrary, translations are by the authors, with assistance from colleagues.

Europe and the Middle East

Some 30 records which specifically state that the Sun completely disappeared are known to us from this source. Many of these come from places lying close to the central line of a wide zone of totality running nearly parallel to a line of latitude and thus the magnitudes are insensitive to changes in $\dot{e}$ and $\dot{n}$. These observations, which are valueless for the purpose of this study, except that they do not contradict any of the results deduced in this paper, are as follows: A.D. 840 (Bergamo), 1133 (Heilsbronn, Reichersberg), 1176 (Antioch), 1239 (Arezzo, Cesena, Florence, Montpellier, Salona, Siena), 1241 (Reichersberg), 1406 (Braunschweig), 1415 (Altaich and Prague) and 1485 (Melk). Many of the descriptions from the above places are remarkably vivid, yet the fully reliable record from Melk in 1485 mentions no more than that the whole Sun was obscured.

The following total eclipse reports are critical: 1375 B.C. (Ugarit), 136 B.C. (Babylon), A.D. 693 (Bagdad), 912 (Cordoba), 968 (Constantinople), 1079 (Seville), 1124 (Novgorod), 1133 (Salsburg), 1239 (Cerrato and Toledo), 1241 (Stade), 1560 (Coimbra), 1567 (Rome) and 1715 (various places in England). Of the numerous reports which specifically deny that the central phase was witnessed, three may be listed which clearly refer to very large eclipses: 1133 (Vysehrad), 1153 (Erfurt) and 1178 (Vigeois). There are numerous other careful descriptions of a partial eclipse, but in every case the central zone passed too far to the north or south of the appropriate place to provide useful sensitivity.

To the above list of observations selected for detailed study may be added a single timed contact of a partial eclipse from Babylon. This eclipse, occurring in 322 B.C., was observed to begin only a few minutes before sunset.

We have ignored two particularly vivid accounts from Kerkrade in 1133 and Altaich in 1544 which describe many of the effects normally associated with the total phase but do not make a direct reference to the degree of obscuration of the Sun. Newton (1972, p. 434) assumed that the eclipse of 1241 was total at Worms, but the term *generalis* (also used in the same annals to describe the annular eclipse of 1191) most probably means that the eclipse was seen over a wide area.

3 May 1375 B.C., Ugarit (B). A detailed linguistic and astronomical discussion of a tablet excavated in the ancient city of Ugarit has been published by Sawyer and Stephenson (1970). A translation of the text is given in the epigraph to the present paper. The description, 'the Sun went down in the daytime', may be most readily understood to refer to a total solar eclipse. The date of this event is given in the text as the first day of the month of Ḫiyar. The equivalent months on the Babylonian and Hebrew calendar are respectively Aiaru and Iyyar, both of which correspond to April–May. Preliminary analysis showed that in the period covered by the texts (1450–1200 B.C.) only three eclipses could possibly have been very large in Ugarit: 15 July 1406 B.C., 3 May 1375 and 5 March 1223.

If we could be certain that the Ugaritic calendar at this very early period corresponded with the Babylonian and Hebrew, we would have no hesitation in assigning the date of the eclipse as 3 May 1375 B.C. However, until it can be shown that this is definitely the case, the suggested date is somewhat questionable. This is, therefore, a class B record.

26 September 322 B.C., Babylon (A). This is recorded on a tablet (B.M. 34093) which is part of an astronomical diary for the period 323/2 B.C. Drawings of both sides of the tablet by T. G. Pinches have been published by Sachs (1955, pp. 53–54: tablet No. 212). Line 23 of the reverse may be translated as follows:

'On the 28th day at 3 *uš* before sunset, a (solar) eclipse ... North west wind. It set eclipsed.'

Earlier in the text the month is given as *Ulul* and the year as the 2nd year of Philip. Using the tables of Parker and Dubberstein (1956), the date reduces to 26 September 322 B.C.

Fotheringham (1935), on the authority of Langdon, stated that the eclipse was said to begin 4 *uš* before sunset but Pinches' drawing clearly reads 3 *uš*, and this has been confirmed by Professor P. J. Huber, who has recently consulted the original tablet on our behalf. The unit of time (*uš*) is the interval required for the heavens to turn through 1°, so that the eclipse was observed to begin 12 min before sunset.

The delay in the actual detection of the eclipse will be considered in Part III below, but it is appropriate to discuss rounding and timing errors here. Stephenson

(1974) analysed 77 Babylonian observations of the interval between the rising or setting of the Sun and Moon near the time of full Moon. These cover the period 323–320 B.C., i.e. they are contemporaneous with the eclipse. About one-third of the measurements are expressed in half integers (the unit being the *uš*), suggesting that measurements were usually, but by no means systematically, rounded to the nearest 0·5 *uš*. Again, analysis of the data reveals that over a period of up to about 1 hr the clepsydra was subject to a mean drift of 15%. We consider that ±0·5 *uš* is a fair estimate of the uncertainty in the eclipse time arising from a combination of both sources of error.

15 April 136 B.C., Babylon (A). Our attention was first drawn to the observation of the total phase of this eclipse in Babylon by Professor Huber. A direct reference to totality is given on B.M. tablet 34034, which is a goal year text containing predictions and a few eclipse observations (lunar and solar). Sachs (1955, p. 198: table No. 1285) has published a drawing of the tablet by Pinches. Huber (personal communication) has dated the tablet from the predictions for Venus and Mars as 136 B.C. He translates lines 24–28 of column 1 of the reverse as follows:

'On the 29th day there was a solar eclipse beginning on the south-west side. After 18 *uš* . . . it became complete (*til*) such that there was complete night at 24 *uš* after sunrise.'

Earlier in the text the month is given as Addaru II. From the tables of Parker and Dubberstein (1956) the date is equivalent to 15 April 136 B.C. The sign *til* is frequently used to express totality in Late Babylonian records of lunar eclipses.

An enquiry to Professor A. J. Sachs yielded a further observation of the same eclipse in an unpublished tablet (B.M. 45745), which Sachs (1955) catalogues as *429. This is part of an astronomical diary for the last part of the year 175 of the Seleucid Era (137/136 B.C.). Lines 13 to 15 of the reverse may be translated as follows (Sachs, personal communication):

'Daytime of the 28th the north wind blew. Daytime of the 29th, 24 *uš* after sunrise, a solar eclipse beginning on the south west side . . . Venus, Mercury and the Normal Stars (i.e. the stars which were above the horizon) were visible; Jupiter and Mars, which were in their period of disappearance (i.e. between last and first visibility) were visible in that eclipse . . . (the shadow) moved from south west to north east. (Time interval of) 35 *uš* for obscuration and clearing up (of the eclipse). In that eclipse, north wind which . . .'

The tablet is extensively damaged, hence the several gaps in the text.

Between them the two observations give by far the most reliable description of a total eclipse which we possess from the ancient world. From Tuckerman (1962), the elongations of the four planets Mercury, Venus, Mars and Jupiter from the Sun were respectively 21° W, 31° W, 18° W and 2° E, so that all would be above the horizon and, as stated, both Mars and Jupiter would be between heliacal setting and heliacal rising. Saturn (166° W) was far below the horizon. The whole description is a testimony of the skill of the Babylonian astronomers.

5 October A.D. 693, Baghdad (B). The chronicle of Elias and Nisibis (a Nestorian bishop, who died about 1049) quotes the following:

'The year 75 (A.H.) began on Saturday, the 2nd day of Ḥiyar (May 2), 1005 of the Greek calendar (A.D. 694); on this day there was a total eclipse of the Sun on Sunday the 5th day of Tesrin I (October 5) at the 5th hour of the day.'

The chronicle has been published in the original Syriac together with a translation into Latin by Brooks and Chabot (1954). Although an error of 1 year must be presumed, the record clearly refers to the eclipse of 5 October 693 (the time of day is roughly correct). Elias names his source as 'Huwarasmi'. From Ginzel (1918), Huwarasmi (*fl.* 833) was one of the Caliph's astronomers and librarian at Baghdad. It seems reasonable to suppose that Huwarasmi obtained his information from a Baghdad source, but as he lived some 150 years after the eclipse, and the context of the record is wanting, there can be no certainty that the observation was made there.

17 June 912, Cordoba (B). This observation is contained in an Arabic chronicle of Cordoba, *Al Muqtabis*, by Ibn Hayyan. We read:

'In this same year (299 A.H.) on Wednesday, the last day of the month Sawal, a total eclipse of the Sun occurred. Darkness covered the Earth and the stars appeared. The greater part of the people believed that the Sun had set below the horizon and got up for the Prayer of Sunset (*Magrib*). The shadow dissipated in a normal time behind the distant horizon'.

An edition of the *Al Muqtabis* in Arabic is published by Antuña (1937) and a translation into Spanish by Guriaeb (1960).

The date given by Ibn Hayyan is precisely correct. The last section of the *Al Muqtabis* (containing the above text) is essentially a chronicle of Cordoba, the capital, of the Arab dominions in Spain at the time. Ibn Hayyan tells us that the eclipse of 27 June 903 was seen in Cordoba. Unfortunately he does not name the place of observation in 912 and as he wrote at least a century after the event (he lived 988–1076), there is a slight doubt concerning the place observation. However, in view of the nature of the chronicle, we presume the place to be Cordoba. The magnitude is certain.

22 December 968, Constantinople (A).

'There was a defection of the Sun at the winter solstice such as never happened before ... The defection was such a spectacle. It was the 22nd day of the month of December at the 4th hour of the day. The sky was clear when darkness was spread over the Earth and all the brighter stars revealed themselves. Everyone could see the disc of the Sun without brightness, deprived of light, and some dull and feeble glow, like a narrow band shining round the extreme edge of the disc. Gradually the Sun going past the Moon (for this appeared covering it directly) sent out its original rays and light filled the Earth again ... At the same time I myself was also staying in Constantinople ...'

(*Leonis Deaconi Historicae*, IV, 11.)

There is an edition of the history in the original Greek and with a translation into Latin in *Corpus Scriptorum Historiae Byzantinae*, Vol. 33 (Niebuhr, 1828). Leo Deaconus spent most of his later life at Constantinople and his history, which covers the period 959–971, is much concerned with the city. The description of the corona is probably unique in a medieval text. There is no question as to magnitude, identification or place of observation as all are explicitly stated (a very rare occurrence).

1 July 1079, Seville (B). Sarton (1947) in his discussion of the work of Samuel ben Judah ben Meshullam (Samuel of Marseille), the 14th-century translator, states that he translated into Hebrew a treatise on a total solar eclipse. This treatise was written by Abū 'Abdallāh Muhammad ibn Mu'ādh of Seville and the subject is the total solar eclipse which occurred on Monday the last day of 471 A.H. According to Sarton, the last day of that year was Wednesday, 3 July 1079. The date of the eclipse is thus correct except in the detail that it was the last day of the year.

We have ordered a copy of the MS. from the Bibliothèque Nationale in Paris. For the present we shall assume that the eclipse was total in Seville. This conclusion seems justified both from the context of Sarton's note and the fact that the treatise was actually written. Further information is eagerly awaited.

11 August 1124, Novgorod (A).

'(6632 AM): In the month of August on the 11th day, before the evening service, the Sun began to diminish and perished completely. Great fright and darkness were everywhere. And the stars appeared and the Moon. And again the Sun began to augment and its face became full again and everybody in the town was very glad.'

(*Novgorodsky I.*)

The above translation is taken from Vyssotsky's (1949) paper on the astronomical records in the Russian chronicles. *Novgorodsky I* is one of the four chronicles of Novgorod, the Russian capital at the time. Much the same description is to be found in *Novgorodsky II* and *IV*. The mention of 'the town' in the last sentence of the text proves beyond any doubt that the total phase was seen in Novgorod. The year 6632 AM is equivalent to A.D. 1124, so that the recorded date is precisely correct.

2 August 1133, Salzburg (A).

'1133. In this year on the 4th day before the Nones of August (August 2) in the heat of midday, the Sun suddenly disappeared (*disparuit*) and a little afterwards it seemed terribly covered over, like a round bag.'

(*S. Rudberti Salisburgensis Annales Breves.*)

This very brief chronicle, which was compiled in the monastery of S. Peter in Salzburg, is published in *Monumenta Germaniae Historica*, Vol. 9 (Pertz, 1851). From Newton (1972, p. 256), after 1060 the entries are in various contemporaneous hands and are probably original. Thus it may confidently be presumed that the observation was made in Salzburg. The sudden disappearance of the Sun implies totality and this is borne out by the subsequent description, which seems to refer to the corona.

2 August 1133, Vysehrad (A).

'1133 . . . On the 4th day before the Nones of August an eclipse of the Sun appeared in a wonderful manner, which, gradually decreasing, diminished to such an extent that a crown like a crescent Moon proceeded to the south part, which afterwards turned round to the east henceforth to the west; at length it was transformed to its original state.'

(*Canonici Wissegradensis contin. Cosmae.*)

This continuation of the chronicle of Cosmas of Prague by an anonymous canon of Vysehrad is published in *Monumenta Germaniae Historia*, Vol. 9 (Pertz, 1851). It covers the period 1126 to 1142. Its author had a particular interest in astronomy and meteorology and records frequent eyewitness accounts of unusual phenomena, in all probability based on his own experiences. Had the total phase enveloped Vysehrad, it seems extremely unlikely that he would have preferred to quote a description of a much less spectacular partial eclipse from elsewhere.

26 January 1153, Erfurt (B). The various Erfurt chronicles show sketches of the annular eclipse of the Sun in 1153 accompanied by a note of the following form:

'A sign appeared in the Sun on the 7th day before the Kalends of February (January 26) in this manner.'

Copies of these sketches are published in Vol. 16 of *Monumenta Germaniae Historica* (Pertz, 1859) and Vol. 42 of *Scriptores Rerum Germanicarum* (Holder-Egger, 1899). The drawings make little attempt to portray the true configuration, but in each case a large partial eclipse with Erfurt lying to the south of the central zone is indicated. However, on any reasonable values of $\dot{e}$ and $\dot{n}$, Erfurt lay close to the northern edge of the wide zone of annularity. Presumably the Sun was viewed by reflection in water (e.g. after rain), which would explain the apparent inversion of the crescent. As the eclipse was generally annular we shall assume, as discussed in Part I above, an uncertainty in the magnitude of 1%.

13 September 1178, Vigeois (A).

'1178. On the 4th day of the week, the Ides of September (September 13), the 28th day of the Moon, on a clear day the Sun suffered an eclipse at about the 5th hour; its disc began to be covered from the east (*sic*) until it was like a two or three day old Moon. The star Venus was visible to the north. After six the brightness returned from the east in the order that it had blackened until the Sun was fully illuminated. Then we saw each others faces as beside a glowing furnace.'

(*Ex Chronico Gaufredi Vosiensis.*)

There is an edition of the chronicle of Gaufredus in *Recueil des Historiens des Gaules*, Vol. 12 (Bouquet, 1781). The chronicle ends in 1182 and here its author is writing as an eyewitness. There can be little question that he observed an eclipse which was not quite total at Vosium (now known as Vigeois). The wording of the last sentence of the above quotation is obscure but we have given what seems to be a reasonable translation. The effect described there probably refers to after-images caused by looking at the solar crescent for too long.

3 June 1239, Cerrato (A).

'In the year of Grace 1239 on the 3rd day before the Nones of June (3 June), the 6th day of the week, the Sun was completely (*totus*) obscured at midday.'

(*Chronicon Cerratensis.*)

The chronicle of Rodrigo de Cerrato, a Dominican monk, is published in Vol. 2 of *España Sagrada* (Florez, 1778→). In describing the eclipse, Rodrigo is writing as a contemporary so that the presumption of Cerrato as the place of observation is strong. By medieval European standards his description is remarkably brief but this need not lower its reliability in any way (as for Melk in 1485).

3 June 1239, Toledo (A).

'The Sun was darkened (obscured) on Friday at the 6th hour and it lasted for a while between the 6th and 9th hours and it lost all its strength (*toda su fuerza*) and it was as though night; and there appeared stars—what a number of stars—and then the Sun grew bright again of its own accord, but for a long time it did not regain the strength it usually has: era 1277.'

(*Annales Toledanos Segundos.*)

The *Annales Toledanos Segundos* is a contemporary chronicle of Toledo, and has been published in its original language (Castilian) in Vol. 23 of *España Sagrada* (Florez, 1778→). The year 1277 is equivalent to A.D. 1239, and from Schroeter the eclipse of 3 June 1239 (which occurred on a Friday), was total in Spain just before noon, so that the identification is certain. As the chronicle is contemporaneous, the probability that the observation was made in Toledo is high.

The record does not explicitly state that the Sun completely disappeared, but it contains four details that are characteristic of the total phase:

(i) the Sun lost *all* of its strength;
(ii) it was as though night;
(iii) many stars were seen;
(iv) the Sun grew bright again of its own accord, but did not regain its original strength for a long time.

Had only one or two of these details been present in the account, then it would have been no more than a plausible assumption to have concluded that totality was witnessed, but the presence of all four renders this conclusion a very safe one.

6 October 1241, (Stade) (A).

'1241 ... There was an eclipse of the Sun on the octave of Michael, namely the 2nd day before the Nones of October (October 6), on Sunday sometime after midday, with stars appearing and the Sun completely (*penitus*) hidden from our sight. And the sky was so clear that no clouds appeared in the air.'

(*Annales Stadenses.*)

The chronicle of the monastery of S. Maria in Stade is published in *Monumenta Germaniae Historica*, Vol. 16 (Pertz, 1859). This chronicle continues until 1256. The above first-person eyewitness account is virtual proof of totality at Stade.

25 May 1267, Constantinople (B).

'At this time, the Moon obscured the Sun when it was in the 4th part of Gemini, at the 3rd hour before midday on the 25th day of May in the year 6775 (AM). It was a total eclipse of about 12 points or digits. Also such great darkness arose over the Earth at the time of mid eclipse that many stars appeared.'

(*Nicephori Gregorae Historiae Byzantinae*, lib. IV, c. 8.)

The history from which the above passage is taken has been published in its original Greek together with a translation into Latin in *Patrologiae Graecae*, Vol. 148 (Migne, 1865). The date of the eclipse is exactly correct and the description clearly refers to the total phase. Nicephoras was a citizen of Constantinople, and had he been living at the time of the eclipse it would have been very reasonable to have concluded that the observation, which he includes in his history of the city, was made there. However, he lived a century after the event (his history continues up to 1351) so that this conclusion, although probable, is by no means certain.

21 August 1560, Coimbra (A) and *9 April 1567, Rome* (A). Clavius had the good fortune to observe two total solar eclipses in a period of seven years.

Descriptions of both eclipses are contained in his *In sphaeram Ioannis de Sacrobosco* (Clavius, 1593). In discussing the variable distance of the Moon from the Earth he refers to his observations of the eclipses of 1559 (actually 1560) and 1567. He describes the former as total, and presumes the latter to be annular. We read:

'I will cite two remarkable eclipses of the Sun, which happened in my own time and thus not long ago, one I observed in 1559 about midday at Coimbra in Lusitania in which the Moon was placed directly between my sight and the Sun with the result that it covered the whole Sun for a considerable length of time and there was darkness in some manner greater than that of night. Neither could one see very clearly where one placed his foot; stars appeared in the sky, and (miraculous to behold) the birds fell down from the sky to the ground in terror of such horrid darkness. The other I saw in Rome in the year 1567 also about midday in which although the Moon was again placed between my sight and the Sun it did not obscure the whole Sun as previously, but (a thing which perhaps never happened at any other time) a certain narrow circle was left on the Sun, surrounding the whole of the Moon on all sides (*undique*).'

Clavius has mistaken the year of the first eclipse, but from Schroeter (1923) the eclipse of 21 August 1560 was total in Portugal and Spain just before noon. He seems to have observed totality in 1560 and 1567 by pure chance since in 1560 he was a student at the University of Coimbra and in 1567 was teaching mathematics at the Collegio Romano in Rome (where he spent most of the remainder of his life).

Unknown to Clavius, the eclipse of 1567 was total on the Earth's surface (except near the sunrise and sunset positions where it was annular). However, computation shows that the width of the belt of totality was approximately 10 km in the vicinity of Rome. It is important to emphasize here that the computed width of a belt of totality at a given place is for our purpose independent of the choice of $\dot{n}$ and $\dot{e}$. Since Clavius describes 'the certain narrow circle' as 'surrounding the whole of the Moon on all sides' it is clear to us that he is referring to the inner corona. The intensity of the corona falls off very rapidly with distance from the solar photosphere. Even if Clavius were on the central line in 1567, no part of the photosphere would be more than 3 arcsec inside the lunar limb so that the inner corona would be particularly luminous.

It is interesting to note that from the above quotation Clavius appears to be unaware that annular eclipses had ever been observed before. This is not surprising. We are aware of only one clear description of the annular phase from any historical record before the year 1600 (probably from Peking, in 1292).

Even in Clavius' day Rome was of substantial size. We have assumed that the observation in 1567 was made in or near the Collegio Romano, which lies in the heart of the city.

3 May 1715, England (A). This eclipse, which was total over much of England, was extensively observed thanks to Halley predicting the approximate path of totality and forewarning astronomers and laymen alike throughout the country. Full details are readily available in a paper by Halley (1715). Near the northern limit of totality the eclipse was observed to be total in Badsworth and Barnsdale and definitely partial at Darrington, whereas near the southern limit it was total in Wadhurst, Norton Court and Cranbrook and partial in Bocton and Brightling. The descriptions from Darrington, Bocton and Cranbrook prove that these places were within a few tenths of a kilometre from the respective edges to the track.

At Darrington the Sun 'was reduced to almost a point which both in colour and size resembled the planet Mars'. A similar description comes from Bocton, where 'before the Sun had quite lost his light on the east side, he recovered it on the west, and that there was a small light left on the lower part of the Sun that appeared like a star'. At the neighbouring Cranbrook the Sun was observed 'to be extinguished but for a moment and instantly to emerge again'. The last pair of observations indicate again the very high resolution which untrained observers can achieve, as both places are virtually on the southern limit.

China

From section 2 above, the two principal sources of useful eclipse records in Chinese history are the annals of the state of Lu and the astronomical chapters of the dynastic histories of the Empire. In the former work, three eclipses, occurring in the years 709, 601 and 549 B.C., are described as total but without any further details. The astronomical treatises of the dynastic histories contain observations of both central and large partial eclipses. The following observations from this source are accompanied by a measurement (or calculation) of the right ascension of the Sun and thus fulfil the criterion of maximum reliability discussed in section 3:

> 198 B.C. (total), 188 B.C. (almost complete), 181 B.C. (total), 147 B.C. (almost complete), 89 B.C. (not complete, like a hook), 80 B.C. (almost complete), 28 B.C. (not complete, like a hook), 2 B.C. (not complete, like a hook), A.D. 65 (total), 120 (almost complete), 360 (almost total), 702 (almost complete), 729 (not complete, like a hook), 754 (almost complete), 756 (total), 761 (total).

In order to avoid unnecessary historical discussion, it can be stated here that the magnitudes of every partial eclipse in the above list except A.D. 120 fell so far short of centrality at the place of observation (the appropriate capital) that

The Accelerations of the Earth and Moon

on any reasonable choice of $\dot{e}$ and $\dot{n}$ the eclipse occurred as stated. This leaves six records which require further analysis. We have included in this survey four additional observations made in 1330 B.C., A.D. 516, 522 and 1221.

14 June 1330 B.C., An-yang (B). Among the oracle bones (ox bones, tortoise shells, etc.) found near An-yang and dating from the Shang dynasty (1500–1000 B.C.) Jao Tsung-i (1959) has identified an inscription which may relate to a large solar eclipse. The record may be translated:

'On the day *ping-shen* the oracle was consulted; on the day *ting-yu*, a sacrifice was made; it was a clear day. On the following day there was a great eclipse of the Sun (*ta shih jih*). It was a clear day.'

The dates *ping-shen* and *ting yn* are respectively the 33rd and 34th cyclical days, so the eclipse occurred on the 35th day.

A computer check showed that only five solar eclipses were total near An-yang during the Shang dynasty and one of these was the eclipse of 14 June 1330 B.C. which occurred on the 35th cyclical day. The deciphering of the oracle bone inscriptions has presented considerable difficulties to sinologists, but the exact agreement in the cyclical day numbers makes the *a-priori* odds at least 10:1 that the phenomenon referred to was an eclipse. We do not propose to go so far as to suppose that the eclipse was total at An-yang (which was the capital for much of the Shang dynasty), but we regard this very early observation as a valuable check.

8 July 709 B.C., Chü-fu (A), *12 September 601, Ying* (B) and *12 June 549, Chü-fu* (A). The *Ch'un-ch'iu* ('Spring and autumn annals'), which is a chronicle of the state of Lu covering the period 722 to 480 B.C., contains 37 records of solar eclipses. This chronicle has been published, together with a translation into English, by Legge (1960). The chronicle is noted for the extreme brevity of its entries and this applies to the eclipse records. The three observations which interest us here are as follows:

'Autumn, 7th Moon (of the 3rd year of Duke Huan), day *jen-chén*, the 1st day of the Moon, the Sun was eclipsed and it was total.'

'Autumn, 7th Moon (of the 8th year of Duke Hsuan), day *chia-tzu*, the Sun was eclipsed and it was total.'

'Autumn, 7th Moon (of the 24th year of Duke Hsiang), day *chia-tzu*, the first day of the Moon, the Sun was eclipsed and it was total.'

In each case the term *chi* is used to imply totality. This expression has the same meaning throughout Chinese history. For the other 34 eclipses in the series no details of magnitude are given. From Legge (1960, *prologomena*, p. 87) the three dates correspond to 8 July 709 B.C., 12 September 601 and 12 June 549. Intercalation at this early period seems to have been somewhat irregular and Legge had to assume an error of one month in 709 B.C. and three months in 601. In the latter case there may simply be confusion between the characters for 7 and 10, which are very similar. However, each cyclical day number is exactly correct. Chinese dates are normally quoted in terms of the 60-day cycle which has been running continuously since well into the 2nd millennium B.C. (1330 B.C.). The dates *jen-ch'en* and *chia-tzu* (which occur in both the 601 and 549 B.C. records) are respectively the 29th and 1st cyclical days. We have checked these dates using a computer program and found them exact (in common with almost every other eclipse date in the Lu annals and in Chinese history generally).

From the discussion in Part I, because these three eclipses were generally total on the Earth's surface, there are no grounds for questioning the magnitudes, but some remarks on the place of observation are necessary. At this period of the Chou dynasty, China consisted of a number of semi-independent states. Theoretically the princes of the various states owed allegiance to the Chou King, but he held no more than a shadowy authority. To prevent anarchy there arose the device of presiding chiefs—the system of one state taking the lead and direction of all the others and exercising what amounted to royal functions throughout the kingdom. From Legge (p. 271) eclipse ceremonies were held at the imperial and state capitals at this period. We read,

'On occasion of an eclipse of the Sun, the Son of Heaven (i.e. the Chou King) should not have his table spread so full as ordinarily, and should have drums beaten at the altar of the land; while princes of states should have drums beaten in their courts.'

Ceremonies at the Lu court during a solar eclipse in which drums were beaten and victims offered at the altar of the land are mentioned in the *Ch'un-ch'iu* on several occasions and hence it may be confidently inferred that the eclipses recorded in the *Ch'un-ch'iu* were observed at *Chü-fu*, the capital of Lu. Similar eclipse ceremonies were probably held at the other state capitals, but we have no record of these as no annals from these states have survived. From Legge (*prologomena*, p. 115), the presiding chief from 612–590 B.C. was the prince of Ch'u. He frequently called the states together so that a report of a major event such as a total solar eclipse occurring in Ying, his capital, might well have reached Lu. We thus assume Ying to be the place of observation in 601 B.C. but assign to class B because we are playing a form of the identification game (see Part I).

7 August 198 B.C., Ch'ang-an (B) and *4 March 181 B.C.*, Ch'ang-an (A). These are recorded in chapter 27 of the *Ch'ien Han Shu* ('History of the Former Han Dynasty'—compiled A.D. 58–76) as follows:

'9th year (of Kao-tzu) 6th Moon, on the day *i-wei* the last day of the Moon there was an eclipse of the Sun and it was complete. It was 13 degrees in *Chang*.'

The date reduces to 7 August 198 B.C.

'7th year (of the Empress of Kao-tzu) first Moon, on the day *chi-ch'ou* the last day of the Moon, there was an eclipse and it was complete. It was 9 degrees in *Ying-shih* (which represents) the interior of the Palace chambers. At that time the empress dowager showed aversion from it and said, "This is for me." The next year it was fulfilled.'

The commentary on the text adds that she died. The date reduces to 4 March 181 B.C.

The Accelerations of the Earth and Moon　　　　　　　　　　　　　　　491

There is nothing in either of the above descriptions to suggest that the observations were made elsewhere than the capital (Ch'ang-an). It is interesting to note that the former eclipse was annular. There does not seem to have been a separate term to describe the ring phase, and we must interpret the term *chi* to imply that the eclipse was annular in Ch'ang-an. The same expression was used to describe the annular eclipses of 616, 1507 and 1527 in China, and the annular eclipse of A.D. 1245 in Korea.

16 December 65, Kuang-ling (B) and *18 January 120, Lo-yang* (A). The source of both records is chapter 28 of the *Hou Han Shu* ('History of the Later Han Dynasty'—compiled 398–445). We read:

'8th year (of the Yung-p'ing reign period), 10th Moon, day *jen-yin*, the last day of the Moon, there was an eclipse of the Sun and it was total. It was 11 degrees in (Nan-)Tou. (Nan-)Tou represents the state of Wu. Kuang-ling as far as the constellations are concerned belongs to Wu. Two years later, Ching, King of Kuang-ling was accused of plotting rebellion and committed suicide.'

The date reduces to 16 December A.D. 65.

'6th year (of the Yüan-ch'u reign period), 12th Moon, day *wu-wu*, the first day of the Moon there was an eclipse of the Sun and it was almost complete (*chi-chin*). On the Earth it was like evening. It was 11 degrees in Hsü-nu. The woman ruler showed aversion from it. Two years and three months later, Teng, the Empress Dowager died.'

The date reduces to 18 January A.D. 120.

Both eclipses were total on the Earth's surface, and from the record it is clear that the eclipse of 120 was on the verge of totality at the capital (Lo-yang). We can be extremely confident that this observation was made at Lo-yang because, around this period in particular, the imperial astronomers were careful to state where an eclipse was observed if it was not seen at the capital. For both the preceding and following entries in the section containing the eclipse records (3 September 118 and 12 August 120) it was explicitly stated that the eclipse was not seen by the imperial astronomers but reported from . . . province. This comment is omitted from the record under discussion and we may conclude that the imperial astronomers saw the eclipse. Particularly because totality is denied, we consider this to be one of the most reliable of the Chinese records.

On the principles of this study, we initially took Lo-yang as the place of observation in A.D. 65 (the imperial capital). Computation ruled out this possibility, leaving the alternatives of deletion, or re-analysis. The only place mentioned in the text is Kuang-ling, albeit the mention is with respect to the prognostication associated with the observation. Why, then, did the astronomers fail to mention that the observation came from the provinces, as we know from computation it must have? We suggest that this is because the place of observation is implicit in the text (Kuang-ling). This interpretation falls short of the reliability of other Chinese records, and we have used computation as an aid in identification, so the data is assigned to class B.

18 April 516, Nanking (B) and *10 June 522, Nanking* (B). Two total solar eclipses in a period of six years are recorded in chapters 6 and 7 of the *Nan Shih* ('History of the Southern Dynasties', compiled 630–650). This history, in common with the histories of several other short-lived dynasties of the period, does not have an astronomical treatise. All of the astronomical records are contained in the annals of the emperors. The imperial annals of the other histories which do not have astronomical treatises also contain astronomical records, but, unlike the *Nan Shih*, no total solar eclipses are chronicled.

Since the eclipse records in A.D. 516 and 522 do not give the R.A. of the Sun, it is impossible to obtain their source, although, as in other dynasties, it is clear that *most* of the observations listed in the annals would be made at the capital by the imperial astronomers. Our experience with the eclipses recorded in the *Hou Han Shu* show that if an observation was not made at the capital, a report was much more likely to come from the extreme limits of the empire than from the vicinity of the capital itself. Thus if our assumption of Nanking as the place of observation in 516 and/or 522 is not justified, our error should be patently obvious in the astronomical analysis. We therefore take it as a reasonable presumption, but by no means a certainty, that the eclipses of 516 (which was annular rather than total) and 522 (which was total) were observed at Nanking, the capital of the period.

28 October 756 (?) and *5 August 761, Ning-hsien* (B). These observations are contained in chapter 32 of the *Hsin T'ang Shu* ('New Book of the T'ang Dynasty' —compiled 1043–1060). We read:

'First year (of the Chih-tê reign period) 10th Moon, day *hsin-szu*, the first day of the Moon, there was an eclipse of the Sun, and it was total. It was 10 degrees in Ti.'

The date reduces to 28 October 756.

'2nd year (of the Shang-yuan reign period), 7th Moon, day *kuei-wei*, the first day of the Moon, there was an eclipse of the Sun and it was complete. All the great stars were visible. It was 4 degrees in *Chang*.'

The date corresponds to 5 August 761.

Both eclipses were total on the Earth's surface. It would normally have been justifiable to assume that the observations were made in Ch'ang-an, the T'ang capital, but this was a period of intense strife. In 755, An Lu-shan, formerly the imperial favourite, rebelled and rapidly captured both Lo-yang and Ch'ang-an, the principal cities of the empire. The emperor Ming Huang and his court fled to Szechuan. For the next 11 years there was turmoil, and when peace was finally restored in 766 the empire had suffered irreparable harm. It thus seems certain that the eclipses of 756 and 761 were not observed in Ch'ang-an.

Our efforts up to the present time have been insufficient to determine the whereabouts of the imperial court astronomers in 756, but we can be more specific in 761. By this time the Crown Prince had been declared emperor, following the abdication of his father, and his movements were presumably

The Accelerations of the Earth and Moon

accompanied by his court, including the imperial astronomers. From the annals of the *Hsin T'ang Shu* (chapter 6), we know that he was in Peng-yuan commandery (identifiable with the modern Ning-hsien) two days after the eclipse, so that he was probably there on the day of the eclipse, and certainly was within 20 or 30 km even if on the march at the time.

23 May 1221, Kerulen River (A). In 1220, Ch'ang-ch'un, the Taoist Master, was commanded by Genghiz Khan to travel from China to Samarkand. A log of the journey (which took 4 years) was kept by one of his disciples, and his account, entitled the *Ch'ang-ch'un chên-jên hsi-yu chi* ('Journey of the adept Ch'ang-ch'un to the West') has been incorporated into the Taoist Canon (Wieger's catalogue, No. 1410). The narrative has been translated into English, with commentary, by Waley (1931).

When Ch'ang-ch'un's party was by the Kerulen River, a total eclipse of the Sun was observed. Waley (p. 66) translates the description as follows:

'On the first day of the fifth month (1221, May 23) just at noon, there was a total eclipse of the Sun, during which the stars were visible. At the time we were on the southern bank of the river. The eclipse worked across the Sun from north-west to south-east.'

Later, while Ch'ang-ch'un was passing the winter at Samarkand, he discussed the eclipse with an astronomer. 'We were by the Kerulen River,' said the Master 'the eclipse was total towards mid-day' (Waley, p. 95). We have consulted a copy of the original text and find that in describing the totality, the term *chi* was used on both occasions.

Wylie (1897) and Curott (1966) made rough estimates of the geographical coordinates of the observers when they saw the eclipse. We have examined in detail Ch'ang-ch'un's route up the Kerulen Valley, using a large-scale map of the area, and give a confident location as somewhere between the points 48° 10′ N, 115° 51′ E and 48° 12′ N, 115° 57′ E. A full discussion of these considerations will be published elsewhere.

PART III: Analysis of the observations

1. *Introduction*

In Part I we established the criteria under which the historically preserved records would be sifted to provide the data-set for actual analysis. In Part II the records were considered under the established criteria, and a final list of 'certified' observations was prepared. It is the purpose of this section to provide the astronomical and numerical analysis of the selected observations. We shall be attempting to find solutions for the acceleration of the Earth's rotation $\dot{e}$ (in sec of time/cy^2), and the acceleration of the Moon in longitude $\dot{n}$ (in arcsec/cy^2), considered as independent parameters. Various linear combinations of these two parameters will also be determined, as this is necessary to finding the independent solution for $\dot{e}$, and to provide astronomically useful parameterizations.

Definitions of some of the terms used will be given, particularly to satisfy the interdisciplinary reader who may not have encountered the astronomical details before. Further details are given regarding what the analysis seeks to determine, and the method of analysis to be employed. The data is analysed in detail, and the iterative steps to the solution are indicated with the intermediate solutions. The nature of uncertainties in the solutions is investigated, and estimates are provided. A detailed comparison is made with the results from other authors. The final solutions are then utilized to compute the 'true' circumstances of several intriguing historical observations which could not be used in the study, and examples are given to show how this procedure can be used to infer historical conclusions from these records. We conclude with a consideration of possible future prospects, the geophysical interpretation of results, and final summary.

2. Definitions

Ephemeris. An ephemeris is a table of, or any method of analytically computing, the positions of a given heavenly body as a function of time (which is the independent argument). Clearly, there are at least as many different ephemerides as there are astronomers to compute them. There are tables, computer programs, and even integrated ephemerides fit to observational data. Whether observational data is included in an ephemeris, or not, the term is herein employed to refer to the theoretical or computed positions, as distinct from the observed positions which constitute 'truth' against which the imperfect theoretical ephemeris is compared and, consequently, improved.

Ephemeris time (ET). This is the time-scale, or clock, defined by the adopted ephemeris. Of course, the ephemeris 'clock' does not exist physically, but given an observation (assumed perfect), and any ephemeris, we can from that ephemeris determine what time corresponds to the observed position by inverse interpolation. This time would be the Ephemeris Time (ET) of that observation, on the given ephemeris. Since time is the independent argument of any ephemeris, the time-scale defined thereby is independent of other time-scales, both civil and astronomical. It is important to remember two points. (i) Each different ephemeris, whether of the same celestial body or not, defines a distinct time-scale. Changing the ephemeris changes the time-scale. (ii) The Ephemeris Second defined by a given ephemeris is not the same length, and does not necessarily have the same epoch (reference date for time = zero), as that defined by any other ephemeris or civil time-scale. The International standard of ET is the Improved Lunar Ephemeis (1954).

Universal time (UT). This is the time-scale defined by the rotating Earth, and is determined by the position of the Greenwich meridian with respect to the Mean Sun. It is important to note that this time-scale is largely independent of any other physical motions in the solar system, or in atomic resonances (see the next definition). It is related to the Sun, and depends upon the solar ephemeris. It is also related to the Moon's orbital motion through the tidal couple, a matter

The Accelerations of the Earth and Moon

to be explored in this study. Since the Earth's rotation rate is not constant, as measured on the ET or AT time-scales, we compare Universal Time (UT) with these other time-scales in order to express the Earth's rotational variations.

Atomic time (AT). Timekeeping's adopted master-reference source has progressed from using the Earth's rotation, to pendulum clocks, to quartz crystal oscillators, and finally to atomic clocks. The atomic clock's rate is governed by transitions in the electron states of atoms, usually caesium or hydrogen. The world reference time-scale is denoted 'AT' and is coordinated around the world by radio transmissions, physical transport of clocks, and other means. We should bear in mind four points. (i) The reference time-scale, AT, is established by agreement, under the *assumption* that it is the best and most reproducible available source. (ii) All other time-scales, including ET and UT are, by this agreement, referenced against it in all comparisons. (iii) The AT second's length was established to be as nearly equal to the length of the Ephemeris Second (defined by the Improved Lunar Ephemeris 1954), as observations of the Moon prior to 1957 could provide. This was done for convenience, and there is no guarantee that this 'equality' between ET and AT will hold forever into the future. (iv) In this study, our reference time-scale must be Ephemeris Time (ET), since our observations were all made before the introduction of Atomic Time, and ET provides the best standard for comparison.

$\Delta T = ET - UT$. This equation expresses the difference, in the reading at the same instant, between the Ephemeris Time clock and the Universal Time clock. Since this function varies slowly with time (over the months, years and centuries), we must associate an 'epoch' or date with any given value of this function. For example, $\Delta T = 40$ sec of time, approximately, during A.D. 1970. In this paper, we shall consider the *Function* ΔT over the time interval of from about 1000 B.C. to the present. Determining this function, which measures the irregularities of the Earth's rotation, is one of the goals of this study. In this sense we are *assuming* that ET is a uniform time-scale (see above). To find a value for the function ΔT, we determine what the UT and ET clocks read at the same instant, that is, when a given event occurred such as an eclipse of the Sun. While an event must be precisely determined to obtain an accurate value of the clock difference ΔT (on ET and UT clocks running at about one sec/sec of time), only an approximate epoch need be associated with a given value of ΔT as this function itself is very slowly varying (sec of time/*year*). The times associated with specific values of ΔT are called 'epochs', and need be only approximately expressed (usually to the nearest year is adequate).

Solution by least-squares filtering. Most scientific analyses provide solutions from observational data using the so-called least-squares filter. This arises from the typical observational equation encountered in most work, which is of the form:

$$\text{Function of Parameter(s)} = \text{Measurement(s)} \pm \text{Error(s)} \qquad (1)$$

Each observational measurement is subject to errors, and estimates of the errors are provided for each observation. The function of observable parameters is provided by theoretical or numerical determination of how the parameters appear in what is being measured or observed (usually expressed by the partial derivatives relating the parameters to the measurements). The set of observational equations of condition (equation 1) usually has more equations than parameters to be determined. These sets of simultaneous, error-corrupted, equations are usually solved by finding the solutions for the parameters, which minimizes the sum of squares of the residuals after the fit, with each measurement given the reciprocal weight of the error estimates squared. This definition is provided here because most other authors have employed this system in reducing these data, but we do *not* use this particular statistical filter.

Observational linear inequalities. As noted in Part I, *observational* errors are negligible in the data as considered in this study. Our errors, which probably exist, take the form of *incorrect equations* of condition. Therefore, equation (1) is not the form of our observational equations of condition. As will be shown in section 4 below, our data provides equations of the form:

$$\text{Lower Bound} \leq \text{Function of Parameters} \leq \text{Upper Bound} \qquad (2)$$

where the Upper and Lower Bounds are provided by the observations. Most other authors have converted the equations of this kind into the form of equation (1), and used least-squares filtering as an approximation in finding solutions. While this is possible, we prefer to adopt the direct approach to solving sets of linear inequalities.

Solution defined for linear inequalities. Strictly speaking, mathematically, systems of linear inequalities may have solution sets, or they may be inconsistent and have none. We define 'solution' for the purposes of this study as the mathematical solution set if it exists. If it does not (as is probable), then we define 'minimum deletion filtering' as the statistical method to be employed. We delete the minimum number of equations which provides a mathematical solution, and adopt it. If several distinct 'minimal' deletion sets exist, then the solution is the union of all such solution sets. Data equation weighting is provided for by assigning class labels (A and B) to the individual equations, preferring to delete the lower class if necessary to provide the solution. This approach is analogous to that used in least-squares filtering, and similarly depends upon having enough 'good' data so that the 'bad' is recognized, and can be deleted. It should be added here that this method was first applied to eclipse observations by Fotheringham (1920).

The identification game. The so-called 'identification game' was defined by Newton (1970), as using *a-priori* computed eclipse paths to assist materially in the selection or rejection of data to be analysed. We wish to clarify this point. Newton, quite correctly in our view, avoided using the identification game as much as possible. We also categorically decline to use it in identifying either the date of an eclipse, or the magnitude, for all data after 1000 B.C. We bring up the point here, because we believe that it is quite proper to use the identification game, *a-posteriori*, to reject obviously erroneous equations. Our minimum deletion does precisely this, in quite an analogous way to the 'rejection sigmas' frequently employed in least squares filtering to reject obvious 'blunder points'. It should be noted that there is a clear distinction between *a-priori* utilization, and that done *a-posteriori*. The former introduces the identification game into the criteria for data selection. The latter assists in the deletion of erroneous equations and is at the heart of most statistical filtering techniques. If we only deleted equations through this device, the explanation would be complete at this point. In practice, however, we found that the *a-posteriori* computations and data comparison identified two observations with wrongly assigned places of observation. Thus warned, we were fortunately able to do further historical and contextual research which plausibly reassigned the place of observation. Observations thereby readmitted to the data set were, to some degree, affected by the identification game, and their weights were reduced to reflect this fact. While we have avoided using the identification game in *a-priori* data selection, we feel it does have a role in *a-posteriori* analysis which is well accepted in statistical filtering.

3. *What the analysis seeks to determine*

We seek two basic Astronomical/Geophysical parameters in this study. They are the secular acceleration of the Moon's longitude ($\ddot{n}$, measured in arcsec/cy^2), and the rotational acceleration of the Earth's rotation ($\ddot{e}$, measured in sec of time/cy^2). By 'secular' acceleration of the Moon's longitude, we mean any and all accelerations of the Moon in its orbital longitude arising from causes other than the theoretically computable gravitational effects. This secular acceleration

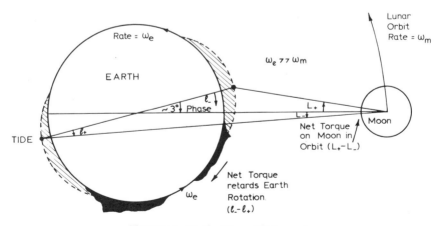

FIGURE 5. Earth–Moon tidal couple

arises, theoretically, from the tidal couple between the Moon and rotating Earth. Figure 5 indicates this effect schematically. The Earth rotates faster (on its axis) than the Moon does (in its orbit) around the Earth. The Earth therefore drags the gravitationally induced oceanic tidal bulges ahead of the Moon (on the nearer Earth hemisphere), and behind the Moon (on the farther Earth hemisphere). This 'phase lag' is small, around 3°. The nearer leading bulge to the Moon tends to accelerate the Moon in its orbit, while the further trailing bulge tends to retard the Moon. The former is larger, and the net effect is to accelerate the Moon in its orbit. If energy is thereby added to the Moon in its orbit, it will move farther from the Earth, slow down, and thereby fall behind its unperturbed position as time progresses. The net tidal torque on the Moon gives rise to a negative acceleration in lunar longitude.

From conservation of momentum, we can see that the Earth must be retarded in its rotation by this torque. The length of day gradually increases, and the value of ET–UT therefore must change as time passes, because the 'Earth clock' is running ever slower. The rotation of the Earth is deaccelerated by this torque, and the effect is termed the 'lunar tidal part' of the (negative) secular acceleration of the Earth's rotation. There are two tides raised by the Sun, an oceanic tide

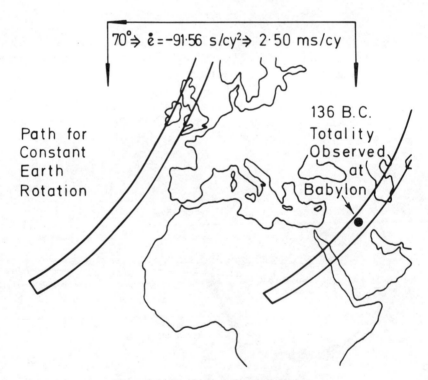

FIGURE 6. Earth's rotational displacement

similar in character to that of the Moon but a quarter the size, and an atmospheric tide. The former provides additional negative acceleration to that from the Moon, and the latter yields a small positive acceleration of the Earth's rotation (semi-diurnal tidal component). The sum of all three tidal accelerations can be termed the 'tidal part' of the Earth's rotational acceleration. It could have turned out that the 'tidal part' was the whole acceleration, but such is not the case. The results of this analysis, as well as of others, support the existence of a large positive non-tidal acceleration term in the Earth's rotation. The sum of all these contributions is the actual Earth acceleration ($\dot{e}$) which this study intends to measure observationally (see Figure 6).

The assumption of constant accelerations is implicit in the foregoing definitions. Theory strongly implies that the tidal components are, on the average over many decades at least, quite constant. This is not proven, however, and the non-tidal acceleration is as yet an open question. It will be seen that under the method of analysis employed here, any non-uniformity in these torques will be identifiable, and can be bounded in magnitude.

A determination of the function ΔT (ET–UT) for the period 1500 B.C. to the present will also, of necessity, be produced by the analysis.

4. *The method of analysis*

The observations of large solar eclipses analysed in this paper provide information on both the Earth's rotational acceleration and the Moon's orbital acceleration. The correlation between these observables, in large solar eclipse observations, is very high. This arises because a change in lunar longitude can almost entirely be cancelled by a corresponding change in the Earth's rotational position. Only small, second-order lunar orbital considerations permit independently distinguishing these two parameters. The sensitivity to be achieved in independently obtaining the parameters is not nearly so high as that which can be achieved in determining one given the other.

We adopt the Improved Lunar Ephemeris (1954) as our standard lunar ephemeris, thereby defining ET_{moon} and Ephemeris Moon. The Newcomb (1895) theory of the Sun defines ET_{sun} and Ephemeris Sun, and we have followed modern practice by making the computations in Ephemeris Time. The Ephemeris Time-Scale defined by this procedure is $ET_{moon} - ET_{sun}$ since our observations all involve the difference between Lunar and Solar longitudes. If we were using occultations of stars by the Moon, the ET scale would be ET_{moon} alone. We make an observational test on the possible systematic effects of the solar ephemeris on ET as defined in this study (through ET_{sun}), but we must admit of possible systematic differences between the results of reducing eclipse data (of all kinds) and occultations or conjunctions because of the presence of the solar ephemeris in the former case, and its absence in the latter. This matter is discussed by Van der Waerden (1961) in some detail, and the method of analysis used in this paper is a simplified and at least roughly equivalent method to his, which in turn relates to techniques employed by Brouwer (1952), De Sitter (1927) and Spencer Jones (1939).

It is critically important to note that changing $\dot{n}$ changes the Lunar Ephemeris and ET_{moon}. We should, therefore, carefully define what we are doing so that others may repeat, test or utilize our results. The term $\dot{n}$ enters the lunar ephemeris in the expression for the mean lunar longitude, to which the other periodic gravitational perturbations are added to obtain the true instantaneous longitude:

$$\text{Mean Longitude} = K + LT + (M + \tfrac{1}{2}\dot{n})T^2 + NT^3 \tag{3}$$

where T is in centuries from 1900. The coefficients K, L, M and N are observed or computed from theoretical gravitational considerations in the motions and shapes of the solid planetary bodies and the Sun. We may presume that they are accurately known. The Improved Lunar Ephemeris (1954) utilizes a value of $\dot{n} = -22\cdot44''/\text{cy}^2$, and this lunar ephemeris is the current astronomical standard. We wish to find an improved value for $\dot{n}$ (as have others), and we are, therefore, proposing to alter the lunar ephemeris. A change in $\dot{n}$ alters the ET_{moon} by an amount easily calculated from a knowledge of the Moon's mean motion in longitude. Such a change to ET implies an equal change in ΔT, and consequently in $\dot{e}$:

$$\delta\dot{e} = 1\cdot97\delta\dot{n} \tag{4}$$

This is nothing more than the mean partial derivative equation between the parameters $\dot{e}$ and $\dot{n}$. In solving for $\dot{n}$, then, we change $\dot{e}$ according to equation (4). For a given $\dot{n}$ solution, we are, for these reasons, interested only in the corresponding departure of $\dot{e}$ from equation (4). It is helpful to note that more negative lunar accelerations (the trend of modern solutions) imply more negative Earth accelerations.

We now turn to a consideration of the parameter $\dot{e}$. The Earth's rotational acceleration, if constant (as will be assumed for the moment), provides a parabolic shape to the function ΔT, as given here:

$$\Delta T_{\text{sec of time}} = A + BT + CT^2 \tag{5}$$

where T is again measured in centuries from 1 January 1900. Once equation (5) is known, $\dot{e}$ follows:

$$\dot{e} = -2C \tag{6}$$

Are we now faced with the determination of three parameters, A, B and C, instead of the one, $\dot{e}$, under conditions where even $\dot{e}$ was going to be difficult in separation from $\dot{n}$? Newton (1970, 1972) and Curott (1966), for example, assume:

$$\Delta T = -\tfrac{1}{2}\dot{e}T^2 \tag{7}$$

though they prefer to use the Geophysical parameter $\dot{\omega}/\omega_e$ which differs from $\dot{e}$ by a constant multiple (see later sections for equivalences to different representa-

The Accelerations of the Earth and Moon

tions commonly found in the literature). This adoption by them is equivalent to assuming that on the adopted lunar ephemeris, the value of ΔT is zero, as well as its derivative, on 1 January 1900. There is no *a-priori* evidence for this assumption, and it turns out that the true epoch is near 1770. Neglecting this consideration provides errors in $\dot{e}$ of 10% for epochs near A.D. 0, and much greater ones for medieval epochs. We shall, therefore, use equations (5) and (6), which is close to the scheme employed most recently by Van der Waerden (1961) and Brouwer (1952).

We must squarely face the problem of the three parameters in equation (5) required to obtain the correct $\dot{e}$ via equation (6). The cited papers have met the equivalent difficulty by using the modern data, such as that reduced by Spencer

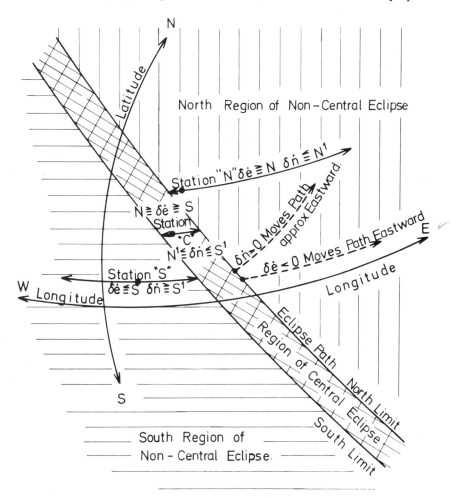

FIGURE 7. Observational equations of condition

Jones (1939) and Brouwer (1952), as well as the ancient data to determine the three coefficients of equation (5). We shall generally follow this approach. There are a number of details inherent in this method which we have glossed over, in the interest of reasonable simplicity and brevity. In this connection, we recommend the detailed theoretical analysis provided by Van der Waerden (1961).

Figure 7 is a schematic representation of a total solar eclipse path, showing three observational stations (north of, within and south of the path respectively), and the corresponding equations of condition implied by observations from each point. Negative changes in $\dot{e}$ move the path directly eastward, altering the longitudes (and not the latitudes) of given points on the eclipse path. This must be so since this parameter changes *only* the value of ΔT, and therefore gives rise to a correction in the rotational position of the Earth. A more negative value of $\dot{e}$ implies a larger positive value of ΔT (equations 5, 6); increasing values of ΔT move the path eastward.

Taking equation (4), and integrating, we have:

$$E = \dot{e} - 1\cdot 97 \dot{n} \qquad (8)$$

which is the mean linear combination of the parameters, or, the 'line of solution' some point on which is the exact determination of the true $\dot{e}$ and $\dot{n}$ considered as independent parameters. The astronomical data best determine E, the linear combination of the parameters, but independent solutions can also be obtained with lesser precision. From equation (8) it is clear that, on Figure 7, an opposite change in $\dot{n}$ moves the path eastward, so the path moves eastward for positive changes in $\dot{n}$ (considered alone). If the path always moved precisely eastward, and by the amount implied in equation (4), for a given change in $\dot{n}$, then the parameters could not be separated at all, and we could solve for E in equation (8) only. Because the lunar orbit is an ellipse inclined to the Earth's equator, equations (4) and (8) apply only to the average of all eclipses, and individual cases vary. The maximum change in the mean ratio of the partial derivatives (1·97 from equation 4) is approximately $\pm 10\%$ for eclipses observed in temperate latitudes. This is, of course, why the sensitivity in separating the parameters is not as high as we would like. We see, then, that changing $\dot{n}$ for different eclipses will move the path along distinct curves, which are roughly parallel to the equator, but which depart from the strict east–west change implied by variations in $\dot{e}$. This fact is illustrated in Figure 7 by showing the path moved along a slightly different curve for a change in $\dot{n}$, whereas changes in $\dot{e}$ move the path directly along a line of equal latitude.

The types of observations fall into two classes, those which imply that the station was within the path (totality, annularity), and those which imply that it was outside (denial of totality). Turning to Figure 7, and considering that it defines the circumstances of a specific eclipse, we can compute four numbers, N, S, N_1 and S_1. These are the changes (from the nominal) in $\dot{e}$ and $\dot{n}$, considered independently, which will bring the path's North and South limits, respectively,

The Accelerations of the Earth and Moon 503

coincident with the observation Station. This set of computations is made by varying the parameters $\dot{e}$ and $\dot{n}$ from the nominal, and noting the changes in the path by computation, thereby obtaining the partial derivatives for the specific eclipse in question. This technique is very common in analysis of this kind, and it is sometimes called the method of finite differences. Having these four numbers, it is possible to determine the equations of condition implied by the observation.

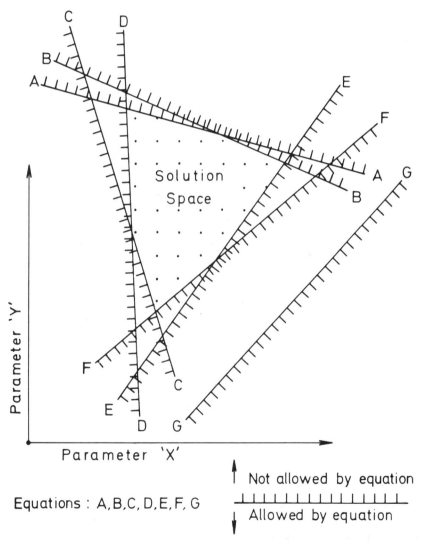

FIGURE 8. Schematic of solution space

If the observation is definitely of centrality, then the station is within the path, at location 'C' in Figure 7. The implied equations of condition are:

$$N \geq \delta\dot{e} \geq S \quad and \quad S_1 \geq \delta\dot{n} \geq N_1 \tag{9a}$$

When centrality is denied, the station is either North or South of the path and we have:

$$S \geq \delta\dot{e} \geq N \quad and \quad N_1 \geq \delta\dot{n} \geq S_1 \tag{9b}$$

where $\delta\dot{e}$ and $\delta\dot{n}$ in the equations constitute the allowable ranges for $\dot{e}$ and $\dot{n}$, implied by the observation.

Each observation considered in this study yields a pair of linear inequalities, of the form (9a) or (9b). The solution to all such equations from the observations considered together, will give us the allowable range (solution space) for the parameters being sought (i.e. $\dot{e}$ and $\dot{n}$). In section 2 we took note of the least-squares filter, so ubiquitous in modern science today. A different technique is more applicable to linear inequalities, and we defined our 'minimum deletion filter' to satisfy our needs in this regard.

Figure 8 is a schematic representation of the method of solution we will use. The parameters are 'X' and 'Y' and the observations have yielded linear inequalities in various linear combinations of X and Y. Equations (2) or (9) show the form these take in our analysis. There may be either one or two inequalities from a given observation, but it will usually turn out that only one is 'critically' near the solution space. Figure 8 therefore represents seven equations 'A' through 'G' as straight lines, with the inequality forbidding values on one side of each line. The solution space is the 'area' shown bounded in Figure 8. Six of the seven schematic equations bound the actual space, while 'G' is consistent with the remainder, but plays no role in the solution. This brings out the first point. Since the solution space may well be very small compared with the separation between some of the bounds, we would expect to find that some of the perfectly good and consistent data has no 'sensitivity' in determining (bounding) the solution. Approximately 40% of the data admitted to this study had this characteristic, as will be detailed later. There was also a large body of data (observations denying totality, particularly in China) which was not included in the first place for this reason.

It may also be noted that if G is an erroneous equation, it has had no effect whatsoever on the solution. Consider the inverse of equation G, where the forbidden area is now above the line instead of below. In this case, there would be no mathematical solution, and equation G would be removed from the solution by the minimum deletion filter, providing there is some redundancy in the other equations (as illustrated). The high stability of this statistical filtering algorithm against errors is indicated by these examples. An erroneous equation failing to intersect the solution space had no effect whatsoever on the solution! This perhaps surprising result depends upon having a sizeable amount of correct

The Accelerations of the Earth and Moon

data equations, but this requirement is demanded equally in least squares and other statistical procedures.

What if an erroneous equation intersects, and therefore bounds, the solution space? Then, of course, unless we can identify it, there will be an effect on the solution. If there is reasonable redundancy in the correct equations, then we can either identify a risky equation, or show that it has little effect on the solution. Figure 8 illustrates the redundancy principle by providing pairs of limits, AB, CD and EF as bounds on the three sides of the solution space. It can be seen that deleting any one equation from each pair has little effect on the solution space. Redundancy provides two benefits. First, as we have already noted, redundancy is necessary for the minimum deletion filter to reject erroneous equations. Otherwise, we might reject a correct equation and retain an erroneous one. Second, redundancy allows us to provide more reliable and conservative limits to the solution space in the presence of erroneous equations which intersect it.

Consider replacing equation E with E inverse. It can be seen that the solution space is reduced in area by an order of magnitude, and now consists of the triangle bounded by equations E inverse, F and B. There is no redundancy for the limit set by E inverse, and removal of this one equation (whether it is correct or incorrect) substantially alters the solution. We therefore adopt the data consistency requirement, to delete any single equation whose absence significantly increases the solution space area. This is analogous to the technique often employed in data consistency checks to the least-squares filter. One removes subsets of the data, and sees if the solution is thereby significantly altered. This procedure, assuming redundancy, provides that no *single* equation, correct or erroneous, will be relied upon as a bound to the solution. The result is that an erroneous equation which bounds the solution space will be removed if it causes a significant change in the solution. If such an equation is correct, we have merely provided a larger solution space, and thereby increased the estimate of uncertainty in the solution. If the error is near other redundant correct limits, then it is not identified, but the effect of its inclusion is minimal compared with the size of the solution space. We have now exhausted all the possible results from an erroneous equation, and find that it cannot have a large effect on the solution. The statistical filter based upon the minimum deletion principle and the related data consistency check provides solutions which will be stable against the presence of errors. In our view, this stability is extremely good, and gives excellent assurance that there can be no significant error bias in the results obtained.

The effect of providing more equations is to reduce the size of the solution space, while removing them increases it. The result of having more data, therefore, is to improve redundancy and decrease the solution space size. As a practical matter, in the kind of data being reduced in this study, the solution space is small compared with the probable spacing of erroneous equations, since they usually arise from assigning the wrong observation location. It is therefore unlikely that an erroneous equation intersects the solution space at all, and even less likely that it will intersect it 'close' to other limits, and therefore be unidentifiable. Practically speaking, it is at least possible, and we consider it reasonably likely, that the errors will have no effect on the solution whatsoever! We hope it has at least been demonstrated that a few erroneous equations cannot seriously bias the results.

In conclusion, we are in a position to define the method by which the solution and its uncertainty bounds will be obtained. First, the simultaneous linear inequalities will be solved by the principle of minimum deletion. The data consistency analysis follows, wherein any single equation which severely restricts the solution space defined without it is removed. The solution will be quoted as the 'centroid' of the solution space, and the uncertainty taken to be a sizeable fraction of the whole semi-diameter of the space (to approximate a standard deviation). We believe that this method is exceptionally stable against the threat of significant bias introduction by errors.

5. *The iterative method of numerical solution*

The data will be numerically reduced by an iterative procedure, with two substeps in each iteration. We have a value of $\dot{n}$ *a-priori* or from the previous iteration, and a corresponding expression for ΔT of the form equation (5). We now intend to vary $\dot{n}$ and $\dot{e}$ in amounts related by equation (4) so that, by finite differences, the equations of condition (9) can be determined for each eclipse and observation site. We must have an expression for ΔT corresponding to each of the two $\dot{n}$ values which will be evaluated to yield the equations of condition. This can be accomplished by modifying the C coefficient of equation (6) to take account of the mean relationship between changes in $\dot{e}$ and $\dot{n}$ indicated by equation (4), with the result:

$$\Delta T_{\text{nominal}} = A_i + B_i T + (C_i - 0\cdot 985\, \delta\dot{n})T^2 \tag{11}$$

where A_i, B_i and C_i are the coefficients of the polynomial corresponding to the value of $\dot{n}$ to be used in this current iteration. For $\delta\dot{n} = 0$ we obtain the nominal elements for the eclipse in $\dot{n}$ and $\dot{e}$, and by choosing a non-zero value for $\delta\dot{n}$, we obtain a value for the altered parameters. From these eclipse computations, the corrections to $\dot{n}$ and $\dot{e}$ required to bring the station onto the North and South limits can be computed, thereby providing the equations of condition (9a) or (9b). The solution space is represented in an $\dot{e}$ versus $\dot{n}$ plot which can be constructed from these equations in a manner similar to Figure 8. From this solution space, we take the new value for $\dot{n}$, considered as an independent parameter, and ignore the solution for $\dot{e}$, as we have only *approximated* the *change* to $\dot{e}$ by using equation (4) in modification of the C coefficient of equations (6) or (11). This is equivalent to using equation (7). If we were prepared to use it, the solution would already be in hand. As can be seen below, this would result in poor solutions

The Accelerations of the Earth and Moon

for $\dot{n}$ as well as $\dot{e}$. In order to use the adopted method, requiring equations (5) and (6), we must add a second step to the iterative procedure, and thereafter continue iterating until a converged solution is obtained. No equations of condition for $\dot{n}$ are deleted until convergence is obtained, since they are dependent upon the epoch of observation.

The second substep in each iteration involves determining the three new coefficients for equation (5), evaluated for the fixed $\dot{n}$ obtained in the first substep. This is easily done through the expedient of making computations for each eclipse using the new $\dot{n}$, and plotting the resulting path limits in a manner similar to Figure 7. The effect of changing ΔT is to move the path directly east–west, and consequently the upper and lower limits to ΔT implied by the observation can be read directly from the plot (in degrees of longitude). Noting that 1° of longitude equals 240 sec in Universal Time, we can provide linear inequalities of the form:

$$\text{Lower Limit} \leq \delta\Delta T \leq \text{Upper Limit} \quad (12)$$

where $\delta\Delta T$ is the departure from the nominal ΔT of equation (11) at the current $\dot{n}$. These bounds on the corrections to ΔT secured from each observation can then be plotted against the epoch (calendar date in years) of the observation. To satisfy equation (5), we fit the best parabola to these limits. The uncertainty in this solution is measured by how much the parabola can be 'bent' without violating the 'limits'. The solution space is the set of all parabolas which can satisfy the observations of the single parameter $\delta\Delta T$ versus time. The resulting parabola is the fit to the *change* required in ΔT and, to complete the iteration, it must be added to the nominal parabola. We then have a new estimate for $\dot{n}$, the corresponding expression for ΔT in the form of equation (5), and the estimate for $\dot{e}$ follows via equation (6).

To proceed with the next iteration, we adopt as *a-priori*, the values of $\dot{n}$ and ΔT from the previous iteration. Equation (13) now uses the new coefficients, and the method of finite differences can be undertaken to determine a new solution space for $\dot{n}$ versus $\dot{e}$, from which only the $\dot{n}$ will be determined as before. This solution space in $\dot{n}$ versus $\dot{e}$ will differ from that obtained previously, because the epoch of zero point and slope for ΔT will have changed, and likewise the relative positions of limits for observations differing in epoch. Observations of nearly equal epochs still provide the same relative limits in the $\dot{e}$ versus $\dot{n}$ solution space. This is an indication that using equation (7), as many other studies have done, provides usable *relative* solutions for data near the same epoch. However, its employment as an approximation makes it quite impossible to deduce any conclusions regarding the *change* in $\dot{e}$ or $\dot{n}$ versus time. More will be said about this later.

As the iterations succeed each other, a point will be reached (assuming convergence) where no further significant changes occur in the parameters. At this point, the procedure ends, and the solutions can be quoted.

6. The data

The observations to be used in the actual numerical processing are summarized in Figure 9. The data which are consistent, but without sensitivity in the solution, are not included. A listing of these observations may be found in Part II, where they are all considered in detail.

In Figure 9, column one is the astronomical* date of the observation. Column two gives the nature of the observation. Columns three and five respectively are the lower and upper bounds to $\delta \Delta T$ provided by each observation. This is measured with respect to the ΔT polynomial and $\dot{n}$ quoted in the final solution (equations 14 and 15). If either of these columns is blank (—), the corresponding limit is so remote from the solution that it is satisfied by any solution within several times the quoted uncertainties, and is therefore irrelevant. Column four is the observation class, A or B, determined in Part II. Column six is the partial derivative of $\delta \Delta T$ with respect to $\dot{n}$ for the particular observation, as measured from the solution ($\delta \dot{n} = \dot{n}_{\text{new}} + 37 \cdot 5$). This permits the reader to make his own computation of limits for any value of $\dot{n}$ he may choose. It is only necessary to multiply column six by $\delta \dot{n}$ and add the result to columns three and five. Column seven is a conservatively estimated probability that the observations is correct as listed (i.e. is not erroneous). Columns eight and nine are the latitude and longitude respectively of the place of observation, longitude measured 0–360° E of Greenwich. Column ten is the common name of the place of observation, and column eleven contains comments.

The two blocks of data in Figure 9 include the observations before A.D. 1500, and after this date, including an artificial check point at 1961 to guarantee computer program consistency. The modern data are affected sufficiently by the 'random' short-term components in the Earth's rotation that they do not exactly fit the mean parabola (nor do we expect them to).

As noted earlier, we must add the modern data into the fit, and we use the results of Brouwer (1952) and further observations up to the present day in Figure 10. Column one is the approximate epoch of the observation(s). The first three points are averages of data found in Brouwer. Column two contains the raw ΔT data, which is referred to the Improved Lunar Ephemeris (1954) and an $\dot{n} = -22 \cdot 44''/\text{cy}^2$. We must first correct these data to an $\dot{n} = -37 \cdot 5$ (our final solution), and this correction is shown in column three, as computed from equation (4). Column four contains the ΔT_o or 'observed' values of Brouwer as corrected, column five gives the computed ΔT from equation (15), and the last column lists the residual $\delta \Delta T_{o-c}$ (observed minus calculated) which is plotted in Figure 11. During previous iterations, it is clear that the corrections were made to the $\dot{n}$ of that iteration, so that the data could be included in the polynomial fit to ΔT.

---

* Astronomical calendar dates are the same as Historical dates for the A.D. period, but differ by one year from B.C. dates: 1 B.C. = 0 astron.; 234 B.C. = −233 astron., etc.

Date Year	Mo	Day	Observed	δΔT_min	C	δΔT_max	2δΔT/δτ	~P(T)	Lat	Long	Place	Comments
-1374	05	03	Total	-550	B	-	-112	.80	35.62	35.78	Ugarit	-1222 Eliminated
-1329	06	14	Large	-	B	-125	-25	.90	36.07	114.33	An-yang	"Great Eclipse"
-708	07	17	Total	-265	A	-	-15	.95	35.53	117.02	Chu-Fu	"Chi"
-600	09	20	Total	-2	B	-	-20	.80	30.34	112.25	Ying	Abandoned Chu-Fu
-321	09	26	Timed	-200	A	+40	-8	1.00	32.55	44.42	Babylon	Began Sunset -12m
-197	08	05	Annular	-	B	-400	-46	.95	34.34	108.90	Ch'ang-an	"Chi"
-180	03	04	Total	-	A	-24	-18	.95	34.34	108.90	Ch'ang-an	"Chi"
-135	04	15	Total	-	A	-145	-11	1.00	32.55	44.42	Babylon	Very Detailed
-65	12	16	Total	-	B	+115	-12	.80	32.42	119.45	Kuang-ling	Abandoned Lo-yang
-120	01	18	Partial	-17	A	-	+17	.98	34.70	112.47	Lo-yang	"Almost Complete"
-516	04	18	Annular	-	B	-215	-24	.50	32.03	118.78	Nanking	"Chi" Annals
-522	06	10	Total	-	B	-24	-9	.50	32.03	118.78	Nanking	"Chi" Annals
-693	10	05	Total	-	B	-182	+6	.70	33.33	44.44	Baghdad	Place Presumed
-912	06	17	Total	-	B	+275	-5	.80	37.88	355.23	Cordoba	Place Presumed
-968	12	22	Total	-440	A	-	-11	1.00	41.02	28.98	Constantinople	Very Detailed
-1079	07	01	Total	-	B	-4	-	.90	37.40	354.02	Seville	Class "A" Candidate
-1124	08	11	Total	-224	A	-	-	1.00	58.50	31.33	Novgorod	Place Certain
-1133	08	02	Partial	-	A	-215	-1.2	1.00	50.062	14.417	Vysehrad	Very Detailed
-1133	08	02	Total	-	A	-215	-1.2	1.00	47.80	13.06	Salzburg	Very Detailed
-1153	01	26	Partial	-58	B	-	-2.5	.90	50.98	11.025	Erfurt	± 1% Magnitude
-1178	09	13	Partial	-	A	-130	+2.5	.98	45.378	1.516	Vigeois	Detailed
-1221	05	23	Total	-	A	+180	-2.5	.98	48.185	115.90	Kerulen R.	Very Detailed
-1239	06	03	Total	-20	A	-245	-	.95	39.87	355.97	Toledo	6 ΔT_max from Cerrato
-1241	10	06	Total	-205	A	-260	-1.5	1.00	53.62	9.48	Stade	Very Detailed
-1267	05	25	Total	-	B	-88	-	.80	41.02	28.98	Constantinople	Source Compiled in 1365
-1560	08	21	Total	-	A	+115	-	1.00	40.21	357.57	Coimbra	Clavius Observer
-1567	04	09	Total	+60	A	-110	-	1.00	41.90	12.48	Rome	Clavius Observer
-1715	05	03	Total	+35	A	-51	+0.3	1.00	-	-	England	Halley Organizer
-1961	02	15	Test	-117	-	-115	-	-	-	-	Europe	Check Point

FIGURE 9. The admissible data

Epoch	ΔT_B	$\delta \Delta T_{\dot{n}}$	ΔT_O	ΔT_C	$\delta \Delta T_{O-C}$
1650	+ 5.	+108.	+113.	+ 51.	+ 62.
1700	-12.	+ 69.	+ 57.	+ 8.	+ 49.
1715	-12.	+ 56.	+ 44.	- 1.	+ 45.
1727	- 8.	+ 48.	+ 40.	- 4.	+ 44.
1738	- 3.	+ 43.	+ 40.	- 7.	+ 47.
1747	0.	+ 39.	+ 39.	- 10.	+ 49.
1761	+ 2.	+ 33.	+ 35.	- 11.	+ 46.
1774	+ 7.	+ 27.	+ 34.	- 13.	+ 47.
1785	+ 8.	+ 22.	+ 30.	- 12.	+ 42.
1793	+ 7.	+ 19.	+ 26.	- 10.	+ 36.
1802	+ 6.	+ 16.	+ 22.	- 8.	+ 30.
1812	+ 5.	+ 13.	+ 18.	- 5.	+ 23.
1820	+ 4.	+ 10.	+ 14.	- 2.	+ 16.
1830	+ 2.	+ 8.	+ 10.	+ 4.	+ 6.
1840	0.	+ 6.	+ 6.	+ 10.	- 4.
1850	+ 2.	+ 4.	+ 6.	+ 17.	- 11.
1860	+ 3.	+ 3.	+ 6.	+ 25.	- 19.
1870	- 2.	+ 2.	0.	+ 35.	- 35.
1880	- 8.	0.	- 8.	+ 44.	- 52.
1890	- 8.	0.	- 8.	+ 55.	- 63.
1900	- 3.	0.	- 3.	+ 66.	- 69.
1910	+10.	0.	+ 10.	+ 72.	- 62.
1920	+20.	0.	+ 20.	+ 92.	- 72.
1930	+23.	+ 2.	+ 25.	+107.	- 82.
1940	+24.	+ 3.	+ 27.	+121.	- 94.
1950	+30.	+ 4.	+ 34.	+137.	-103.
1960	+35.	+ 6.	+ 41.	+155.	-114.
1970	+40.	+ 8.	+ 48.	+173.	-125.

FIGURE 10. Modern ΔT data

7. The iterative fit to the data

We begin the process of iteration towards the solution by adopting the Improved Lunar Ephemeris (1954) and its 'Spencer Jones' lunar acceleration of $-22\cdot44''/\text{cy}^2$. For ΔT we take as our initial *a-priori* the polynomial to be found in the Explanatory Supplement to the Ephemeris (1961) page 87, specifically:

$$\Delta T = 24\cdot349 + 72\cdot318T + 29\cdot950T^2 \tag{13}$$

where T is in centuries from 1900 (the term in parameter 'B' is deleted). Iteration one, step one, consisted of determining the limits on $\dot{n}$ and $\dot{e}$ implied by using $\delta\dot{n}$ values of zero, and $-20''/\text{cy}^2$, and plotting these on a sheet of graph paper. Equations (13) and (11) provide ΔT. Equation (6) provides the necessary estimates of $\dot{e}$ as required. The procedure is to use the ΔT polynomial and each of the two $\dot{n}$ values individually, as the 'nominal' values for each observation. Solution limits are then obtainable in $\dot{e}$ for each $\dot{n}$, and the two sets of values define two points on each equation of condition. The solution space is exhibited by simply plotting these linear inequalities in $\dot{e}$ and $\dot{n}$ on graph paper similarly to Figure 8. The result of this step was to bound the limits on $\dot{n}$ between $-28\cdot80$ and $-37\cdot50$ with mean value $-33\cdot15$, for an $\dot{e}$ of $-79\cdot1$. We now have a new $\dot{n}$ to iterate with, but as the solution space changes with the ΔT *polynomial*, and not just $\dot{e}$ alone as we have seen above, we move on to the second step of iteration one.

Step two in iteration one consisted of taking the new $\dot{n}$, and the same ΔT polynomial, and computing the $\delta\Delta T$ limits implied by each observation. So far, in the ΔT polynomial, we have corrected only the coefficient of T^2, C in equation (11). We shall now solve for A, B and C, using all data. This is accomplished by plotting the $\delta\Delta T$ limits versus epoch of the observations (-1374 to 1972) and fitting the best parabola. The fitting was done by hand, using transparent overlays of parabolas of various C values. The polynomial solution for ΔT corresponding to $\dot{n} = -33\cdot15''/\text{cy}^2$ was $54\cdot13 + 115\cdot89T + 41\cdot55T^2$.

The specific numbers from iteration one are not important or interesting in themselves. However, there are important remarks concerning the results of iteration one. First, we found that the medieval data were inconsistent with the remainder. The new $\dot{n}$ was therefore obtained from the ancient class A data. Step two of the first iteration resulted in a poor fit to -321, $+65$, -1374 and $+522$, but all class A data were consistent. The effect was for the medieval data to 'pull the solution down' to the point where our credulity was strained.

In step one of the second iteration, we found that the new ΔT polynomial resulted in larger-than-expected changes to the $\dot{n}$ versus $\dot{e}$ solution space. The second iteration did not quite converge, but the third had only very small additional corrections (about 1%), and we will now give the final iteration results. For $\dot{n}$:

$$-30\cdot0 \geq \dot{n} \geq -45\cdot0; \quad \dot{n} = -37\cdot5 \pm 5 \text{ arcsec/cy}^2 \tag{14}$$

where both the maximum range and mean solution with standard deviation are given. The uncertainty quoted is fairly conservative in view of the fact that it is two-thirds of the whole range. For ΔT corresponding to the given $\dot n$:

$$\Delta T = 66\cdot 0 + 120\cdot 38T + 45\cdot 78T^2 \pm 100 \text{ sec of time};$$
$$\text{or } \pm 0\cdot 5 \text{ sec of time } T^2 \tag{15}$$

where the first uncertainty is the standard deviation of the departure of the polynomial from the instantaneous values, and the second is the uncertainty on the T^2 coefficient applicable to the long-term average behaviour. The epoch of zero derivative is 1770, and T is in centuries from A.D. 1900. The epoch of the *a-priori* equation (13) is 1780, and it has not changed significantly. Equation (13) is from Spencer Jones (1939), in effect, and this indicates once again that the astronomical method of data reduction via equations (5) and (6) is correct, and apparently largely invariant in epoch over changes in $\dot n$. Equation (11) can be used to map the results of (15) to other values of $\dot n$.

The solution for $\dot e$, considered as an independent parameter, may be obtained from (15) and (6):

$$\dot e = -91\cdot 56 + 10 \text{ sec of time/cy}^2 \tag{16}$$

As an independent parameter, the uncertainty is obtained from the estimate in equation (14) applying equation (4), that is, it must be approximately doubled. One form of the solution for linear combination of $\dot e$ and $\dot n$, with the correspondingly reduced error estimate, is contained in equations (15) and (11). This may also be expressed in the form:

$$\dot e - 1\cdot 97\dot n = -17\cdot 7 \pm 1\cdot 0 \text{ sec of time/cy}^2 \tag{17}$$

where $\dot e$ is in sec of time, $\dot n$ in arcsec, per cy^2.

Another way to present the data is to provide a plot of the allowable range of $\delta\Delta T$. This is the departure from the solution for ΔT in equation (15), implied by each observation (Figure 11). In effect, it is a display of the observed minus calculated ΔT values for each observation. The final solution parabola, equation (15), is simply the abscissa on the plot. In previous iterations, a similar plot was generated, with the *initial* polynomial as the abscissa. The new polynomial correction was deduced by fitting a parabola through these residuals. Figure 11 thereby illustrates the limits on the final solution, as well as the method by which the corrections to ΔT were obtained in each iteration.

In Figure 11, data which provide exact, or hard limits, are marked differently from the small subset of the data which provided estimated uncertainties. The mean of all medieval data from this study and that of Newton (1970) is also plotted, to indicate that the mean solution (abscissa) passes above the instantaneous values of $\delta\Delta T$ during the medieval period. It can be seen from the Figure that the modern $\delta\Delta T$ data residuals taken from Figure 10 are quite consistent with the modern eclipse observations from 1715, 1567 and 1560. This indicates that there cannot be a very large systematic difference between the occultation

The Accelerations of the Earth and Moon

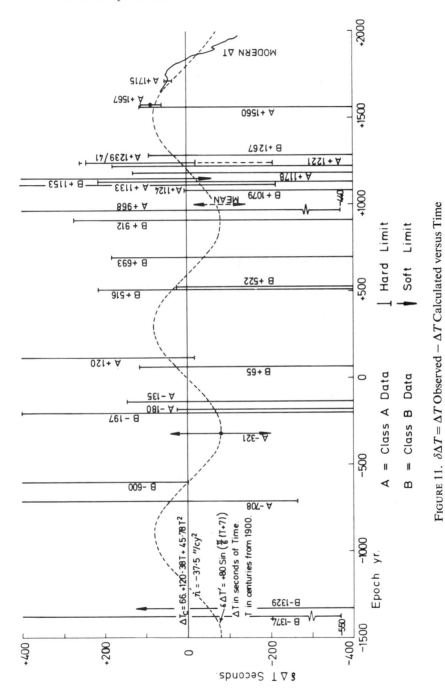

FIGURE 11. $\delta \Delta T = \Delta T$ Observed $- \Delta T$ Calculated versus Time

data which provided the modern curve and the eclipse data. In effect, this shows that our reliance upon $ET_{moon} - ET_{sun}$ instead of ET_{moon} alone does not introduce any significant problems. The medieval discrepancy noted by Newton (1970, 1972) does not appear in our results, and it should be emphasized that there are numerous redundant limits, both above and below the curve, which rule out the very large variation noted by Newton (1972); see Figure 12. These points will be discussed in more detail below.

Figure 11 also indicates that between A.D. 900 and the present the departure of the instantaneous Earth rotation from the mean has been systematic, and long-period rather than wildly fluctuating. The mean fits the instantaneous data within 200 sec of time (worst case) over the entire historical period, and we have a good picture of the variations after A.D. 900. It will be emphasized again later, that these results appear to rule out significant departures of either Earth rotational acceleration or lunar acceleration from the mean during the historical period. We will attempt to reconcile the results of other recent authors who considered the ancient data with the opposite conclusion in a later section.

We would again emphasize here the standard astronomical elements to be employed in the reproduction or use of these results. The Improved Lunar Ephemeris (1954) provides the standard ephemeris, changing only the $\dot{n}$ by equations (14) and (3). The solar ephemeris should be that of Newcomb, operated on Ephemeris Time. The only remaining expression required is that of equation (15) for ΔT, which provides $ET - UT$. Mapping to other lunar accelerations may be accomplished as noted above.

8. *Errors and the estimates of uncertainties*

In the final analysis, only the consistency, or lack of it, between the many observations provides hard evidence to support confidence in the solutions. The consistency of the data utilized in this study, as demonstrated by the quality of fit, could not be better. Each and every pre-A.D. 1500 observation is satisfied by the solution, and none had to be deleted either for being inconsistent, or for providing an unsupported (non-redundant) critical limit. In other words, the whole data set provided a solution without deleting any data! The post-A.D. 1500 data depart from the mean solution, but we can see that this results from their higher precision and the dominance of short-term Earth rotational variations at recent epochs.

It might be suggested that the apparent consistency is too good, and that we should have expected to find more errors. One answer to this is that we did find two errors in the assignment of observation location, 601 B.C. and A.D. 65. We were fortunate to amend these, thus warned, through additional historical analysis. Since it is roughly equally probable that erroneous equations will be either consistent with, or inconsistent with, the solution, we can guess that *a-priori*, we probably have an equal number of 'consistent' errors. These, of course, cannot be identified from Figure 11 or the list of consistent, but low-sensitivity, observations included in Part II but not numerically analysed in this section. Only further historical research or independent means could be used in this case. However, as we have seen above, neither consistent, nor inconsistent, erroneous equations can have any effect on the solution if they fail to intersect the solution space. We have, in a sense, identified four probable errors; two corrected after discovery, and two which cannot be discovered numerically. This is in the range of the number of errors we would expect to find in this size data set, given the estimated probabilities on each observation noted in Figure 9. Further errors, which intersect the solution set and provide bounds, are unlikely for the reasons considered earlier. The quality of the final solutions, however, does not depend even upon this assumption, because the redundancy in the limits is quite adequate.

Beyond the question of redundancy, which was adequate, we undertook other data consistency checks. We considered the question whether there was a single historically preserved eclipse observation which was inconsistent with the solution obtained here. The historically preserved data falls into three categories. First, a large body of data not admitted to the study because the magnitude was too small, or the identification was questionable, or for other basic reasons. Second, a body of data admitted or admissible to the study, under our criteria, but which failed to have any apparent sensitivity in the solution. It was therefore not included in the numerical analysis. Third, the data actually used in the solution, Figure 9. We have already seen that the third set is *perfectly* consistent with the solution. We undertook computation to check the second set, as the new solution may have 'moved' enough to provide an inconsistency. Neither the Chinese 'non-total' observations which were not listed or analysed in Part II, nor those observations which were analysed historically, contain a single inconsistency.

Analysis of the first set is more subtle, but these observations fall into two groups. First, those which cannot be used under any reasonable criteria. Second, those which fail to pass our stringent requirements, but which might pass somebody else's. The former group is of interest only historically, though we can use the solution obtained here to provide some interesting, if not terribly significant, deductions along these lines (see section 9). The latter group involves, primarily, observations of partial eclipses which we did not include, but which Newton, for example, did or would have, for least-squares processing. There is not a single inconsistency in this group, either.

There is, therefore, not a single, astronomically usable, historically preserved eclipse observation known to us which is inconsistent with the solution. Another analyst might find a case or two which, when he interpreted the text, provided an inconsistency. This would be a matter of historical judgment. What we are emphasizing is that our selection criteria have not been used to 'sweep the inconsistencies under the rug', thereby creating an artificial consistency and consequent false confidence in the solution.

We now turn to a consideration of the errors, and determination of the uncertainties to be applied to equations (15) and (17). It was emphasized earlier that the uncertainty in making the discrimination between 'total' and 'not total' was negligible for the purposes of this study. It appears that observers of the same eclipse can make this discrimination with a precision and resolution of better than 1 km. However, the irregularities in the lunar limb, and the resulting narrow band of 'Bailey's beads' along the path edge, must be considered. This raises the actual uncertainty on path edges to the range of 2–4 km, or about 10 sec in ΔT. This we have ignored, and it will be assumed that path edges defined by 'totality' are hard, exact limits.

It is also of interest to note that the linear inequalities implied by this kind of observation are either true, or false. They do not have error bounds in the traditional sense. Despite this 'exactness' of the equations, the solution space still has a definite and finite extent, and this is the estimate of the uncertainty in the solutions obtained. For example, in equations (14) and (16), we merely used a sizeable fraction of the solution space semi-diameter as the estimated uncertainty. Considered as independent parameters, this is the correct way to treat uncertainties in $\dot{e}$ and $\dot{n}$. For equation (15), the situation is more complex.

Plots of the form in Figure 11 are used in finding the solution set, as has been explained above. To investigate the maximum range of polynomials which can fit this data, we proceeded as follows. The soft limits were largely ignored, and the modern data were replaced by a bound of ±100 sec of time at epoch 1800. The polynomial was then 'bent' the maximum amount until it ran up against two class B limits, or one class A. That is, we would automatically throw out a class B limit. Exceptions to this rule were the data before 1000 B.C., where the limits were treated as soft, but inviolable beyond a reasonable amount. The estimated maximum variation in the polynomial was ±1·5 sec of time/cy².

If this were the only factor involved, we would now have the desired uncertainties for equations (15) and (17). While each limit is quite hard, for the epoch of the observation, the short-term variations in ΔT have the effect of increasing the limit uncertainties when a mean, long-term solution, is desired. To account for this, the analysis of the preceding paragraph was repeated, after moving each limit back by 100 sec of time, to widen the possible polynomial variation. The result was approximately to double the allowance variation:

$$\Delta T \text{ Polynomial Acceleration Variation} = \pm 3\cdot 0 \text{ sec of time/cy}^2 \qquad (18)$$

This is a very conservative estimate of the maximum allowable variation in the solution, and the adoption of one-third of this amount as a standard deviation for equations (15) and (17) seems reasonable. We quote, therefore, two distinct uncertainties for equation (15). First, 100 sec of time applies in the fit of this mean polynomial at any point against the instantaneous values of $\delta \Delta T$ (which include the short-period variations). This implies that, on the average, we cannot expect the polynomial to fit the real function any better than this even for recent epochs because of the small, but discernible, short-term variations in Earth rotation. Second, an uncertainty of 1·0 sec of time/cy^2 is deduced for the mean Earth acceleration. This applies only for a given fixed value of $\dot{n}$ and, as we have seen, the estimated uncertainty on Earth acceleration considered as an independent parameter is much larger (equation 16). Over the short term, then, the 100 sec of time uncertainty dominates, while over longer periods of many centuries, the 1·0 sec/cy^2 dominates in the estimate of the polynomial goodness-of-fit.

Despite apparently favourable theoretical analysis of the minimal role to be played by a limited number of erroneous equations, we should still at least consider the possibilities for systematic errors. We do not expect to have any errors in eclipse magnitude, or identification (dating). These would not appear to be biased and, for reasons already considered, such errors if they do occur should be eliminated by the statistical filtering procedures. In any least-squares fit to smaller magnitude eclipse observations such as in Newton (1972), the risk of population-induced bias is very real. This was illustrated in Figures 3 and 4 of Part II and in the surrounding text. The question at issue here, is whether population bias can produce bias in our solution. Bias would arise in this method if all or most of our observations came from the same side of a population area, so that *errors* in place assignment tended in the *same* direction. We have no reason to suspect that this problem arises, and conclude that population bias does not have any significant effect. If, on the other hand, one averages over many places of observation of eclipses where the *magnitude* uncertainty dominates, the population bias enters the process directly. Our dominant error source is in *identifying* the place of observation, because we have deleted all of the partial eclipse data. This provides freedom from population bias effects. In conclusion, we have not been able to identify any other significant source of error bias, and tentatively conclude that this method of analysis is stable in this regard.

9. *Comparison with other recent determinations*

We now turn to a comparison of results from this paper with those of other recent studies, primarily Newton (1970, 1972). Figure 12 lists the relevant numbers. Column one is the type of data, where the first nine and last two entries come from Newton. Column two is the approximate epoch of the solution. Column three is the residual $\delta \Delta T_{o-c}$ (observed minus calculated) in the same sense as Figure 11, and elsewhere in this paper. The 'observed' value comes from Newton or this paper, and the 'calculated' value is from equation (15). Column four is the standard deviation as quoted by the source; as column three, in sec of time. Column five is column three divided by column four, the ratio of the actual residual to the standard deviation. Column six is the same as column five, except the residual is computed from the weighted mean instead of equation (15) for medieval data.

Data	Mean Epoch	$\delta \Delta T_{o-c}$	$\delta \Delta T_o$	$F(\sigma)$	$F'(\sigma)$
Babylonian L.E.T.	-600	+150	700	+0.2	-
Hipparchus EQ.	-140	-1550	3300	-0.5	-
Amalgest OCC.	-100	+880	700	+1.2	-
Mediterranean L.E.T.	-80	-800	430	-1.8	-
Theon S.E.T.	+360	+1530	680	+2.5	-
Islamic EQ.*	+840	-920	1060	-0.9	-0.8
Islamic S.E.T.*	+930	-230	85	-2.7	-1.5
Islamic L.E.T.*	+940	-300	170	-1.8	-1.2
Hakemite OCC.*	+1000	-60	50	-1.2	+0.8
$\delta \Delta T$ This Study *	+1000	-80	50	-1.6	+0.4
*Weighted Mean	+1000	-100	30	-2.3	-
ΔT Polynomial	+1000	0	50	-	+2.0
A.A.O. Solution (1970)	+1000	-120	40	-3.0	-0.5
M.C. Solution (1972)	+976	-500	100	-5.0	-4.0

* Included in the Weighted Mean.

From this study. All other data from R. Newton (1970, 1972).

FIGURE 12. Comparison of data from other sources

Newton quotes his Earth accelerations as $\dot{\omega}/\omega_e$ in cy^{-1} which is related to $\dot{e}$ by the equation:

$$\dot{e} = 3 \cdot 15 \times 10^9 \, \dot{\omega}/\omega_e \tag{19}$$

Newton uses equation (7) to define both ΔT and $\dot{e}$, while it is the contention of this paper that equations (5) and (6) must be employed. Newton's results must therefore be compared in the parameter ΔT, not $\dot{e}$ or its equivalent $\dot{\omega}/\omega_e$. As can be seen from equation (11), this must be done at the same value of $\dot{n}$, and we utilize the result of this paper from equation (14). We 'map' Newton's value of $\dot{e}$ to the lunar acceleration of $-37 \cdot 5$, by equation (4), and then compute his 'observed' ΔT at the epoch in question by equation (7). Differencing this result against the corresponding value obtained from equation (15) yields the observed minus calculated $\delta\Delta T$ residuals given in column three of Figure 12.

Examination of column five of Figure 12 permits a determination of whether or not there are obvious systematic biases between the various observables processed by Newton, and the results of this paper. The data from Newton (1970) include Lunar Eclipse Times, Equinoxes, Timed Occultations and Solar Eclipse Times. In the first block of the Figure, ancient observations, we find that the Theon solar eclipse times depart from the mean by +2·5 standard deviations, but the remaining data are balanced and acceptably random. In our view, Newton (1970) was optimistic regarding the accuracy of Theon's clocks, but this a minor consideration. On balance, all the ancient observations are in excellent agreement, with no obvious biases between the various data types considered by Newton and this paper. The data in the second block of Figure 12, Newton's (1970) medieval observations, are below ours by modest but consistent amounts of approximately two standard deviations. We do not imply by this that his data are questionably biased; if anything, it suggests that our mean polynomial may be a bit high during this period.

To investigate this situation, we formed the weighted mean of the data marked (*) in Figure 12. This medieval mean included the data of Newton (1970), plus the best estimate of the instantaneous medieval $\delta\Delta T$ from our data in Figure 11. The latter were obtained from the dotted sine curve plotted on Figure 11, justification for which was found in both the very modern and ancient data near epoch zero A.D. It turns out that the weighted mean is consistent with the sine curve estimate of the medium-term variation in $\delta\Delta T$, adding strength to this proposed 'modelling' of the variation which is considered in more detail later. We do not suggest that our data make a very strong estimate of this medieval value, and reliance is more strongly based in the data of Newton (1970), which is why we employ the weighted mean as the best estimate for comparative purposes. Column six provides the standard deviation ratios as measured from the weighted mean of all data, and we see that most of the originally apparent bias has vanished. Compared with the mean, the several data types are consistent, and separated by reasonable multiples of the standard deviations.

We draw two conclusions from these points. First, it has been shown that *all* of the ancient and medieval data, including our mean polynomial and the reinterpreted results of Newton (1970), are internally consistent. Second, it follows from this that most, if not all, of the 'medieval variation' in the accelerations discussed by Newton from his data was the spurious result of his method of analysis (i.e. the choice of equation (7) instead of (5) and (6) for the definition of $\dot{e}$). We do not intend this in any way to be a criticism of the remainder of his (1970) analysis. As Figure 11 and the other considerations amply testify, his reductions of the several distinct data-types available in the Astronomical Records are in accord, and free of any obvious systematic biases. We repeat once again, that these are consistent with our mean (and quite constant) accelerations.

We would suggest that the question of variable accelerations in the historical period is very nearly answered in the negative, and further investigation of this point will now be undertaken.

Newton (1972) analyses the large solar eclipses from the medieval period. He fits by least squares all of the astronomically usable data, involving all eclipse observations whether of large or small magnitudes. As we have previously noted, this process may be subject to the effects of population bias. The fourth and last block of entries in Figure 12 comprises Newton's final (1970) and (1972) solutions, respectively, for the medieval data only. In column six we have the standard deviation ratios compared with the weighted mean. It can be seen that the (1970) solution from several sets of observations is quite reasonably consistent, as we have already noted. The (1972) reduction is biased, by four standard deviations, in comparison with his own results as well as those of this paper. In fact, this solution is quite inconsistent with two total eclipse observations actually contained within his own data sets, i.e. 968, 1241; which also appear in this paper in Figure 9, with 1124 and 1153. We believe that population bias may be to blame. The numerical analysis of Newton's valuable and extremely well-considered data should, perhaps, be redone, if a filter can be devised which eliminates or reduces the apparent bias. Until this is done, or the analysis given here is disproved, we must suggest that the results of Newton (1972) be laid aside. In particular, we should be very careful about drawing any conclusions from them regarding possible variations in the accelerations.

Beyond Curott (1966), who uses eclipse data (some of which also appear here), there is probably no other paper which contributes materially to confidence in these results. Fotheringham (1920), De Sitter (1927), and Van der Waerden (1961) among others, used eclipse data in their analyses. In our view, none of this eclipse data would pass the requirements of any modern study. In fact, there is not a single large solar eclipse observation in any of these three papers which appears in this investigation. There are many valuable, even critically important, contributions in these works, but no usable data appear there, in our judgment. We shall therefore make further numerical comparisons of studies including ancient data only between ourselves, Newton and Curott.

Figure 13 lists several author's results for $\dot{n}$ considered as an independent parameter, as opposed to considerations of ΔT given a fixed $\dot{n}$. The consistency in the determinations from ancient data is very good, as has been noted by innumerable other authors. The crucial question is whether or not the lunar acceleration has changed during historical time.

Figure 13 also lists determinations from modern data. They fall into two groups, the first two observations are old (1930s), and the last three are recent (1970s). Only the older determinations are inconsistent with the remaining ones, both from modern and ancient data. It seems quite apparent that the older papers dealing with the modern astronomical data are, for some reason, biased. Furthermore, a constancy in the lunar acceleration is much more easily explained theoretically than a variation (Munk and MacDonald, 1960). It should also be noted that these several consistent determinations use largely independent data sets. Newton (1970) is based primarily on ancient and medieval, non-large solar eclipse data. Our results come primarily from central solar eclipses, few of which have significant weight in Newton. Van Flandern (1970) and Morrison (1973) utilize very modern occultations on the Atomic Clock. Oesterwinter and Cohen

FOTHERINGHAM	-30.8 (-34.)‡	
DE SITTER	-37.7 ± 4.3	
NEWTON (Ancient; 1970)	-41.6 ± 4.3*	EARLY
NEWTON (Medieval; 1970)	-42.3 ± 6.1*	
STEPHENSON (Thesis)	-34.2 ± 1.9	
THIS PAPER	-37.5 ± 5.0*	
SPENCER JONES	-22.4 ± 1.1	
CLEMENCE	-17.9 ± 4.3	
VAN FLANDERN	-52 ± 16*	MODERN
OESTERWINTER	-38 ± 8*	
MORRISON	-42 ± 6*	
Weighted Mean of all (*)	-40.75 ± 2.5	

* Comparable Recent Independent Determinations

‡ As corrected by Newton (1970)

FIGURE 13. Values of $\dot{n}$ deduced from early or modern observations

(1972) use meridian circle data back to 1912 on the best available clocks applicable in each period since. The agreement between these independent determinations is excellent, and we feel it is appropriate to compute the weighted mean of these results marked (*) in Figure 13. The result is $\dot{n} = -40.75$ arcsec/cy^2 ± 2.5, considered as an independent parameter. All of the determinations are well within their respective standard deviations of this mean.

We therefore believe that there is less justification than was formerly supposed for believing that the present value of the lunar acceleration in longitude is at variance with that in effect during ancient times. In fact, we now believe that there is strong evidence that this acceleration has been substantially constant over the historical period.

10. Some random historical considerations

The most intriguing thing about this kind of study is the concern we must have with the attitudes of ancient, and modern, men as they observed and reported these phenomena. It is necessary, but tragic, to pass over so many historically preserved statements simply because they are not usable under the contraints adopted in this analysis. We should like to devote a brief part of this paper to some random considerations of the historcial observations which we did not, for one reason or another, utilize in the analysis.

By far the most famous eclipse observation is that reported by Pappus, Cleomedes and Ptolemy, and attributed to the great Astronomer Hipparchus. It is said that he used an eclipse which was total at the Hellespont and 0·80 eclipsed at Alexandria, to determine the parallax of the Moon. Fotheringham (1920) and others have identified this as the eclipse of 20 November −128, since this was in Hipparchus' lifetime. Unfortunately, the references do not tell us which eclipse was actually utilized, and the identifications of 15 August −309 and 14 March −189 are also plausible, as noted by Newton (1970). We had thought of including this eclipse before the strict criteria were established for this analysis, and had to abandon the eclipse because identification was certainly questionable (see Newton, 1970, for an excellent and detailed discussion of these matters). On the solution obtained in this paper, the correct identification would be 15 August −309, if totality at the Hellespont is required. This eclipse was total over the entire region for all $\dot{e}$ and $\dot{n}$ within the quoted uncertainties. The eclipse of −189 was total over the whole of the Troad (south of the Hellespont) and came within 50 km of the Hellespont itself. Totality over such a wide region of Greek City States in the Troad could hardly have escaped Hipparchus' notice, and designation of the Hellespont raises a problem with this identification. The eclipse of −128 was near the end of Hipparchus' life, and was total about 50 km north-west of the Hellespont, on a track parallel to it. It is still just possible that either of these was 'his' eclipse, but we shall probably never be certain.

The eclipse of −309 was also observed by the Tyrant Agathocles on board one of his ships sailing from Syracuse to Carthage (Diodorus Siculus). The analytic question involved in the use of this eclipse is to determine whether Agathocles sailed north, or south, to clear Sicily. Since the observation is definitely of a very large eclipse, either total or verging on totality, our results indicate that he probably sailed north through the straits of Messina.

The so-called eclipse of Plutarch has been considered by the older authors, who try to identify this eclipse by analysing the clearly fictitious literary passage in which the reference occurs. We have always felt that the *a-priori* way to analyse such records is simply to assume, in the case of a detailed record, that the author personally witnessed it at his home. Such a presumption falls far too short of certainty for the purposes of this Astronomical analysis, but it does provide, in our view, a fairly high probability of success. Computation of the extremely narrow belt of totality for the best identification applicable to this eclipse, 20 March +71, does in fact include Plutarch's home city of Chaeronea. This could be coincidence, or accident, but the probability of this happening by chance is rather small. We believe, therefore, that Plutarch very probably witnessed this eclipse at home, and later included it in his literary efforts as writers invariably do with their personal experiences. In view of this, his account probably is the first written record of the solar corona seen during an eclipse.

The poet Archilochus (ca. 650 B.C.) has composed a vivid poem including a detailed eclipse record. This 'datum' has been considered by previous authors under the assumption that he witnessed it at home. This had to be either at Paros, as a young man, or Thasos as an older man. Our solution implies Paros as the place of observation on 6 April −647, and the uncertainties in the solution are not enough to admit of any alternative. We can conclude, therefore, that Archilochus saw the referenced eclipse, as a young man, at home. It apparently inspired his later poem.

The Chinese observation of 14 June −1329 was identified for the first time in this paper through analysis of the cyclical day number provided with the record (day 35 in a cycle of 60 days). Aside from its use in the analysis as a class B observation, we can conclude two things with high (0·90) reliability. First, the Chinese cyclical day calendar was correctly synchronized in −1329, and has presumably been carried forward to the present day without error. The tortoise shell on which the inscription appears, and all the other records preserved there, are thereby dated; the eclipse record to the day. This provides the earliest established date in the Chinese Calendar. Of course, the establishment of calendrics for ancient civilizations is one of the desired applications for results of the kind reported in this paper.

The Ugaritic eclipse of 3 May −1374 is now positively identified as the only total eclipse visible in Ugarit during the historically admissible period defined by the analysis of Sawyer and Stephenson (1970). It is also the best fit to the calendar month cited in the record (first day of Ḫiyar) of any eclipse reaching a reasonable magnitude in Ugarit. We can now definitely say that this is the date of that tablet, and it becomes the earliest date established for the city of Ugarit.

An eclipse is reported at sunrise in Athens on 14 January 484. Unfortunately, we could not use it because sunrise/sunset magnitudes can be questioned. Stephenson analyses this record, and many others, in his unpublished Thesis. It turns out that the eclipse was total in Athens, at sunrise, and the sun rose totally eclipsed, a fairly rare event. Similarly, for Rome on 17 July −187, we have the report of a great darkness, after which three days of prayer were ordered The prayers were justified, as Rome witnessed a total eclipse on this date, as confirmed by our solution.

Saint Luke refers to an eclipse, and describes it, in his gospel (24: 45). No matter which of the two cities with the name Antioch he resided in, he saw a total or magnitude 0·995 eclipse on 24 November +29. See the analysis of Sawyer

(1972a) which uses the computations of Stephenson's Thesis (1972); that result is not significantly changed by the new solution and is confirmed. Another Biblical eclipse description is attributed to Joshua from the Old Testament. Sawyer (1972b) gives the date for this as 30 September −1130, and the place as Gibeon. The solution given in this paper places Gibeon within the path of this total eclipse, but barely so.

We have quickly summarized a few high points in the consideration of ancient historical eclipses, using the determined solution to evaluate these records for calendrics and identification. Detailed analysis of the more critical points is to be undertaken in the immediate future, and will be submitted for later publication. We are sure that the high-interest potential of these subjects will not let the smallest question long go unanswered, if an answer is achievable. For those interested in this kind of analysis, we suggest that the solutions to the Astronomical elements presented in this paper are probably the best available for this purpose at the present time. The results were obtained from eclipses, and therefore should be best suited to historical extrapolation for the purpose of more detailed and precise analysis of them than has been possible in the past.

We would further be derelict in our duty if we did not issue a resounding call to historians, Assyriologists, Sinologists and any others who may have come across references to the Sun, or the Moon, or eclipses, or any other phenomena of a kind which might be Astronomically identifiable. The authors would very much appreciate receiving information on any such data or potential observations, and will cooperate in the reduction of these data (if suitable) to the improvement of the solutions needed for Astronomical purposes. Particularly, anyone having access to, or experience with, the Sumerian Texts (ca. 2000 to 3000 B.C.), could make a strong contribution to science by providing the lead to even a single eclipse observation from this period.

11. *Possible future prospects*

In believing, as we do, that the method utilized in this paper is best suited to extracting high-resolution information about ΔT from central solar eclipse observations, we would like to obtain more data so as to refine the details of the $\delta \Delta T$ behaviour, particularly during the medieval period. We believe that there is a way to do this effectively. Now that the $\delta \Delta T$ function is at least reasonably well defined for the medieval period, which it was not before, we can compute more refined eclipse paths than previously available. It is clear that the vast bulk of observations are not useful in the present analysis. We must have totality observed, or denied convincingly, to be of use. Furthermore, these observations must come from near the edge of a path to be of great sensitivity in refining $\delta \Delta T$ during this recent period. We intend, therefore, to compute refined paths for every suitable eclipse between A.D. 900 and A.D. 1600 and plot these on the European map. After adding the locations of all monasteries and centres of culture during this period which are available from the historical atlas (such as

Menke, 1880), we can determine the sites where observations of the right kind are likely to be found. Allowing the proper uncertainty to the elements, we determine thereby a list of dates and places corresponding to possible observations. Inquiry will then be made, particularly in Spain and Portugal, to determine the existence of local annals which may contain the required information. We believe that much of the European information has not yet been brought to light, and there is real hope of extending the data set in this way.

The Chinese occultations provide another source of data. Occultations have been reduced in the past by consideration of the timed contacts (the time of day associated with the event). These observations are usable in this form, as Newton (1970) has shown, but they can be used in another way. An occultation is like an eclipse, except the track is the width of the whole Moon. This is why the sensitivity is not nearly as great as for central eclipses. However, out of hundreds of occultation observations, there will be a few 'grazes' where the observation site is very near the lunar limb. If several of these can be found by analysis and computation, they could provide another source of untimed, and therefore precise, observations.

There is also the possibility, suggested in the previous section, of finding some even more ancient data than those employed for the first time in this analysis. Even one or two solid eclipse observations from Sumeria (ca. 2500 B.C.), would considerably refine (or confirm) the long-term mean acceleration solution for the Earth's rotation, and possibly the lunar acceleration as well.

The results of all such analysis, and elaborations on matters raised but not adequately discharged in this chapter, are to be considered in a book intended for publication in about a year's time.

12. *Geophysical interpretation*

We have already considered, indirectly at least, several geophysical implications to be drawn from the results of this analysis. The data of Newton (1970) have been reconsidered in the light of equations (5) and (6) instead of (7), thereby revealing consistency between his reductions and ours. It has been shown that all of the modern analyses, involving many independent sources of data both modern and ancient, are now quite consistently indicating unchanged lunar and Earth accelerations (tidal plus non-tidal). This is contrary to the position taken by Newton (1970, 1972) and many others. This observation is also consistent with modern theory (Munk and MacDonald, 1960).

I. The sum total of all modern, medieval and ancient data of all kinds indicates sensibly constant lunar and Earth accelerations, in accordance with theory, and despite conclusions to the contrary reached in other recent papers

The above refers to the total, observed accelerations in each case. Since the various analyses have determined values for both $\dot{e}$ and $\dot{n}$ considered as independent parameters, we can ask whether or not the tidally induced accelerations on the Earth add up to the observed acceleration $\dot{e}$. From conservation of angular momentum, we can compute the contribution to $\dot{e}$ from $\dot{n}$. This has been analysed by many authors, but we prefer the solution of Newton (1970), which includes the best estimate of the solar tidal contribution:

$$\text{Tidal}_{m+s}(\dot{\omega}/\omega_e) = 1\cdot 147 \times 10^{-9}\dot{n} \text{ (Units cy}^{-1}) \tag{20}$$

Newton (1968) also derives a value for the solar atmospheric tidal acceleraton:

$$\text{Solar atmosphere tide } (\dot{\omega}/\omega_e) = +2\cdot 7 \times 10^{-9} \text{ cy}^{-1} \tag{21}$$

Other contributions to the Earth acceleration which are known, theoretically, are very small and will be neglected here (i.e. Munk and MacDonald, 1960; Dicke, 1966). We may, therefore, compute the non-tidal acceleration from the observed Earth acceleration:

$$\text{Non-tidal} = \text{Observed} - \text{Atmospheric} - \text{Tidal}_{m+s}(\dot{\omega}/\omega_e) \tag{22}$$

Since the estimate of non-tidal Earth acceleration depends upon the $\dot{n}$ as well as the observed Earth acceleration, we should make comparisons at a fixed $\dot{n}$. It can be shown that:

$$\text{Non-tidal } (\dot{\omega}/\omega_e) = \text{Observed non-tidal } (\dot{\omega}/\omega_e) - 0\cdot 525\delta\dot{n} \tag{23}$$

where the left-hand term is for some new $\dot{n}$, given the observed value at the nominal $\dot{n}$ ($\delta\dot{n}$ = New $\dot{n}$ − Nominal $\dot{n}$).

We will now find the best estimates of the Earth's non-tidal acceleration from Curott (1966), Newton (1970) and this paper. In Figure 14, column one is the data source, column two the approximate epoch, column three lists the $\dot{n}$ solution associated with the estimate of $(\dot{\omega}/\omega_e)$ appearing in column four. When we say 'raw' $(\dot{\omega}/\omega_e)$, we are referring to the fact that the other authors used equation (7) in their determination, and we will need first of all to correct these results to the epoch 1770 determined in this study, from the A.D. 1900 implied by equation (7). We first correct to an $\dot{n} = -37\cdot 5$ in column five, and to epoch 1770 in column six via the equation:

$$\dot{\omega}/\omega_e \text{ Corrected} = ((\text{Epoch-}1900)/(\text{Epoch-}1770))^2 \dot{\omega}/\omega_e \text{ Raw} \tag{24}$$

We now have the desired basis of comparison for the different solutions. Column seven is the observed non-tidal component from equation (22) and is associated with an $\dot{n} = -37\cdot 5$ to facilitate the consistency comparison between the various sources. To have the absolute value of this non-tidal acceleration at the $\dot{n}$ actually obtained by the given author, we must apply equation (23), and the results are displayed in column eight. The second block of Figure 14 provides the determination resulting from the $\dot{n}$ mean best estimate noted in Figure 13, with equation (17)

The Accelerations of the Earth and Moon

providing the corresponding $\dot{e}$. The approximate standard deviations applying to the individual determinations and the mean are also indicated.

The first thing which strikes our eyes is the absence of any medieval variation in the data of Newton as herein re-interpreted in column 6. This is just another

Author	Epoch	$\dot{n}$	Raw $\dot{\omega}/\omega_e$	$\dot{\omega}/\omega_e$ $\dot{n}=-37.5$	Corrected* $\dot{\omega}/\omega_e$	Non-Tidal $\dot{n}=-37.5$	Non-Tidal Author $\dot{n}$	Tidal Energy
Newton (1970)	−200	−41.6	−27.7	−25.1	−28.4	+12.1	+14.3	6.5
Newton (1970)	+1000	−42.3	−22.5	−19.5	−28.0	+12.5	+15.0	6.6
Curott (1966)	∼0	−22.4	−16.0	−25.4	−29.1	+11.4	−	−
This Study	∼0	−37.5	−29.0	−29.0	−29.0	+11.5	+11.5	5.9
Std. Dev. of Above	−	±5	±3	−	(±1.)	(±1.)	±3.	±1.
Mean Best Estimate	−	−40.75	−30.5*	−	−	+12.0	+13.7	6.2
Std. Dev. of Mean	−	±2.5	±1.5	−	−	−	±2.0	±1.

*Corrected to Epoch of 1770 A.D. (See Text)
$\dot{\omega}/\omega_e$ in $cy^{-1} \times 10^9$
Tidal energy in watts $\times 10^{-12}$

FIGURE 14. Comparison of corrected results of recent studies ($\dot{\omega}/\omega_e$)

way of displaying this fact, from that adopted above in the comparison of results in the parameter ΔT. The consistency of relative results (column 7) is extremely good. Even the determinations of non-tidal Earth acceleration, considered as an independent parameter (column 8), all fall within a reasonable range. We believe that not only has the constancy of the non-tidal acceleration of the Earth been demonstrated, but that a reasonable estimate of its magnitude is obtained as well.

II. The Earth's non-tidal acceleration has also been sensibly constant over the historical period, and the results from all ancient and medieval data are consistent.

As can be seen from Figure 11, the mean polynomial (and hence the constant acceleration in $\dot{e}$) does not perfectly represent the observations. The modern data are up to 100 sec above the mean near A.D. 1700 and rather more below it in 1970. The variation in medieval times near A.D. 1000 is again negative by about 100 sec of time. Looking across the whole panorama of data, we can see that it is extremely unlikely that the short-term variations in ΔT from the mean (constant) acceleration have exceeded 200 sec of time. This is the rather tight-bound, placed on short-term variation.

III. The departure of instantaneous ΔT from the mean, constant acceleration, is bounded by 200 sec of time, and this is the limit on integrated short- and medium-term variations over the whole historical period.

A bound has been placed on the short-term variations of ΔT, but many would be interested in whether or not there is any modellable pattern to the actual observed variations. While rather speculative at this point, we do wish to consider this possibility. Figure 11 provides sufficient data density from about A.D. 900 to the present time for the preliminary evaluation of consistency in the departures of the instantaneous $\delta \Delta T$ from the mean. We have a very fine mean for the epoch near A.D. 1000, primarily from the data of Newton (1970) as considered above. Our data provide very tight limits at A.D. 1079 and A.D. 1239. The modern eclipse data circa A.D. 1567 fill in this region, and the Halley eclipse of 1715 provides the consistency tie-in, joining us to the modern astronomical data which are available continuously up to the present time. Over this interval of some 1000 years, we can see that the instantaneous values depart quite systematically from the mean, rather than wildly fluctuating. A good fit to this systematic variation is provided by the function:

$$\delta \Delta T' = +80 \sin((\pi/6)(T+7)) \tag{25}$$

where $\delta \Delta T'$ is the correction to equation (15) and T is in centuries from 1900. This additional correction remains consistent with all of the observations, and reduces the 'strain' against the class A limits for A.D. 120, and -180. It also provides an ideal fit to the Babylonian timed contact in -321. While we emphasize again the speculative nature of this suggested modelling of the medium-term $\delta \Delta T$ variation, it does seem markedly to improve the quality of fit both to modern detailed data as well as the sparser historical limits. Perhaps suggesting this model will lead some geophysics theoretician to suggest an explanation which, if convincing, would go a long way towards strengthening this speculation.

It is interesting to summarize the proposed theoretical explanations which have been offered by others to explain the observed magnitudes of the various geophysical parameters. As we have noted, all of the recent determinations of the lunar acceleration cluster around and determine a value near -40 arcsec/cy^2, which implies a tidal energy dissipation of approximately $6 \cdot 2 \times 10^{12}$ watts. The constancy of this estimated value over the historical period is consistent with theoretical analyses such as Munk and MacDonald (1960) and Dicke (1966). Finding the required energy dissipation in the shallow seas is difficult, but now appears to be possible. The recent analysis of Hendershott (1972) obtains values between 3 and 4×10^{12} watts. Smith and Jungels (1970) obtain 3 to 5×10^{12} watts from observational considerations of the gravimetrically determined tidal phase lag. Figure 14 indicates that the range determined from the astronomical observations is between 5 and 7×10^{12} watts. Considering the difficulties involved in the analyses noted here, the agreement between the observational and theoretical determinations is reasonably good. We may hope that the next few years provide further convergence on the 'correct' answers.

The next major consideration is the determination of the constancy of the Earth's non-tidal acceleration and its magnitude, which is in the range 10 to 15×10^{-9} cy^{-1}. The large estimate of Newton (1970) has been re-interpreted and falls in this range, as noted above. Yukutake (1972) obtains approximately 5×10^{-9} cy^{-1} from the effect of the changing geomagnetic dipole as observed in the palaeomagnetic data. Dicke (1969) uses the results of Curott's (1966) eclipse analysis to determine the viscosity of the lower mantle. In so doing, he provides a plot (his Figure 2) which gives tidally induced Earth accelerations for various isostatic relaxation times in the N = 2 mode. He determines that the relaxation constant is 1×10^{-3}/year. Applying this to the sea-level data of Redfield and Rubin (1962), Dicke obtains about 6×10^{-9} cy^{-1} in the non-tidal Earth acceleration from changing sea level. Dicke uses Curott, who in turn assumed $\dot{n} = -22\cdot44$, but the estimate of relaxation time does not seem to depend heavily on this assumption. The sea-level variation and geomagnetic variation, between them, appear capable of producing a non-tidal Earth acceleration of $+11 \times 10^{-9}$ cy^{-1}. This is well within the range of 10 to 15×10^{-9} cy^{-1} implied by the recent astronomical determinations quoted in this paper. There appears to be hope at least, that these observationally determined parameters (sea level and geomagnetism) can point the way to theoretical understanding of the non-tidal Earth acceleration estimate provided by astronomy.

13. *Final summary of results*

The data of this paper, plus the reconsidered results of other recent analyses, imply the constancy of the tidal, and non-tidal, Earth and lunar accelerations and provide reasonably accurate and reliable estimates of their magnitudes. These observational results are now internally consistent, and consistent with the suggestions of theoretical analysis. Previous claims that the accelerations have changed significantly in the historical period have been denied, and we conclude quite the opposite. The bound on the departure of instantaneous, and/or non-constant acceleration contributions to $\delta\Delta T$, is set at 200 sec of time, which is a very tight limit indeed. The numerical results from this paper included the following.

From this analysis

Lunar acceleration $\dot{n}$ as an independent parameter: $\dot{n} = -37\cdot5 \pm 5$ arcsec/cy^2
(equivalent to rate of retreat of Moon from Earth): $= +5\cdot8$ cm/year
Earth acceleration $\dot{e}$ as an independent parameter: $\dot{e} = -91\cdot56 \pm 10$ sec/cy^2
Or: $\dot{\omega}/\omega_e = -29\cdot0 \pm 3 \times 10^{-9}$ cy^{-1}
Or: $\delta\text{LOD} = 2\cdot50 \pm 0\cdot3$ millisec/cy
ΔT (ET − UT for $\dot{n} = -37\cdot5$): $\Delta T = 66\cdot0 + 120\cdot38\,T + 45\cdot78\,T^2$ sec
for T in cy from 1900 and: ± 100 sec, or $\pm 0\cdot5\,T^2$, or $\pm 1\cdot0$ sec/cy^2

Linear combination of $\dot{e}$ and $\dot{n}$: $\dot{e} - 1.97\,\dot{n} = -17.7 \pm 1.0$ sec/cy^2
Bound on $\delta\Delta T$ (Instantaneous − Mean ΔT): 200 sec maximum, 100 typical
Speculative profile of medium-term $\delta\Delta T$: $\delta\Delta T' = +80\sin((\pi/6)(T+7))$ sec for T in cy from 1900
Non-tidal Earth acceleration $\dot{\omega}/\omega_e$: $(\dot{\omega}/\omega_e)_{\text{non-tidal}} = +11.5 \pm 3 \times 10^{-9}$ cy^{-1}

Better estimates for some of these parameters can be obtained from the averaging of other recent independent determinations with the above.

From the data of all recent studies

Lunar acceleration $\dot{n}$ as an independent parameter: $\dot{n} = -40.75 \pm 2.5$ arcsec/cy^2
Earth acceleration $\dot{\omega}/\omega_e$ as an independent parameter:
$$\dot{\omega}/\omega_e = -30.5 \pm 1.5 \times 10^{-9} \text{ cy}^{-1}$$
Non-tidal Earth acceleration $\dot{\omega}/\omega_e$: $= +13.7 \pm 2.0 \times 10^{-9}$ cy^{-1}
Lunar tidal energy dissipation: $= 6.2 \pm 1.0 \times 10^{12}$ watts

These mean values of the geophysical parameters are recommended as the best available for use in geophysical interpretation. For the prediction of ancient astronomical data, we suggest that the solutions of this paper are the best choice, and the likeliest to be free of systematic error.

The data and theory presented in all recent papers have been shown to be reasonably consistent. This provides hope that the situation in this field is now resolving itself in the direction of observational consistency and theoretical understanding.

Acknowledgments

We wish to express our gratitude to the following colleagues for much assistance and many valuable discussions. Mr. A. C. Barnes, School of Oriental Studies, University of Durham; Dr. T. J. Saunders, Department of Classics, University of Newcastle upon Tyne; Professor P. J. Huber, Department of Mathematical Statistics, Federal Technical University, Zurich; Professor A. J. Sachs, Department of the History of Mathematics, Brown University, Providence, Rhode Island, USA; Dr. T. C. Van Flandern, United States Naval Observatory, Washington, DC; Professor H. C. Urey, University of California at San Diego, LaJolla, California, USA; Dr. J. H. Lieske, Mr. R. N. Wimberly and Mr. W. S. Sinclair, Jet Propulsion Laboratory, Pasadena, California, USA. Additionally, many colleagues from diverse fields have supplied assistance in their specialities, who are far too numerous to mention individually.

P. M. Muller's research is supported by: Jet Propulsion Laboratory, Contract NAS7-100; Leverhulme visiting Fellowship of the Royal Astronomical Society, London (1973–74). He is primarily responsible for Parts I and III.

F. R. Stephenson's research is supported by the Earl Grey Memorial Fellowship of the University of Newcastle upon Tyne. He is primarily responsible for Part II.

References

Antuña, M. M. (Ed.) (1937). *Ibn Hayyan—Al Muqtabis*, Paris

Bouquet, M. (Ed.) (1781). *Recueil des Historiens des Gaules et de la France*, vol. 12, Paris

Brinkman, J. A. (1968). *A Political History of Post-Kassite Babylonia*, Pontifical Biblical Institute, Rome

Brooks, E. W. and Chabot, J. B. (Eds.) (1954). *Eliae Metropolitae Nisibeni Opus Chronologicum*, Louvain

Brouwer, D. (1952). A study of the changes in the rate of rotation of the Earth. *Astron. J.*, **57** (1201), 125–146

Celoria, G. (1877a). Sull' eclissi solare totale del 3 giugno 1239. *Memorie del Reale Istituto Lombardo di Science e Letteri, Classe di Science Mathematiche e Naturali*, **13**, 275–300

Celoria, G. (1877b). Sugli eclissi solari totali del 3 giugno 1239 e del 6 ottobre 1241. *Memorie del Reale Istituto Lombardo di Science e Letteri, Classe di Science Mathematiche e Naturali*, **13**, 367–382

Clavius, C. (1593). *In Sphaeram Ioannis de Sacrobosco Commentarius*, Sumptibus fratrum de Gabiano, Lugduni

Curott, D. R. (1966). Earth deceleration from ancient solar eclipses. *Astron. J.*, **71** (4), 264–269

De Sitter, W. (1927). On the secular accelerations and the fluctuations of the longitudes of the Moon, the Sun, Mercury and Venus. *Bull. Astron. Inst. Netherlands*, **4** (124), 21–38

Dicke, R. H. (1966). The secular acceleration of the Earth's rotation and cosmology, pp. 98–163 in *The Earth–Moon System* (Ed. B. G. Marsden and A. G. W. Cameron), Plenum Press, New York

Dicke, R. H. (1969). Average acceleration of the Earth's rotation and the viscosity of the deep mantle. *J. Geophys. Res.*, **74** (25), 5895–5902

Dubs, H. H. (1938, 1944 and 1955). *The History of the Former Han Dynasty* (3 vols.), Waverley Press, Baltimore

Dubs, H. H. (1951). The date of the Shang period. *T'oung Pao*, **40**, 322–355

Explanatory Supplement to the Astronomical Ephemeris and the Nautical Almanac (1961). H.M. Stationery Office, London

Fitzgerald, C. P, (1942). *China, a Short Cultural History*, Cresset Press, London

Florez, H. (1778→). *España Sagrada* (50 vols.), Madrid

Fotheringham, J. K. (1920). A solution of ancient eclipses of the Sun. *Monthly Notices Roy. Astron. Soc.*, **81**, 104–126

Fotheringham, J. K. (1935). Two Babylonian eclipses. *Monthly Notices Roy. Astron. Soc.*, **95**, 719–723

Ginzel, F. K. (1884). Astronomische untersuchungen über finsternisse. *Sitzungsberichte der Kaiserlichen Akademie der Wissenschaften, Wien, Math.-Naturwiss. Classe*, **88**, 629–755; **89**, 491–559

Ginzel, F. K. (1899). *Spezieller Kanon der Sonnen und Mondfinsternisse*, Mayer and Muller, Berlin

Ginzel, F. K. (1918). Beitrage zue kenntnis der historische sonnenfinsternisse und zur frage ihrer verwendbarkeit. *Abhandlungen der Koniglichen Akademie der Wissenschaften, Berlin, Phys.-Math. Klasse*, **4**, 3–46

Guriaeb, J. E. (1960). 'Al-Muqtabis' de Ibn Hayyan. *Cuadernos de Historia de Espana*, **32**, 316–321

Halley, E. (1695). *Phil. Trans. Roy, Soc.*, **19**, 160–175

Halley, E. (1715). Observations of the late total eclipse of the Sun. *Phil. Trans. Roy. Soc.*, **29**, 255–266

Hendershott, M. C. (1972). The effects of solid Earth deformation on global ocean tides. *Geophys. J.*, **29**, 389–402

Holder-Egger, O. (Ed.) (1899). *Scriptores Rerum Germanicarum*, vol. 42, Hahn, Hanover

Ho Peng Yoke (1966). *The Astronomical Chapters of the Chin Shu*, Mouton, Paris

Improved Lunar Ephemeris (1954). Joint Supplement to the American Ephemeris and Nautical Almanac. U.S. Government Printing Office, Washington, D.C.

Jao Tsung-i (1959). *Yin-tai Cheng-pu Jen-wu T'ung-kao* (Oracle Bone Diviners of the Yin Dynasty), Hong Kong Univ. Press

Jones, H. S. (1939). The rotation of the Earth, and the secular accelerations of the Sun, Moon and planets. *Monthly Notices Roy. Astron, Soc.*, **99**, 541–558 (see also under Spencer Jones, H. 1939)

Lecomte, L. (1696). *Nouveau Memoires sur l'Etat Present de la Chine*, Anisson, Paris

Legge, J. (1960). *The Chinese Classics*, 2nd ed., vol. 5, Hong Kong Univ. Press (1st edn. 1861)

Liu, Chao-yang (1945). Chia ku wen chih jih erh kuan ts'e chi lu (Mention of the solar corona on the oracle bone inscriptions). *Yu-chou*, **15**, 15–16

Menke, T. (1880). *Spruner-Menke Hand-Atlas*, J. Perthes, Gotha

Migne, J.-P. (Ed.) (1865). *Patrologiae Graecae*, vol. 148, Paris

Morrison, L. V. (1973). Rotation of the Earth from AD 1663–1972 and the constancy of G. *Nature*, **241**, 519–520

Munk, W. H. and MacDonald, G. J. F. (1960). *The Rotation of the Earth*, Cambridge University Press

Muratori, L. A. (Ed.) (1723→). *Rerum Italicarum Scriptores*, (25 vols.), Milan

Needham, J. (1959). *Science and Civilisation in China*, vol. 3, Cambridge University Press

Neugebauer, P. V. (1932). Spezieller kanon der sonnenfinsternisse. *Astronomische Abhandlungen*, **8**, Nr. 4

Newcomb, S. (1895). Tables of the Sun and inner planets. *Astronomical Papers for the American Ephemeris and Nautical Almanac*, VI, Washington, D.C.

Newton, R. R. (1970). 'Ancient Astronomical Observations and the Accelerations of the Earth and Moon', Johns Hopkins Press, Baltimore and London, 309 pp. Summary in *Astrophysics and Space Science*, **16**, 179–200 (1972)

Newton, R. R. (1972). *Medieval Chronicles and the Rotation of the Earth*, Johns Hopkins University Press, Baltimore and London, 825 pp.

Niebuhr, B. C. (1828). *Corpus Scriptorum Historiae Byzantinae*, vol. 33, Weber, Bonn

Oesterwinter, C. and Cohen, C. J. (1972). New orbital elements for Moon and planets. *Celestial Mechanics*, **5**, 317–395

Oppolzer, T. R. von (1887). *Canon der Finsternisse*, Kaiserlichen Akademie der Wissenschaften, Wien. Reprinted as *Canon of Eclipses*, Dover, New York (1962)

Parker, R. A. and Dubberstein, W. H. (1956). *Babylonian Chronology 626 BC–AD 75*, Brown University Press, Providence, R.I.

Pertz, G. H. (Ed.) (1826→). *Monumenta Germaniae Historica, Scriptores* (32 vols), Hahn, Hanover

Redfield, A. C. and Rubin, M. (1962). The age of salt marsh peat and its relation to recent changes in sea level at Barnstable, Massachusetts. *Proc. Nat. Acad. Sci., U.S.*, **48**, 1728

Rowton, M. B. (1946). Mesopotamian chronology and the 'Era of Menophres'. *Iraq.*, **8**, 94–110

Sachs, A. J. (1948). A classification of the Babylonian astronomical texts of the seleucid period. *J. Cuneiform Studies*, **2**, 271–290

Sachs, A. J. (1955). *Late Babylonian Astronomical and Related Texts*, Brown University Press, Providence, R.I.

Sarton, G. (1947). *Introduction to the History of Science*, vol. 3, part 1, Published for Carnegie Institute of Washington by Williams and Williams

Sawyer, J. F. A. and Stephenson, F. R. (1970). Literary and astronomical evidence for a total eclipse of the Sun observed in Ancient Ugarit on 3 May 1375 BC, *Bull. School Oriental and African Studies*, **33**, 467–489

Sawyer, J. F. A. (1972a). Why is a solar eclipse mentioned in the Passion Narrative (Luke 23. 44–45). *J. Theological Studies*, **23**, 124–128

Sawyer, J. F. A. (1972b) Joshua 10. 12–14 and the solar eclipse of 30 September 1131 BC. *Palestine Exploration Quarterly* (1972), pp. 139–145

Schroeter, J. F. (1923). *Spezieller Kanon der Zwntralen Sonnen und Mondfinsternisse*, Dybwad, Kristiania

Smith, S. W. and Jungels, P. (1970). Phase delay of the solid Earth tide. *Physics of Earth and Planet Interiors*, **2**, 233–238

Spencer Jones, H. (1939). The rotation of the Earth, and the secular accelerations of the Sun, Moon and planets. *Monthly Notices Roy. Astron. Soc.*, **99**, 541–558

Stephenson, F. R. (1972). Some geophysical, astrophysical and chronological deductions from early astronomical records. Ph.D. Thesis, School of Physics, University of Newcastle upon Tyne, England

Stephenson, F. R. (1974). Late Babylonian Observations of 'Lunar Sixes'. In 'The Place of Astronomy in the Ancient World' (Ed. F. R. Hodson), *Proc. Roy. Soc.* (in press)

Tuckerman, B. (1962). Planetary, solar and lunar positions, 601 BC to AD 1. *Mem. Am. Phil. Soc.*, **59**, viii, 333 pp.

Tung Tso-pin (1945). *Yin Li P'u* (On the Calendar of the Yin Dynasty.) Lichuang

Van der Waerden, B. L. (1961). Secular terms and fluctuations in the motions of the Sun and the Moon. *Astron. J.*, **66** (3), 138–147

Van Flandern, T. C. (1970). The secular acceleration of the Moon. *Astron. J.*, **75** (5), 657–658

Vyssotsky, A. N. (1949). Astronomical records in the Russian Chronicles. *Medd. fran. Lunds Astr. Obs.*, Historical Papers, 22

Waley, A. (1931). *The Travels of an Alchemist*, Routledge, London

Wylie, A. (1897). *Chinese Researches*, Shanghai

Yukutake, T. (1972). The effect of change in the geomagnetic dipole moment on the rate of the Earth's rotation. *J. Geomagnetism Geoelec.*, **24**, 19–47; or pp. 228–230 in *Rotation of the Earth* (Ed. P. Melchior and S. Yumi), I.A.U. Symposium 48, Reidel, Dordrecht, Holland

DISCUSSION

RUNCORN: I wonder whether tidal effects on the Earth's rotation could be distinguished from non-tidal effects, by examining the Earth's and Moon's acceleration as determined from ancient astronomical observations?

MULLER: This depends on accuracy of the observations. The accuracy of the observations which we believe are valid depends on accurate timing, description of totality and location of observer. Newton's (1972) determinations are often inadequate in at least one of the three regards, so they cannot be used in calculating the Moon's acceleration, although Newton's results are correlative with ours.

O'HORA: Wouldn't it be difficult to distinguish an annular from near total eclipse of an irradiating Sun in the first place?

MULLER: The criteria for selecting valid eclipse observations in ancient records require specific statements of totality or non-totality. Partial eclipse data would be used only if the beginning and end were accurately timed. The information could then be used similar to that from a simple observation of total eclipse.

The data examined to date give an accurate value for deceleration of the Earth of i.e. 2·5 msec/cy because the value agrees from one investigator to another, although each used different means of calculating it. However, the value for the Moon's acceleration is not as well-determined. Our effects are aimed at conclusive independent determination of $\dot{n}$.

Another point is that the observed total (orbital) acceleration of the Moon and the observed total rotational acceleration of the Earth which we have determined are independent. From the observed acceleration of the Moon, one can tell how much of the Earth's rotational change is due to all tidal causes (by conservation of momentum). The difference between the tidal effects and the total change in the Earth's acceleration is then due to non-tidal causes.

RUNCORN: Could we detect these latter effects from present eclipse data? Could we describe these changes graphically for the historical period?

MULLER: At this time one could separate the tidal and non-tidal components, but because the non-tidal component was so large (almost half the total acceleration of the Earth's acceleration) and the variation relatively small, one could not at this time describe these short-term accelerations. It does seem that these variations occurred with a period of 1200 years and amplitude of 80 sec of time in Δt but the present data do not permit certainty.

Our data do not require a sharp change in Earth's acceleration about A.D. 1100 whereas Newton's do; the end result of Newton's work is to suggest a more important non-tidal contribution than we regard as plausible. The best we can do at present is to determine the maximum departure of the short-term variation from constant mean acceleration of the Earth and to recognize that the departures may not be random.

RUNCORN: There is no geophysical need for the non-tidal variation to be quadratic, although the acceleration due to tidal causes has been uniform since about 1000 B.C.

MULLER: Any non-constant parameter would appear as a residual departure from the quadratic equation.

CONCLUSIONS

Changes in the Earth's angular momentum and moment of inertia can be computed from palaeontological data, ancient and modern astronomical observations, and geophysical considerations of tidal friction and physical constants. The different methods give results which tend to agree, although additional data are needed for confirmation. The palaeontological data are of central importance as they provide the only known means of directly measuring the Earth's angular momentum and moment of inertia in the distant past.

Palaeontological considerations

The palaeontological data of Berry and Barker are in agreement with existing data and were collected independently before the publication of the data for Phanerozoic bivalves by Pannella, MacClintock and Thompson. Johnson and Nudd's Carboniferous coral data are also in agreement with published data. However, concern was expressed regarding their method of converting number of days/month to days/year by extrapolating months/year from the Devonian, and for using their result to suggest short-term variation in the rate of dissipation of the Earth's rotational energy. Mohr's stromatolite data suggest a 25-day Middle Precambrian month and 800–900 days/year. This contrasts with earlier work by Pannella. Acceptance of Mohr's data would indicate a reversal in the Precambrian of the trend for the number of days/synodic month to increase from the present back through the Phanerozoic. Truswell and Eriksson establish the tidal origin of Transvaal stromatolites and thus confirm the presence of the Moon in the Lower Proterozoic. Changes in the acceleration of gravity may accompany changes in the Earth's rotation. Creber's account of the effect of gravity on the growth of trees suggests that a gravitational influence on biological growth rhythms may be worth pursuing in fossils. A general survey of the suitability of invertebrates to growth analyses by the Termiers indicates that many taxa not yet studied are potentially reliable chronometers, notably the stromatoporoids. The Termiers also emphasized the importance of sexual rhythms as physiological controls of growth rhythms.

Increasing reliability of growth increment measurements is a consequence of studies of growth rhythms in living organisms. Clark presented convincing evidence of daily growth periodicities in bivalves kept in the laboratory. His time-lapse photography of specimens regularly adding increments supports his belief that the maximum number of increments added by individuals in a laboratory population of *Pecten* approaches the number of days elapsed. Evans' work and Pannella's studies correlate tidal cycles and growth increment patterns in selected bivalves. Evans' data also show that some species of bivalves have no geophysically significant increment patterns; awareness of associated structural differences will help provide standards for selection or rejection of species regarding their temporal reliability. Similarly, Buddemeier and Kinzie's coral studies show that regularly spaced increments in *Porites* do not necessarily have regular periodicities. However, lunar growth periodicities are also found in *Porites*, and these are geophysically important.

Thompson's laboratory studies of *Mercenaria* confirm *Mercenaria* growth periodicities described by the Yale group of Rhoads, Pannella and MacClintock. Moreover, they further establish a link between a behavioural process (gaping) and increment production. Thompson also believes that her work establishes the existence of an internal biological clock in *Mercenaria*, independent of the environment. This conclusion should be the subject of interesting discussion between the endogenous and exogenous schools of biorhythm workers. Hall has found latitudinal, depth and substrate correlative growth patterns in the bivalves *Tivela* and *Callista*. The latitudinal variations can be used to suggest ancient pole positions. His work also discusses the obvious but often overlooked fact that growth in bivalves declines with age. On the one hand, his work supports the chronometric reliability of some bivalves in their early growth stages and, on the other hand, serves as clear warning that growth periodicities are not necessarily continuous throughout life. Also notable are Dolman's automated techniques of increment study and analysis. These should speed the collection and refinement of growth data. His results in *Cardium* further establish daily growth in bivalves. The techniques will prove of interest to palaeoecologists wishing to correlate growth rhythms with environmental fluctuations. Whyte has studied subdaily growth variations in the same species. His recognition of structural and compositional differences between nocturnal and daylight growth should also be of interest to molluscan physiologists. Such detailed analyses by Whyte are important sequels to the structural analyses undertaken by Barker, an investigator whose interesting work is sometimes unfortunately overlooked. Yet such detailed analyses, along with the techniques of Dolman, are potentially able to refine correlations of structure and periodicity to a high degree.

Rosenberg and Jones presented further evidence for a chemical alternative to structural analysis of growth periodicities. Suggestion of tidal periodicities of calcium and sulphur distribution in living molluscs may eventually enable the measurement of the changing number of days/month and days/year with the electron microprobe. Recognition of repeating stromatolite periodicities potentially extends the chemical record back to the Precambrian. Diagenesis of organisms and fossil metabolism also may eventually be studied with microprobe techniques.

Growth studies are of value to the archaeologist. Coutts uses the season of growth last recorded in bivalve shells dumped in Maori Indian Middens to determine migration patterns of aboriginal New Zealanders. In turn, Coutts' growth increment counts further suggest periodic growth in some molluscs.

Conclusions 537

Pannella has summarized the growth increment work since Wells. Most important is his recognition and discussion of different types of daily and sub-daily increments. One can now begin to distinguish between the affect of tidal and day–night cycles on increment pattern. He and Evans recognize the difference results in an interference pattern of solar and lunar daily increments. The determination of these differences is important in accurately determining the number of days/fortnight, month and year in the distant past.

Rosenberg's discussion of the increment and the series provides a standard for describing and classifying growth rhythms. The temporal increment is a functional unit for use in the geophysical classification of the chronometric reliability of organisms. Rosenberg also believes it to be a functional unit enabling the comparison of growth structures between species for evolutionary purposes. Comparison of increments with similar periods and use of the environmental and endogenous information contained in the increment enables discussion of the process of selection of the increment and consequently of the growth rhythm.

Geophysical considerations

Creer is the first to correlate palaeontological growth rhythm data with changes in the Earth's geomagnetic polarity bias and reversal frequency, a correspondence which may help differentiate the relative importance of tidal and non-tidal causes of changes in the Earth's rotation. This is all the more important considering Hipkin's belief that, on tidal considerations alone, it is impossible to predict the past rate of the Earth's rotation. Hipkin's description of the many kinds of tides may also help biologists to recognize different kinds of tidal cycles registered in accretionary skeletons.

Jacobs and Aldridge explore viscous core–mantle coupling as a cause of the Earth's deceleration but there is at present no known means of determining its importance relative to other possible causes, such as tidal friction and magnetic or topographic coupling on the basis of growth rhythms.

Changes in the gravitational and Hubble's constants have also been considered in association with changes in the Earth's rotation. Wesson has explored the possible changes in G and g, and concludes the Earth is undergoing a secular expansion, with G constant. But there may be a cosmoslogical effect in operation with tidal friction to slow the Earth. The botanists' analysis of the effect of gravity on the growth of wood (see Creber) may eventually provide a means of testing Wesson's ideas. Weinstein and Keeney's theoretical examination of Hubble's constant discusses possible variation of physical constants and expansion of the universe in relation to palaeontological data and history of the Earth–Moon system.

Proverbio and Poma compared modern determinations of astronomical time measured simultaneously on different continents and find discrepancies which they believe are caused by different plates of the Earth's crust moving at different rates. They also believe that the movements correlate with variations in the rate of the Earth's rotation and suggest they are due to variations in atmospheric circulation as well as tidal forces. Tarling considers the inverse, concluding that the changing distribution of the continents has had little effect on the Earth's rotation except as the distribution might affect depth of the oceans, hence the value of tidal friction.

Gribbin explores the correlation between variations in solar radiation and variations in the Earth's rotation. Solar radiation varies with sunspot cycles and tides induced in the Sun by movement of the planets Jupiter, Mercury, Venus and Earth, so it may be necessary to consider dynamics of a large part of the solar system to explain fully changes in the Earth's angular momentum and consequent changes in biological growth periodicities. O'Hora and Morrison have summarized the astronomical and atomic time-scales. O'Hora discusses recent changes in the Earth's rotation which occur with regular periods, and maintains that the causes of these changes cannot be simplistically determined from a comparison of TAI and UT. Both Morrison and O'Hora trace the history of increasingly accurate measurements of variations in the Earth's rotation. They indicate that increasingly accurate measurements will help to separate the tidal from the non-tidal causes of the changing rotation rate.

Muller and Stephenson's analysis of ancient eclipse observations indicates a deceleration of the Earth about $2\cdot50 \pm 0\cdot3$ millisec/cy within the past 3000 years. Their conclusions differ from previous calculations in that they indicate a constant deceleration. Muller's novel analysis of Stephenson's scholarly research by the technique of linear inequalities is aimed towards independent determination of $\dot{n}$ and $\dot{e}$. Their work is important in that it prepares the field for separate observation of tidal and non-tidal effects of deceleration within the historical period. These results can then be extrapolated back to the distant past with increasing confidence, and compared with palaeontological data and considerations of tidal friction and other geophysical parameters to determine more precisely the history of the Earth–Moon system.

AUTHOR INDEX

A contributor's references to his own publications are not repeated here.

Accad, Y.   324, 325, 454
Adam, N. V.   313
Adelsberger, U.   434
Akimoto, S.   306
Allan, J. R.   224
Anderson, C. N.   400, 402
Anderson, D. L.   322
Aschoff, J.   152, 262
Ash, M. B.   354, 358, 363, 370
Atuña, M. M.   483
Audus, L. J.   85

Backus, G. E.   343
Bailey, I. W.   83
Ball, R. H.   341, 342, 343, 344
Bambach, R. K.   185
Barber, N. F.   202, 225
Barghorn, E.   44
Barker, R. M.   xiii, xiv, 10, 17, 150, 158, 163, 181, 194, 223, 257, 274, 276, 277, 300, 301, 302, 380
Barnes, D. J.   xiv, 136, 137, 270, 271, 272
Barnes, G. E.   151
Barnothy, J. M.   382
Bauer, L. A.   287
Bauld, J.   49, 257, 273
Bender, D. L.   448
Benkova, N. P.   313
Berkner, L. V.   404
Berry, W. B. N.   xiii, 150, 194, 276, 277, 300, 301, 302, 380
Bertotti, B.   356, 357
Bevelander, G.   106, 178
Binford, L. R.   244
Birch, F.   309, 346, 361
Blackman, R. B.   202, 211
Bogdanov, K. T.   324, 331
Bondi, H.   347, 348, 349
Boquet, M.   485
Bostrom, R. C.   386
Bott, M. H. P.   35

Boucot, A. J.   22
Bouquet, M.   485
Bowden, P.   403
Boyden, C. R.   185, 188, 189
Bracy, D. R.   402
Brans, C.   354, 357, 358, 359, 363, 364, 371
Brill, D.   356
Brinkman, J. A.   476
Brock, T. D.   49, 257, 273
Brooks, E. W.   483
Brosche, P.   323, 400
Brouwer, D.   288, 450, 500, 501, 502, 508
Brown, F. A.   155, 159, 160, 267
Bryan, G. W.   223
Bucha, V.   341
Bullard, E. C.   344, 405
Bullen, K. E.   305, 310
Burek, P. J.   294
Butler, R.   371, 372

Campbell, D. A.   13, 14
Cartwright, D. E.   274
Cazenave, A.   436, 454
Celoria, G.   474, 476, 477
Chabot, J. B.   483
Challinor, R. A.   414, 415, 436, 440, 454
Chaloner, W. G.   79
Chapman, S.   325
Cherevko, T. N.   313
Clark, G. R.   xiv, 8, 120, 121, 163, 195, 196, 254, 259, 263, 267, 277, 278
Clavius, C.   487, 488
Cloud, P., Jr.   43, 44, 50
Coe, W.   169
Cohen, C. J.   450, 521
Cooley, J. W.   202
Cooper, C.   287
Counselman, C. C.   382
Cox, A.   294, 298, 313, 341
Craig, G. Y.   10, 12

539

Crain, I. K.   313
Crain, P. L.   313
Creer, K. M.   399
Crenshaw, M. A.   181
Curott, D. R.   460, 493, 500, 520, 526, 527, 529

Dahl, A. L.   91, 94
Danjon, A.   414, 415
Darwin, G.   320, 331, 332, 333, 399
Davenport, C. B.   163, 263
Davidson, E.   81
Defant, A.   323
Degens, E. T.   264
Delaunay, C.   399
Dicke, R. H.   354, 356, 357, 358, 359, 361, 362, 363, 367, 370, 371, 372, 377, 381, 409, 453, 526, 528, 529
Digby, J.   81
Digby, P. S. B.   178, 181
Dirac, P. A. M.   354, 356, 358, 377, 381
Dixon, D. S.   94
Dodd, J. R.   223
Doell, R. R.   298
Dolman, J. W.   274, 277, 278, 280
Dooley, J. C.   361
Dubberstein, W. H.   482
Dubs, H. H.   477, 478, 479
Duell, B.   419
Duell, G.   419
Dugal, L. P.   151, 265

Egyed, L.   399
Einstein, A.   357
Eisma, D.   185
Elsasser, W. M.   381
Essen, L.   431, 432
Eugster, H.   66
Evans, J. W.   xiv, 98, 150, 178, 186, 196, 197, 245, 269, 322
Ewart, A. J.   85

Farrell, W. E.   323
Farrow, G. E.   xiv, 12, 13, 14, 17, 98, 150, 163, 177, 181, 182, 184, 185, 192, 195, 196, 198, 263, 269, 270, 277, 278
Finch, H. F.   434
Finzi, A.   358
Fitch, J. E.   169
Fitzgerald, C. P.   477
Fleming, N. C.   401
Florez, H.   486

Fotheringham, J. K.   460, 474, 475, 476, 481, 520, 521
Freedman, C.   344, 405
Fujisawa, H.   306

Gans, R. F.   347
Garrett, C. J. R.   322
Gebelein, C. D.   59, 68, 272
Gellman, H.   344, 405
Gerstenkorn, H.   331
Gignoux, M.   37
Ginsburg, R. N.   61
Ginzel, F. K.   460, 476, 477, 483
Goldich, S.   44
Goldreich, P.   310, 367, 381, 408
Goreau, T. F.   97, 270
Govett, G. J.   66
Groves, G. W.   323, 324
Guinot, B.   362, 455
Guriaeb, J. E.   483

Haddon, R. A. W.   306, 310
Hall, A.   163, 192
Hall, C. A.,   245, 260
Hall, R. G.   432
Hallam, A.   10, 12, 185, 402
Halley, E.   285, 288, 460, 465, 466, 488
Hanney, C. E. A.   81
Hansen, W.   324
Hare, P. E.   264
Harland, W. B.   298
Harrigan, J. F.   144
Hartig, R.   83
Hartmann, W. D.   97
Harwit, M.   358
Hastie, J.   297
Haurwitz, B.   325
Hays, J. D.   304, 401
Heiskanen, W.   323, 400
Helmberger, D. V.   404
Hendershott, M.   321, 324, 528
Hendrix, W. O.   287
Hess, H. H.   403
Hicken, A.   297
Hide, R.   294, 298, 338, 339
Higgins, G.   346
Higham, C. F. W.   244
Hinteregger, H. F.   448
Hipkin, R. G.   xiv, xv, 28, 254, 270, 271, 272, 276, 277, 302, 360
Hoffman, P. F.   59, 68, 235
Holder-Egger, O.   485

Author Index 541

Horodyski, R. S. 58
House, M. R. xiv, 12, 17, 150, 163, 177, 181, 184, 195, 262, 263, 270, 277, 278
Howse, D. 434
Hoyle, F. 354, 358, 359, 363, 371, 372, 453

Ingalls, R. E. 354, 358, 363, 370
Irving, E. 293, 294, 297

James, H. 49
Jao Tsung-i 489
Jeffreys, H. 326, 329, 360, 365, 366, 367, 370, 398, 400
Johnson, A. H. 294
Johnson, J. G. 22
Johnson, L. R. 294
Jordan, P. 354, 357, 358, 359
Jungels, P. 528

Kahle, A. B. 341, 342, 343, 344, 454
Kant, I. 399
Kapp, R. O. 354
Kaula, W. 328
Keeney, J. 360, 361, 363, 371, 372
Kennedy, G. C. 346
Kennedy, W. J. 163, 192
Kennedy, W. Q. 403
Khramov, A. N. 294, 295, 297, 313
Kidwai, P. 81
King, J. W. 417, 419, 422, 436, 440
Kitazawa, K. 341
Klein, O. 356
Knopoff, L. 386
Koike, H. 244, 246, 263
Komissarova, R. A. 295, 297
Kozai, Y. 328
Kreger, D. 184
Krempf, A. 270
Krishnaswami, S. 136
Kronberg, P. 371, 372
Krotkov, R. 356
Kulp, J. 37
Kummel, B. 22
Kuznetsov, M. V. 324, 331, 400, 454
Kuznetzova, L. V. 331, 400, 454
Kvenfolden, K. A. 58

Lacross, R. T. 219
Lagus, P. L. 322, 400
Lamar, D. L. 9, 51, 53
Lamb, H. H. 417, 423
Lambeck, K. 436, 454

Lange, I. 287
Laporte, L. 287
Larson, P. R. 81
Larson, R. 294
Law, L. K. 297
Lecomte, L. 478
Leeds, A. 386
Legge, J. 489, 490
Lemessurier, xiv, 120
Levington, J. S. 185
Lilley, F. E. M., 298, 313
Lindzen, R. S. 325
Liu Chao-yang, 477
Ljustih, J. N. 386
Lloyd, R. M. 61
Logan, B. W. 59, 60
Lyttleton, R. A. 347, 348, 349

Ma, T. Y. H. 270
McBurney, C. B. M. 244
MacClintock, C. xiii, xiv, 17, 22, 44, 57, 91, 120, 121, 131, 150, 152, 153, 156, 157, 158, 163, 171, 178, 181, 186, 192, 194, 196, 265, 298, 338, 360, 380
MacDonald, G. 53, 286, 310, 323, 324, 326, 333, 362, 364, 381, 385, 398, 400, 404, 405, 407, 408, 442, 454, 455, 521, 525, 526, 528
MacDonald, N. J. 419
MacDougal, J. 371, 372
McGann-Lamar, J. 53
McGregor, A. 57
McGugan, A. 300, 301
MacIntyre, I. G. 136
McKenzie, D. P. 362, 364, 365, 367, 370, 408
Magarik, V. A. 324, 331
Malin, S. R. C. 345
Mansinha, L. 406, 442
Maragos, J. E. 144
Marceau, F. 181
Markowitz, W. 432, 448, 455
Marshall, L. C. 404
Martin, C. F. 327, 450
Mason-Jones, A. J. 85
Mazzullo, S. J. xiii, 28, 194, 196, 202, 278, 300, 302, 380
Melchior, P. 323
Menke, T. 474, 525
Merifield, P. M. 9, 51, 53
Migne, J. P. 487
Miller, G. R. 299, 302

Milne, E. A.   353, 354
Mintz, Y.   386
Moberly, R.   224
Monty, C. L. V.   60, 61, 67, 68
Moore, W. S.   136
Moores, E. M.   303, 304
Morrison, L. V.   333, 363, 364, 370, 382, 436, 449, 460, 521
Morton, B. S.   181, 267
Mueller, I. I.   408
Munk, W. H.   274, 286, 310, 322, 323, 324, 326, 333, 381, 385, 398, 400, 404, 405, 407, 408, 442, 454, 455, 521, 525, 526, 528
Muratori, L. A.   476

Nairn, A. E. M.   294
Nakahara, H.   106, 178
Narlikar, J. V.   354, 358, 359, 363, 371, 372
Nečesaný, V.   83
Needham, J.   478
Neugebauer, P. V.   460
Neville, A. C.   8, 262, 263
Newcomb, S.   338
Newell, E. R.   185
Newton, R. R.   302, 325, 329, 331, 361, 436, 452, 460, 465, 467, 468, 470, 474, 475, 477, 484, 496, 500, 512, 514, 517, 518, 519, 520, 521, 522, 525, 526, 527, 528, 529
Niebuhr, B. C.   483
Nixon, J.   344, 405
Njoku, E.   81

Oehler, D. E.   58
Oesterwinter, C.   450, 521
O'Hora, N. P. J.   414, 416, 455
Olsen, R. M.   419
Onaka, F.   85
Oppolzer, T. R., von   460
Orton, J. H.   10, 12, 13, 184, 185, 263
Ozawa, I.   323

Palincsar, J. S.   267
Palmer, A. R.   22
Palmer, J. D.   149, 150, 267
Pannella, G.   xiii, xiv, 13, 17, 22, 44, 48, 49, 52, 57, 91, 120, 121, 131, 135, 136, 150, 152, 153, 156, 157, 158, 160, 163, 171, 178, 181, 186, 192, 194, 196, 198, 199, 265, 270, 278, 298, 299, 300, 314, 338, 380, 301, 302, 360

Pariiskii, N. N.,   331, 400, 454
Parker, R. A.   482
Parker, R. H.   264
Parry, J. V. L.   431, 432
Parry, L. G.   294
Pavlov, N.   386, 391, 393
Pechersky, D. M.   294
Pekeris, C. L.   324, 325, 454
Penny, C. J. A.   416
Perfect, D. S.   430
Pertz, G. H.   474, 476, 484, 485
Petersen, G.   178, 263
Peterson, D. N.   294
Pettingill, G. H.   354, 358, 363, 370
Pitman, W. C.   294, 304, 401
Pratt, D. M.   13, 14, 163
Prokhovnik, S. J.   354

Rao, K. P.   267
Redfield, A. C.   529
Reiter, R.   419
Rhoads, D. C.   xiv, 13, 153, 159, 163, 186, 270, 278
Rich, W. H.   185
Richardson, W. N.   94
Rishbeth, H.   405
Robards, A. W.   81, 82, 85
Roberts, D. G.   401
Roberts, W. O.   419
Rochester, M. G.   287, 338, 360, 381, 399, 400, 431, 454, 455
Rodianov, V. P.   295, 297
Rona, P. A.   401
Rosenberg, G. D.   xiv, 13
Ross, C. A.   22
Rowton, M. B.   476
Rubin, M.   529
Runcorn, S. K.   xiii, xv, 9, 57, 135, 301, 302, 338, 359, 377, 378, 381, 401, 406, 453

Sachs, A. J.   479, 481, 482
Sakuma, A.   328
Salanki, J.   181, 182
Sandquist, A.   371, 372
Sarton, G.   484
Saunders, I.   344
Sawyer, J. F. A.   523, 524, 481
Schatzman, E.   415
Scheibe,   434
Schmalz, R. F.   240
Schopf, J. M.   81

Author Index

Schopf, J. W. 58
Schopf, T. J. M. 224
Schroeter, J. F. 460, 487
Schwarz, C. R. 408
Sciama, D. W. 357
Sclater, J. G. 408
Scrutton, C. T. xiii, xiv, xv, 28, 31, 35, 44, 94, 135, 196, 199, 254, 289, 302, 338
Semikhatov, M. 44
Shapiro, I. I. 354, 358, 363, 370, 382
Shinn, E. 61
Sidorenkov, N. S. 454
Simpson, J. F. 313
Sims, P. 49
Singleton, R. C. 202
Sitter, W. de 327, 460, 500, 520, 521
Smith, A. G. 298, 402
Smith, D. E. 448
Smith, H. M. 432, 434
Smith, S. V. 136
Smith, S. W. 528
Smith, W. B. 354, 358, 363, 370
Smylie, D. E. 406, 442
Spencer, D. W. 264
Spencer Jones, H. 286, 287, 288, 326, 399, 442, 449, 460, 500, 502, 511, 512, 521
Sreeramulut, 94
Stacey, F. 309
Stanley, S. 128
Steiner, J. 358
Stephen, A. C. 10, 13, 184, 185, 186
Stephenson, F. R. 452
Stewartson, J. 346
Stoyko, N. 386, 434, 440
Struever, S. 244
Sündermann, J. 323, 400
Swallow, J. 240, 400
Swan, E. F. 13
Swanson, F. J. 240

Tait, P. 399, 404
Talent, J. A. 22
Taylor, G. I. 400
Taylor, J. D. 96, 163, 192
Thomas, D. B. 430
Thompson, M. N. xiii, 17, 22, 44, 57, 91, 150, 158, 171, 194, 298, 338, 360, 380
Thomson, W. 399, 404
Thorson, G. 158
Tinsley, B. M. 382

Toomre, A. 310, 367, 408
Tucker, R. H. 434
Tukey, T. W. 202, 211
Tung Tso-pin 477
Tyler, S. 44

Ulrych, T. J. 219
Umamaheswarara, M. 91,
Urey, H. C. 288, 327, 328, 338, 381, 406

Valentine, J. W. 303, 304
Van der Waerden 460, 499, 501, 502, 520
Van Flandern, T. C. 327, 449, 460, 521
Vaughan, T. W. 29, 38
Vestine, E. H. 287, 341, 342, 343, 344, 405, 454
Vogt, P. R. 402
Vyssotsky, A. N. 484

Waley, A. 493
Walter, M. R. 49, 67, 257, 273
Wareing, P. F. 81
Wedemeyer, E. H. 346
Weide, M. L. 244
Weinstein, D. H. 360, 361, 363, 371, 372
Wells, J. W. xiii–xiv, 9–10, 27–28, 29, 31, 35–36, 38, 44, 94, 103, 120, 135, 196, 253, 254, 270, 279, 289, 300, 301, 302, 338
Wershing, H. F. 83
Wesson, P. 377
Weymouth, F. W. 10, 163, 169, 185
Wheeler, J. A. 356
White, D. J. B. 82
Whitfield, R. P. 27, 29, 38
Whittington, H. B. 22
Wilbur, K. M. 178, 181
Wilcock, B. 298
Will, C. M. 358
Winstanley, D. 420, 423
Wise, D. U. 401
Wood, K. D. 422, 424
Wunsch, C. 321
Wylie, A. 479, 493

Yukutake, T. 326, 340, 341, 529

Zahel, W. 324, 325

TAXONOMIC INDEX

(Bold numbers denote Figures)

Acropora palmata 38
Actinostroma **96**
Agathiceras suessi 99
Anomia 15
Archaeolithothamnium 94, **95**
Arctocephalus fosteri 249
Argopecten circularis 110, 111, 112
Argopecten gibbus 110
Argopecten irradians **107**, 109
Astartella concentrica **20**
Ateyapecten 98
Aulophyllum fungites 39

Bangia 94

Callista chione 164, 165–169, 170, 172
Callocardia morrhuana 13
Cardium edule, see also *Cerastoderma edule* 13, 192–194, 196, 206–218, 223, 226–234
Cardium lamarcki 178
Cerastoderma edule, see also *Cardium edule* 98, 177–186, 188, 189, **256**, 267, 268, 269, 270
Cerastoderma glaucum 178
Chaetetes **97**
Chione californiensis 11, 18
Chione cancellata **16**, 18
Chione stutchburyi 244–246
Chione succincta 18
Chione undatella 11–13, **16**, 18, 223, 227
Chlamys hastata hastata 110
Chlamys hastata herica 110
Clinocardium nuttalli 120, 121–128, 132–133, 178, 245, 267, **268**, 269
Codakia orbicularis 18
Collenia **93**
Conchocelis 94
Conocardium herculeum **275**, 276, 280
Conocardium 20, 275
Conophyton **91**
Crassatella **275**

Crassatella lincolnensis 18
Crassatella vadosus 19
Cyathaxonia cornu 39

Dibunophyllum bipartitum 39
Dictyota dichotoma 91, 94
Dosinia 15

Echinoconchus 94
Endophyllum archiaii 271
Eudyptula m. minor 247
Evechinus chloroticus 247

Glycymeris lacertosa 19
Gruneris biwabika 44

Heliophyllum halli 28
Hinnites multirugosus 110
Hydrocorallia **96**

Idonearca vulgaris 19
Ischyrospongia **96**

Jordanidia solandri 247

Kellia suborbicularis 11

Lepidodendron 85
Lima gigantea 19
Lingula 94
Lithostrotion junceum **30**, 39
Lithostrotion martini **30**, 31, **32**, **33**, 35, 37, 38–39
Lithostrotion minus **30**, 39
Littorina 99
Lonsdaleia 29
Lonsdaleia floriformis 39
Lucina subundata 19
Lyriopecten 98

Macrocallista 15
Macrocallista hornii 18

Megateuthis 223, 235, 236–237
Mercenaria campichiensis **255**
Mercenaria mercenaria 11, 13, 15, **16**, 120, 121, 151–160, 178, 181, 186, **258**, 265, **266**, 267, **268**, 269, 273–274
Meretrix 15
Meretrix lusoria 246
Meretrix splendida 19
Merlia 99
Montastrea annularis 272
Mya arenaria 13
Myalina subquadrata 20
Mytilus 15
Mytilus californianus 264, 267
Mytilus edulis 181, 267

Nucula 15
Nucula proxima 186

Orionastraea 29
Orionastraea phillipsi 39
Ostrea 15
Ostrea edulis 181
Ostrea virginica 267

Pachyptila d. depolata 247
Pachyptila turtur 247
Pecten 15, 104
Pecten diegensis 104, **105**, 106, 109, 111, 112, 113, **115**, 121, 278

Pecten jacobaeus 182
Pecten maximus 98
Pecten vogdesi 106, **107**, 109, 111, 112
Pelecanoides urinatrix chathamensis 247
Penitella penita 120, 121, 128–133, 186
Pinus strobus 75, **77**
Populus alba 79, **80**
Porites lobata 136–145
Prasopora **98**
Protothaca staminea 11, 13, 15, 128, 178
Puffinus g. gavia 247
Pustula 94

Scleractinia 96
Septocardia 19
Siliqua patula 185
Spisula 223, 227, 231, 232, **233**

Tadaorna variegata 247
Tellina tenuis 10, 186
Thracia 96
Thyrsites atun 247
Thysanophyllum 40
Tivela stultorum 10, 15, 164, 169, 171, 172
Tridacna squamosa 158, **259**, **268**, 269, 270

Venus 15

Xenoxylon latiporosum **76**

SUBJECT INDEX

(Bold numbers denote Figures or Tables)

Absolute time, from fossils xiii, 27
Abū'Abdallāh Muhammad ibn Mu'ādh 484
Abu Dhabi (stromatolites) 60
Activity charts 153–156
Adductor Muscles, see Gaping rhythms
Adriatic Sea (bivalves) 164, 168, 169, **170**
Africa (solar activity and weather) 420
Agathocles 476, 522
Alaska (bivalves) 120
Algae 89, 91, 94
  (in stromatolites), see also Palisade structures 43, 44, 45, 49–50, 54, 57–59, 89–94, 100
  (symbiotic zooxanthellae) 135, 144, 158
Aliasing 194, 197, 208–209
Alizarin Red 137–143
Al Muqtabis 483
Alps (solar activity and atmosphere) 419
Amino acids 264
Angiosperms 79, **80**, 81–83, 85
Angle of deviation, trees 85
Angular momentum, see under factors affecting momentum such as Coupling or Moment of inertia, or Rotation rate of Earth
  conservation of xiv–xv, 50–54, **51**, 289, 300, 305, 326, 327, **330**, **332**, 341, **379**, 381, 440, 453, 498, 526
Annales Stadenses 486
Annales Toledanos Segundos 486
An-yang 477, 489
Archaeology and growth of bivalves 243–252
Archilochus 523
Arithmetic mean spectra, coherence of 211–213, 216
Artificial growth increments (trees) 81
Ashfell Sandstone 31, **39**, **40**
Asymmetrical increments, definition **258**, 259, **268**, 269, **275**, 276

Astrolabe 387
Astrology, Chinese, see Eclipse, solar, Chinese
Astronomical date vs. historical date 508
Atmosphere, altitude of pressure surfaces 400
  circulation of 286, 325, 337, 339, 386, 387, 398–399, 419–422, 454
  evolution of 404
  pressure gradient force in 339
Atmospheric equator 419
Australia (stromatolites) 59, 60, 68
Austro-Hungarian Empire 474
Auxin Theory 79–85
Axis of Earth's rotation, see also Chandler wobble, Markowitz wobble 327, 337, 338, 339, 345, 360, 362, 408
  inclination of, and tree growth 79

Bacteria (in stromatolites) 43
Bahamas (corals) 38
  (stromatolites) 67
Baily's Beads 464, 516
Baltic (tides) 323
Banded iron formations (stromatolites) 44–50, 66–67
Bankhouses Limestone **39**
Basalts and damping coefficient 404
Bay of Bengal (tides) 400
Belemnites 223, 235
Bermuda (stromatolites) 68
Biocheck 4, **166**, 168–169, 170–173, 260
Biological clock (test for) 3, 79, 112–113, 149–160, 262, 263, 267
  evolution of 8
Bivalves xiv, 9–22, 96, 98, 103–133, 149–189, 192–194, 196, 206–218, 223, 226–234, 244–246, 254–260, 263–270, 273–280
Biwabik 44, 50, 56
Boetsap **59**, 60, 65–66
Botany Limestone **39**

547

Boundary Problem 4, **5**, 6
Brachiopods 28, 94
Bryozoa 97
Bulawayan (stromatolites) 52, 58
Bureau International de l'Heure (BIH) 386, 416, 430, 448
Burning of the books under Shih Huang-ti 477

Caesium Beam Resonator, *see also* Time, atomic 431
Calcium physiology, algae 49–50
 bivalves 121, 125, 131, 151, 158, 177–186, 223–234, 240, 263–270
 corals 27, 135–144, 270–272
California (bivalves) 11–14, 120–121, 169, 171
Cambium (and tree growth) 75, 85
Canada (bivalves) 121, 125
 (stromatolites) 223, 235, 273, 380, 381
Canonici Wissegradensis Contin Cosmae 485
Carbon (in stromatolites) 60, **62**, 65, 66
Carboniferous (bivalves) 20
 (corals), *see also* Viséan, Namurian 29–40
 (lycopods) 85
Catskill Delta 98
Cell size (trees) 75, **77**, **78**, 79
Cephalopods, *see also* Belemnites, Ammonoids 99, 223, 235, **260**, 276
Chandler wobble 290, 360, 414, 430, 442
Ch'ang-ch'un chên-jên hsi-yu chi, Journey of the adept Ch'ang-ch'un to the West 493
Ch'ang-ch'un, The Taoist Master 493
Chi 466, 467, 489
Ch'ien Han Shu History of the Former Han Dynasty 490
Chih-tê Reign 492
China, Dynasties Chin (Tsin) 471, 477
 Chou 477, 490
 Han 471, 477, 478, 479, 490, 491
 Manchu (Ch'ing) 478
 Shang 477, 489
 Southern 492
 Sung 471
 T'ang 471, 479, 492
Ching, King of Kuang-ling 491
Chin Shu History of Chin (Tsin Dynasty) 477
Christmas Islands (corals) 138, 143

Chronicon Cerratensis 485
Chronometric species, definition 8
Ch'un-ch'iu, Spring and Autumn Annals 489, 490
Circumpolar Westerlies 420
Clavius 487, 488
Cleomedes 475, 522
Climate, *see also* Atmosphere *and* Solar Activity 381, 413–424, 436, 440
Compression wood **82**, 83–85
Conchiolin, *see also* Calcium physiology, bivalves 125, 130, 131, 158, 177–178, **179**, 263–270
Connecticut (bivalves) 269
Constants, physical 377–382, 353–359
 as dimensional units 358
 as dimensionless units 356–358
 electron, mass 359
 fine structure 356, 359
 Hubble's 354, 356, 371, 377–382
 number of particles in Universe 356
 Planck's 359
 positron, mass 359
 ratio of atomic period to age of Universe 356
 ratio of gravitational to electrical force 356
 strong and weak coupling 356, 359
Continental drift, *see* Plate tectonics
Convection within Earth, *see also* Core, Mantle, Coupling and Fluids 288, 294, 298, 310, 362, 365, 367, 406–409
Corals xiii–xiv, 27–40, 94, 135–147, 254, 270–272
Core, *see also* Coupling, core-mantle
 adiabatic gradient 346
 conductivity in 341, 343
 density gradient in 328, 346
 as dynamo 298, 339, 345
 expansion of 288, 328, 338, 362, 406–407, 453
 fluid velocity near surface 342
 as fluid, westward drift 287, 344, 345
 free stream velocity in 343
 melting point gradient 346
 ohmic dissipation in 346
 Precambrian 51
 rotation of 287, 345, 405
 steady state 346–349
 stratification 339, 345
 turbulence in 346
 viscosity of 346

Subject Index

Coriolis force (and Ekman number) 346
Corona of Sun, appearance during totality 462, 464, 516
  observation by Clavius 487
  observation by Plutarch 462
  observation by Leo Deaconus 483
Corpus Scriptorum Historiae Byzantinae 483
Cosmas of Prague 485
Cosmic rays 415, 419
Cosmological drag force (H_p) 377
Cosmology, theories 353–372, 377–383
Coupling, core–mantle xv, 287–290, 326, 327, 406, 407
  conservative or dissipative 346
  electromagnetic 287, 338, 341, 362, 405, 406, 454
  topographic 294, 298, 338, 454
  viscous 338, 346–349, 454
Coupling, Earth–atmosphere 337, 362, 405, 406, 454
  Earth oceans 337, 362
Cretaceous (algae) 94
  bivalves 19
  sea level 401

Damping coefficient of mantle 404
De Facie 475
Degassing 401, 403, 404, 405, 409
Density distribution in Earth 305–309, 310, 328, 346, 407
Depth of oceans 22, 35, 37, 299, 302–304, 323–324, 336, 362, 400–404, 412, 528–529
Devonian (bivalves) 20, 98
  cephalopods 276
  corals xiii, 10, 27, 28, 29, 31, 271
  stromatoporoids 97
  trees 75, 81
Diagenesis 97
  belemnites 235
  bivalves 15, 17
  corals 29, 31
  stromatolites 48, 61, 65, 68
Dinantian 29
Doublet, definition 259
Duluth Gabbro 44

Earth, elasticity of, *see also* Earth, shape of, *and* Tides, solid Earth 53–55, 322–325, 362, 385, 386, 390, 398, 430, 455
  expansion of xv, 288–289, 304–305, 308, 328, 354, 359, 361, 362, 363, 371–372
  obliquity of ecliptic **330**, 331, **332**
  orbital eccentricity of 378
  revolution around Sun xiii, 31, 289, 291, 378, 379, 380
Earth, shape of, *see also under* Tides and equatorial tidal bulge, *and* equilibrium tide model, of solid earth, *and under* Earth, expansion of
  and equatorial bulge changes *see* Earth, shape of and J_2
  and J_2 harmonics 309–310, 328–329, 362, 363, 364–372
  axes of figure 408
  no strain conditions 367–368
Earth, spheroidal free oscillation 322
Earthquakes and rotation rate 365, 366, 372, 440
Eclipses, lunar 135 B.C. (of Hipparchus) 460
Eclipses, solar 288, 341, 361, 362, 450–453, 459–530
Eclipses solar (named by date and place or observer)
  1406 B.C. (Ugarit) 481
  1375 B.C. (Ugarit) 459, 480, 481, 509, 511, 513, 523
  1330 B.C. (An-yang) 489, 509, 513, 523
  1223 B.C. (Ugarit) 481
  1130 B.C. (of Joshua) 524
  1063 (Sic) B.C. (Babylon) 476
  763 B.C. (Eponyn Canon) 475
  709 B.C. (Chü-fu) 477, 488, 489, 490, 509, 513
  647 B.C. (Archilochus) 523
  601 B.C. (Ying) 488, 509, 513, 514
  549 B.C. (Chü-fu) 477, 488
  322 B.C. (Babylon) 470, 480, 481, 509, 511, 513, 528
  310 B.C. (of Agathocles) 476, 522
  198 B.C. (China) 488, 490, 509, 513
  188 B.C. (China) 478, 522
  188 B.C. (Rome) 476, 523
  181 B.C. (Ch'ang-an) 488, 509, 513, 528
  147 B.C. (China) 478, 488
  136 B.C. (Babylon) 480, 482, **498**, 509, 513

Eclipses, solar (*cont.*)
129 B.C. (Hipparchus) 450, 452, 475, 522
89 B.C. (China) 488
80 B.C. (China) 488
28 B.C. (China) 478, 488
2 B.C. (China) 488
29 A.D. (St. Luke) 523
65 A.D. (Kuang-ling) 488, 491, 509, 511, 513, 514
71 A.D. (Plutarch) 475, 522–523
120 A.D. (Lo-yang) 478, 488, 491, 509, 513–528
360 A.D. (China) 488
484 A.D. (Athens) 476, 523
516 A.D. (Nanking) 489, 492, 509, 513
522 A.D. (Nanking) 489, 492, 509, 511, 513
616 A.D. (China) 491
693 A.D. (Bagdad) 480, 482, 483, 509, 513
702 A.D. (China) 488
729 A.D. (China) 488
754 A.D. (China) 488
756 A.D. (China) 488, 492
761 A.D. (Ning-hsien) 488, 492
840 A.D. (Bergamo) 480
912 A.D. (Cordoba) 480, 483, 509, 513
968 A.D. (Constantinople) 480, 483, 509, 513, 520
1039 A.D. (Europe) **473**
1079 A.D. (Seville) 480, 484, 509, 513, 528
1093 A.D. (Europe) 474
1124 A.D. (Novgorod) 480, 484, 509, 513, 520
1133 A.D. (Europe) 474
1133 A.D. (Heilsbronn, Reichersberg) 480
1133 A.D. (Kerkrade) 481
1133 A.D. (Salzburg) 480, 484, 509, 513
1133 A.D. (Vysehrad) 480, 484, 485, 509
1147 A.D. (Europe) 474
1153 A.D. (Erfurt) 480, 485, 509, 513, 520
1176 A.D. (Antioch) 480
1178 A.D. (Vigeois) 480, 485, 509, 513
1187 A.D. (Europe) 474
1191 A.D. (Europe) 474, 481
1221 A.D. (Kerulen River) 489, 493, 509, 513
1239 A.D. (France, Italy, Spain) 472–474, 475, 480, 485, 486, 509, 513, 528
1241 A.D. (Stade) 474, 480, 481, 486, 509, 513, 520
1245 A.D. (Korea) 491
1263 A.D. (Europe) 474
1267 A.D. (Constantinople) 486, 487, 509, 513
1406 A.D. (Braunschweig) 480
1415 A.D. (Altaich and Prague) 480
1485 A.D. (Melk) 480
1507 A.D. (China) 491
1527 A.D. (China) 491
1544 A.D. (Altaich) 481
1560 A.D. (Coimbra) 480, 487, 509, 512, 513
1567 A.D. (Rome) 480, 487, 509, 512, 513, 528
1715 A.D. (England) 480, 488, 509, 512, 513, 528
1918 A.D. (Western US) 464
1925 A.D. (New York, 86th Str and Broadway) 465
1961 A.D. (Europe) 509
1973 A.D. (Kenya) 464
angle of track to equator, importance 475
annular 461, 467, 480, 502
Assyrian 475, 482, 483
atmospheric scattering of light during 462
Babylonian 470, 471, 476, 479–482, **498**, 509, 511, 513, 528
and Bailey's Beads 464, 516
centrality 461, 504
Chinese 466, 471, 477–479, 488–493, 509, 511, 513, 523
class A and B definition 469, 480
corona, appearance during totality 462, 464, 516
date of observation, importance of 468, 469, 474, 517
duration of, *see* Magnitude
and equations of condition 496, **501**, 504
frequency 462
Greek 471, 475–476, 522–523
identification game and 496–497
location of observation, importance of 468, 469, 470, 475, 517
magnitude (light curve) 462–469, 517

Subject Index

Eclipses, solar (*cont.*)
 magnitude of observation, importance of 468, 469, 475, 517
 medieval European 471–475, 476–477, 480, 483–488, 509, 513, 519, 520, 528
 Middle East 459, 470, 471, 475, 480–484, 486
 Moon's shadow, speed during 464
 observations, ancient methods 466
 observations, subjectivity and reliability of, 462–467, 517
 partial 460, 461, 481, 470
 partial vs. total 460–462
 and pollution of atmosphere 465
 population bias 472–475, 517
 and principle of rotational displacement xiv, 450–453, 497–499
 Roman 466, 471, 475–476
 shadow geometry 450–453, 460–462
 stars, appearance during 464, 465
 statistical evaluation of, by least squares 473, 495–496
 statistical evaluation of, by linear inequalities 496, 503–521
 totality, criteria of 464, 465, 466, 467, 468
 Ugaritic xiv, 459, 480, 481, 509, 511, 513, 523
Ectoprocta, *see* Bryozoa
Eddington Numbers, *see also* Constants, physical, 356
Ekman Number 346
Electromagnetism and bivalve growth 159, 162
Electron microprobe 223–240
Elias of Nisibis 482
Eocene (bivalves) 19
Eotovos experiment 359
Epochs, definition 494, 446
España Sagrada 486
Evolution, of biological clocks 8
 of photoperiodism in trees 81
 of sclerosponges 97
Ex Chronico Gaufredi Vosiensis 485
Extrapallial fluid, pH of 151, 158, 178, **180**, 181

Facies, *see* Growth, and substrate, and tides
Fanning Islands (corals) 138, 142
Fast Fourier Transform 202, 204–206

Faunal diversity and plate tectonics **303**, 304
Figure of Earth, *see* Earth, shape of
Fistuliporids, *see* Bryozoa
Flamsteed, First Astronomer Royal 434
Florida (bivalves) 269
 stromatolites 68
Fluids, axisymmetric oscillations of 339, 345–349
 inertial range 345
 steady-state considerations 346–349
 viscosity of, *see also* Coupling, *or* Core 346–349
Fourier analysis 45, 79, 202–220, 225, 227–235
Fraunhofer gratings 277
Free pendulum clocks 434
Fugacity, oxygen 49

Galapagos Islands (solar activity and climate) 420
Galaxy, rotation of Earth's rotation in 313
Gaping rhythms in bivalves 121, 125, 131, 132, 151–160, 181, 263–270
Gastropods 99
Gaufredus 485
Gauss coefficient **340**, 341, **342**
Gelatinous fibres 82–83
Geocentric dipole formula 287
Geological Society of London Scale (radiometric ages) 298
Geomagnetic centre and pole 310–313, 344–345
Geomagnetic field, secular variations **287**, 288, 294, 326, 339–345, 405, 407, 408, 454
Geomagnetic polarity bias 294–298, 339, 341, 529
 and core-mantle topography 294, 298
 Devonian interval 297
 Early Cenozoic interval 297
 Kiaman interval 294, 297
 Mesozoic 297
 reversal frequency 294–298, 313
 Silurian interval 298
 stability of, *see* Geomagnetic polarity bias, reversal frequency
 stochastic model of 313
 and temperature anomalies in core 298
Geomagnetism, and coupling, *see* Coupling, core–mantle

Geomagnetism and coupling (*cont.*)
   and Earth's dynamo 287, 293, 298, 339
   and gauss coefficient **340**, 341, **342**
   and Lorentz force 339
   and solar activity 419, 422
Geothermal flux 298, 304–309, 364, 365, 366, 367
Geotropism, *see* Gravity, biological influence *and* Growth and gravity
German Empire 474
Ghenghiz Khan 493
Gibberelic acid 81
Glaciation 302, 403
Gravimetry and tides 322, 325, 328
Graviperception of plants, site of 85
Gravity xv, 289, 319, 322, 323, 324, 325, 326, 328, 353–372, 375, 377–382, 430, 448, 449, 453
   biological influence 81–85, 159, 162, 355
   and expansion of Earth xv, 289, 328, 353–354, 359, 361, 363, 364–372
   G, limits of change 354, 359, 361, 363, 381, 382
   G vs g 355
   and lunar occultation of stars 382
   temporal vs spatial variability 354–356
   and temperature changes 361, 375, 381
Great Britain (bivalves) 178, 182–186, 192, 194, 196, 206, 211–216, 269, 278
   corals 29–40
Great Limestone 39
Great Scar Limestone 39
Greenland (solar activity and atmosphere) 419
Growing season, *see* Climate, *or* Solar activity
Growth, *see also* Taxonomic Index, *or* Subject Index under common group names (e.g. Bivalves, Corals)
   and ageing (bivalves) 164–173, 185, 197, 206
   ammonoid cephalopods 99
   and anaerobic environment (algae) 49, 94
   and biological clock (bivalves), experimental test for 3, 79, 112–113, 149–160, 262, 263, 267
   of bivalves, and archaeology 243–252
   boring rhythms (bivalves) 128–133
   boundary problems 4, **5**, 6

   and breeding rhythms 28, 38, 91, 94, 97, 98, 99, 142, 144
   and calcium physiology (algae) 49–50, (bivalves) 121, 125, 131, 151, 158, 177–186, 223–234, 240, 263–270 (corals) 27, 135–144, 270–272
   and carbonate saturation of seawater 240
   and CO_2 concentration of seawater 182, 240
   and conchiolin deposition, *see also* Growth and calcium physiology 125, 130, 131, 158, 177–178, **179**, 263–270
   in constant conditions (bivalves) 149–160
   continuity of, *see also* Growth, and random environmental disturbances 104, 106, 113–114, 178, 184, 209–211, 219–220, 262–263, 270, 278
   co-ordinate measurement (stromatolites) 44–45
   density increments in corals 136–145
   and depth, *see also* Growth, and tides 12–13, 15, 22, 50, 58–61, 65–68, 98, 157–160, 165–173
   and diagenesis 15, 17, 29, 31, 48, 61, 65, 68, 97, 235, 254, 271
   distinctness of increment boundaries 4–6, 109, 125, 128, 131, 134, 192, **193**, 195–196
   and dolomite layering (stromatolites) 60, 64, 65–70, 235
   and electromagnetism 159, 162
   endogenous vs exogenous 3, 79, 112–113, 149–160, 263, 262, 267
   and fault displacements 164
   and feeding, *see also* Growth and gaping 110, 144, 245, 265, 269
   and gaping rhythms 121, 125, 131, 132, 151–160, 181, 263–270
   and gravity (bivalves) 159, 162
   and gravity (plants) 81–85
   increment, (definition of) 3
   and iron in stromatolites 44, 49, 50, 66–67, 235, 238–240
   latitudinal variations 13, 14, 29, 38, 75, 79, 163–173, 182, 278
   and light 262, 273, 274, 280
   and light (algae) 91, 94
   and light (bivalves) 110–116, 149–160, 177–182, 192, 194, 196, 209, 218, 267, 269, 274

Subject Index

Growth, (cont.)
   and light (corals) 27, 28, 144, 270, 271
   and light (stromatolites) 44, 50, 68
   and light (trees) 81
   line 259, 265
   and longevity (bivalves) 185
   and magnesium concentration (stromatolites) 235, 238–240
   and molluscan provinces 164
   and oxygenation of ocean 110
   and oxygen fugacity 49
   and oxygen uptake 185
   period of the increment, definition 2
   periodicities, *see references to* Growth and environmental cycles such as tides, light, temperature
   and pH of extrapallial fluid 151, 158, 178–181, 265
   and pH of seawater 67, 240
   phase shifts, *see also* Growth and switch over 113, 121, **124**, 125
   and photosynthesis (algae) 50
   psuedo periodicities 2, 128, 143–144, 184
   and random environmental disturbances 12, 13, 14, 60, 68, 104, 106, 109, 184, 209, 211, 218, 274, 278
   rate of, definition 3
   and seasons, *see also* Growth, and temperature and light
   and seasons (algae) 89, 91
   and seasons (bivalves) 10, 12, 98, 125, 128, 163–172, 178, 182, 184–186, 206–207, 219, 232, 233, 244–246, 272, 274, 278
   and seasons (cephalopods) 276
   and seasons (corals) 27, 29, 138–144
   and seasons (stromatolites) 57, 61, 67, 68, 257
   and seaons (trees) 75, 79
   secretion, period of definition 2
   secretion, rate of definition 3
   series, definition 3
   silica concentration (stromatolites) 44, 49, 50, 66, 67, 235, 238–240
   and statistics of increment measurements (mode, mean, etc) 28, 45–49, 194–202
   and substrate (bivalves) 12, 14, 25, 121, 128, 130–133, 165, 169, **170**, 171–173
   and substrate (corals) 29, 31, 38, 40
   and substrate (stromatolites) 44, 49, 60–61, 65–68, 91
   and substrate (stromatoporoids) 96–97
   sulphur concentration (molluscs) 226–227, 230, **231**, 232–235
   suppressed (bivalves) 184
   and switch-over (interference patterns) 182–184, 192, 194, 218, **258**, **266**, **268**, 269
   and temperature, *see* Growth, and seasons
   and temperature (bivalves) 12, 13, 14, 110, 128, 168, 172, 173, 182, 184, 245, 269
   and temperature (corals) 27, 29, 38, 144
   and temperature (stromatolites) 44, 53
   terminology of 1–8, 259–262
   and tides (algae, other than stromatolites) 91, 94
   and tides (bivalves) 10, 11, 13, 14, 20–22, 98, 110, **111**, 120–133, 150, 152–160, 181, 182–186, 192–194, 196, 209, 210–220, **255**, **256**, 257, **259**, 262, 265–270, 273–276, 278
   and tides (corals) 10, 28, 35, 144, 270–272
   and tides (stromatolites) 44, 50, 56, 57–70, 89, 91, 94, **261**, 272–273
   and tides (stromatoporoids) 96–97
   of trees 75–85
Gulf of Alaska (solar activity and atmosphere) 419
Gunflint 273, 380, 381
Gymnosperms 75, **76**, **77**, **78**, 79, 81–83, **84**, 85

Haematite in stromatolites, *see* Growth and iron in stromatolites
Halley, Edmund, 285, 288, 460, 465, 466, 488
Harmonics, Fourier analysis 202–208
Hawaii (corals) 137
Hearne Formation 235
Heat flow, *see* Geothermal flux
Heat pulse due to lunar orbital change 53–54
Hennops river formation 60
Herbaceous plants (gravity and growth) 85
Hipparchus 450, 452, 460, 475, 522

Holy Roman Empire 474
Hormones in trees, see Indolyl acetic acid and Gibberelic acid
Hou Han Shu, History of the Later Han Dynasty 491, 492
Hsin T'ang Shu New Book of the T'ang Dynasty 492
Hsü Chih, The Astronomer Royal 477
Hubble's constant 354, 356, 377–382
Hunter-gatherers (New Zealand) 243–252
Huwarasmi 483
Hydrodynamics, see under Fluids, or Tides or Core
Hypsometric curve **401**

Ibn Hayyan 483
Identification game 496–497
Increment, definition of 3
  period of definition 2
  temporal, definition 6
Indolyl acetic acid 81, 83
Inertia, origin of 377
In Sphaeram Ioannis de Sacrobosco 487
Intertropical Convergence Zone (ITCZ) 419, 420
Ionosphere 406
Irish Sea, tidal friction in 400
Isotopes and climatology 403, 419, 424
Italy (bivalves) 164, **170**, 172

J_2, see under Earth, shape of
Jew Limestone 39
Jupiter and solar activity 422, 424
Jurassic (belemnites) 223
  bivalves 19
  gymnosperms 79, 85
Juvenile water, production of, see Degassing

Kaapvaal Craton 58
Kao-tzu, Empress of 490
Kelvin 366
Kepler's Laws of Motion 378, 289
Koestler's 'Ghost in the Machine' 150
Kuang-ling 491
Kymograph trace 181

Laser ranging 448
Latitude and growth, see Growth and latitude
Latitude and variation of lunar tidal torque, see Tides, latitudinal variations
Least squares filtering of eclipse data 473, 495–496
Length of day (l.o.d.), see also Rotation rate of Earth 285, 286, 290, 309, 428, 437, 442, 446, 450, **451**, 529
Length of daylight 87
Leo Deaconus 483
Leonis Deaconi Historicae 483
Le Pichon 385
Lignin 82–83
Linear inequalities 496, 503–521
Line Islands (corals) 139
Loran-C system of navigational radio transmission 432
Lorentz force 339
Love number 324, 331, 455
Lu, annals of 477, 488, 489, 490
Lunar acceleration, see Moon, acceleration
Lunar occultations of stars and planets 382, 449, 471, 477, 525
Lunar parallax, see Moon, distance from Earth
Lunar Sixes 471
Lycopods 85

Mach's Principle 377
Magnetic anomaly stripes on sea-floor 294
Man-days of Maori occupancy 244, 246–252
Mantle (bivalve), see also Calcium physiology 113–114, **115**, 121, 125, 151, 158, 263, **264**
Mantle (Earth's), see also Coupling, core–mantle 51, 287
  convection in 288, 294, 298, 310, 362, 365, 367, 406–409
  damping coefficient 404
  density zones **307**
  and geothermal flux 304–309
  lateral inhomogenities 310
  ohmic dissipation in 346
  olivine–spinel transition **308**
  phase changes, temperature-induced 305–309
  rotational pole of 345
  seismic zones **306**
  thermal considerations 294, 304–309

Subject Index

Mantle (Earth's) (*cont.*)
 viscosity, *see also* Mantle, and thermal considerations, *and* Coupling, core–mantle 338, 346–349, 362, 529
Maori Indians 243–252
Markowitz wobble 360
Mars, rotation rate 338
Mass, inertial (anisotropy of) 359
Maximum entropy power spectrum 219
Maxwell stresses 346
Mean Sun 494, 427
Melting of Earth (during Precambrian) 53, 398
Mercury 382, 422, 449, 464, 482
Meteorites, and mass change in Earth and Moon 405
Mexico (bivalves) 11, 12, 13, 169
Micrometer eyepiece, digitized **195**, 196
Middle East (solar activity and weather) 420
Ming-huang, Emperor of T'ang 492
Minimum deletion filtering; defined 496, 504
Mink formation 56, 273, 280
Minnesota (stromatolites) 44
Missing increments, *see* Growth, continuity of
Mississippian (bivalves) *see also* Carboniferous 276, 280
Modulation, in Fourier analysis 197, 208, 209–211
Moment of Inertia of Earth xv, 51, 53–54, 289, 294, 301–302, 304–315, 327–329, **330**, 337–338, 359–363, 378–381, 385, 386, 405–409, 442, 453, 454
 affects on rotation, distinguished from tidal friction 289, 300, 304, 309, 310, 313, 315, 327–328, 449, 497–499, 534, 525–530
 calculation from palaeontological data 289, 290, 301, **302**
 changes, *see under factors affecting moment of inertia, such as* Coupling, *or* Earth, expansion of, *or* Gravity
Monumenta Germaniae Historica 474, 476, 484, 485, 486
Moon, acceleration in longitude (ṅ) 286, 287, 288, 326–334, 448–450, 460, 497, 499–521, 529–530
 distance from Earth, *see also* Moon, recession rate *and* Moon, acceleration in longitude xv, 53, 460

eccentricity of orbit 50, 289, 320, 327, 329, 330, 332, 378, 381
inclination of orbit 50, 329, 330, 331, 332
longitude of ascending node of orbit 440
occultations of stars and planets by 382, 449, 471, 477, 525
origin of xv, 273
perigree, mean longitude of 442
precession of orbital plane 326–327
recession rate (from eclipse observations) 529
shadow, speed during solar eclipse 464

Namurian 29
Nan Shih, History of the Southern Dynasties 492
Neptune, and solar activity 422
Newtonian mechanics 286, 360, 448
Newton's Law of Gravitation 355
New York (corals) 38
New Zealand (bivalves and archaeology) 243
Nicephoras 487
Nicephori Gregorae Historiae Byzantinae 486
Nomarsky phase microscopy 131
Non-dipole field, *see* Geomagnetic field
Normal stars 482
Normal wood 82
North Sea, tidal friction in 400
Novgorodsky 484

Obliquity of ecliptic, 330–333
Oceans, inertial oscillations in 339
 depth of, and rotation rate 22, 35, 37, 299, 302–304, 323–324, 336, 400–404, 412, 528–529
 volume of water changes 401, 403
Oligocene (bivalves) 18
Oolites (in stromatolites) 44, 50, 56, 60
Ophiolites 407
Optical microdensitometer 193, 195
Oracle bones of Shang Dynasty 477, 489
Ordering, of increment types 257, 260, 262, 272–273
Ordovician (sclerosponges) 97
Oregon (bivalves) 120, 121, 269
Organic layer, definition 259
Organic line, definition 259
Orogenies 304

Ozone, as atmospheric energy source 325

Palaeocene (bivalves) 19
Palaeoclimatology 422, 424
Palaeogeography 22, **33**, **34**, 37, 38, 311–313, 324–325, 399–404
Palaeomagnetism, *see* Geomagnetism, Geomagnetic polarity bias
Palaeotemperatures from bivalves 244
Palisade structures in stromatolites 60, 61, **62**, **63**
Pangaea 304, 310–313, **402**
Pappus 475, 522
Paradox of growth rhythms 8
Patrologiae Graecae 487
Patterns, growth (definition) 4, 259
Periodograms 207
Permian (ammonoids) 99
Permian pole and tree growth 81
Pethai Group 223, 235
pH of extrapallial fluid 151, 158, 178, **180**, 181, 265
  of seawater 67, 240
Phaeophyta, *see* Algae
Photoelectric Transit Instrument (PTI) 387
Photographic Zenith Tube (PZT) 387, 430, 448
Photomicrography, time lapse 113–114, **115**
Photon frequency 377
Photoperiodism, *see also* Growth, and light 81
Photosynthesis (algae) 50
Pinches, T. G. 479, 481 482
Pith, excentric (gymnosperms) 83
Planetary alignments and solar activity 422–424
Plate tectonics 98, 288, 294, 304, 310–313, 364–367, 385–395, 401–404, 406–409
Pleistocene (bivalves) 18, **165**, 169, 172
  (corals) 142
Pliocene (bivalves) 18, **164**, **165**, 169, 172
Plutarch 462, 475, 522–523
Poincare equation 339
Pole, geomagnetic, *see* Geomagnetic field
  Permian, from trees 81
  rotational, measurements of motion *see also* Axes 430
Pollution, and bivalve growth 245

Polanyi, M. and the Laplacean delusion 277
Power, in Fourier analysis, def. 204
  leakage of in Fourier analysis 204, 205, 208
Precambrian 5
  periods of, from radiometric dating 288
  (stromatolites) 43–55, 57–70, 223, 257, 272, 280
Precessional torque 326–327
Presiding Chiefs of the Chou Dynasty 490
Proton, mass of, 357
Proudman–Taylor theorem 339, 347
Pseudoperiodicities 2, 128, 143–144, 184
Ptolemy 475, 522
Pumping rates in bivalves, *see also* Gaping 267

Quartz crystal clocks 431, 432

Radau equation 328
Reaction wood 81, **82**, 83–85
Rebound acceleration of bivalve calcification 266
Recombination (Fourier analysis) 216, **217**, 218
Recueil des Historiens des Gaules 485
Red tone 83
Reducing vs oxidizing environment (Precambrian) 49, 50
Relativity 355, 356, 372
Repeatability of chemical series 225–226
Rerum Italicarum Scriptores 476
Revolution of Earth around Sun xiii, 31, 289, 291, 378, 379, 380
Reynolds Number 345
Rheology of Earth 53, 54
Rhodophyte, *see* Algae
Ridges, oceanic 401, 403
Right ascension of the Sun (R.A.) 479, 492
Rodrigo de Cerrato 486
Rotation rate of Earth 20, **21**, 22, 35, **36**, 37, **51**, **52**, 298–302, 326, **330**, **332**, 333, 337, **340**, 341, **342**, **343**, **379**, 386, 428, 432–442, 445–455, 529–530
  causes of variability, distinguished by fossils 289
  Cenozoic **21**, 280, 300, **301**, 379
  Cenozoic anomaly 300

Subject Index

Rotation rate of Earth (*cont*.)
 history of measuring irregularities 428, 430, 432, 434
 maximum accuracy from fossils 219
 Mesozoic 21, 22, 280, 299–300, **379**
  anomaly 299–300, **301**
 Palaeozoic xiii, 10, **21**, 22, 35, **36**, 37, **379**
  anomaly 299–301
 Precambrian **51, 52**, 273, 280, **379**, 381
 (Earth's origin) 367, 370
 seasonal variations 337, 386, 414, 428, 434, 448
Rotholz 83
Royal Greenwich Observatory 424, 430

Sahara (and climate) 420
Saint Luke 523
Salt balance in oceans 403
Samuel of Marseille 484
Satellite tracking (and tides) 325, 328, 331
Saturn and solar activity 422
Scanning electron microscope (stereoscan) 113, 114, 121, 178, 257, 276
Scriptores Rerum Germanicarum 485
Sea-floor spreading 294, 304, 403
Sea-level changes 22, 35, 37, 299, 302, 303–304, 362, 401, 528, 529
Seasonality, relative 246
Seasons, timing of 219
Seismicity 365, 366, 372, 440, 442, **437**
Seneca 466
Series, definition 3
Shang-yuan reign 492
Shell structure (bivalves), *see also* Conchiolin 96, 178, **179**, 192, **256, 264**, 269, 275
Shih Huang-ti-Emperor of China 477
Silica (in stromatolites) 44, 49, 50, 66, 67, 223, 235, 238–240
Silurian (corals) 28
Simple increment, definition 259
Sine rule 85
Sixty-day cycle 489
Skye (belemnites) 223
Solar activity and atmosphere circulation 325, 413–424, 436–442, 454–455
Solar system vibration of 313
South Africa (stromatolites) 58–60
South China Sea (tides) 322
Specific dissipation function 53–54

Sphaerolithic structures 89, 96
Sponges (incl. Stromatoporoids) 96, 97
Square data window 204–206
S. Rudberti Salisburgensis Annales Breves 484
Statistics of growth increment measurements 28, 45–49, 194–202
Stromatolites 43–50, 52, 57–70, 89–94, 223, 235, 238–240, 251, 255, 260, **261**, 262, 272–273, 280–281
Stromatoporoids 96, 97
Strong equivalence principle 356
Sunspots, *see* Solar activity
Symmetrical increments, definition 259, 269
Synodic month, definition 322
Syzygy (astronomical) 159

Tabulae (corals) 40
T'ai-ho Reign of Ming Ti 477
Temperature and cosmology 361, 375, 381
Temporal increment, definition 6, 8
Temps Universal Définitif (Tu Def) 430
Teng, Empress Dowager's death 491
Tension wood 81, **82**, 83, 85
Terminology of growth 1–8, 259, 260, 262
Tethys Sea, *see also* Pangaea **402**
Tetracycline 106, 109
Thysanophyllum Limestone 40
Tides, Atlantic 321, 323, 324, 322
 atmospheric (gravitational and thermal) 319, 323, 325, 331, 366, 362, 404, 499, 526
 Baltic 323
 Bay of Bengal 400
 body, *see* Tides, solid Earth
 charts 324, 400
 depth variations 22, 35, 37, 299, 302–304, 319, 322, 323, 324, 325, 334, 361, 362, 398, 399–404, 412, 528, 529
 diurnal, physical considerations 320, 321, 322, 323, 324, 329, 331, 332
 and equatorial tidal bulge 54, **289**, 291, 302, 319, 320, 322, 362, 364–372, 398, 498
 equilibrium tide model 319–322, 323, 329
 and free oscillation of Earth 323
 friction (including dissipation) 35, 37, 53–54, 287, 289, 290, 291, 299, 300,

Tides (*cont.*)
  301, 302–304, 319, 322–334, 338, 341, 360–363, 370, 380, 381, 385, 400, 453, 454, 525–527, 528, 529, 530
  and geography of coastline, *see* Tides, depth variations
  Irish Sea 400
  and isostacy 328
  latitudinal variations 399
  long period, physical considerations 320, 321, 332
  and Love number 324, 331, 455
  neap, physical considerations 322
  North Sea 400
  oceanic, acceleration by 400
    and loading 322, 324, 325
    (resonant) amplification in 321, 322, 323, 400
  Pacific 322
  partial, physical considerations 320, 321, 322, 323, 324, 325, 329, 331, 332
  planets, tidal 422, 424
  Precambrian 50–55, 57–70, 381, 398
  and sea-level changes, *see* Tides, depth variations
  semi-diurnal, physical considerations 320, 321, 322, 324, 325, 329, 331, 332
  solar, effect on Mercury 382
  of solid Earth 319, 320, 322, 323, 385, 400, 404, 455
  speed of 320–322
  spring, physical considerations 322
  on Sun 422–424
  torques, solar and lunar 35, 51, 52, **289**, 302, 324, 325, 326, 329, 331, 333, 334, 378, 380, 382, 399, 400, 453–455, **497**, 498, 499
  types, notation 320–322
Til definition 482
Time, comparisons xiv, 286
  ET–UT 449, 450, **451**, 495, 498, 499, 506–521, 527–530
  TU2DEF – IAT **435**, 436
  UT – AT 434, 448
  UT(0) – UTC$_{BIH}$ 387
  UT2 – AT 415, 436, **437**, 440
  UT2 – GA2 **433**, 434
  UT2 (RGO) – UT2Def **429**, 430
  UT2 – UT1 386
Time instruments, ancient 466, 471, 482
  modern 387, 430, 431, 432, 434, 448

Time scales, Atomic Time (AT) xiv, 431, 432, 445, 448, 495
  biological, *see* Rotation rate of Earth, *and under* Growth, periodicities
  Ephemeris Time (ET) xiv, 431, 432, 445, 446, 494
  ET* 449
  ET$_{moon}$ – ET$_{sun}$ 499, 514
  GA$_2$ 432
  Greenwich Mean Time (GMT) 427
  Improved Lunar Ephemeris, 494, 495, 511
  Mean Solar Time 427, 445
  second, (SI unit) 432, 446
  Sidereal Time (ST) 427, 428
  Universal Time (UT) xiv, 362, 427–431, 445, 494–495
  UT1 430, 448
  UT2 434
Torques, *see* Tides, torques
Trabeculae 138–145
Tracheids 75, **77**
Transvaal Dolomite 58–70
Triangular data window 205, 206
Triassic (bivalves) 19, 22
Troposphere 419
Tyne Bottom Limestone 39

Unification of China (2221 B.C.) 477
Universe, age of 356, 357, 375
  expansion of xv, 353–372, 377–382
  number of particles in 356
Uranus and solar activity 422
UŠ, definition 481

Ventersdorp 58
Venus, Chinese sightings of 477
  rate of rotation 338
  and solar activity 422
Very long baseline interferometry (VLBI) 436, 448, 360
Viscosity, *see under* Fluids, *or* Coupling *or* Core
Viséan 29, 31, 35, 37, 38, 40
Visual transit instrument (VTI) 387
Volatiles, release of, *see* Degassing
Volcanism, and gravity determinations 358–359
  and mass loss from Earth and Moon 405
  and stromatolites 58, 66

Subject Index

Warring states period 477
Water clock (clepsydra) 471, 482
Wegener 288
White dwarfs 358
Wieger's catalogue 493
Wolkberg 58
Wood, quality affected by gravity 83
Wu, state of 491

X-Radiography (*see also* Electron microprobe) 135–145

Xylem (and tree growth) 75, 79, 81

Year, constancy of length xiii, 31, 289, 291, 360
Yüan-ch'u Reign 491
Yung-p'ing Reign 491

Zooxanthellae 135, 144, 158
Zwartkops **59**, 60, 65–66